20th Australasian Weeds Conference 2016

AA005468

Perth, Western Australia
11-15 September 2016

Editors:

Rod Randall
Catherine Borger
Sandy Lloyd

ISBN: 978-1-5108-3408-8

Printed from e-media with permission by:

Curran Associates, Inc.
57 Morehouse Lane
Red Hook, NY 12571

Some format issues inherent in the e-media version may also appear in this print version.

Printed by Curran Associates, Inc. (2017)

For permission requests, please contact Weed Society of Western Australia (WSWA) at the address below.

Weed Society of Western Australia (WSWA)
Department of Agriculture
Baron-Hay Court
South Perth 6151
Australia

Phone: (08) 9368 3760
Fax: (08) 9474 3814

info@wswa.org.au

Additional copies of this publication are available from:

Curran Associates, Inc.
57 Morehouse Lane
Red Hook, NY 12571 USA
Phone: 845-758-0400
Fax: 845-758-2633
Email: curran@proceedings.com
Web: www.proceedings.com

TABLE OF CONTENTS

BIOSECURITY

IWM NON-CHEMICAL CONTROL METHODS

IWM

WONS AND OTHER WEEDS

WEED CONTROL

INVASION CURVE

WATTLE WE DO ABOUT WATTLES?

HERBICIDE RESISTANCE

DRONES, UAVS AND ROMOTE SENSING

KEYNOTE ADDRESSES

WEED POTENTIAL AND ASSESSMENT

BIOLOGICAL CONTROL

Do as I say, not as I do: benchmarking weed hygiene practices across NSW

Sonia Graham[1], Nick Gill[2], Rebecca Cross[1], Viveka Simpson[1] and Eli Taylor[1]
[1] School of Social Sciences, University of New South Wales, Kensington, New South Wales 2052, Australia
[2] AUSCCER, University of Wollongong, Wollongong, New South Wales 2522, Australia
(sonia.graham@unsw.edu.au)

Summary Numerous industry and government guidelines describe and recommend weed hygiene practices to limit common pathways for weed spread. Yet despite these guidelines poor weed hygiene continues to undermine the efficacy of weed control efforts and enables the spread of noxious weeds into and across NSW. This paper presents the results of a study that interviewed over 50 individuals with responsibilities for managing weed spread; private landholders, public land managers, weed contractors and agricultural transport operators were interviewed about their knowledge, practice, and support of good weed hygiene practices. A further 40 individuals were invited to participate, but many declined because they do not engage in weed hygiene. None of the recommended practices discussed in the interviews were undertaken by all applicable interviewees. The weed hygiene practices that were most likely to be undertaken involved using wash down bays for machinery by landholders and weed contractors, providing weed identification guides to staff by public land managers and weed contractors, and using wash down bays for vehicles for agricultural transport operators. The weed hygiene practice that was least likely to be used across all interview groups was washing down boots and clothing. While participants knew about most of the recommended weed hygiene practices, they provided a large range of explanations for not prioritising weed hygiene in their day-to-day activities. If weed hygiene is to become part of everyday farm practice in NSW then effort needs to be invested in addressing the widespread view that the costs far outweigh the benefits.

Keywords Weed control, farm hygiene, rural Australia.

Do new weeds introduced from outside New Zealand still pose a threat to our farmers?

T.K. James[1], M.R. McNeill[2] and P.D. Champion[3]
[1] AgResearch, Ruakura Research Centre, Private Bag 3123, Hamilton 3240, New Zealand
[2] AgResearch, Lincoln, Private Bag 4749, Canterbury 8140, New Zealand
[3] National Institute of Water & Atmospheric Research, PO Box 11-115, Hamilton 3251, New Zealand
(trevor.james@agresearch.co.nz)

Summary Despite New Zealand being considered among world leaders in biosecurity at our international borders, there are concerns that pathways leading to new weed introductions still continue to exist. Our hypothesis is that potential weeds continue to enter New Zealand via a variety of pathways. Several weed incursions during the last ten years are detailed. Many incursion pathways have been identified and will be described in more detail.

Briefly, 2007/08: 31 new to NZ weed species found in imported coco peat, 2009/10: 10 new species found in footwear examined at the airport, 2013/14: 23 new weed species found in soil collected off shipping containers at our ports, 2013: black grass (*Alopecurus myosuroides* Huds.) seed as a contaminant in grass seed, 2015: Noogoora bur (*Xanthium strumarium* L.) as a contaminant in imported maize for milling, 2016 velvetleaf (*Abutilon theophrasti* Medik.) introduced to circa 260 farms as a contaminant on fodder beet (*Beta vulgaris* L.) seed. Additionally, a number of new to New Zealand potential aquatic weeds have been introduced via illegal aquarium imports.

New species of weeds are continually entering New Zealand by a variety of pathways and while border biosecurity intercepts the majority of these weeds, it is reasonable to expect that there will be unwanted introductions. Without vigilance and early intervention these weeds could be rapidly spread through natural and human assisted pathways. Some practical steps for improving on-farm vigilance and biosecurity as it relates to unwanted weeds will be discussed.

Keywords Biosecurity, incursion, new weeds.

Growing our biosecurity culture for property and landscape resilience

Belinda J. Callanan
98 Venz Road, Nobby, Queensland 4360, Australia
(belinda@th9.com.au)

Summary Weeds have a big impact on our capability, our livelihoods and the environment. For a long time now, weed management has been considered the realm of a select few. But has the climate for weeds also triggered a time for cultural change?

Our awareness of biosecurity risks is growing. So it's not surprising that biosecurity measures are becoming a way of life that permeates everything we do. Biosecurity is now considered at a range of critical points - whether you're operating a property, negotiating land access agreements, moving vehicles and equipment or even delivering a construction project. So how can this growing biosecurity culture help to improve our land management and create resilience within our properties and landscapes?

Biosecurity protects our interests. And managing biosecurity doesn't have to be hard… even the simple things can make a big difference. We'll take a look at some of the progressive biosecurity approaches being implemented across private properties, industry and the community - from increasing awareness via social media, to designating movement corridors, negotiating contract conditions and implementing practical hygiene measures at the farm gate. We'll also share some case studies that demonstrate how preventing and managing weed and other biosecurity risks in a practical way yields results for properties and the landscape.

Keywords Weed management, biosecurity, resilience, capability, practical methodologies, case studies.

Control of tropical soda apple: five years on

Reece Luxton[1], Rod Ensbey[2] and Elissa van Oosterhout[2]
[1] Clarence Valley Council, Grafton, New South Wales 2460, Australia
[2] NSW DPI, Grafton, New South Wales 2460, Australia
(reece.luxton@clarence.nsw.gov.au)

Summary Tropical soda apple (*Solanum viarum* Dunal), first identified in Australia in the NSW Northern Rivers Region during 2010, has been the subject of a control program for the past five years. Through a concerted effort led by NSW DPI and Local Control Authorities (local government), the weed has been contained to a number of sites. This paper seeks to highlight the successes achieved, and reflect on the lessons learnt from the review of progress to date.

To oversee the coordinated management of the control program, a tropical soda apple taskforce was established in 2011 and continues to this day. The taskforce is made up of members from the NSW Farmers Association, Local Government, the Local Land Services and NSW Department of Primary Industries, and seeks to identify funding opportunities and recommend improvements in the control program.

The paper will identify the development of a strategic plan, on-ground control efforts, education and awareness programs, and monitoring and evaluation. It will briefly discuss the use of the National Livestock Identification Scheme for identification of new incursions, as well as current approaches utilised for tackling the weed including use of community based social marketing (CBSM).

Keywords Tropical soda apple, strategy, taskforce, NSW DPI, Local Government.

Can a plant's cultural status and weed history provide a generalised weed risk score?

Roderick P. Randall
Invasive Species, Biosecurity and Regulation, Department of Agriculture and Food, South Perth, Western Australia 6151, Australia
(rod.randall@agric.wa.gov.au)

Summary Previous editions of *A Global Compendium of Weeds* (Randall 2012), whilst providing references relat ing to a plant's weed history, have not provided any real guidelines about the level of risk these species pose.

Using the extensive dataset behind the *Compendium,* including data indicating a plant's means of entry to new environments, dispersal pathways and impacts this paper considers the possibility of a generalised weed risk score.

Of the 40,874 referenced weeds listed in the database 9855 have a score indicating their level of weed risk. With further data collection more species can easily attain a score, and for species with no score the system is simple to use even without the underlying dataset.

Keywords *Global Compendium of Weeds*, database, risk, introduction, naturalised, environmental, agricultural, score, ranking.

INTRODUCTION

For several decades researchers have looked at various groupings of plant biological characteristics to determine those most useful in assessing a species' weed potential (Richardson *et al.* 1990, Rejmánek 1995, Rejmánek and Richardson 1996, Reichard and Hamilton 1997). Later publications looked at various other combinations of criteria to try and predict weediness, such as taxonomy (Daehler 1998, Diez *et al.* 2009), impacts and environment (Pyšek *et al.* 2012), weed groupings (Morin *et al.* 2013), comparing naturalisation rates (Duncan *et al.* 2004) and distribution traits (Pyšek *et al.* 2009), among others.

This paper considers the question: Could a database containing a relatively small range of dispersal, cultural (use) and weed status attributes for plants produce a generalised weed risk score?

Phases For a plant species to become a problematic weed in a new region it must progress through three phases:

1. **Enter** and establish a population in the new environment
2. **Disperse** throughout that environment normally from the initial population/s
3. **Impact** on the environment or human activities within its new range

The entry of a species into a new environment is considered an incursion. Once a species has become established and starts to spread without human assistance it can be considered naturalised and a cycle of dispersal, establishment, naturalisation and dispersal continues throughout the new environment.

After naturalisation the impacts may be immediate and costly. Species of *Striga* or *Orobanche* can have huge financial ramifications, restricting access to some markets just by the mere knowledge of their introduction into a previously clean cropping region. Most weeds, however, do not begin to accumulate noticeable impacts until years after their incursion, normally when public awareness starts to occur (Figure 1). With effective surveillance systems in place detection should hopefully occur in the first incursion phase. Detection in later phases usually means eradication is no longer an option and management systems need to be developed to cope with the new species.

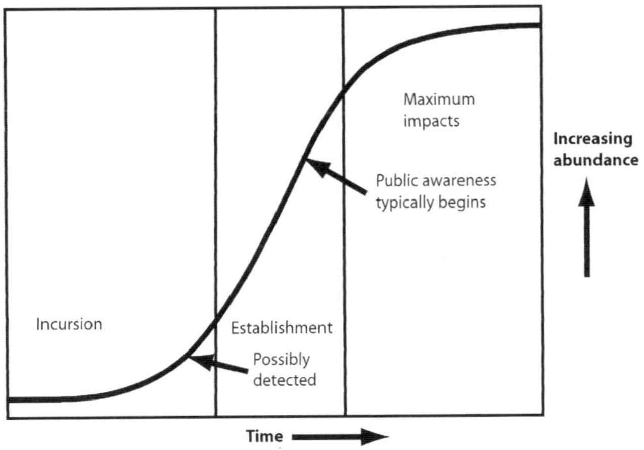

Figure 1. A typical 'Invasion Curve' with the phases of an organism's Incursion (Entry), Establishment (Dispersal) and Maximum impacts.

Rather than relying on a number of biological or ecological traits this proposed system considers the outcomes of a plant species' interaction with humans and its observed behaviour around the world. This approach is supported in part by a number of studies that concluded that the single most useful characteristic of a plant in predicting weediness is its behaviour as a 'Weed Elsewhere' (Gordon *et al.* 2008, Daehler and Carino 2000, Daehler *et al.* 2004, Mack 1996, Parker *et al.* 2007, Rejmánek *et al.* 2005). The 'Weed Elsewhere' question is not a function of a plant's biology or ecology but rather how a species' behavior is perceived by humans.

Consequently a number of the database characteristics used in this study are drawn from attributes that are closely allied to human behaviors, observations and values. This includes categories like 'intentionally grown' or 'considered to have medicinal values' and other categories of human values where these plants interact and impact, their uses and how they are dispersed by people.

MATERIALS AND METHODS

The Department of Agriculture and Food, Western Australia currently maintains a database that contains over 1.2 million plant names referentially linked by their taxonomy to 4.62 million bibliographically referenced records (Randall 2016). These referenced records contain data on hundreds of thousands of plant species, including their cultivated and naturalised status, dispersal mechanisms and their weed status from various regions around the world.

For the purposes of this system a species incursion risk is determined by (1) its ability to enter new environments, (2) to spread and establish over wider areas by its dispersal options and lastly (3) by its impacts in these environments. By matching selected database categories to these three phases and placing a value on each category a relatively simple scoring system was devised with the intention of producing a generalised weed risk score.

Not all categories were given the same value. Within the 'entry' phase the score value allocated was based on the percentage of known weeds as a proportion of the total species number within each category. Essentially the weediest categories were weighted with a higher score (Table 1). Within the 'dispersal' phase all categories were given the same score as there is a much higher percentage of weeds in each of these categories. In the 'impact' phase the 'invasive' category was valued the same as 'environmental' and 'agricultural' for reasons discussed later.

Fifteen categories, five in each phase (Table 1), were selected from those available in the plants database (Randall 2016) with the end result that 189,996 plant species were represented in at least one of the fifteen categories.

The database categories were selected to try and separate the more significant end of the weed spectrum rather than try to generate an all inclusive ranking of the many thousands of 'minor weeds', many of which may never generate the level of risk to warrant an eventual score.

Within each phase the scoring is additive, with the subsequent scores of the three phases multiplied. The maximum score any phase can add up to is four, with the maximum score therefore being 64:

<u>WEED RISK SCORE</u> = <u>ENTRY</u> (A + B + C + D + E) x <u>DISPERSAL</u> (F + G + H + I + J) × <u>IMPACT</u> (K + L + M + N)

For a species to generate a weed risk outcome it needs at least one score in each of the three phases, otherwise its outcome will be zero. Essentially the system assumes that if a species, has little ability to enter new regions, does not manage to easily disperse or has no definable impacts then it would not be considered a significant weed.

Scores were then generated for 9855 eligible species, which were then exported and graphed to try and determine the most effective allocation of the scores for a range of four risk outcomes.

A number of different cut offs between each category were originally tested in a graph of the risk ranking versus species distribution (Figure 2). The final distribution cut offs selected were those shown in Figure 3 (the blue lines) which closely follows the logarithmically reducing relationship between the categories, as illustrated in Figure 2.

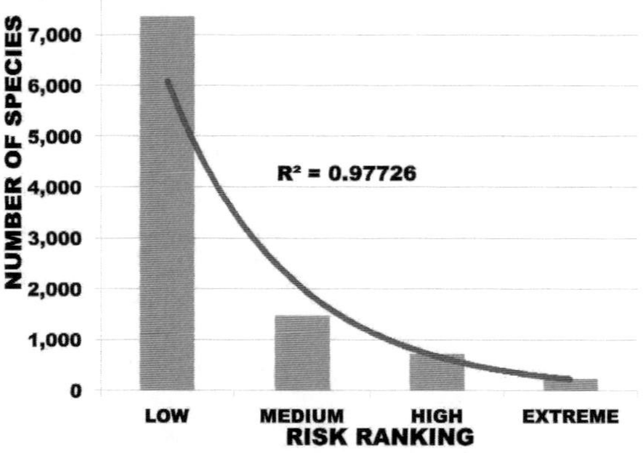

Figure 2. Risk ranking versus number of species per rank.

Table 1. The system categories, definitions, total species numbers from database, weed percentages and allocated values for each.

PHASE	CATEGORY		No. of species	% weeds	Value
ENTRY	Introduction via the most important pathways (human mediated)				
A	Pastures	includes fodder, forage, silage, hay, straw	3,170	67.0	1.3
B	Forestry	timber, pulp, firewood, shade, utility	1,422	61.8	1.2
C	Herbal	plants considered of medicinal use	32,842	37.5	0.6
D	Crops	edible plants	26,270	31.1	0.6
E	Ornamentals	grown by people for all sorts of reasons	149,636	13.1	0.3
DISPERSAL	Significant dispersal mechanisms				
F	Human dispersed	Sown as crops, garden plants, sold, bartered Intentional dispersal via human mediated pathways including machinery and vehicles	53,614	30.8	0.8
G	Animal dispersed	Includes all records of dispersal by animals	4,953	80.6	0.8
H	Wind or Water	When recorded as dispersed by either	4,004	89.1	0.8
I	Contaminants	Contaminants in seed for sowing, hay, fodder, other agricultural products	5,777	100	0.8
J	Escapees	Plants that escape from rehab sites, crops, orchards, gardens, collections etc	5,902	100	0.8
IMPACT	The most significant or potentially significant impacts				
K	Agricultural weed	Known weed of an agricultural crop	15,927	100	1
L	Environmental weed	Impacts are reported on the natural environment	5,471	100	1
M	Noxious weed	Quarantine species or Prohibited species as well as Declared, Noxious or Pest plants	4,238	100	1
N	Invasive	Species that can impact significantly on the environment or agricultural (primary production) operations, are seen to be actively spreading, often with impacts increasing	6,076	100	1

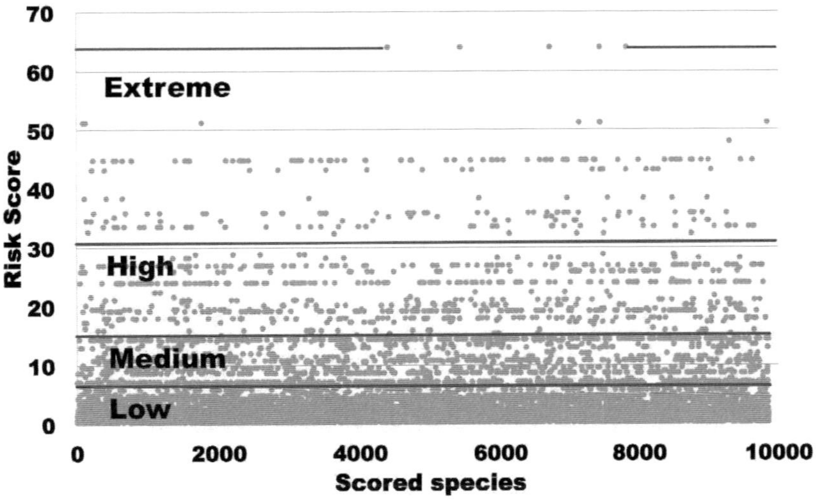

Figure 3. Risk score versus number of scored species (n = 9855) showing allocated risk ranks (blue lines).

Table 2 lists these cut off scores and includes the numbers and percentages of species in each rank category and as a percentage of all known weeds.

There is also an effective fifth rank outcome, that of 'Unscored', which includes all those species for which insufficient data exist to allow a score to be calculated. This outcome contains 180,141 species with at least one category captured in at least one phase.

RESULTS

The number of species in each of the four risk categories is outlined in Table 2. The 235 'Extreme' risk species identified by this process are all well known global weeds with the top 13 of this 'extreme' group containing very well known invasive species (Appendix 1).

Some of the 180,141 'Unscored' species could potentially generate a risk ranking with further research and therefore could be considered Data Deficient. However some species may never generate a risk score, these species could potentially be considered Non-Weedy species.

The only available measure of the likelihood of a species being Data Deficient, or a Non-Weed, is to consider the total number of records for that species. The more records a species has then potentially the lower is its likelihood of being data deficient. The implication of this is that species with no calculated risk rank, no weed references and numerous non-weed related records could effectively be considered non-weedy.

To determine just how many species may meet this potential Non-Weedy outcome the number of database records for all species with at least one phase score in the system, and no weed references (140,298), were exported and graphed to determine their distribution in a similar manner to the risk scored species (Figure 4).

While a more definitive analysis is needed, preliminary work implies that those species referenced 30 or more times with no actual weed references in the *Plant Database*, could be considered as candidate 'non-weeds'.

In terms of numbers of species, 30 or more references encompasses 1622 candidates, shown as points above the blue line in Figure 4. This also implies that the remaining 138,676 species from Table 2 could be considered data deficient.

Whilst interesting, this aspect of the *Plants Database* requires more statistical analysis and verification and clarification of the species names, including the determination of any synonymy and correction of spelling errors.

Table 2. Allocated Risk Ranks and species statistics for each.

Risk Score	Risk Rank	No. of species	% of species scored (n = 9855)	% of all known weeds (n = 40,874)
Zero	Unscored	180,141	–	–
>0–6	Low	7,437	75.4	18.19
7–15	Medium	1,472	14.9	3.60
16–31	High	712	7.2	1.74
32–64	Extreme	235	2.3	0.57

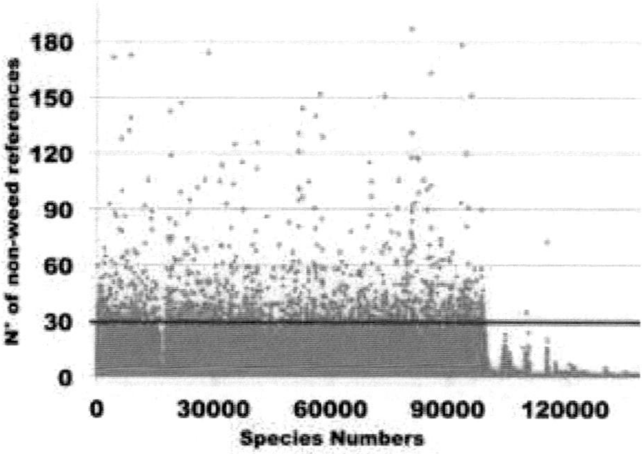

Figure 4. Number of references (for no weed reference species) versus species numbers (n = 140,298). Blue line = 30 records (vertical axis truncated at 200 records).

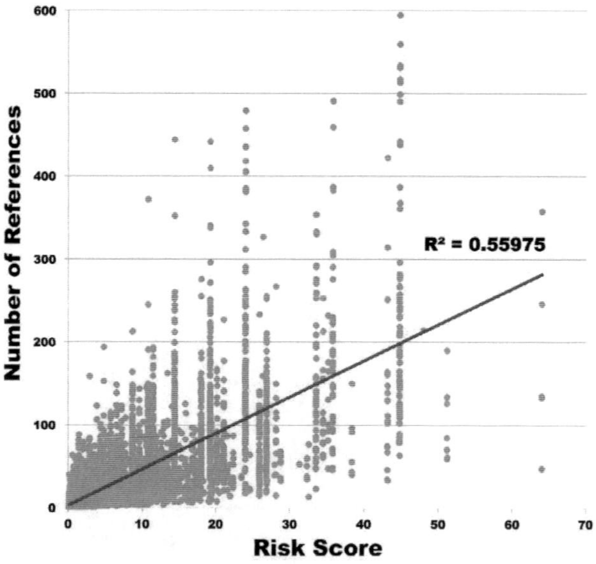

Figure 5. Species risk score versus the number of Weed References in the *Plants Database* (n = 9855).

That so many species are ranked in the database is an indication of the amount of relevant data held. Currently the database contains over 396,000 weed related records and another 4.26 million records relating to various other aspects of the species it holds.

With such a large number of weed related records it was considered that a potential relationship may exist between the number of weedy records for any species in the database and its ranked outcome (Figure 5). While there is a reasonably strong relationship between the risk score and the number of weed references for the dataset, there is still a significant degree of variation. The effect of cascading references (reference not a primary source) would have an impact on this approach.

The numerous non-weedy data sources also play a part in providing viable outcomes but a similarly plotted figure comparing instead the total number of references versus risk score showed no clear relationship.

DISCUSSION

Definitions and Categories Determining the correct category for a species from the literature can be difficult, sometimes impossible, so the descriptive term used by the authors is generally left unchanged. Therefore the score used in this system will more likely reflect the utility of the descriptive term used rather more than its formal definition might suggest. This is an issue that currently affects much of invasive species science and while a widely cited definitions paper has been written, those definitions are not well applied (Richardson *et al.* 2000).

'Invasive' is a term that is widely used and often misrepresented in the literature. Quite often the terms agricultural or environmental weed would be more informative and appropriate. Currently 'Invasive' scores the same as these more defined impacts as the majority of sources in the literature do not apply the term correctly.

Many plant species reported as environmental weeds could often be more correctly considered reports of naturalisation. Environmental impacts are harder to ascertain than agricultural and a great deal of work needs to be done to describe and quantify the full extent and costs of the environmental impacts of weeds. For the most part however species reported as environmental and agricultural weeds have been accepted as such unless the criteria as explained in the source indicates a clear mismatch.

Dispersal The vast majority of plant propagules are dispersed over relatively small distances, normally in metres to tens of metres (Cain *et al.* 2000, Horn *et al.* 2001, Nathan and Muller-Landau 2000, Tackenberg *et al.* 2003). This makes the human mediated dispersal of plants and plant propagules a significant class of global dispersal and is the reason why most natural means of dispersal were ignored.

Unsurprisingly a large proportion of species in each of the dispersal categories are also recorded as weeds. The same score was allocated to each as no single dispersal means is particularly important but multiple means of dispersal do mean an increased level of propagule pressure (Lockwood *et al.* 2005, Simberloff 2009). Furthermore demand for ornamental species is often driven by fashion, advertising and pricing, human foibles that have led to many successful plant incursions (Dehnen-Schmutz *et al.* 2007).

Unscored Species Many species in the database are yet to have a risk score calculated but may well do so as further data are incorporated. While the *Plants Database* (Randall 2016) is certainly not complete, it does have weed risk scores determined for 9855 weed species which represents a significant saving in time and resources for potential users of this system. It should also be remembered that many of these species may never meet enough criteria to enable them to attain a risk score, these being non-weedy plants.

For any species not yet scored, and no *Plants Database* available, Appendix 2 provides an example of three species scored with an example/template table that could be used to collate the information and tally up the scores. Any species currently unscored could have its risk rank determined using this table; the real effort of course lies in searching through references for the relevant data to complete the category scores within each phase.

The issue of non-weedy species being potentially identifiable by this system will be further investigated.

Benefits One of the biggest issues of the majority of 'ranking' or 'prioritisation' systems is that most have no species banked (assessed) in the system for others to use right away. With many thousands of species already assessed this proposed ranking system is likely to meet the requirements of most users without their having to research any, or many, species. As data collection in the database continues the numbers of species with ranked risks will continue to increase over time.

Consider the Climate While the generalised weed risk score provides a good indication of a species' weed potential globally, any potential risk should be considered in respect of a species' ability to naturalise in the region of concern. If a suitable climate exists and there are no cultural, agronomic or management restraints against the establishment or spread of the suspect species then the systems risk rank outcome could be considered appropriate and acted on accordingly.

However if there are no suitable climates available then the likelihood of the assessed species establishment in the region is likely to be low. There are very few plant species that are globally invasive and those are all well documented and hence already captured in this system.

A simple system that could be employed to modify the risk score based on an analysis of its preferred climate would be to apply a discount based on the degree of the climate match. For example a good climate match would mean no discounting of the risk score, whereas a moderate climate match could reduce the score by 25%, a low match by 75% and a poor climate match could reduce the score by 95%.

The degree to which a species' climate requirements are met in a new environment should therefore provide a level of discount that could be applied to its generalised risk score.

For a state like Western Australia, which covers over 30° of latitude across a range of climates, it is unlikely many species scores would be discounted; there is generally always somewhere a plant could survive and thrive. However in a region like Tasmania many tropical species may find their risk ranks heavily discounted.

Microclimates It is also important not to overlook potentially valuable microclimates. Many plants can become weeds in production glasshouses in regions where they would not ordinarily establish or survive in the prevailing weather conditions outside. Under such circumstances the costs of such weeds to industry and consumers can be high.

A system like this will not identify a potential weed where no records of weediness exist. Even known weeds that have no incorporated references for agricultural or environmental impacts will remain unscored in this system. The intent is to highlight and rank the worst known weeds rather than try and rank all possible weedy species, known and unknown. That this system can determine risk rankings for 33% of the world's known weeds is testament to the many and varied impacts that so many plant species have globally.

Overall this system could provide a great many users with a ready made, data populated, tool to assist in the ranking of their current weed species, representing significant savings in time and resources.

The potential savings for individuals and government, and the reduced downstream impacts to the environment and agriculture, by removing just a single 'extreme' risk species from a rehab program, cropping or landscaping project could be incalculable.

Any future editions of the *Global Compendium of Weeds* will also contain this system's risk ranking to provide readers with more data on species weed potential.

ACKNOWLEDGMENTS

The full results of this project are available as an Excel spread-sheet which contains all the category scores, phase scores, final risk scores and rank outcomes for 9855 plant species. This supplementary data spreadsheet is available on the conference USB drive along with the rest of the conference proceedings.

Many thanks to the thousands of authors whose work has gone into and made entirely possible this potential tool and lastly to the very useful suggestions from the reviewer.

REFERENCES

Cain, M.L., Milligan, B.G. and Strand, A.E. (2000). Invited special paper: long-distance seed dispersal in plant populations. *American Journal of Botany* 87 (9), 1217-27.

Daehler, C.C. (1998). The taxonomic distribution of invasive angiosperm plants: ecological insights and comparison to agricultural weeds. *Biological Conservation* 84, 167-80.

Daehler, C.C. and Carino, D.A. (2000). Predicting invasive plants: prospects for a general screening system based on current regional models. *Biological Invasions* 2 93-102.

Daehler, C.C., Denslow, J.L., Ansari, S. and Kuo, H. (2004). A risk assessment system for screening out harmful invasive pest plants from Hawaii's and other Pacific islands. *Conservation Biology* 18, 360-8.

Dehnen-Schmutz, K., Touza, J., Perrings, C. and Williamson, M. (2007). A century of the ornamental plant trade and its impact on invasion success. *Diversity and Distributions* 13, 527-34.

Diez, J.M., Williams, P.A., Randall, R.P., Sullivan, J.J., Hulme, P.E. and Duncan, R.P. (2009). Learning from failures: testing broad taxonomic hypotheses about plant naturalization. *Ecology Letters* 12 (11), 1174-83.

Duncan, R.P., Randall, R.P. and Williams, P.A. (2004). Comparative naturalization rates in the exotic floras of Australia and New Zealand. Paper presented at the fourth Southern Connections Conference, 19–23 January, University of Cape Town, Cape Town, South Africa.

Gordon, D.R., Onderdonk, D.A., Fox, A.M., Stocker, R.K. and Gantz, C. (2008) Predicting invasive plants in Florida using the Australian Weed Risk Assessment. *Invasive Plant Science and Management* 1 (2), 178-95.

Horn, H.S., Nathan, R. and Kaplan, S.R. (2001). Long-distance dispersal of tree seeds by wind. *Ecological Research* 16, 877-85.

Lockwood, J.L., Cassey, P. and Blackburn, T. (2005). The role of propagule pressure in explaining species invasions. *TRENDS in Ecology and Evolution* 20 (5), 223-8.

Mack, R.N. (1996). Predicting the identity and fate of plant invaders: emergent and emerging approaches. *Biological Conservation* 78, 107-21.

Morin, L., Paini, D.R. and Randall, R.P. (2013). Can global weed assemblages be used to predict future weeds? *PLoS ONE* 8 (2): e55547. doi:10.1371/journal.pone.0055547

Nathan, R. and Muller-Landau, H.C. (2000). Spatial patterns of seed dispersal, their determinants and consequences for recruitment. *TREE* 15 (7), 278-85.

Parker, C., Caton, B.P. and Fowler, L. (2007). Ranking nonindigenous weed species by their potential to invade the United States. *Weed Science* 55, 386-97.

Pyšek, P., Jarošík, V., Hulme, P.E., Pergl, J., Hejda, M., Schaffner, U. and Vilà, M. (2012). A global assessment of invasive plant impacts on resident species, communities and ecosystems: the interaction of impact measures, invading species' traits and environment. *Global Change Biology* 18, 1725-37. doi:10.1111/j.1365-2486.2011.02636.x

Pyšek, P., Jarošík, V., Pergl, J., Randall, R., Chytrý, M., Kühn, I., Tichý, L., Danihelka, J., Chrtekjun, J. and Sádlo, J. (2009). The global invasion success of central European plants is related to distribution characteristics in their native range and species traits. *Diversity and Distributions* 15 (5), 891-903.

Reichard, S.H. and Hamilton, C.W. (1997) Predicting invasions of woody plants introduced into North America. *Conservation Biology* 11 (1), 193-203.

Rejmánek, M. (1995). What makes a species invasive. *In* Pyšek, P., Prach, K., Rejmánek, M. and Wade, P.M. (eds), 'Plant invasions', pp. 3-13. (SPB Academic Publishing, The Hague, Netherlands).

Rejmánek, M. and Richardson, D.M. (1996). What attributes make some plant species more invasive? *Ecology* 77, 1655-60.

Rejmánek, M., Richardson, D.M., Higgins, S.I., Pitcairn, M. and Grotkopp, E. (2005). Ecology of invasive plants: state of the art. *In* Mooney, H.A., McNeely, J.A., Neville, L., Schei, P.J. and Waage, J. (eds), 'Invasive alien species: a new synthesis', pp. 104-61. (Island Press, Washington, DC).

Randall, R.P. (2016). 'Plants Database'. Department of Agriculture and Food, Western Australia. Unpublished database (4.62 million records).

Randall, R.P. (2012). 'A Global Compendium of Weeds, Second Edition', 1124 pp. (Department of Agriculture and Food, Western Australia).

Richardson, D.M., Cowling, R.M. and Le Maitre, D.C. (1990). Assessing the risk of invasive success in *Pinus* and *Banksia* in South African mountain fynbos. *Journal of Vegetation Science*, 1 629-42. doi: 10.2307/3235569

Richardson, D.M., Pyšek, P., Rejmánek, M., Barbour, M.G., Panetta, F.D. and West, C.J. (2000). Naturalization and invasion of alien plants: concepts and definitions. *Diversity and Distributions* 3 (6), 14-93.

Simberloff, D. (2009). The role of propagule pressure in biological invasions. *Annual Review of Ecology, Evolution and Systematics* 40, 81-102.

Tackenberg, O., Poschlod, P, and Bonn, S. (2003). Assessment of wind dispersal potential in plant species. *Ecological Monographs* 73 (2), 191-205.

Appendix 1. Top 13 Ranked Weeds (the next 75 species all score the same at 44.8).

Species	Global Risk Score
Gleditsia triacanthos L.	64
Leucaena leucocephala (Lam.) De Wit.	64
Parkinsonia aculeata L.	64
Prosopis glandulosa Torr.	64
Robinia pseudoacacia L.	64
Acacia farnesiana (L.) Willd.	51.2
Acacia nilotica (L.) Willd. ex Delile	51.2
Calotropis procera (Aiton) W.T.Aiton	51.2
Pithecellobium dulce (Roxb.) Benth.	51.2
Prosopis juliflora (Sw.) DC.	51.2
Prosopis pallida (Humb. & Bonpl. ex Willd.) Kunth	51.2
Ziziphus mauritiana Lam.	51.2
Trifolium pratense L.	48

Appendix 2. Generic Weed Risk Ranking System as applied to three species.

Phase Categories (No = Zero)	***Malva sylvestris* L.** **tall mallow**	***Abies pinsapo* Boiss.** **Spanish fir**	***Emex australis* Steinh.** **doublegee**
ENTRY			
Pastures (Yes = 1.3)	1.3	0	0
Forestry (Yes = 1.2)	0	1.2	0
Crop or Edible plant (Yes = 0.6)	0.6	0	0.6
Herbal/Medicinal (Yes = 0.6)	0.6	0.6	0
Ornamental (Yes = 0.3)	0.3	0.3	0.3
Sum	2.8	2.1	0.9
DISPERSAL	×	×	×
Human Dispersed (Yes = 0.8)	0.8	0.8	0.8
Animal Dispersed (Yes = 0.8)	0.8	0	0.8
Contaminant (Yes = 0.8)	0.8	0	0.8
Escapee (Yes = 0.8)	0.8	0	0
Wind and/or Water (Yes = 0.8)	0	0	0.8
Sum	3.2	0.8	3.2
IMPACT	×	×	×
Agricultural Weed (Yes = 1.0)	1	0	1
Environmental Weed (Yes = 1.0)	1	1	1
Noxious Weed (Yes = 1.0)	1	0	1
Invasive (Yes =1.0)	1	1	1
Sum	4.0	2.0	4.0
Weed Rank Score = Entry × Dispersal × Impact	**35.84**	**3.36**	**11.52**
Generic Risk Ranking	**Extreme**	**Low**	**Medium**

Stevia ovata – not so sweet

Melissa J. Setter[1], Stephen D. Setter[1], Simon J. Brooks[2] and Shane D. Campbell[2]
[1] Department of Agriculture and Fisheries, PO Box 20, South Johnstone, Queensland 4859, Australia
[2] Department of Agriculture and Fisheries, PO Box 187, Charters Towers, Queensland 4820, Australia
(melissa.setter@daf.qld.gov.au)

Summary Native to elevated areas of the tropical Americas, *Stevia ovata* Willd (candyleaf) does not have a record of weediness overseas. Closely related to *Stevia rebaudiana* Bertoni (a commercially grown sweetening agent), *S. ovata*, with its pretty white flowers and soft sweet-scented leaves, may not seem like much of a threat. However, affected land managers in north Queensland's Atherton Tablelands know better. The woody multi-stemmed perennial shrub has infiltrated open habitats, including cattle grazing lands, and road and power line corridors. Its rate of spread and formation of virtual monocultures is a growing cause for concern. *S. ovata*'s observed weedy behaviour in these situations, coupled with the fact that this region has the only known infestations in Australia, has led to its local declaration by the Tablelands Regional Council, and the formation of the *S. ovata* Working Group.

Keen to quash further *S. ovata* infestation, this group of concerned stakeholders initially requested research into the ecology and control of the plant, and were a driving force behind a successful multi-agency delimitation exercise which took place in June 2015. Biosecurity Queensland took up the research challenge and investigated several aspects, including seed longevity, seed dispersal, growth rates, reproductive age and size, lifecycle trends and control options.

This paper presents an overview of *S. ovata* and its history of in Australia, and describes our research into its ecology and control.

Keywords Candyleaf, ecology, seed, control, herbicides.

INTRODUCTION

Stevia ovata belongs to the tribe Eupatorieae of the family Asteraceae. Also in this tribe are the genera *Mikania* Willd. and *Chromolaena* D.C which contain species that are also priority weeds in Queensland (DAF 2016). A native to the southern United States and tropical Central America, plants of the genus *Stevia* are often found at elevations of 1000–3000 m (Grashoff 1972).

Stevia ovata was first reported in Australia in 2007 near Ravenshoe on the Atherton Tablelands inland from Cairns at an altitude of around 900 m. Despite the plant's limited known distribution in Australia, analysis of the climate in this species native range suggests it has the potential to infest far larger and cooler areas of eastern and southern Australia (Murphy *et al.* 2009). When it first came to the attention of Local Government weed managers, surveys found plants to be within a relatively restricted area. This situation gave rise to the possibility of containing the weed's spread and realistically reducing the population with our research findings providing the tools.

Research into the following aspects have been either completed or initiated:

- Germination requirements,
- Age to reproductive maturity,
- Seed longevity in soil (in Wet and Dry Tropics of north Queensland),
- Seed longevity in water (fresh/salt), and
- Herbicide screening and rate refinement.

A summary of general observations observed during the research along with the key findings are presented in the following sections.

GENERAL OBSERVATIONS

These observations were made in the field and in shade house experiments with both showing very similar results.

The plant itself originates as a single-stemmed herb which is easily identifiable by its distinctive white flower clusters. Flowering occurs around May (late autumn), and seed maturation commences around June (early winter) As the flowers are relatively easily seen in the open dry sclerophyll setting, surveying is best achieved during this period, though treatment is most effective before flowers are fully formed.

Seedlings can flower as early as three months after germination and at heights as small as 8 cm. Plants can exceed 3 m in height reaching densities of up to 30 plants m^{-2} to the exclusion of most other understory species.

Stevia ovata stems die back to ground level around September with re-shooting and seedling emergence occurring after the first substantial summer rain. This aspect of the species biology means that there are periods of each year when control operations are not possible and detection is extremely difficult. After its annual dieback, the following year's growth often emerges with multiple stems, increasing in number until it obtains a dense, shrubby appearance. Some have over 30 stems, with a large, strong root mass. Reproduction is not limited to seeds with vegetative reproduction occurring from stem sections placed on soil.

RESEARCH OVERVIEW

Germination requirements Freshly collected achenes (hereafter seeds) were placed in temperature/light controlled incubators to determine suitable conditions for germination. Conditions implemented included set temperature (25°C, 24 h), cycling temperatures (30°C, 12 h/20°C, 12 h) and set light (24 h), cycling light (12 h/12 h). Results indicate fresh seed is readily germinable and highly viable at approximately 75%. Varying temperature and light regimes had little to no effect.

Seed longevity in soil Fresh seed enclosed within mesh packets were buried at the Centre for Wet Tropics Agriculture (South Johnstone near 17° 36'S, 145° 59'E, altitude 15 m) and the Tropical Weeds Research Centre (Charters Towers near 20° 05'S, 146° 16'E, altitude 310 m). The seeds were placed on the soil surface and at depths of 2.5–3, 10 and 20 (Charters Towers only) cm below the surface, a common method for assessing soil seed bank persistence (Brooks and Setter 2012). Packets were retrieved at set intervals for viability testing. Most viable seed buried at 0, 2.5–3 or 10 cm in soil was exhausted within one year. No viable seed was found after three years buried in soil in either trial site.

This indicates that *S. ovata* has a relatively short-term persistent seed bank, according to the classification system developed by Thompson *et al.* (1997). However, seed could be retained on the stems even after the plant had died back. Some of this 'standing seed' was viable after one year.

Seed viability after immersion Longevity was determined by placing fresh seed within porous mesh packets and immersing them in fresh, brackish or salt water in fish tanks in the laboratory at the Centre for Wet Tropics Agriculture. Replicate seed packets were retrieved from the tanks at set intervals between 1 and 128 days for viability testing. Seed placed in fresh water continued to germinate in the water for the first four weeks before being depleted. Seed in brackish and saltwater did not germinate while immersed but were found to be viable after 14 weeks. Whilst immersion in fresh, brackish and saline water is no barrier to dispersal, germination and establishment is ultimately dependent on viable seed landing in a suitable niche.

Seed buoyancy In laboratory trials of seed buoyancy, lots of 50 *S. ovata* seeds placed in beakers of fresh water were observed regularly until all the seed sank. Some beakers were also constantly agitated using a laboratory sample rocker. In all beakers, less than 50% of the seed was floating after 10 hours, less than 10% of seed was floating after 48 hours and nearly all seeds had sunk after 126 hours.

These findings suggest that *S. ovata* has the potential to spread considerable distances down fast flowing tropical creek and river systems. Immersion in different water types is no barrier to viable seed dispersal and the seed can remain buoyant in moving water. Downstream spread is more likely to be limited by the seed reaching a site to establish and the speed of the water, rather than distance or salinity levels.

Seed size and dispersal Although *S. ovata* is in the Eupatorieae tribe, where many species are described as primarily wind dispersed, Cortes-Flores *et al.* (2013) list this species, and several other *Stevia* spp. as having an epizoochorous dispersal syndrome. While the length of the seed is 2.44 mm (n = 100), *S. ovata* has an extremely short pappus. Based on the weights of four lots of 50 seeds the seed is also light with an average weight of 0.086 mg. Due to the form of the seed and shrubby stature of plants, most wind dispersal is likely to be in the vicinity (metres to tens of metres) of the parent plants. Water dispersal and adhesion of the light seed are seen as more likely to lead to long distance dispersal. In Mexico, 92.4% of the 79 native *S. ovata* populations reported by Soejima *et al.* (2001) contained agamospermous polyploid plants; the remaining 7.6% were diploid populations. If plants in the local incursion are apomictic like most of those in Mexico, then a single *S. ovata* could produce viable seed without pollination and start a new infestation.

Age to reproduction Under controlled conditions within shade houses at both the Centre for Wet Tropics Agriculture (South Johnstone) and the Tropical Weeds Research Centre (Charters Towers), newly germinated seedlings were observed for growth rates and reproductive capacity. Cohorts were planted monthly to examine variations due to seasonality. Minimum age to reproductive maturity was recorded at three months with flowering and seeding were found to be synchronised irrespective of planting month, and season-driven. The older cohorts obtained heights of up to three metres with multiple stems.

Herbicide screening A comprehensive herbicide screening trial has been completed in a dense infestation (near 17° 44'S, 145° 25'E, altitude 690 m) south of the town of Ravenshoe. We applied 12 foliar herbicides as listed in Table 1 at two rates, to 1800 plants, using high-volume, low-pressure application methods. Pulse wetter at 200 mL 100 L^{-1} and Spraymarker dye at 200 mL 100 L^{-1} were used with each treatment (excluding control). Treatments were applied using Croplands Swissmex 20-litre backpack sprayers with hand-wands to deliver an overall foliar spray to the point of run-off.

The large-scale field trial identified several herbicides that caused high mortality of *S. ovata*. For example, fluroxypyr (Starane™ Advanced) and triclopyr/picloram/aminopyralid (Grazon™ Extra) were both very effective. Aminopyralid/fluroxypyr (Hotshot™) also gave high mortality but was not quite as effective as the other two. These chemicals were effective at rates in line with the Queensland Environmental Weeds Permit (APVMA 2016).

Some herbicides were quite ineffective, including glyphosate (Roundup® 360), glyphosate/ metsulfuron-methyl (Roundup® 360/Brushoff® 600) and di-sodium methyl arsonate (DSMA Clear 250). At present, this permit (APVMA 2016) contains only a few recommendations for *S. ovata*, however this research will lead to a specific *Stevia* permit in the future, with more options.

Rate refinement The three effective herbicides listed above were further tested in another field experiment to determine their efficacy at lower rates, bringing some in line with those used for co-occurring weeds such as *Chromolaena* odorata L. King & Robinson (Siam weed) and *Lantana camara* L. The results are currently being analysed but all appear to have been quite successful at the lower rates thus reducing overall costs to management.

Other control options Low-volume, high-concentration application methods for the three herbicides listed previously were also tested in the same infestation as the screening and refining trials, and show very promising results. Further replicated trials have commenced to verify the effectiveness of this method and refine the application rates.

'Splatter' techniques are especially useful for difficult terrain, where access with regular foliar spray equipment may be impossible (Brooks *et al.* 2014). A cut-stump method using 4.47 g a.i. L^{-1} aminopyralid and 44.7 g a.i. L^{-1} picloram (Vigilant II™ gel) has also been trialled which may prove useful in controlling individual plants while surveying in isolated/ rough terrain.

Table 1. Rates of herbicides used in initial herbicide screening.

Product	Active ingredient| (g a.i. L^{-1})	Low rate applied (g a.i. L^{-1})	High rate applied (g a.i. L^{-1})
Roundup® 360	glyphosate 360	4.50	9.00
Grazon Extra™	triclopyr 300	1.50	3.00
	picloram 100	0.50	1.00
	aminopyralid 8	0.04	0.80
Starane Advanced™	fluroxypyr 333	2.00	4.00
Hotshot™	fluroxypyr 140	0.98	1.96
	aminopyralid 10	0.07	0.14
Garlon™ 600	triclopyr 600	1.80	3.60
Tordon™ 75D	2,4-D 625	1.95	3.90
	picloram 75	0.49	0.98
Lontrel™	clopyralid 300	0.38	0.75
2,4-D™ 625	2,4-D 625	2.50	5.00
DSMA™	di-sodium methyl arsonate 250	7.50	14.00
Unimaz™	imazapyr 250	1.25	2.50
Roundup® 360 + Brushoff®	glyphosate 360	4.50	4.50
	metsulfuron-methyl 600	0.03	0.06
MAT™ 28	aminocyclopyrachlor 212	0.10	not applied
	metsulfuron methyl 600	0.76	

Due to an annual capacity to regenerate from ground or below ground level, slashing/cutting without the use of herbicides does not appear to be a useful control measure for *S. ovata*. Field observations of plants after fire have also shown re-shooting from the below ground root mass.

CONCLUSION

The preliminary results of the ecological and herbicide trials provide useful information and a number of tools for land managers to deal with this weed before it becomes a greater problem. They should be confident that they can target a conspicuous plant at a set time of year with effective control measures and a short-lived seed bank. With adequate resourcing, this small incursion can be effectively treated in the short to medium term before this species becomes a more widespread problem that requires control at a much larger cost. However, our studies reinforce the weedy potential of *S. ovata*, as demonstrated by its high fecundity and dominance of pasture situations. The window for detection and control of this species is short and the potential for seed dispersal particularly through vectors such as water, roads and grazing stock is high.

ACKNOWLEDGMENTS

We would like to thank Dr Joe Scanlan for reviewing the manuscript and the owners of Wooroora Station for site access. We would also like to thank Joe Vitelli for herbicide selection and rate advice, Faiz Bebawi for scientific advice and Richard Boyne and Kirsty Gough for technical assistance.

REFERENCES

Australian Pesticides and Veterinary Medicines Authority (2016). Minor use permit 11463 – Environmental weeds in non-crop areas. http://permits.apvma.gov.au/PER11463.PDF (accessed 7/6/16). Australian Government, Canberra.

Brooks, S.J and Setter, S.D. (2012). Soil seed bank longevity information for weed eradication target species. *Pakistan Journal of Weed Science Research* 18, 73-83.

Brooks, S.J., Gough, K.L. and Campbell, S.D. (2014). Refining low-volume, high-concentration herbicide applications to control *Chromolaena odorata* (Siam weed) in remote areas. *Plant Protection Quarterly* 29, 71-7.

Cortés-Flores, J. Andresen, E., Cornejo-Tenorio, G. and Ibarra-Manríquez, G. (2013). Fruiting phenology of seed dispersal syndromes in a Mexican Neotropical temperate forest. *Forest Ecology and Management* 289, 445–54.

DAF (2016). Declared Plants, Department of Agriculture and Fisheries. https://www.business.qld.gov.au/industry/agriculture/species/declared-pests/weeds (accessed 7/6/16).

Grashoff, J.L. (1972). A systematic study of the North and Central American species of *Stevia*. Ph.D dissertation, University of Texas, Austin, Texas.

Murphy, H., Metcalfe, D., Kriticos, D. and Lawson, T. (2009). Pro-active management of weeds based on their potential distributions under changing climatic conditions. Proceedings of the 9th Queensland Weed Symposium, eds T. Baker, S. Campbell, T. Grice, S. Kinnear and C. Sandral, pp. 61-7. (Queensland Weed Society, Brisbane).

Soejima, A., Yahara, T. and Watanabe, K. (2001). Distribution and variation of sexual and agamospermous populations of *Stevia* (Asteraceae: Eupatorieae) in the lower latitudes, Mexico. *Plant Species Biology* 16, 91-105.

Thompson, K., Bakker, J.P. and Bekker, R.M. (1997). 'Soil seed banks of NW Europe: methodology, density and longevity'. (Cambridge University Press, Cambridge, UK).

Many eyes on the prize: the role of citizen scientists in active weed surveillance

Melinda J. Laidlaw, Ailsa Holland and Gordon Guymer
Queensland Herbarium, Department of Science, Information Technology and Innovation, Toowong, Queensland 4066, Australia
(Melinda.Laidlaw@dsiti.qld.gov.au)

Summary Engaging and training citizen scientists to detect incursions of new and emerging weed species through active surveillance has proven an effective plant biosecurity strategy in Queensland. Between 2013 and 2015, approximately 550 community members attended Weed Spotter training sessions across Queensland where they were trained to detect declared pest plants in the field and submit specimens for formal identification. During the same three year period, more than 3000 weed specimens were submitted to the Queensland Herbarium for identification, 45% of which were subsequently incorporated into the Herbarium's collection and databases including Australia's Virtual Herbarium. Data from both specimen backed and non-specimen backed data are also accessible via the Atlas of Living Australia.

Submission of weed specimens by the community between 2013 and 2015 has allowed 383 incursions of 88 priority weed species to be notified to Biosecurity Queensland and local government officers, as well as other weed and land management organisations. In addition to priority weeds, members of the Weed Spotters Network have assisted in detecting newly naturalised and doubtfully naturalised species for Queensland and for Australia. Since 2013, 25 new naturalisations and 31 new 'doubtful naturalisations' have been detected including six species with no previous record in Queensland and eight species not previously known to occur in Australia. By detecting emerging and priority weeds through active surveillance, Weed Spotters are making a valuable contribution to biosecurity in Queensland.

Keywords Active surveillance, citizen science, early detection, Weed Spotters Network Queensland.

INTRODUCTION

Queensland is a large state (1.7 M km^2) with a significant weed problem, costing the economy in excess of $600 million a year in control and lost productivity costs (Biosecurity Queensland 2016). Addressing new weed threats quickly ensures that the control and management of weeds is targeted and cost effective. Discovering and correctly identifying new weed occurrences are vital first steps. Plant identification services are provided to Queenslanders by the Queensland Herbarium, however the size of the state and the scale of the weed threat mean that plant scientists, botanists and biosecurity staff need the help of the community to spot and identify new and emerging weeds. The Weed Spotters Network Queensland project aims to establish and support a weed incursion detection capability focused on the early detection of new and emerging weeds through harnessing and fostering community interest and skills in collecting and identifying weeds.

The Weed Spotters Network Queensland is a citizen science collaboration between the Queensland Herbarium, Biosecurity Queensland and the community. Local government support for the network is provided via the Land Protection Fund. The Queensland Herbarium, part of the Department of Science, Information Technology and Innovation (DSITI) is the Queensland Government's centre for research and information on Queensland plants, animals, fungi, algae and ecosystems. Herbarium botanists, ecologists and taxonomists identify and verify weed specimens submitted for identification and provide data to scientists, government agencies and the public. Where new naturalisations (species detected as becoming weeds for the first time), new occurrences of declared weeds or potential new weed species are detected, the Queensland Herbarium notifies Biosecurity Queensland and local government agencies. Biosecurity Queensland, part of the Department of Agriculture and Fisheries (DAF), provides leadership and support for invasive plants and animal management in Queensland through the development of policy, legislation, pest risk assessments, research, state-wide coordination of pest programs, communication, education and awareness programs. Biosecurity Queensland works directly with local governments and the community to manage new naturalisations, new occurrences of declared weeds, and potential new weed species where they are identified through a risk assessment process.

The Weed Spotters Network Queensland project coordinator is based at the Queensland Herbarium and manages communication across the state, provides training and educational resources and raises awareness of weed issues through the network's monthly bulletin, media releases, website and publications. Training is delivered to participants at the regional level and covers weed ecology, plant identification and plant collecting techniques to assist Weed Spotters to locate, identify and collect weed specimens. The Network coordinator is also the regional coordinator for Brisbane. Other volunteer regional coordinators are located in 21 regions around the state and provide assistance to Weed Spotters in their local area with the collection, identification and sending of specimens to the Queensland Herbarium. They also provide feedback to local Weed Spotters about new and emerging weeds found locally and assist with accessing government information.

Participants in the Weed Spotters Network come from diverse sectors of the community but share a common interest in weeds for either scientific, economic, or environmental reasons. The current membership includes landholders, industry representatives, regional natural resource management groups, Landcare, Bushcare, scientists, slasher drivers, weed contractors, gardeners, biosecurity and local government officers. Once trained, Weed Spotters send specimens of potential, new and emerging weeds and accompanying data to the Queensland Herbarium for identification, data capture and mapping. Analysis of this data leads to a better understanding of new weed incursions in Queensland and tracks how far and how quickly weeds are spreading. This information is crucial for developing early response strategies for new weeds, and for the ongoing control and containment of established weeds, a key objective of the project.

RESULTS

The Weed Spotters Network has been engaging and training citizen scientists to detect incursions of new and emerging weed species through active surveillance since 2008. Network membership has continually grown and now exceeds 1000 members (Figure 1). Between 2013 and 2015, 550 participants attended Weed Spotter training sessions on weed ecology, plant identification and plant collecting techniques to assist them to locate, identify and collect weeds across Queensland. To support this training, Weed Spotters have received over 30 editions of the Weed Spotters bulletin via email, providing them with information on weed spread pathways, biosecurity legislation changes and identification training for over 100 declared weed species.

Between 2013 and 2015, more than 3000 weed specimens were submitted to the Queensland Herbarium for identification. Almost half of these specimens (45%) and their associated data were subsequently incorporated into the Queensland Herbarium specimen collection and associated Herbrecs database. Data from weed specimens not incorporated into the collection is captured in Queensland's Wildnet (Wildlife Online) database. As a result, data on weed species, distribution and spread is accessible to the public, governments and scientists via tools such as Australia's Virtual Herbarium, the Atlas of Living Australia and Queensland's Wildlife online. Specimens are also extensively used for plant molecular and taxonomic studies by botanists and university researchers from around the world.

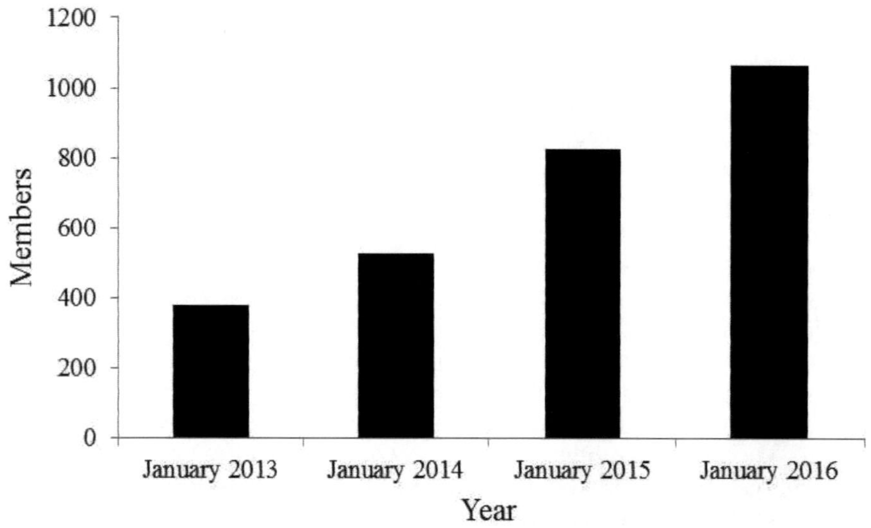

Figure 1. Weed Spotters Network Queensland membership 2013–2016.

Between 2013 and 2015, 383 incursions of 88 priority weed species were reported to Biosecurity Queensland and other weed management agencies as a result of specimens submitted to the Herbarium for identification (Table 1). These included 109 specimens of 28 Class 1 and 89 specimens of 27 Class 2 species declared under Queensland's *Land Protection (Pest and Stock Route Management) Act 2002*, including 98 specimens of 27 species also listed as Weeds of National Significance (WONS). These declared species are prioritised as targets because of their potential to cause economic, environmental and social impacts in Queensland. These incursions were notified to management agencies including Biosecurity Queensland and local governments for management. In addition, since 2013, 25 species have been detected as new naturalisations and a further 31 species have been detected as new 'doubtful naturalisations'. Six of these species were not previously known to occur in Queensland and eight species were not previously known to occur anywhere in Australia.

DISCUSSION

Weed Spotters are trained individuals and groups with the knowledge, skills, contacts and support to obtain accurate and timely weed identifications. Correct identification empowers actions in the community including weed control and management programs, the issuing of local government weed control notices and the pursuit of community project funding opportunities. The provision of advice, data and resources assists communities wishing to prepare funding applications, educational publications and programs aimed at increasing community knowledge and changing local attitudes towards weeds. The support of scientists, botanists and other weed specialists allows Weed Spotters to become informed biosecurity advocates within their communities. An online survey of Weed Spotters Network members in July 2015 asked Weed Spotters what they felt were the benefits of being part of the network:

- 93% of respondents told us that as a result of participation in the network, they know how to collect and submit a weed specimen to the Queensland Herbarium for identification
- 94% know who to contact to report a declared or potential weed in their region
- 100% of respondents felt that being a part of the Weed Spotters Network helped them learn more about weeds in their region and across the state
- 97% told us they were now actively looking out for new and emerging weeds in their region.
- 89% said they talk to their community, friends and/or family about the importance of early weed detection and
- 97% said that participation in Weed Spotters made them feel they contribute to keeping Queensland free of some of the world's worst weeds.

Weed Spotters have an increased understanding of invasion ecology. They understand that the time lag between when a weed species first invades and when it begins to have serious impacts can be exploited to save on weed control costs and lost production, as well as to protect our native ecosystems. They also appreciate the value of accurate identifications, data and mapping in supporting on ground actions. Weed Spotters understand the important role of herbaria and biological collections, the value of plant and biodiversity science and the applied benefits to our biosecurity of understanding our state's alien flora. As a result, the Weed Spotters Network has proven to be an effective aid to both plant science and biosecurity in Queensland and Australia.

REFERENCES

Biosecurity Queensland (2016). Impacts of weeds. [accessed 20 June 2016] https://www.daf.qld.gov.au/plants/weeds-pest-animals-ants/weeds/impacts-of-weeds.

Table 1. Weed incursions notified by the Queensland Herbarium 2013–2015.

	Number of notifications				Number of species			
Weed category	2013	2014	2015	**Total**	2013	2014	2015	**Total**
Class 1	34	37	38	**109**	17	15	20	**28**
Class 2	33	15	41	**89**	16	7	20	**27**
Class 3	6	1	0	**7**	3	1	0	**3**
Weeds of National Significance (WONS)	22	22	54	**98**	11	8	25	**29**
Northern Australian Quarantine Strategy (NAQS)	16	4	8	**28**	8	3	5	**13**
National Environmental Alert List (NEAL)	13	8	6	**27**	5	4	3	**6**

Introductions at what cost? Can insurance provide a new tool to manage weed incursions at the border?

Susie Hester[1,2] and Paul Downey[3]
[1] UNE Business School, University of New England, Armidale, New South Wales 2351, Australia
[2] Centre of Excellence for Biosecurity Risk Analysis, University of Melbourne, Victoria 3010, Australia
[3] Institute for Applied Ecology, University of Canberra, Canberra, Bruce, Australian Capital Territory 2601, Australia
(shester@une.edu.au)

Summary While international trade has many benefits, a significant side effect has been the creation of multiple pathways for the introduction and spread of non-native species across the globe. In Australia, a variety of tools, legislation and policy mechanisms have been developed and implemented to try and limit these accidental introductions, with limited success. Part of the problem is that import requirements do not provide an incentive for an importer to consider the impact on Australia of a potential invasive species introduction – typically importers do not bear any of the costs associated with managing an invasive they have inadvertently introduced. Instead, any management costs would be borne by the government and/or affected industries.

In this paper we explore the concept of introducing insurance contracts to the importing process – requiring importers to insure against the risk that their imported product either introduces a new invasive species or becomes invasive. Insurance companies would assess the risk of an incursion associated with a particular import and derive a premium accordingly. Importers would be able to reduce premiums through actions that result in a lower risk of incursion, such as further cleaning of packaging and containers pre-arrival, or additional testing of imported plants for invasive characteristics. As such, the scheme would both provide incentives for importers to reduce the likelihood of a weed introduction, and contribute to covering the cost of management should an incursion occur.

Keywords Insurance, border quarantine, importer behaviour, incentives, economic instruments.

Biological-based weed management, past successes and future opportunities

Andy W. Sheppard
CSIRO, GPO Box 1700, Canberra, Australian Capital Territory 2601, Australia
(andy.sheppard@csiro.au)

Summary Beyond the historical use of grazing animals, biologically based solutions to widespread weeds started over 100 years ago with the first classical biological control programs against *Opuntia* cacti and lantana. These two programs illustrated the fickle nature of biological control and weed management. The first was largely under complete control within a generation while the second still remains a problem without highly effective agents. Nonetheless biological control was the main form of ecological weed management until the arrival of chemical herbicides in the 1950s and its application continued grow until a peak into the 1990s, despite chemicals becoming the main basis for weed management in agricultural settings. 16 years into the new millennium, it is now the turn of chemical options to fade as understanding of their environmental and health impacts are recognized and as widespread weeds evolve resistance to most of the remaining active ingredients. This could be the new dawn for biological-based weed management as even agrochemical companies push all new investments into biopesticide options. Biological control continues to generate benefits, not limited by resistance, and ground breaking new genetic approaches like RNAi and gene-drive open new possibilities for tackling the current generation of intractable weeds species and resistant genotypes. I look back at past successes in bio-based weed management and also look to the future and the possibilities for managing weeds in ways we wouldn't have dreamed of even at the start of this millennium… a chance for biologists to control the weeds once and for all?

Habitat and landscape associations of sea spurge (*Euphorbia paralias* L.) invasion of coastal dunes across NSW

Ben Gooden[1], Liam Kelly[1], Peter J. Turner[2] and Kris French[1]
[1] Centre for Sustainable Ecosystem Solutions, University of Wollongong, Wollongong, New South Wales 2522, Australia
[2] Pest and Ecological Management Unit, National Parks and Wildlife Service, NSW Office of Environment and Heritage, PO Box 1967 Hurstville, New South Wales 1481, Australia
(bgooden@uow.edu.au)

Summary Containing the spread of new weed incursions is a key pillar of contemporary weed management and asset protection. The first step in containing a novel weed is to determine its range limits and habitat features associated with spread at the edges of its distribution. The aim of this study was to: (1) document the spatial extent of sea spurge invasion of coastal dunes across southern NSW and (2) determine whether its invasion is associated with particular habitat features of coastal dunes.

The presence or absence of sea spurge was recorded from 581 beaches between the Victorian-NSW border and Port Macquarie, using a combination of field surveys, interviews with local land managers, and inspection of herbarium specimens. Overall, sea spurge populations were detected at 64% of beaches, with the northernmost population detected near the township of Forster. However, the likelihood of detecting sea spurge infestations declined exponentially with decreasing latitude, from >90% of beaches in Bega Valley Shire, to ~40% in Shoalhaven and Illawarra and ~7% of beaches northwards of Sydney. Principal components analysis revealed that invasion was not significantly associated with dune length, orientation or distance between adjacent dunes, indicating that all beaches are likely to share a similar risk of invasion. However, given the low rates of invasion north of Sydney, it is likely that containment and local eradication of sea spurge on the mid-north coast of NSW will be effective to arrest its spread further north in the long term.

Keywords Coastal dune, invasion front, new weed incursions, weed containment.

Eradication in the tropics: constantly changing and adapting

Mick Jeffery[1] and Simon Brooks[2]
[1] Biosecurity Queensland, PO Box 20, South Johnstone, Queensland 4859, Australia
[2] Biosecurity Queensland, PO Box 187, Charters Towers, Queensland 4820, Australia
(mick.jeffery@daf.qld.gov.au)

Summary The National Tropical Weeds Eradication Program targets *Limnocharis flava*, *Mikania micrantha* and three *Miconia* species in Queensland and northern New South Wales. Over the 15 year life of the project, improvements have been made across a range of areas including surveillance methodologies, data collection and analysis, and community engagement. A summary of progress to eradication for each of these species is presented.

Keywords Eradication, *Limnocharis flava*, *Miconia* spp., *Mikania micrantha*.

INTRODUCTION

Biosecurity Queensland, a business group of the Queensland Department of Agriculture and Fisheries, is responsible for managing the National Tropical Weeds Eradication Program (NTWEP). This is a nationally cost-shared program that commenced in late 2001 and targets five species of highly invasive weeds which originate from tropical America: *Limnocharis flava* (L.) Buchenau, *Miconia calvescens* DC., *Miconia nervosa* (J.E.Smith) Triana, *Miconia racemosa* (Aubl.) DC., and *Mikania micrantha* Kunth.

SURVEILLANCE

How often? As eradication is the management objective, it is vital that any surveillance activities detect plants before they become reproductive. Through 15 years of successes and failures, the NTWEP has developed a range of methods to ensure we have the best possible chance of halting any future replenishment of the soil seed bank.

Firstly, scientific research has been undertaken to establish the time to reproductive maturity for most of the species within the program (Brooks and Setter 2014). For example, *M. calvescens* can produce fruit after four to five years (Murphy and Brooks 2010), while *L. flava* can flower in 46 days (Brooks *et al.* 2008). Surveillance visits for each of these species need to be scheduled accordingly. The program allows for two or more visits to an infestation within the time to reproduction period, which means every 16 months for *M. calvescens* and every three weeks for *L. flava*.

The aim is to have enough visits to an infestation to take into account:

(a) that detectability will improve with increasing plant size; and
(b) that field teams usually do not detect 100% of the plants in a complex rainforest landscape every time they traverse through. This needs to be balanced with cost so that unnecessary 'over' surveillance does not occur.

The NTWEP has increased the site revisit frequency over the past five years for most species with positive results. In 2011, the average revisit interval for *M. calvescens* sites was 29 months which had reduced to 16 months by the end of 2014 (S. Brooks unpublished data). This has correlated with a reduction in mature plant detections from 2.5/100 ha searched in 2010–11 to only 0.6/100 ha searched in 2015–16.

Technological edge In 2008 the NTWEP commenced using newer model GPS units which could provide an accurate location under dense rainforest in steep terrain (Brooks and Jeffery 2010). Prior to that, there was no accurate way to relocate previously treated sites. Not only did these GPS units provide very accurate plant locations for the first time, but they also allowed the ability to record a tracklog of where each survey team member surveyed each day (Figure 1).

These tracklogs are utilized for several purposes, including: the ability to check the thoroughness of surveillance; the creation of Geographical Information System (GIS) polygons to record the exact daily surveillance area completed; and a means of estimating the average time required to survey for each species in different terrain types. The latter information has been used to generate very accurate estimates of resources and related budget bids.

The NTWEP GPS units are also pre-loaded with locational data of previously found plants. This provides a visual reminder to teams undertaking surveillance to be extra vigilant when passing through these locations.

The other technological development has been the utilization of UAVs (un-manned aerial vehicles) to detect plants from the air. Eventually this will

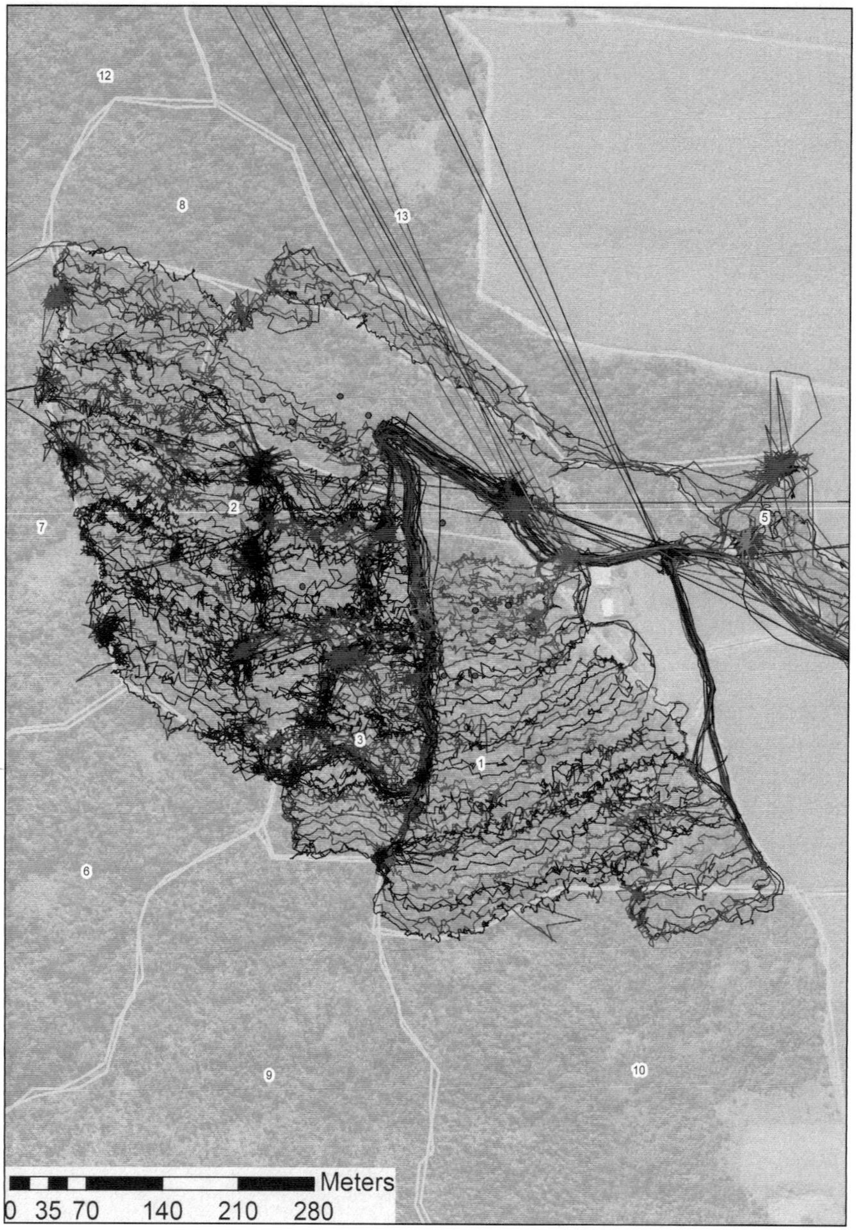

Figure 1. GPS tracklogs of ground survey coverage for *M.calvescens* within pre-determined GIS surveillance polygons.

replace manned helicopter flights (see separate paper at this conference). Over the past three years CSIRO have developed a UAV which has been trialed in north Queensland rainforests to detect the rainforest invader, *M. calvescens*. The UAV carries a camera payload which captures images every two seconds. In a small comparative trial, the UAV flew an area which had received helicopter surveillance two weeks prior and detected a much larger number of plants than a helicopter with three spotters on board. The NTWEP has recognized the greater detection ability, reduced cost and much higher safety of UAVs, and will move all of its aerial surveillance to drones during 2016-17.

Where to search? For known infestations, the NTWEP targets surveillance to the most likely areas of dispersal. Bird and wind dispersal kernels are generated from all detections of mature plants of the *Miconia* species or *M. micrantha*, and extensive downstream searches are undertaken for all species. Ground teams will search all surrounding suitable habitat out to radius of 1 km for *M. calvescens* (Murphy *et al.* 2008) and 200 m for *M. micrantha.*

Searching more broadly across north Queensland, or northern NSW in the case of *M. calvescens*, is reliant on a community engagement program which utilizes a range of techniques to educate both trained weed officers, and the general public (Brooks and Galway 2008) to help determine the extent of the incursions. For the more easily identifiable species such as *M. calvescens*, television commercials and social media have been used to reach a large audience. The NTWEP maintains a collection of live specimens of all species which is taken around the region to 'toolbox talks' of field staff from a wide variety of land management agencies, including National Parks, local government and Green Army. These specimens are also on display at all regional agricultural and garden events so that the public can see and touch the plants.

DATA MANAGEMENT

Consistent and comparable A wide range of data is collected on GPS units and data sheets which are later entered into the Tropical Weeds Database.

To assist with comparability across all sites, a static grid of one hectare management units has been overlaid on GIS mapping. All data recorded within these units are now pooled and progress to eradication is statistically analyzed every six months using plant presence and absence from each unit. This has allowed the NTWEP to compartmentalize large infestations into smaller management units, and will allow portions of sites to progressively meet the criteria for area freedom. An infestation is classified as eradicated if, after a period equating to the species' seed bank longevity (with no known input into the seed bank from adult plants), consistent monitoring has not detected seedling emergence for at least five consecutive years. This strict eradication criterion requires the Program to revisit sites for some species for over 20 years.

To provide an accurate mapping of plant distribution, any new plants of any species located at least 30 m away from a previously known site are recorded as a new site. This chosen spatial scale provides an excellent representation of plant density and distribution on GIS, but also provides fine scale locational data which is used to guide field teams via their GPS units. Each time teams then revisit these sites, they record whether any plants are present or not, remembering that the recording of nil presence is an essential part of showing progress to eradication.

The database is also interrogated monthly to inform surveillance scheduling. By checking visitation records to each management unit, it is possible to reprioritize surveillance for the month ahead to ensure all sites are visited at the required frequency to minimize the occurrence of reproductive plant relapses.

PROGRESS TO ERADICATION

Limnocharis flava The only aquatic species targeted in the NTWEP, *L. flava* occurs in locations which provide relatively easy access and detection. The visitation frequency for this species is every three weeks. Viable seed has been retrieved from one infestation for the past 12 years (S. Brooks unpublished data). No new sites have been detected since May 2015.

Of the known infestations, we are dealing mostly with the residual seedbank, and over 63% of management units had no seedling regeneration during 2015 (Table 1). The estimated year of complete eradication of *L. flava* from Australia is 2029.

Mikania micrantha This vine species is the most problematic to detect as it co-habits a rainforest landscape with hundreds of other vine species. The NTWEP has a specialist team of experienced *M. micrantha* spotters which is dedicated to the detection and removal of this species alone.

As this species has a high light requirement, it tends to dominate cleared areas, rainforest edges and canopy gaps. Two category five cyclones have struck the region in the past ten years and have provided optimum conditions for *M. micrantha* growth. Nonetheless, the NTWEP has reasserted control over this elusive invader, with excellent progress being made across the region. In Ingham, where *M. micrantha* was first detected in 1998, no plants have been detected since February 2014. In the other main area of distribution, Mission Beach, no germination was detected at 63% of known management units during 2015 (Table 1). This species appears to have a shorter seed longevity than the other species in the Program, but will still take at least until 2030 for eradication to be declared.

Miconia calvescens The rainforest tree, *M. calvescens* is the largest component of the NTWEP, with a surveillance area of over 4000 hectares per annum. As the fruit of this species is bird dispersed, most of the surveillance and control occurs in dense tropical rainforest on steep slopes, a large portion of which is within the Wet Tropics World Heritage Area.

This species had a limited legal trade prior to its declaration in 1996. Consequently, it has taken over 20 years to track down all of the plants that were purchased and planted in tropical gardens in north Queensland and subtropical gardens in northern New South Wales (Brooks and Jeffery 2010). With a seed longevity of at least 16 years (Brooks and

Setter 2012), this species will take the longest time to eradicate from Australia, with an estimated end point of 2039.

The species has a number of features which will assist its eradication. Firstly, it is easily identifiable due to its very large leaves with bright purple undersides. This not only assists field staff to detect it in the rainforest, it also allows it to be spotted from the air once it is over two metres in height, and is readily identifiable by the general public. The NTWEP regularly uses multi-media campaigns targeting this species due to the high level of positive reports we receive from the public (Brooks and Galway 2008) and the fact that many infestations originate from cultivated garden specimens (Brooks and Jeffery 2010). Secondly, the species has the longest pre-reproductive period of all the species, of at least four years (Murphy and Brooks 2010). This allows on ground and aerial surveillance a number of opportunities to detect plants before they set fruit.

An increase in surveillance resources which commenced in 2011 and improved again in 2015 has seen a much higher visit frequency to *M. calvescens* infestations. This has led to a decline in mature plant detections, now down to 0.6 plants per 100 hectares searched. At the majority of infestations the NTWEP is now just dealing with the persistent seed bank. During 2015, no germination was detected at 43% of sites (Table 1) and 99.5% of plants that the ground teams found were seedlings.

Table 1. Number and status of management units (1 ha) for each target species at the end of 2015. Units are in monitoring phase when plants are not detected for a year or more.

Species	Total units	% in monitoring
Limnocharis flava	62	63
Miconia calvescens	771	43
Miconia nervosa	28	39
Miconia racemosa	45	44
Mikania micrantha	83	63

Miconia nervosa* and *M. racemosa These two rainforest shrub species are each restricted to single locations in Far North Queensland. The densities of both species have reduced significantly, with only the occasional mature plant detected each year. The seed banks are in decline (S. Brooks, unpublished data), with a large portion of each of these locations demonstrating no seedling regeneration in 2015 (Table 1). As both of these species are co-located within *M. calvescens* infestations, teams search a far wider area surrounding these small infestations to confirm no further spread beyond the prescribed dispersal buffers. Both of these species are tracking well towards eradication, but a long seed life will mean it will be 2034 before area freedom is declared.

PROGRAM REVIEW

Independent assessment Every three years the NTWEP hosts a full week visit by a team of independent experts who are tasked to critically assess all aspects of Program performance. This process has been welcomed by the NTWEP as an opportunity to strengthen aspects that may require improvement, and to recognize progress and continual evolution of the program.

The last program review in 2013 was the first to set the same strict performance milestones for each species which would trigger an immediate review by the national Consultative Committee for Exotic Pest Incursions if any is not met each year. These milestones focus on three main aspects: (a) ensuring infested area decreases over time; (b) total extent is not expanding; and (c) that reproductive plant detections decline on a year to year basis.

For a program that has been running since 2001 and will not see the first of its species reach eradication before 2029, these performance milestones are necessary to ensure the large investment of time and resources allocated to each of these five species is achieving the intended results.

ACKNOWLEDGMENTS

The authors would like to acknowledge all of the eradication program staff and regional collaborators including National Parks and Local Government staff that work enormously hard to achieve the ultimate goal.

REFERENCES

Brooks, S.J. and Galway, K.E. (2008). Processes leading to the detection of tropical weed infestations during an eradication program. Proceedings of the 16th Australian Weeds Conference, eds R.D. van Klinken, F.D. Panetta and J.C. Scanlan, pp. 424-6. (Weed Society of Queensland, Brisbane).

Brooks, S.J. and Jeffery, M. (2010). Status of *Miconia calvescens* and the eradication program in Australia. Proceedings of the International Miconia Conference, eds L.L. Loope, J-Y. Meyer, B.D. Hardesty and C.W.Smith. (Maui Invasives Species Committee and Pacific Cooperative Studies Unit, University of Hawaii, Hawaii).

Brooks, S.J and Setter, S.D. (2012). Soil seed bank longevity information for weed eradication target species. *Pakistan Journal of Weed Science Research* 18, 73-83.

Brooks, S.J. and Setter, S.D. (2014). Issues and solutions for researching weed eradication target species. Proceedings of the 19th Australasian Weeds Conference, ed. M. Baker, pp. 255-8. (Tasmanian Weed Society, Hobart).

Brooks, S.J., Weber, J.M., Setter, S.D. and Akacich, B.A. (2008). Seed production and maturation of *Limnocharis flava* (L.) Buchenau in the field and glasshouse. Proceedings of the 16th Australian Weeds Conference, eds R.D. van Klinken, F.D. Panetta and J.C. Scanlan, pp. 180-2. (Queensland Weeds Society, Brisbane).

Murphy, H.T. and Brooks, S.J. (2010). The ecology of *Miconia calvescens* in Australia. Proceedings of the International Miconia Conference, eds L.L. Loope, J-Y. Meyer, B.D. Hardesty and C.W. Smith. (Maui Invasives Species Committee and Pacific Cooperative Studies Unit, University of Hawaii, Hawaii).

Murphy, H.T., Hardesty, B.D., Fletcher, C.S., Metcalfe, D.J., Westcott, D.A. and Brooks, S.J. (2008). Predicting dispersal and recruitment of *Miconia calvescens* (Melastomataceae) in Australian tropical rainforests, *Biological Invasions* 10 (6), 913-24.

Emerging environmental weed issues for Western Australia

Greg J. Keighery
Department of Parks and Wildlife, Locked Bag 104, Bentley Delivery Centre, Perth, Western Australia 6983, Australia
(greg.keighery@dpaw.wa.gov.au)

Summary Weeds are dynamic, as are the issues and resources to address these weeds. The Western Australian Herbarium currently makes over 100 changes to the names and naturalization status of plant species in WA per year and is currently attempting to obtain clear definitions of what should be added (or deleted) as naturalized plants to the WA/Australian flora.

The Department of Parks and Wildlife has completed several major prioritizations of environmental weeds across the State to attempt to utilize resources efficiently at a regional level. However, since these reviews a large number of Ecological Communities are being listed by the Commonwealth in Western Australia. Several of these have weed issues that have not been addressed in the past (*Limonium* species, tall wheat grass). Also changing land practices in the rangelands and the shift to more intensive farming in tropical Australia is changing the potential for more environmental weeds and conflicts with agricultural imports in these areas.

Community based and indigenous groups are increasingly becoming the front line in weed management as the government consists more of a declining and aging weed research/management cohort, where there is less active and passive surveillance of weeds in the field. The engagement and education of these groups needs to be a higher priority.

Finally increasing numbers of local native species are behaving as serious environmental weeds (*Typha orientalis* C.Presl.) under a changing climate and land uses. These are potentially challenging to control in natural ecosystems where native species are by default protected.

Keywords Environmental weeds, Western Australia, emerging issues.

Incorporating passive surveillance into management of weeds

Oscar Cacho[1] and Susie Hester[1,2]
[1] UNE Business School, University of New England, Armidale, New South Wales 2351, Australia
[2] Centre of Excellence for Biosecurity Risk Analysis, University of Melbourne, Victoria 3010, Australia
(shester@une.edu.au)

Summary Information on the location of pests and diseases often comes from members of the public, industry groups, plant or animal health professionals and their networks, all of whom report suspected detections at their discretion. Surveillance that leads to these 'passive detections' is activated and maintained through public- or industry-funded awareness campaigns. While several recent studies have demonstrated the benefits of passive surveillance across a range of pests and diseases, and while public awareness campaigns have become an integral part of invasion management programs, there is still much to learn about this type of surveillance. This includes the 'return on investment' in the activities that lead to passive surveillance, the size and types of investment required to achieve the desired level of reporting from the public, which parts of the community are more likely to report and whether these reports are reliable. This paper discusses recent findings on passive surveillance and puts forward a general framework for assessing the cost-effectiveness of passive surveillance.

Keywords Passive surveillance, general surveillance, community engagement, cost-effectiveness.

Weed invasions and native plant extinctions: a six-threshold framework

Paul O. Downey[1,2] and David M. Richardson[2]
[1] Institute for Applied Ecology, University of Canberra, Australian Capital Territory 2601, Australia
[2] Centre for Invasion Biology, Department of Botany and Zoology, Stellenbosch University, Private Bag X1, Matieland 7602, South Africa
(paul.downey@canberra.edu.au)

Summary Despite weeds being widely acknowledged as a major threat to biodiversity, there is growing debate, regarding the nature of the weed threat – in particular whether the outcome is likely to be the widespread extinction of native plants. The debate has raised questions on whether the threat posed by weeds to native plants has been overstated. We provide a conceptual framework to guide discussion in which the threat posed by weeds is considered in the context of a progression from no impact through to extinction.

We define six thresholds along this 'extinction trajectory' – global extinction being the final threshold. Although there are no documented examples of either 'in wild' (Threshold 5) or global extinctions (Threshold 6) of native plants that are attributable to weeds, there is evidence that native plants have crossed or breached other thresholds along the extinction trajectory due to weeds. Management interventions focussing on Thresholds 1 to 3 (a declining population through to the local extinction of a population) are likely to halt progress along the extinction trajectory.

Several other factors may be masking where native species are on the trajectory; these include a lack of appropriate data to accurately map the position of native species on the extinction trajectory, and the timeframe required to definitively state that extinctions have occurred. The critical issue for conservation managers is the underlying trend, because interventions cannot be implemented after extinctions occur. Thus the lack of evidence for extinctions attributable to weeds does not mean we should ignore the broader threat.

Keywords Biological invasions, conservation, extinction trajectory, threshold breaches, invasive plant species, declining populations.

REFERENCE

Downey, P.O. and Richardson, D.M. (2016). Alien plant invasions and native plant extinctions: a six-threshold framework. *AoB Plants* 8, plw047. (DOI 10.1093/aobpla/plw047)

Long term impact of harvest weed seed destruction on annual ryegrass

Catherine P.D. Borger[1] and Glen P. Riethmuller[2]
[1] Department of Agriculture and Food Western Australia, 75 York Road, Northam, Western Australia 6401, Australia
[2] Department of Agriculture and Food Western Australia, PO Box 432, Merredin, Western Australia 6415, Australia
(catherine.borger@agric.wa.gov.au)

Summary A long term trial (2003 to 2013) investigated annual ryegrass seed production in a crop rotation in Merredin, WA. The weed control methods in the trial included herbicides, or herbicides combined with weed seed destruction at harvest. Data from this trial was used to investigate varying scenarios of harvest weed seed destruction using the Weed Seed Wizard (WSW) model. The WSW demonstrated that herbicides plus weed seed destruction at harvest could reduce the annual ryegrass seedbank to very low levels over eight years. Use of herbicides alone did not reduce the seedbank to the same extent over the entire 11 year span of the trial. Harvest weed seed destruction in alternate years reduced the annual ryegrass seedbank to very low levels compared to use of herbicides alone.

Keywords *Lolium rigidum*, Weed Seed Wizard, model, seed production, residue burning, seedbank.

INTRODUCTION

Herbicide resistance is an increasingly severe problem throughout the world and successful weed control relies on the implementation of integrated weed management (IWM) programs. However, there is a general perception that IWM is difficult and expensive in the short term, compared to chemical weed control. It is important to present growers with IWM programs that are simple to implement and inexpensive in the short term, as well as financially rewarding in the long term.

A good example of a simple and inexpensive IWM technique is weed seed destruction at harvest, which can be achieved using a harvest seed destructor or chaff cart, or through residue burning (Walsh and Newman 2007, Walsh *et al.* 2013). Harvest weed seed destruction can destroy up to 85% of annual ryegrass (*Lolium rigidum* L.Gaud.) seed present at harvest time (Walsh *et al.* 2013, Walsh and Powles 2014). However, in a farm business it may not be possible to use harvest weed seed destruction in every year. Weed seed destruction using a seed destructor or chaff cart causes a delay to harvest, which may be unacceptable in those years where weather conditions will cause damage to the grain. Windrow burning may fail to effectively destroy weed seeds in some years, due to low residue levels or poor burning conditions (Walsh and Newman 2007).

A recent paper demonstrated that annual seed destruction at harvest can eradicate annual ryegrass over an 11 year period (Borger *et al.* 2016). However, the long term impact of occasional weed seed destruction at harvest has not been investigated. Computer simulations of weed growth under varying agronomic scenarios are an excellent way to formulate IWM programs, and demonstrate their long term efficiency and financial viability to growers. A useful model to assess IWM programs is the Weed Seed Wizard (WSW) (Department of Agriculture and Food Western Australia 2016). This model uses site specific soil type, weather records, and farm management records to simulate how different agronomic practices can affect density and seed production of a range of weed species. The current research utilised field data from the study by Borger *et al.* (2016) in the WSW model to investigate harvest weed seed destruction. We hypothesised that a standard crop rotation using herbicides alone would not adequately control annual ryegrass over an 11 year period, a system utilising herbicides plus weed seed destruction at harvest would eradicate annual ryegrass, and a system utilising weed seed destruction in alternate years would reduce but not eradicate annual ryegrass.

MATERIALS AND METHODS

The Weed Seed Wizard model The WSW is a computer simulation tool that allows the user to set weed species, crop and soil parameters, climate conditions from weather records and agronomic management events. The output interface produces results that include (but are not limited to) crop yield, size of the weed seedbank, and an economic analysis of the system. Full details of the model are included in Renton *et al.* (2008).

Field trial data to parameterise the model A trial was conducted from 1987 to 2013 at the Department of Agriculture and Food Western Australia (DAFWA)

Merredin Research Station (Borger *et al.* 2016). Crop species were sown at row spacings of 9, 18, 27 or 36 cm, and the crop residue was burnt or retained at the end of each winter annual growing season. Non-selective, pre-emergent and in crop herbicides were applied to control weeds, which predominately included annual ryegrass and Indian hedge mustard (*Sisymbrium orientale* L.). A chaff cart was used in 2003 to 2006. The herbicides successfully controlled wild mustard but the annual ryegrass remained in crop. Annual ryegrass seed production was assessed from 2003 to 2013. Full details of the results of this experiment from 2003 to 2013 are found in Borger *et al.* (2016).

Scenarios in the model The scenarios in the WSW model were based on the results of the field trial. Scenario 1 in the WSW model used data from the crop residue retained treatments and Scenario 2 used data from the residue burnt treatments. Both Scenarios assumed a row spacing of 18 cm.

The initial parameters loaded into the model were 'Southern WA grain region'. Under 'Scenario Setup' tab, the timeframe was 1 December 2002 to 31 December 2013. The initial annual ryegrass seedbank was set to 1500 seeds for Scenario 1 and 500 seeds for Scenario 2. This was because the annual ryegrass seedbank had reached higher levels in the plots where residue was retained (the seedbank had built up from 1987 to 2002). The soil type was 'Southern region – Silty clay loam', as the closest approximation to the actual soil type of the field trial, which was a mottled eutrophic red chromosol (Borger *et al.* 2016). A weather record was created from data generated by the DAFWA Merredin Research Station weather station (Department of Science Information Technology and Innovation 2016).

Each agronomic event (seeding, harvest etc.) and herbicide spray event that occurred in the field trial in Borger *et al.* (2016) was added to the WSW scenarios (Table 1).

Each harvest event in the WSW included date, crop species and harvest type. For Scenario 1, the harvest type was 'Chaff cart' for 2003 to 2006. However, the percent of weed seeds dropped in the field peas (in 2005) was increased from the default value of 20% to 80%, as Matthews *et al.* (1996) found that only 20% of the annual ryegrass seed was captured by a chaff cart in a pea crop. From 2007 to 2013, the harvest selected was 'All chaff spread/normal harvest'. However, crop lifters were used to lift/harvest a greater proportion of annual ryegrass seed heads, and so the percent of annual ryegrass seed dropped was adjusted from 99% to 96% in cereal and canola crops and left as 99% in the chickpea crop, based on the data in Borger *et al.* (2016). For Scenario 2, the harvest type was 'Narrow windrow and burning', as the closest approximation to burning all crop residue that was done in the initial field trial (Borger *et al.* 2016). However, the seed dropped was altered to 35% in the shorter field pea and chickpea crops. This was again based on the field pea data in Matthews *et al.* (1996), where only 65% of the annual ryegrass seed entered the harvester. 'Advanced Events' and 'Seed Drop' were selected in the WSW to indicate abscission of annual ryegrass seeds in the fallow of 2008.

Scenario 3 was used to investigate weed seed destruction at harvest in alternate years. Scenario 3 was the same as Scenario 2, but in every second year, 'Narrow windrow and burning' was replaced with 'All chaff spread/normal harvest'.

RESULTS

Over the 11 year span, the number of annual ryegrass seeds in the seedbank after harvest was greatest under Scenario 1, followed by Scenario 3 and Scenario 2 (Table 2). The number of seeds predicted by the WSW was similar, but not identical to the actual number of seeds observed by Borger *et al.* (2016). However, it should be noted that the WSW reports the total number of seeds in the soil seedbank, whereas the seeds observed in Borger *et al.* (2016) was only the annual ryegrass seed produced at harvest each year.

The income (and potential income) generated by each cropping system was greatest in Scenario 1 (Table 3). Income was similar in Scenario 2 and 3. The annual and total cost of weeds was greatest in Scenario 1, followed by Scenario 3 and 2 (total weed cost of \$104, \$35 and \$28 ha^{-1} over 11 years).

DISCUSSION

The prior research by Borger *et al.* (2016) noted that residue burning in each year (plus herbicide use) reduced annual ryegrass seed production to 0 seeds m^{-2} by 2011 (Table 2). The result from Scenario 2 in the WSW was similar, with the annual ryegrass seedbank reduced to 2 seeds m^{-2}. Again, it should be noted that the WSW accounts for all the seeds in the seedbank, not just the seeds added to the seedbank at harvest. Therefore, it is reasonable that the WSW seed number should be slightly higher than that observed by Borger *et al.* (2016). By comparison, herbicide use alone (no residue burning) did not reduce annual ryegrass seed production to zero, which was confirmed by Scenario 1 in the WSW. However, the WSW predicted a 'blow out' at the end of Scenario 1 (500 seeds in the 2013 seedbank) due to no in-crop herbicide. In reality, the Sakura herbicide used in 2013 was highly effective

Table 1. The crop sowing details, herbicide options and harvest date entered into the model, based on the agronomic events that occurred in the field trial detailed in Borger *et al.* (2016).

Sowing date, crop species (and number of viable seeds sown)	Herbicide, date of application (and percent annual ryegrass kill rate, residual kill rate and half-life of the herbicide at the first mention of each product)	Harvest date
4/6/03: wheat (*Triticum aestivum* L.) (255 seeds m^{-2})	1/5/03: Glyphosate 450 (85%, 0%, 0 days)	5/12/03
	28/5/03: Glyphosate 450	
	4/6/03: SpraySeed/Trifluralin (85%, 90%, 20 days)	
	29/7/03: Spear (80%, 0%, 0 days)	
3/6/04: wheat (218 seeds m^{-2})	2/6/04: Spray.Seed/Trifluralin	17/12/04
	20/7/04: Spear	
2/6/05: field pea (*Pisum sativum* L.) (62 seeds m^{-2})	15/4/05: SpraySeed (85%, 0%, 0 days)	11/11/05
	2/6/05: SpraySeed/Trifluralin	
	25/7/05: Select (80%, 0%, 0 days)	
8/6/06: wheat (280 seeds m^{-2})	25/1/06: Glyphosate + Hammer (85%, 0%, 0 days)	9/11/06
	10/5/06: Glyphosate + Hammer	
	8/6/06: SpraySeed/Trifluralin	
	8/6/06: Dual Gold/Atrazine (50%, 90%, 30 days)	
	21/8/06: Decision (80%, 0%, 0 days)	
26/6/07: barley (*Hordeum vulgare* L.) (247 seeds m^{-2})	26/6/07: SpraySeed/Trifluralin	6/11/07
	13/8/07: Decision	
5/5/08: chemical fallow	5/5/08: SpraySeed	1/12/08
	29/6/08: Glyphosate 450	
	26/8/08: Glyphosate 450	
	1/9/08: SpraySeed	
15/6/09: oilseed (*Brassica napus* L.) rape (129 seeds m^{-2})	15/6/09: SpraySeed/Trifluralin	9/11/09
3/6/10: wheat (217 seeds m^{-2})	3/6/10: SpraySeed/Trifluralin	9/11/10
	2/8/10: Achieve (80%, 0%, 0 days)	
7/7/11: wheat (198 seeds m^{-2})	7/7/11: Glyphosate + Hammer	30/11/11
	7/7/11: Boxer Gold (20%, 95%, 30 days)	
	26/7/11: Achieve	
18/6/12: chickpea (*Cicer arietinum* L.) (55 seeds m^{-2})	15/6/12: SpraySeed/Simazine (85%, 50%, 20 days)	10/12/12
	24/7/12: Select	
28/5/13: wheat (240 seeds m^{-2})	13/5/13: Glyphosate 450	26/11/13
	28/5/13: SpraySeed	
	28/5/13: Sakura (40%, 95%, 30 days)	

and annual ryegrass seed production did not increase (Table 2). However, the WSW confirmed the findings of Borger *et al.* (2016) that herbicides cannot eradicate annual ryegrass, whereas herbicides combined with harvest weed seed destruction can reduce annual ryegrass seed production to very low levels in eight years. These results indicate that the WSW is an excellent tool for modelling IWM programs, to help growers explore methods to eradicate annual ryegrass.

The total cost of weeds over the 11 year span was \$28 ha^{-1} in Scenario 2 (residue burnt). The weeds reduced the income (compared to the potential income) by less than \$1 ha^{-1} over the last four years of the trial. The total cost of the weeds was \$104 ha^{-1} in Scenario 1 where annual ryegrass seed production at harvest was not reduced to 0 seeds m^{-2}. Actual income (and potential income) was greater in Scenario 1 in spite of the higher weed burden, because residue burning reduced yield (Borger *et al.* 2016). However, on an actual farm, residue burning could be replaced with other methods of weed seed destruction at harvest that do not reduce yield (Walsh *et al.* 2013).

Scenario 3 demonstrated that the annual ryegrass seedbank could be reduced to very low levels even where it is not possible to utilise harvest weed seed

Table 2. The actual number of annual ryegrass seeds m^{-2} produced at harvest in Scenario 1 and 2 from Borger *et al.* (2016), and the number of seeds m^{-2} in the seedbank after harvest predicted by the WSW in Scenario 1, 2 and 3.

	Scenario 1		Scenario 2		Scenario 3
Year	Actual seeds	WSW seeds	Actual seeds	WSW seeds	WSW seeds
2003	296	394	117	152	152
2004	312	288	117	119	147
2005	558	586	221	129	151
2006	18	105	5	29	52
2007	54	23	23	5	8
2008	0	33	0	8	13
2009	319	121	152	17	49
2010	24	37	1	3	8
2011	162	41	0	2	8
2012	50	81	0	4	17
2013	1	500	5	9	85

Table 3. The year, grain price, potential income (income from the potential yield that could be achieved in the absence of weeds), actual income, weed cost (i.e. the difference between the actual income and the potential income), and the total income or weed cost over the 11 year span, for each Scenario.

		Scenario 1			Scenario 2			Scenario 3		
Year	Price (\$ t^{-1})	Potential income (\$ ha^{-1})	Actual income (\$ ha^{-1})	Weed cost (\$ ha^{-1})	Potential income (\$ ha^{-1})	Actual income (\$ ha^{-1})	Weed cost (\$ ha^{-1})	Potential income (\$ ha^{-1})	Actual income (\$ ha^{-1})	Weed cost (\$ ha^{-1})
2003	216	800	783	16.9	655	650	4.7	655	650	4.7
2004	197	377	372	5.2	349	347	1.8	349	347	1.8
2005	407	848	808	40.1	856	839	16.9	859	839	19.3
2006	242	644	635	9.9	641	639	2.2	642	639	2.6
2007	347	154	153	0.7	114	114	0.1	114	114	0.3
2008	*	*	*	*	*	*	*	*	*	*
2009	433	387	382	4.4	383	382	1.0	384	382	1.8
2010	257	278	276	1.9	278	278	0.3	278	278	0.8
2011	227	502	501	1.1	434	434	0.1	434	434	0.2
2012	393	53	53	0.3	63	63	0.0	63	63	0.1
2013	316	727	703	23.7	575	574	0.9	578	574	4.0
Total (\$ ha^{-1})		**4770**	**4666**	**104**	**4348**	**4320**	**28**	**4356**	**4320**	**35**

* Data not available as no crop was sown in 2008.

destruction in every year. Further, there was little change to the total cost of weeds (\$35 ha^{-1} over 11 years) compared to Scenario 2.

ACKNOWLEDGMENTS

This research utilised the Weed Seed Wizard model, which was a national initiative funded by the Grains Research and Development Corporation.

REFERENCES

Borger, C.P.D., Riethmuller, G. and D'Antuono, M. (2016). Eleven years of integrated weed management: long-term impacts of row spacing and harvest weed seed destruction on *Lolium rigidum* control. *Weed Research* doi: 10.1111/wre.12220.

Department of Agriculture and Food Western Australia (2016). 'Weed Seed Wizard. Version 7.1.7' (https://www.agric.wa.gov.au/weed-seed-wizard-0, verified 29 April 2016).

Department of Science Information Technology and Innovation (2016). 'Daily climatic data: patched point dataset & DAFWA weather stations' (http://srig-web/ppd/, verified 27 Jan 2016).

Matthews, J., Llewellyn, R., Jaeschke, R. and Powles, S. (1996). Catching weed seeds at harvest: a method to reduce annual weed populations. Agronomy – science with its sleeves rolled up. 8th Australian Agronomy Conference, eds D.L. Michalk and J.E. Pratley, pp. 684-5. (Australian Society of Agronomy, Toowoomba, Queensland).

Renton, M., Diggle, A. and Peltzer, S. (2008). Understanding, predicting and managing weed seedbanks in agricultural systems with the Weed Seed Wizard. Hot topics in the tropics. 16th Australian Weeds Conference, eds R.D. van Klinken, V.A. Osten, F.D. Panetta and J.C. Scanlan, pp. 77-9. (Queensland Weeds Society, Cairns, Queensland).

Walsh, M. and Newman, P. (2007). Burning narrow windrows for weed seed destruction. *Field Crops Research* 104, 24-30.

Walsh, M., Newman, P. and Powles, S. (2013). Targeting weed seeds in-crop: a new weed control paradigm for global agriculture. *Weed Technology* 27, 31-6.

Walsh, M.J. and Powles, S.B. (2014). High seed retention at maturity of annual weeds infesting crop fields highlights the potential for harvest weed seed control. *Weed Technology* 28, 486-93.

Use of cover crops as weed management tools in sugarcane fallow

Emilie F. Fillols and Timothy N. Staier
Sugar Research Australia Ltd, Meringa, Queensland 4865, Australia
(efillols@sugarresearch.com.au)

Summary In the Wet Tropics, new cost effective weed management strategies in sugarcane must be developed to reduce the reliance of herbicides and comply with the 60% reduction in pesticide pollutant loads under the ReefPlan program.

Two replicated trials investigating the weedicide potential of several cover crops and blends that could be used in sugarcane fallows were implemented in fallow blocks at Meringa, Queensland.

Trials were designed as split plots with three replicates. Five treatments were compared: *Vigna unguiculat a* (L.) Walp cultivar Ebony, *Dolichos lablab* L. cultivar Rongai, a blend of *V. unguiculata* and *D. lablab* (50/50), a blend of *V. unguiculata, D. lablab* and *Canavalia ensiformis* (L.) DC.) (40/40/20) and a blend of *V. unguiculata*, *D. lablab* and *Echinochloa esculenta* (A.Braun) H.Scholz variety Shirohie (40/40/20). Each treatment was tested at two sowing rates (green manure rate/twice green manure rate) and managed with and without herbicide control.

Cover crop and weed coverages were calculated (using photographic assessment) every second weed and dry biomass and nitrogen available for the following crop were measured after the cover crop was sprayed out.

Results show that the best cover crops to reduce weed coverage were either *V. unguiculata* alone or mixed 50% with *D. lablab*. Adding *C. ensiformis* or *E. esculenta* to the mix did not significantly improve the ground coverage. The high sowing rate tended to increase the speed of legume establishment and ground coverage and slightly decrease the amount of weeds (no significant differences). Managing weeds (with herbicide) was unnecessary and even slightly decreased the legume biomass. The amount of available nitrogen provided by the leading treatments was around 140 kg N ha^{-1}.

Keywords Cover crop, lablab, cowpea, sugarcane, green manure.

Annual ryegrass seed set reduced by desiccation and swathing of canola

Glen P. Riethmuller[1], Abul Hasham[2] and Catherine P.D. Borger[2]
[1] Department of Agriculture and Food, PO Box 432, Merredin, Western Australia 6415, Australia
[2] Department of Agriculture and Food, 75 York Road, Northam, Western Australia 6401, Australia
(glen.riethmuller@agric.wa.gov.au)

Summary Field studies were conducted at Mt Barker in 2010 and 2011 and Katanning in 2012 and 2013 to investigate techniques to reduce annual ryegrass viable seed set. The Mt Barker experiments included swathing with a commercially available sprayer kit attached and desiccation of the standing crop while at Katanning only desiccation was tested. Results in 2010 showed, on average, desiccation with diquat reduced the viable ryegrass seed production by 65%, and on average, desiccation or spraying on the swather with glyphosate reduced the viable ryegrass seed production by 45%.

The 2011 results showed swathing with above 2.4 L ha^{-1} of glyphosate at either 30% or 60% seed colour change of the canola reduced ryegrass seed viability by an average of 43% compared to the average of the nil and swathing alone while desiccation with above 2.4 L ha^{-1} of glyphosate at either 30% or 60% seed colour change of the canola reduced ryegrass seed viability by an average of 73% compared to the average of the nil and swathing alone.

In 2012, desiccation with Roundup Attack had no effect on yield or ryegrass viability at different rates at 20% or 50% seed colour change. In 2013, desiccation with Weedmaster DST at different rates and different water rates had no effect on yield but there was less ryegrass seed produced at 4.0 L ha^{-1} at 40 and 120 L ha^{-1} of water at 37% seed colour change.

Keywords Canola, desiccation, swathing, ryegrass.

Harvest weed seed control: the influence of harvester set up and speed on efficacy in south-eastern Australia wheat crops

John C. Broster[1], Michael J. Walsh[2] and Allison J. Chambers[1]
[1] Graham Centre for Agricultural Innovation (Charles Sturt University and NSW Department of Primary Industries), Charles Sturt University, Locked Bag 588, Wagga Wagga, New South Wales 2678, Australia
[2] IA Watson Grains Research Centre, University of Sydney, 12656 Newell Highway, Narrabri, New South Wales 2390, Australia
(jbroster@csu.edu.au)

Summary Herbicide resistant weeds are a major impediment to Australian grain production and one of the newer methods used to reduce their impact is the collection and/or destruction of weed seeds at harvest, harvest weed seed control (HWSC). An important factor in many HWSC systems is the proportion of weed seed exiting the harvester in the chaff fraction, with chaff carts, chaff tramlining, chaff lining and the Harrington seed destructor (HSD) only targeting this fraction. Therefore, this study was aimed at determining the proportion of annual ryegrass seed collected by harvester that subsequently exits in the chaff fraction under commercial harvest conditions. Dyed annual ryegrass or oat seed was introduced into the front of harvesters during the 2014 and 2015 harvests and these seed were then recovered from collected harvest residues. Five harvesters at separate locations were evaluated in 2014 with the proportion of introduced annual ryegrass seed collected in the chaff portion ranging from 27.9% and 69.6% (mean 47.1%) while between 1.2% to 9.2% (mean 3.7%) of seed was found in the grain portion. The proportion of oat seed recovered was 43.7% in the grain and 4.0% in the chaff. In 2015 using a single harvester at one location setup for maximum HWSC efficacy less than 5% of annual ryegrass seed was found in the straw fraction. It was determined that increasing harvester speed did not influence the proportion of annual ryegrass seed in the straw portion but did increase the proportion of wheat seed exiting in this fraction from 0.2% to 5.0%. This research shows that harvester set up and operation is important in the efficiency of HWSC as well as harvest efficiency.

Keywords Herbicide resistance, HWSC.

INTRODUCTION

A great proportion, if not most, of the crops across the cereal growing regions of south-eastern Australia contain populations of herbicide resistant annual ryegrass (*Lolium rigidum* Gaudin) and many of these populations are resistant to multiple herbicide modes of action (Broster *et al.* 2011, Boutsalis *et al.* 2012, Broster *et al.* 2013). As herbicides are the dominant weed control method the risk of herbicide resistance evolution is high posing a significant threat to the remaining effective herbicides. This has resulted in both farmers and researchers trying to reduce this reliance on herbicides by investigating alternate methods of weed control.

An increasingly popular alternative to herbicides is the suite of HWSC systems. These include narrow windrow burning, chaff lining, chaff tramlining, chaff carts and the Harrington seed destructor (HSD) (Walsh *et al.* 2013). HWSC targets weed species with a potential weakness, in that they retain a large portion of their seed at maturity (Walsh and Powles 2014). As such, while harvest can be a major factor in spreading weed seeds across a paddock, (Blanco-Moreno *et al.* 2004) it can also be utilised to reduce seed bank inputs.

An important factor in many HWSC systems (e.g. chaff carts, chaff lining and HSD) is that they only target the proportion of weed seed exiting the harvester in the chaff fraction. Therefore, this study was aimed at determining the proportion of annual ryegrass seed collected by the harvester that subsequently exits in the grain, straw and chaff fractions under commercial harvest conditions.

MATERIALS AND METHODS

2014 Harvest In 2014 sampling was undertaken in wheat crops on five different farms in southern New South Wales in December. The farmer's harvester (Sites 1 and 3 John Deere 9660 STS; Site 2 New Holland CR9080; Site 4 John Deere CTS; Site 5 Case 8120) was used with the machine set up as for the rest of that paddock, however harvester speeds and settings were not recorded. The harvest height at all sites was 15 cm, the optimal height for HWSC.

The plots were 20 m long and the width of the harvester front. Either 10,000 ryegrass or 500 Winteroo oat seeds that had been dyed (to differentiate from native seeds) were introduced into the front elevator of the harvester for the full length of the plots. Winteroo oats were chosen as the wild oat proxy as they were a similar size to wild oat seed and they accepted the dye better than the wild oats. Large bags were used to catch the grain and chaff portions of the harvested material. To minimize interference with air flow the chaff collection bags were made from shade cloth.

2015 Harvest In 2015 sampling was undertaken at a single location using one harvester fitted with an integrated HSD operating at the different ground speeds (4, 6 and 8 km h^{-1}) but with no other setting changes. Plot length was 50 m but sampling only took place in the last 20 m with the first 30 m used to obtain the required ground speed. As in 2014, 10,000 dyed ryegrass seeds were introduced into the front elevator of the harvester over the 20 m sampling length.

The straw fraction of the harvested material was caught on shade cloth which was rolled out as the harvester moved. To reduce the amount of material to be processed for annual ryegrass seed recovery much of the larger sized straw material was raked out of the collected sample in the field.

Sample processing Chaff and straw samples were processed in the laboratory to allow the collection and subsequent counting of annual ryegrass seed. In 2014 the amount of ryegrass seed in the uncollected straw fraction was determined as the difference between that introduced into the harvester and that found in the grain and chaff portions. In 2015 any wheat seed found in the straw portion was also collected and the percentage of grain loss calculated for each plot.

Crop yield was calculated from dry matter cuts in 2014 and from a yield monitor in 2015.

RESULTS

2014 Harvest The mean wheat yield at the five sites was 2.72 t ha^{-1} (1.85–3.67) with a mean harvest index of 37.3% (32.6–44.3). The straw fraction was 42.7% (30.5–52.3) of the total dry matter production and the chaff fraction the remaining 20% (13.3–25.2).

There were differences between harvesters in the proportion of annual ryegrass and oat seed recovered in the grain (difference between minimum and maximum – annual ryegrass 775%; oats 185%) and chaff (annual ryegrass 250%; oats 675%) fractions of the harvested material (Figure 1). There was no significant correlation between wheat yield or harvest index and the proportion of ryegrass or oat seed recovered in either the grain or chaff fraction (data not shown).

On average 3.7% (1.2–9.2) of annual ryegrass seed was recovered in the grain fraction and 47.1% (27.9–69.6) in the chaff fraction. The harvester with the highest level of ryegrass seed in the grain fraction (Site 1) had the lowest percentage in the chaff fraction. Conversely, the site with the highest percentage of ryegrass seed in the chaff fraction (Site 4) had the lowest percentage of seed recovered in the grain fraction, although the amount of seed recovered from the grain fraction was no different to that of the other sites (except Site 1).

On average 43.7% (32.3–60.6) of the oat seed was recovered in the grain fraction and 4.0% (1.5–10.2) in the chaff fraction. With the exception of Site 1, similar to the ryegrass, the sites with a low percentage of oat seed recovered in one fraction (grain or chaff) had a higher percentage of seed recovered in the other fraction (Figure 2).

A significant ($r = 0.874$; $P < 0.05$) positive correlation was recorded between the percentage of ryegrass and oat seed found in the chaff portion but none of the other correlations (grain:grain, chaff:grain or grain:chaff) were significant

2015 Harvest The mean crop yield in 2015 was 2.45 t ha^{-1}. There was a significant difference in crop yield for the three harvest speeds with a lower yield ($P < 0.01$; lsd = 0.26) recorded at 6 km h^{-1} (2.16 t ha^{-1}) compared with both 4 km h^{-1} and 8 km h^{-1} (2.67 and 2.54 t ha^{-1} respectively).

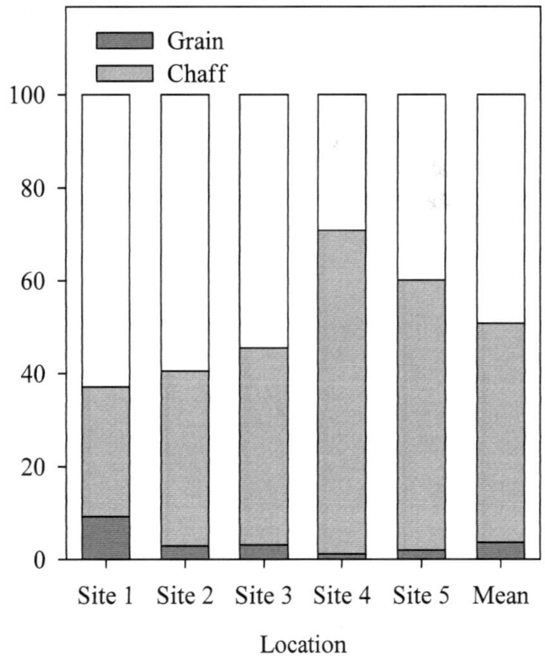

Figure 1. Differences between harvesters in 2014 for the amount of ryegrass seed found in grain ($P < 0.001$; lsd = 2.54) and chaff ($P < 0.001$; lsd = 16.3) fractions, the remainder of the seed was assumed to be in the straw fraction.

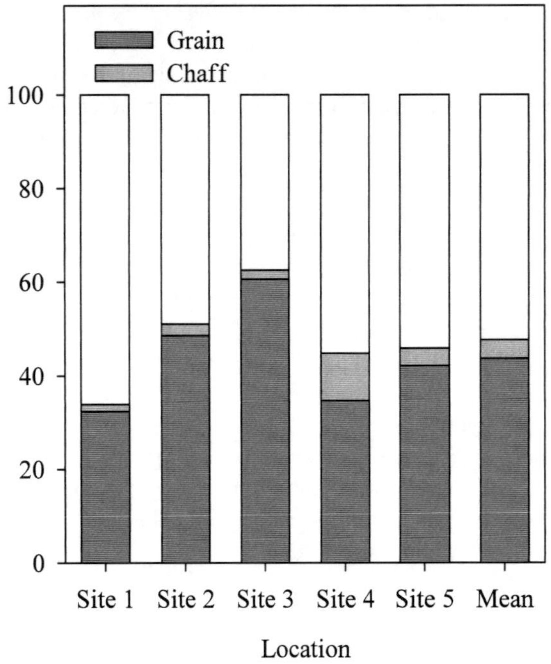

Figure 2. Differences between harvesters in 2014 for the amount of oat seed found in grain ($P < 0.001$; lsd = 6.27) and chaff ($P < 0.001$; lsd = 3.76) fractions, the remainder of the seed was assumed to be in the straw fraction.

Overall 3.38% of the ryegrass seed introduced into the harvester was recovered with no difference between the three harvest speeds ($P > 0.05$) (Figure 3a).

However, increasing the harvest speed from 6 to 8 km h^{-1} resulted in a significant increase in the amount of wheat seed found in the straw fraction, from 6.2 kg ha^{-1} (0.98%) to 127.4 kg ha^{-1} (5.03%) ($P < 0.001$; lsd = 37.04), there was no difference between 4 and 6 km h^{-1} in the percentage of wheat seed found (Figure 3b).

DISCUSSION

The harvester used in 2015 was set-up to maximize the efficiency of the attached IHSD. This is shown by the low percentage (3.4%) of ryegrass recovered in the non-targeted straw fraction, compared to the predicted average level of 49.2% (introduced minus grain and chaff) from the previous year. While part of this difference may have been due to the shade cloth bags used to capture the chaff portion interfering with air flow through the harvesters it is considered unlikely that this would affect nearly 50% of the introduced seed especially as large amounts of chaff were still being captured in these bags.

Harvester operation is of vital importance in maximizing the efficiency of HWSC systems, especially if the system chosen is one that focuses only the chaff fraction e.g. HSD, chaff carts, chaff lining or chaff tramlining. However, the differences between annual ryegrass and oat seed show that the most effective harvester operation for HWSC may also be species dependent (Figures 1 and 2).

When the settings are optimal for wheat, as was the case in 2015, the harvest speed was shown not to influence the amount of ryegrass seed present in the straw fraction (Figure 3). Increasing the harvester speed did result in increased grain loss (in the straw) especially when the harvester was operating at full capacity, in this case 8 km h^{-1}.

Those systems that target only the chaff fraction (e.g. HSD and chaff carts) also remove less nutrients (none for the HSD) than narrow windrow burning, a system which captures both the chaff and straw fractions of the harvested material. Narrow windrow burning is often the system chosen when farmers commence using a HWSC system due to its low setup cost. This is, however, offset by the increased labour component when burning the windrows.

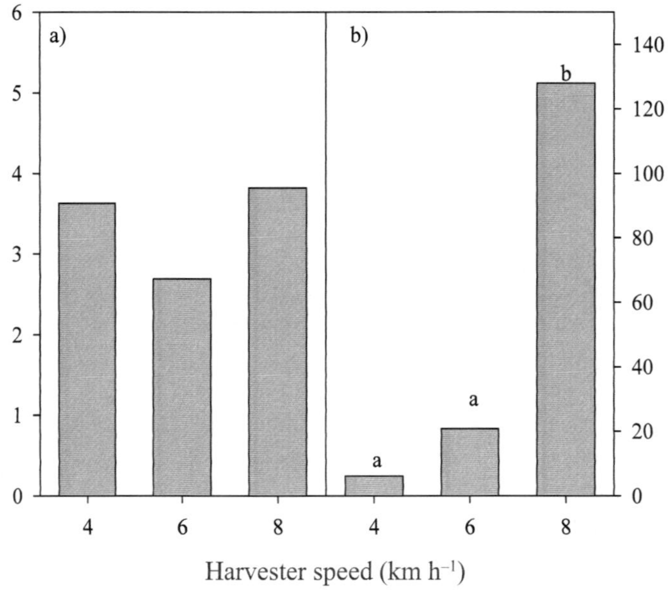

Figure 3. Percentage of annual ryegrass seed (a) and amount of wheat seed (kg ha^{-1}) (b) recovered from straw fraction of harvested material at three harvest speeds in 2015.

ACKNOWLEDGMENTS

This work was funded by a Grains Research and Development Grant (UWA00146). The authors would like to acknowledge the assistance of both casual staff and students from Charles Sturt University who assisted in the collection and processing of the samples. This work would not have occurred without the cooperation of several growers (Des Mason, Mark Kreutzberger, Andrew Harding, Ian Mason, Roy Hamilton and David Humphris) who generously made available their crops and harvesters for this research.

REFERENCES

Blanco-Moreno, J.M., Chamorro, L., Masalles, R.M., Recasens, J. and Sans, F.X. (2004). Spatial distribution of *Lolium rigidum* seedlings following seed dispersal by combine harvesters. *Weed Research* 44, 375-87.

Boutsalis, P., Gill, G.S. and Preston, C. (2012). Incidence of herbicide resistance in rigid ryegrass (*Lolium rigidum*) across southeastern Australia. *Weed Technology* 26, 391-8.

Broster, J.C., Koetz, E.A. and Wu, H. (2011). Herbicide resistance levels in annual ryegrass (*Lolium rigidum* Gaud.) in southern New South Wales. *Plant Protection Quarterly* 26, 22-8.

Broster, J.C., Koetz, E.A. and Wu, H. (2013). Herbicide resistance levels in annual ryegrass (*Lolium rigidum* Gaud.) and wild oats (*Avena* spp.) in south-western New South Wales. *Plant Protection Quarterly* 28, 126-32.

Walsh, M.J., Newman, P. and Powles, S.B. (2013). Targeting weed seeds in-crop: a new weed control paradigm for global agriculture. *Weed Technology* 27, 431-6.

Walsh, M.J. and Powles, S.B. (2014). High seed retention at maturity of annual weeds infesting crop fields highlights the potential for harvest weed seed control. *Weed Technology* 28, 486-93.

Influence of annual ryegrass seed retention height on harvest weed seed control (HWSC) and harvest efficiency

Michael J. Walsh[1], John C. Broster[2], Charlotte Aves[3] and Stephen B. Powles[4]
[1] I.A. Watson Grains Research Centre, University of Sydney, 12656 Newell Highway, Narrabri, New South Wales 2390, Australia
[2] Graham Centre for Agricultural Innovation (Charles Sturt University and NSW Department of Primary Industries), Charles Sturt University, Locked Bag 588, Wagga Wagga, New South Wales 2678, Australia
[3] Melbourne School of Land and Environment, University of Melbourne, Nalinga Road, Dookie, Victoria 3647, Australia
[4] Australian Herbicide Resistance Initiative, School of Plant Biology, University of Western Australia, 35 Stirling Highway, Crawley, Western Australia 6009, Australia
(m.j.walsh@sydney.edu.au)

Summary HWSC is an Australian innovation where systems have been invented to specifically target the weed seed bearing chaff fraction during crop harvest. This approach to weed control is a response to escalating frequencies of herbicide resistant crop weeds that was developed because there is the opportunity to target the high seed retention levels of these weeds at crop maturity. However, seed retention is not complete and weed seeds are frequently retained at lower heights than crop seed heads. The aim of this study was to examine the seed retention height of annual ryegrass through the canopy of wheat crops at maturity. During the 2013 harvest period annual ryegrass and wheat plant material was collected at five heights (0, 10, 20, 30, and 40 cm) each within four 1.0 m^2 quadrats in 25 wheat fields across the WA wheatbelt. On average 66% of annual ryegrass seed was retained above a 10 cm harvest height. However, a harvest height aimed at grain collection only (30 cm) would have reduced annual ryegrass seed collection and HWSC potential to 40%. The major proportion of wheat biomass (66%) occurs above 30 cm however, reducing harvest height from 30 to 10 cm would mean the collection of an additional 14% of the total biomass production. Thus, operational cost of harvesting lower for HWSC is the processing of just an additional 14% of crop biomass. This will impact to some extent on the speed of harvest but in contrast the gains from the adoption and use of HWSC are substantial.

Keywords Herbicide resistance, harvest height.

INTRODUCTION

Seed retention at maturity has been identified as a weed control target for the problematic annual weed species of Australian crop production systems. The major weeds, *Lolium rigidum* Gaudin (annual ryegrass), *Raphanus raphanistrum* L. (wild radish), *Bromus* spp. (brome grass) and *Avena* spp. (wild oats) retain significant proportions of total seed production, above a low harvest height (15 cm) at wheat crop maturity (Walsh and Powles 2014). During wheat harvest, the seed of these weeds are collected and processed and then exit the harvester, predominantly in the chaff fraction, where they are evenly redistributed across the paddock. Preventing inputs to the weed seedbank has been a focus for weed control for many years and therefore, harvest is now recognized as a weed control opportunity. Subsequently, HWSC systems have been developed specifically to target weed seeds during commercial wheat crop harvest to prevent inputs to the seed bank (Walsh *et al.* 2013).

Approaches used to target weed-seed bearing chaff include: collection and subsequent burning (chaff cart), concentration in a narrow windrow with straw residues for subsequent burning (narrow windrow burning), collection in bales along with straw residues (bale direct system) and mechanical destruction during harvest (Harrington seed destructor). When implemented effectively, these systems all deliver similarly high levels of weed seed destruction (>85%). Therefore, growers can choose an HWSC system to suit their farming operation.

The developed HWSC systems effect weed control by targeting the weed seed bearing chaff fraction and when implemented correctly are all similarly effective. The overall efficacy of these systems then is reliant on the proportion of total seed production that is collected during the harvest operation. The height and distribution of weed seed through the crop canopy at maturity impacts on the efficacy of HWSC. Thus the aim of this study was to establish the distribution through the canopy of mature wheat crops of retained annual ryegrass seed at widely dispersed WA wheatbelt locations.

MATERIALS AND METHODS

At the start of the 2013 harvest, (Nov.–Dec.) wheat crops at 25 locations across the WA wheatbelt were sampled to establish the height of annual ryegrass seed retention at crop maturity. Sampling was conducted within the first one to two weeks past crop maturity. In each wheat crop a 1.0 m^2 quadrat was place around annual ryegrass plants, these along with the surrounding wheat plants were then sampled at five heights commencing at 40 cm above the soil surface and then at 10 cm intervals to the soil surface. The soil surface within quadrat area was swept with a brush to collect any seed, seed heads and plant material that had fallen from wheat and annual ryegrass plants. Four quadrat samples were collected at each location. Before processing, the collected plant samples were oven dried at 70°C for 48 hours and then weighed. Any wheat heads present at each sample height were threshed and weighed to determine crop yield and proportion of yield at each sampling height. Annual ryegrass samples were weighed to determine dry matter production, threshed and then the seed produced was counted to determine seed production at each sampling height. The percentage of annual ryegrass seed collected above each of the harvest heights was calculated for each site.

RESULTS

Wheat crop sampling at 10 cm intervals from 40 cm downwards identified a uniform distribution of retained annual ryegrass seed through the wheat crop canopy at harvest (Figure 1). The 40 cm sampling height resulted in the collection of 24% of total seed production. Each successive 10 cm decrease in sampling height realised the collection of approximately 15% additional retained seed.

With the major proportion of total wheat biomass (61%) located above the 40 cm sampling height there were only reduced amounts of biomass collected at lower 10 cm sampling increments. As was observed for annual ryegrass seed distribution there were consistent proportions of wheat biomass collected at each of the 10 cm sampling intervals below 40 cm. However, in contrast to annual ryegrass seed retention only 8% of total biomass was collected at each of the 10 cm sampling increments. Therefore, at the 10 cm height the cumulatively collected proportion of wheat biomass was just 24% more than biomass collected at 40 cm.

Incremental sampling identified an average harvest height of 30 cm that would likely have been used to harvest the sampled wheat crops in 2013. Wheat grain was almost entirely located in the upper canopy (i.e. above 30 cm). Although on average 91% of wheat

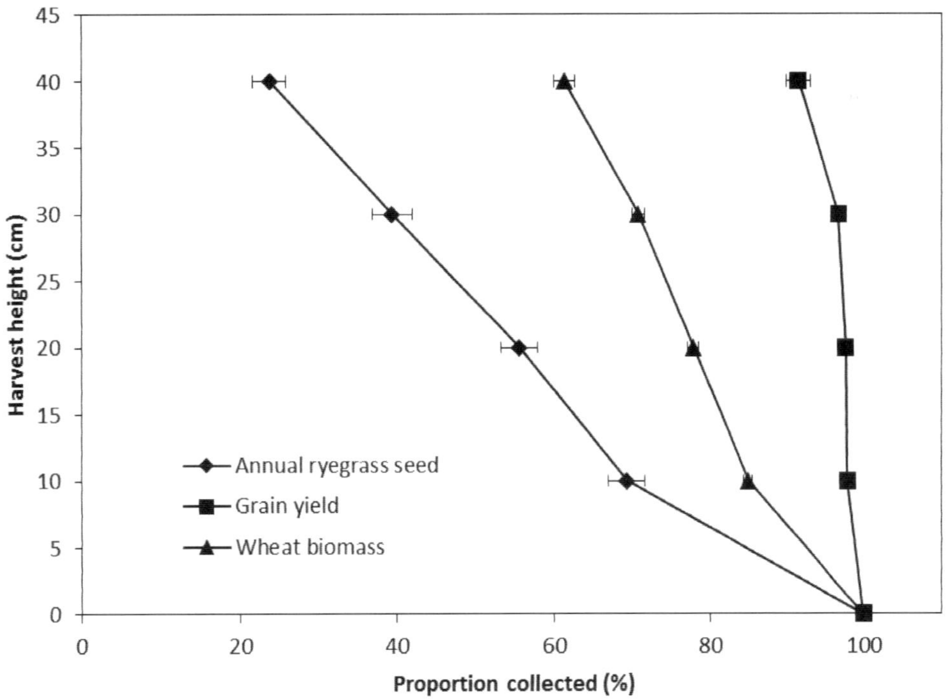

Figure 1. The cumulative proportion of annual ryegrass seed, wheat biomass and wheat grain collected at decreasing harvest heights.

grain was collected at the 40 cm sampling height, an additional 6% of yield was collected at the 30 cm height. There was no grain collected at the 10 or 20 cm sampling heights. However, there was some head and seed loss that prevented the entire grain production from being collected.

High average seed retention levels were recorded for annual ryegrass present in wheat crops across the WA wheatbelt during the initial stages of the 2013 harvest. In annual ryegrass infested areas of wheat crops at 25 locations across the WA wheatbelt there was an average of 66% of total annual ryegrass seed production retained above a low 10 cm harvest height (Table 1). Although, there was large variation around this mean seed retention (35–89) the median of 69 indicates that the majority sites had higher than average seed retention levels.

Annual ryegrass seed production levels clearly highlighted the seedbank replenishment/establishment potential for annual ryegrass plants surviving to maturity. The mean plant density for the sampled areas was 9 plants m^{-2} with a seed production of 218 seed plant^{-1}. This would potentially result in a seedbank contribution of 2008 seed m^{-2}, more than sufficient for the establishment or continuation of an annual ryegrass infestation (Table 1).

Table 1. Annual ryegrass plant density, seed production and seed retention at wheat crop maturity at 25 locations across the WA wheatbelt in 2013. Numbers in brackets are the standard errors of the mean of four replicates.

Location	Plant density (plants m^{-2})	Seed Production (seed m^{-2})	Seed retention >10 cm (%)
Mullewa	16 (5.3)	3287 (1134)	56 (6.4)
Kulin	20 (0.9)	3527 (924)	80 (4.9)
Eradu	23 (0.1)	4865 (1160)	50 (13)
Gibson	4 (1.3)	346 (83)	46 (13.9)
Bruce Rock	6 (0.7)	577 (193)	84 (3.1)
Irwin	3 (1.0)	1091 (628)	67 (8.2)
York	7 (0.5)	2366 (678)	50 (3.3)
Wagin	4 (0.6)	951 (191)	71 (6.0)
Pingelly	4 (0.6)	879 (184)	78 (7.8)
Darkan	9 (0.7)	2147 (713)	68 (6.6)
Hyden	4 (1.2)	1136 (222)	75 (7.7)
Holt Rock	7 (0.1)	494 (156)	85 (8.2)
Latham	14 (6)	3941 (1080)	70 (8.7)
Muradup	6 (0.8)	1188 (511)	64 (2.7)
Yarawindah	3 (0.4)	926 (367)	69 (6.7)
Corrigin	3 (0.4)	429 (120)	35 (3.8)
Tenindewa	3 (0.4)	817 (162)	80 (3.1)
Koorda	16 (3.4)	1467 (494)	78 (12.5)
Mingenew	4 (1)	798 (234)	57 (13)
Callingiri	35 (3.4)	4144 (709)	89 (2.7)
Coorow	6 (1.5)	983 (271)	36 (19)
Scadden	5 (1.3)	487 (128)	59 (4.6)
Wyalkatchem	4.3 (1.4)	2206 (321)	60 (14.6)
Merredin	3.8 (0.2)	1872 (890)	81 (4.7)
Kojonup	34.7 (2.8)	6177 (977)	72 (4.0)
Average	**9.2 (0.9)**	**2008 (218)**	**66 (2.5)**

DISCUSSION

Across 25 WA wheat crops in 2013 66% of annual ryegrass seed production would have been collected at a 10 cm harvest height. Additionally it was determined that annual ryegrass seed was found to be retained uniformly through the wheat crop canopy profile, therefore any increase in harvest height consistently reduced the amount of seed collected. Cross-sectional sampling down through the crop canopy determined a uniform distribution of retained annual ryegrass such that every 10 cm cross section contained 15% of annual ryegrass seed production. Thus every 1 cm increase in harvest height would have resulted in a 1.5% reduction in annual ryegrass seed collection. Therefore, a 30 cm harvest height as identified as optimal for grain only collection in 2013 would have resulted in just 39% annual ryegrass seed collection.

With the majority of wheat crop biomass located above 30 cm then lowering harvest height does not substantially increase the amount of crop biomass that needs to be processed during harvest. Similar findings have been reported for wheat crops in NSW and Victoria (Broster *et al.* 2015). A barrier to the adoption and effective use of HWSC systems has been the perception that lower harvest heights result in the collection of excessive amounts of crop biomass.

Across the 25 wheat crops sampled in this study 66% of total biomass was located above 30 cm wheat only harvest height. Reducing this height to 10 cm resulted in just 14% additional biomass being collected.

ACKNOWLEDGMENTS

This work was funded by a Grains Research and Development Grant (UWA00146). The authors would like to acknowledge the assistance of both casual staff and students from University of Western Australia who assisted in the collection and processing of the samples and the farmers who allowed the sampling to take place.

REFERENCES

Broster, J.C., Walsh, M.J., Aves, C. and Powles, S.B. (2015). Harvest weed seed control: ryegrass seed retention levels in south-eastern Australia wheat crops. Proceedings of the 17th Australian Agronomy Conference. Hobart, Tas, eds T. Acuna, C. Moeller, D. Parsons and M. Harrison. (Australian Society of Agronomy, Hobart)

Walsh, M.J., Newman, P. and Powles, S.B. (2013). Targeting weed seeds in-crop: a new weed control paradigm for global agriculture. *Weed Technology* 27, 431-6.

Walsh, M.J. and Powles, S.B. (2014). High seed retention at maturity of annual weeds infesting crop fields highlights the potential for harvest weed seed control. *Weed Technology* 28, 486-93.

Narrow row spacing suppresses weeds more effectively than wide row spacing in mungbean

Bhagirath Singh Chauhan
The Centre for Plant Science, Queensland Alliance for Agriculture and Food Innovation (QAAFI), Gatton/Toowoomba, Queensland 4343/4350, Australia
(b.chauhan@uq.edu.au)

Summary Mungbean (*Vigna radiata* (L.) Wilczek) is a nutritionally important food and feed legume crop. Its production in Australia is concentrated in the northern grains regions; characterised by summer dominant rainfall. In general, the crop is grown in wide rows of up to one metre and therefore, it is prone to heavy weed infestation. Weeds can be controlled using herbicides; however, continuous use of herbicides with similar modes of action could result in the evolution of resistance in weeds. Therefore, weed management tactics, which can make crops more competitive to weeds, are required. A field trial was conducted in the summer season of 2015 at the research farm of the University of Queensland, Gatton, Queensland, to determine the effects of row spacing (25, 50 and 75 cm) and weed infestation period [no weeds (weed-free); weeds grown up to 3 and 6 weeks after crop planting; and weeds grown up to crop harvest] on weed growth and mungbean yield.

There were 60–70% fewer weeds emerging three weeks after crop planting in 25 to 50 cm rows, compared to 75 cm rows. In the weed-free condition, as well as in low to moderate levels of weed infestation, the grain yield of mungbean was lower when grown on 75 cm rows.

Keywords Non-chemical weed control, row spacing, seeding rate, crop competition.

The economics of tactical mechanical weed control in broadacre farming systems

James Hagan[1], Andrew Zull[1], Michael Widderick[2] and Steven Davies[3]
[1] Queensland Department of Agriculture and Fisheries, 203 Tor St, Toowoomba, Queensland 4350, Australia
[2] Queensland Department of Agriculture and Fisheries, Leslie Research Centre, 13 Holberton Street, Toowoomba, Queensland 4350, Australia
[3] Department of Agriculture and Food Western Australia, 20 Gregory Street, Geraldton, Western Australia 6530, Australia
(james.hagan@daf.qld.gov.au)

Summary With increasing pressure being placed on ever smaller available chemical modes of control, and effective chemical options significantly increasing in price as herbicide resistance removes prior control options, the difference in upfront cost of chemical and mechanical control is shrinking. For example newer formulations such as (pyrasulfotole) Velocity®, isoxaflutole (Balance®), and pyroxasolfone (Sakura®), can all cost over \$30 ha^{-1} for a single pass at popular rates and potentially double this when taking into account the higher rates and use of double knocks to control hard to kill weeds. In addition to this upfront cost, over-reliance on a few working modes of action increases the risk of further resistance developing, thus the addition of non-chemical methods of control are a key in the war against weeds.

Farm businesses across Australia are finding that when used tactically, mechanical control of weeds is capable of increasing the profitability of their no-till farming systems, by setting up paddocks for long term control with low weed burdens, whilst not drastically compromising the benefits of their no-till operations.

Using a combination of real world examples and economic modelling, this paper will examine the economics of tactical mechanical weed control in broadacre no-till farming systems. The focus will be on the experiences of West Australian and Queensland farm businesses, specifically on dealing weed species which are currently hard to kill due to herbicide resistance or tolerance, and how they have incorporated mechanical methods of control to complement their farming systems.

Keywords Strategic tillage, machinery, herbicide resistance, mechanical weed control, economics.

Field evaluation of Australian wheat genotypes for competitive traits and weed suppression

James M. Mwendwa[1], William B. Brown[1], Shamsul Haque[1], Graeme Heath[1], Hanwen Wu[2], Jane C. Quinn[1], Jeffrey D. Weidenhamer[3] and Leslie A. Weston[1]
[1] Graham Centre for Agricultural Innovation (NSW Department of Primary Industries and Charles Sturt University), Wagga Wagga, New South Wales 2678, Australia
[2] Department of Primary Industries, Plant Biosecurity, Wagga Wagga, New South Wales 2650, Australia
[3] Department of Chemistry, Geology and Physics, Ashland University, Ashland, Ohio 44805, United States of America
(jmwendwa@csu.edu.au)

Summary In 2014 and 2015, replicated field trials were performed at commercial paddocks in moderate to low rainfall zones at Wagga Wagga and Condobolin NSW, respectively. In 2014, a total of 11 winter wheat cultivars (*Triticum aestivum* L.) representing four major breeding family lines grown in Australia were evaluated with 13 cultivars assessed in 2015. At each site, crop and/or weed growth were monitored at various stages of growth: early season (tillering), vegetative, grain filling, harvest and post-harvest. Significant differences between wheat cultivar and location were observed for crop biomass, early vigour, leaf area index (LAI), weed number, weed biomass, canopy architecture and yield in both 2014 and 2015. Differences in weed suppression were largely impacted by crop architecture and phenology early in the growing season. Cultivar competitive traits were also influenced by both genotype and environmental factors, as shown by clear differences in cultivar performance, yield and weed suppression among both locations. Cultivars Condo and Espada were superior performers for yield and weed suppression in both locations and years analysed. This data supports the concept that choice of wheat cultivar can prove to be a cost effective means of weed management.

Keywords Weed suppression, canopy architecture, phenology, propagules, weed seedbank.

INTRODUCTION

Herbicide resistance in both grasses and broadleaf weeds is on the rise across Australia, with an increasing number of cropping weeds experiencing resistance to multiple herbicides (Owen *et al.* 2013). Globally, weeds have evolved resistance to 22 of the 25 known herbicide modes of action and to 160 different herbicides (Heap 2016).

Highly competitive wheat cultivars typically have the ability to access better light, nutrients, and water resources in a limited space, thus suppressing the growth and reproduction of neighbouring weed species (Bertholdsson 2011, Worthington *et al.* 2015). In Greece, the use of competitive wheat cultivars alone has reduced the use of herbicides by 50% for weed management (Travlos 2012, Andrew *et al.* 2015). However, in the past century, the competitive ability of wheat has been typically reduced by selection based on yield potential. Older cultivars or landraces have been shown to be more competitive with weeds than the higher yielding, semi-dwarf modern cultivars (Bertholdsson *et al.* 2012).

To realise the potential of competitive crop cultivars as a tool in integrated weed management, a quick and simple-to-use protocol for assessing their competitive potential is required as it is likely that selection will not be based on a single trait, but will need to capture the combined effect of multiple traits (Bertholdsson 2011, Andrew *et al.* 2015). In this study, both field and laboratory experiments were performed with the overall objectives of 1) assessing the competitive traits of selected superior Australian winter wheat cultivars which are well adapted for the southern faming region, 2) assessing the impact of environmental factors associated with location and year, including soil moisture and temperature, on weed suppressive ability of wheat, 3) identifying and quantifying key wheat metabolites associated with weed suppression and 4) evaluating weed suppression by remaining wheat stubble post-harvest.

MATERIALS AND METHODS

In 2014 and 2015 field trials were sown at two locations in moderate to low rainfall zones at Wagga Wagga (572 mm) and Condobolin (449 mm) NSW, respectively. Plots were seeded with six replications in a randomized complete block design. Eleven wheat cultivars representing four major genetic families of winter wheat commercially grown in Australia were selected for evaluation, plus one cultivar of winter cereal rye (*Secale cereale*) as a known suppressive control. In 2015, two additional cultivars, Trojan and Federation, were included. Trojan is a recently released cultivar and Federation is an older heritage cultivar bred and released in 1901 and widely used until 1970. Soils were typical Aeolian fine red clays with clay content varying with soil depth.

At sowing, replicated soil samples were taken from each block to evaluate the weed seedbank in the glasshouse. At Condobolin, the crop was sown on 6th and 15th May at 33 cm spacing while at Wagga Wagga the crop was sown on 20th and 22nd May at 25 cm spacing for 2014 and 2015 respectively. A knife point and press wheel planter was used. Fertilizer was applied at 70 kg ha^{-1} MAP (Incitec Pivot Fertilisers). Equal plant density of each cultivar was established in each trial (target 120 plants m^{-2}) by assessment of seed weight per volume when seeding to reduce confounding effects for enhanced evaluation of cultivar competitiveness. No pre- or post- emergent herbicides were applied at either site during wheat production.

All data collection was performed at the critical plant developmental stages of stem elongation, flowering, maturity as well as post-harvest stubble residue assessment. Data collected each year included crop phenological characteristics, percentage light interception using light Ceptometer (AccuPAR LP-80 Ceptometer, Decagon Devices®) to measure PAR (Photosynthetically Active Radiation) both above and below the crop canopy, NDVI (Normalised Difference Vegetative Index) readings (GreenSeeker® 505 hand-held sensor and Trimble Recon PDA) to monitor crop biomass production, biomass cuts of crop and weeds (g m^{-2}), visual vigour ratings (0 = poor, 10 = excellent), yield (kg ha^{-1}) and post-harvest weed suppression(0 = poor, 10 = excellent).

Crop and weed biomass data were collected within a 50 × 50 cm quadrat in each plot at the soil surface with two subplots collected per plot. Grain harvest was performed before 15th December in each year and each location, using a small plot harvester. Yield was measured as harvested cereal grain. Statistical analysis of data was performed by ANOVA for randomised block experiments with six replicates using GenStat (VSN, 2016); significant differences were separated using LSD (0.05).

RESULTS AND DISCUSSION

At each location, cultivar differences were significant for crop growth parameters, including early vigour, leaf area index, NDVI, and crop biomass. Weed numbers and weed biomass were also significantly affected by crop cultivar at each location. Figure 1 depicts cultivar differences in early vigour and weed suppression at Wagga Wagga. Tables 1 and 2 reflect weed suppression provided by various wheat cultivars in terms of weed biomass taken at approximately 100 days after crop emergence (DAE) at both locations in 2014 and 2015.

Based on the results from 2014 and 2015 growing season, Condo, Janz Cl and Espada demonstrated enhanced crop competitive ability against weeds in contrast to other cultivars. At Wagga Wagga in 2015 (Figure 1 and Table 1), Federation, Condo and Janz Cl cultivars demonstrated strongest early vigour in terms of wheat growth at 57 days after establishment (DAE) and subsequently provided the greatest weed suppression at 110 DAE. Similar trends in weed suppression were also observed in 2014; Janz Cl, Espada and Condo were highly weed suppressive (Table 1).

Crop competitive ability can either be specified in terms of crop tolerance against weeds or growth inhibition of weeds by resource competition (Bertholdsson 2010). Previous studies have found that cultivars with higher early vigour are generally capable of extracting more soil moisture which in turn enables them to maintain lower canopy temperatures on warm days (Zerner *et al.* 2008) which is essential in dryland broadacre farming. Certain wheat cultivars under ideal conditions will produce acceptable yields while suppressing weed populations (Worthington *et al.* 2015, Andrew *et al.* 2015).

Condo and Espada were amongst the three most weed suppressive cultivars and also produced the highest grain yields (4.0 and 3.9 t ha^{-1} in 2014 and 5.4 and 5.6 t ha^{-1} in 2015 respectively; data not presented). Recent studies have also shown that certain cultivars successfully reduced the economic burden of weeds over time by resisting yield loss (Vandeleur and Gill 2004) while reducing weed seedbank inputs through successful suppression. Lemerle *et al.* (2001) also observed that cultivars which showed competitive yield advantage also suppressed *L. rigidum.*

At Condobolin in 2014, Mace was highly weed suppressive but was not significantly different from Espada, Condo and Janz Cl. Gregory was significantly less weed suppressive than other cultivars evaluated (Table 2). In 2015, Trojan, Condo, Scout and Espada were most weed suppressive. The ability to suppress weeds appears to be strongly cultivar dependant (Bertin *et al.* 2003, Wu *et al.* 2001). However, environmental factors such as moisture availability and daily temperature also play a key role in influencing the performance of specific genotypes in a particular season as shown by locational performance differences.

The competitive ability of wheat is influenced by a range of plant attributes such as height, tiller number, and light interception by the canopy or light interception at the soil surface. If one can limit light interception at the soil surface, the germination and establishment of weeds can often be significantly reduced. In recent experimentation, increasing plant height improved bread and durum wheat's ability to tolerate and suppress oats. Several plant traits associated with early wheat vigour (early canopy cover, greater leaf width and tiller number) were also

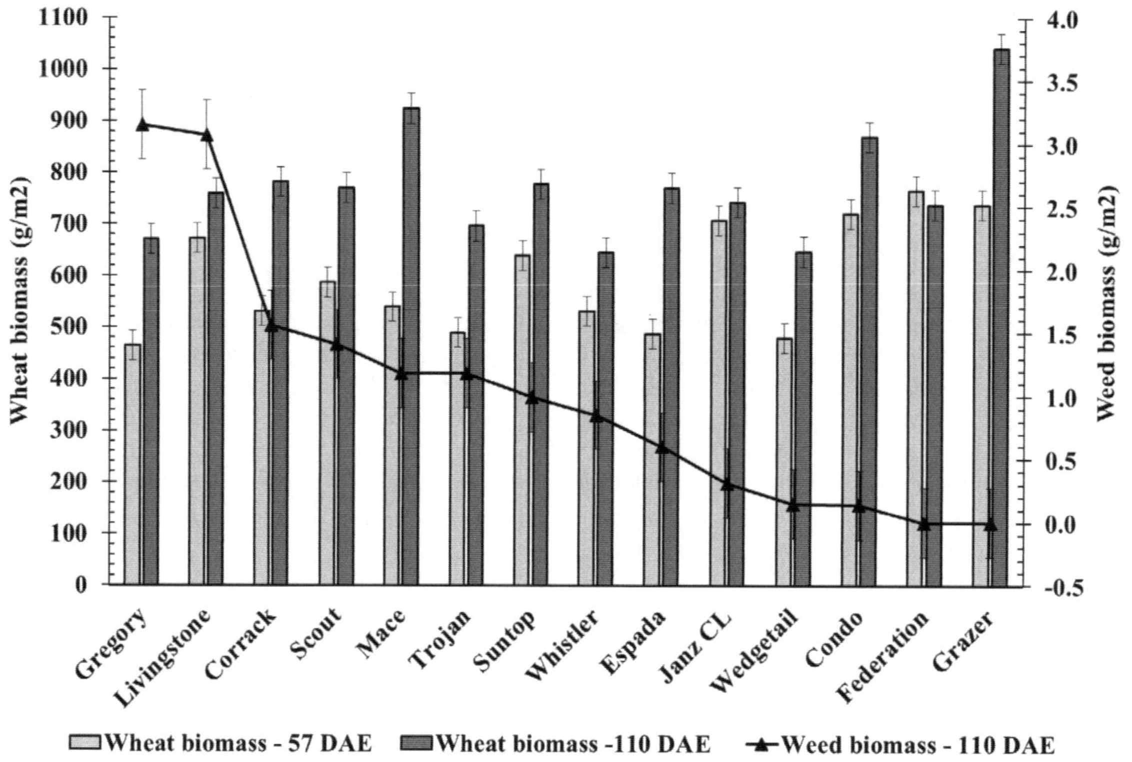

Figure 1. Wheat cultivar biomass taken 57 DAE (LSD 151; P <0.001) and 110 DAE (LSD 106; P <0.001) and weed biomass (LSD 2.9; P <0.5) taken at 110 DAE at Wagga Wagga in 2015 arranged by weed biomass from highest to lowest.

Table 1. Weed biomass (g m^{-2}) taken 130 and 110 DAE at Wagga Wagga in 2014 and 2015 respectively.

2014			2015		
Rank	Cultivar	g m^{-2}	Rank	Cultivar	g m^{-2}
1	Rye	1.1	1	Rye + Federation	0.0
2	Janz Cl	2.5	2	Condo	0.1
3	Espada	3.3	3	Janz Cl	0.3
4	Condo	5.1	4	Espada	0.6
5	Gregory	5.7	5	Trojan + Mace	1.2
6	Livingstone	7.3	6	Livingstone	3.1
7	Mace	13.5	7	Gregory	3.2
P – value		**NS**			*
LSD 0.05		**10.9**			**2.9**

Table 2. Weed biomass (g m^{-2}) taken 133 and 109 DAE at Condobolin in 2014 and 2015 respectively.

2014			2015		
Rank	Cultivar	g m^{-2}	Rank	Cultivar	g m^{-2}
1	Mace	0.03	1	Trojan	4.9
2	Espada	0.04	2	Condo	7.6
3	Condo	0.06	3	Livingstone	11.6
4	Janz	0.06	4	Espada	11.9
5	Livingstone	0.07	5	Federation	12.4
6	Rye	0.23	6	Janz	21.1
7	Gregory	0.70	7	Gregory	44.3
P – value		*			*
LSD 0.05		**0.5**			**30.5**

positively correlated with crop competitive ability (Vandeleur and Gill 2004, Zerner *et al.* 2008).

High leaf area index is known to be important in light interception (Figures 2 and 3). Federation, Condo and Grazer rye had the highest leaf area index of all crops and cultivars in Wagga Wagga, and this trait is most closely associated with early vigour growth and crop competitive ability (Vandeleur and Gill 2004, Zerner *et al.* 2008), but is not always selected for in current commercial lines and cultivars. Previous studies have also reported that early leaf area development is an important contributor to weed competitive ability (Coleman *et al.* 2001).

In Wagga Wagga in 2015 cultivar light interception was highly positively correlated to leaf area index at 57 days after crop emergence ($r^2 = 0.97$, $P < 0.001$; Figure 2). A similar relationship was also seen at Condobolin in 2015 (Figure 3) with cultivar light interception being highly positively correlated to leaf area index at 107 DAE ($r^2 = 0.89$, $P < 0.001$). However, after this time there was no significant correlation between these traits, suggesting crop height and other growth characteristics were important in light interception. For example, Federation and Condo cultivars in Wagga Wagga at 57 DAE were weed suppressive, having less than 50% of photosynthetically active radiation reach the soil surface compared to Wedgetail which had over 70% reach the soil surface. This suggests that the canopy of Federation and Condo successfully intercepted more light than did Wedgetail while allowing less to reach the soil surface, thereby influencing weed establishment (Table 1 and 2).

Plant height was also identified as one of the traits most commonly associated with competitiveness (Vandeleur and Gill 2004). However, timing of emergence also influences light interception, as the same weed species may be relatively tall or short depending on the emergence time relative to the crop. It is important to consider these cultivar competitive traits at breeding and sowing.

We have demonstrated that crop phenology early in the season may be particularly important in impacting overall weed suppression and the subsequent weed seedbank at the time of harvest. Initial studies over two years in two locations in the moderate to low rainfall zone show that the choice of wheat cultivar can prove to be a cost effective means of weed management and may potentially impact weed propagule numbers in the subsequent seedbank.

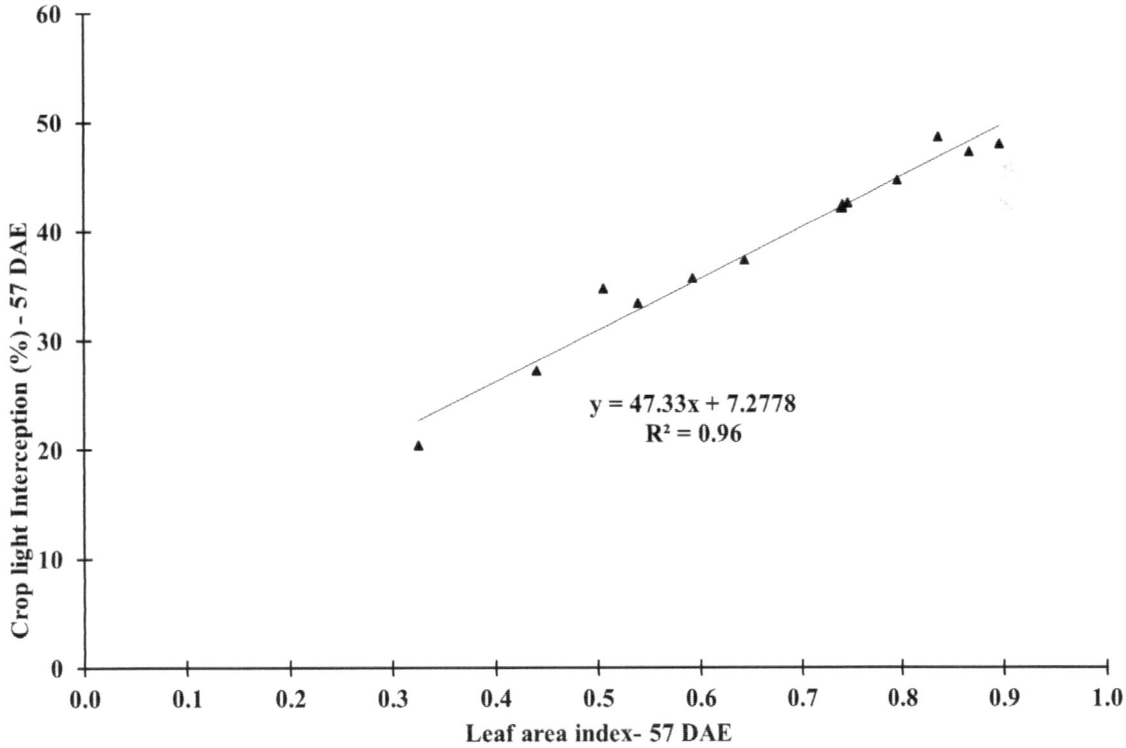

Figure 2. The relationship between wheat cultivar ground surface PAR light interception and leaf area index at 57 DAE in Wagga Wagga in 2015 ($r^2 = 0.97$; $P < 0.001$).

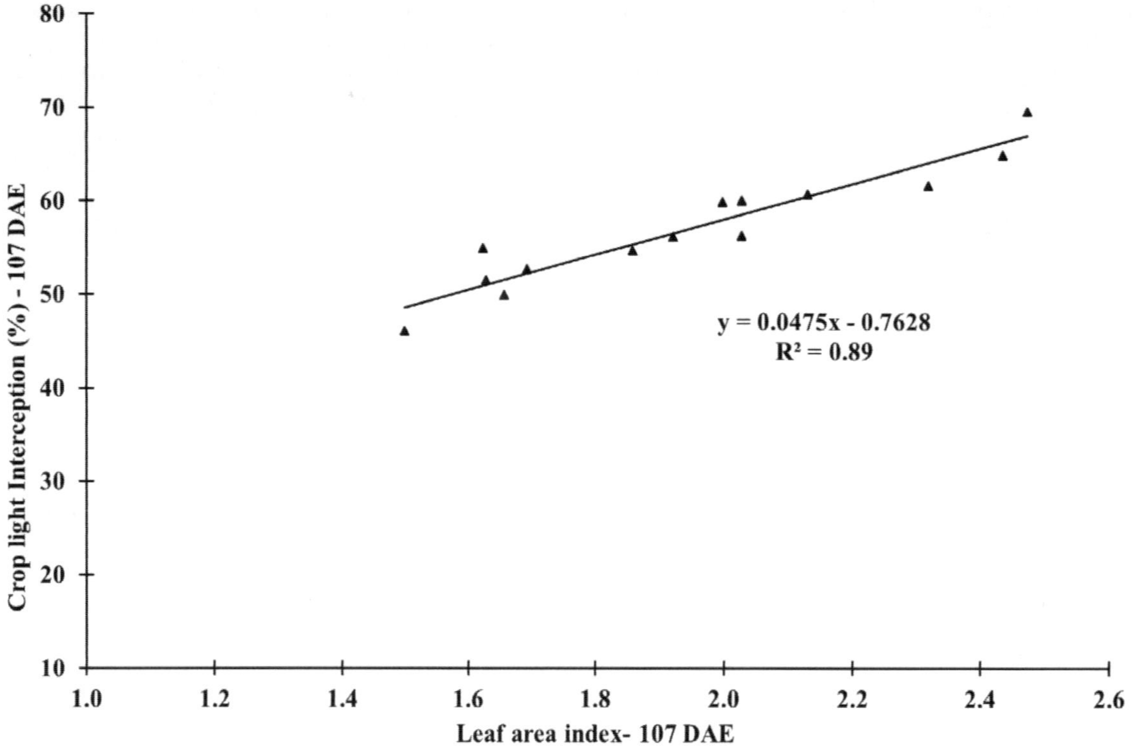

Figure 3. The relationship between wheat cultivar ground surface PAR light interception and leaf area index at 107 DAE in Condobolin in 2015 ($r^2 = 0.89$; $P < 0.001$).

ACKNOWLEDGMENTS

We acknowledge the financial support of the Grains Research and Development Corporation (Projects UCS 00020, 00022, 00023) and technical support of the CSU weed research team in performing these studies.

REFERENCES

Andrew, I. K. S., Storkey, J. and Sparkes, D. L. (2015). A review of the potential for competitive cereal cultivars as a tool in integrated weed management. *Weed Research* 55 (3), 239-48.

Bertholdsson, N.O. (2011). Use of multivariate statistics to separate allelopathic and competitive factors influencing weed suppression ability in winter wheat. *Weed Research* 51 (3), 273-83.

Bertholdsson, N.O., Andersson, S.C. and Merker, A. (2012). Allelopathic potential of *Triticum* spp., *Secale* spp. and *Triticosecale* spp. and use of chromosome substitutions and translocations to improve weed suppression ability in winter wheat. *Plant Breeding* 131 (1), 75-80.

Bertholdsson, N.O. (2010). Breeding spring wheat for improved allelopathic potential. *Weed Research* 50, 49–57.

Bertin, C., Yang, X. and Weston, L.A. (2003). The role of root exudates and allelochemicals in the rhizosphere. *Plant and Soil* 256 (1), 67-83.

Coleman, R.K., Gill, G.S. and Rebetzke, G.J. (2001). Identification of quantitative trait loci for traits conferring weed competitiveness in wheat (*Triticum aestivum* L.). *Crop and Pasture Science* 52 (12), 1235-46.

Heap, I. (2016). www.weedscience.com

Lemerle, D., Verbeek, B. and Orchard, B. (2001). Ranking the ability of wheat varieties to compete with Lolium rigidum. *Weed Research* 41 (3), 197-209.

Owen, M., Preston, C. and Walker, S. (2013). Resistance rising across Australia. GRDC groundcover Supplement on Herbicide resistance, May–June 2013, pp. 5-6.

Travlos, I.S. (2012). Reduced herbicide rates for an effective weed control in competitive wheat cultivars. *International Journal of Plant Production* 6 (1), 1-14.

Vandeleur, R.K. and Gill, G.S. (2004). The impact of plant breeding on the grain yield and competitive ability of wheat in Australia. *Crop and Pasture Science* 55 (8), 855-61.

Worthington, M., Reberg-Horton, S.C., Brown-Guedira, G., Jordan, D., Weisz, R. and Murphy, J.P. (2015). Relative contributions of allelopathy and competitive traits to the weed suppressive ability of winter wheat lines against Italian ryegrass. *Crop Science* 55 (1), 57-64.

Wu, H., Haig, T., Pratley, J., Lemerle, D. and An, M. (2001). Allelochemicals in wheat (Triticum aestivum L.): variation of phenolic acids in shoot tissues. *Journal of Chemical Ecology* 27 (1), 125-135.

Zerner, M.C., Gill, G.S. and Vandeleur, R.K. (2008b). Effect of height on the competitive ability of wheat with oats. *Agronomy Journal* 100 (6), 1729-34.

Metribuzin tolerance of wheat varieties

Harmohinder S. Dhammu[1], Mario F. D'Antuono[2] and David F. Nicholson[1]
[1] PO Box 483, Northam, Western Australia 6401, Australia
[2] Locked Bag 4 Bentley Delivery Centre, South Perth, Western Australia 6983, Australia
Department of Agriculture and Food Western Australia (DAFWA), Australia
(harmohinder.dhammu@agric.wa.gov.au)

Summary Metribuzin 75 g a.i. ha^{-1} is registered on all wheat varieties as an early post-emergent for control of toad rush (*Juncus bufonius* L.). The objective of the study was to investigate the tolerance of new or commonly grown wheat varieties in Western Australia to the higher rates of metribuzin for effective management of weeds like brome grass (*Bromus* spp.), barley grass (*Hordeum* spp.), ryegrass (*Lolium rigidum* Guadin) and wild radish (*Raphanus raphanistrum* L.).

Six field trials under weed free conditions were conducted from 2013 to 2015, three on sandy loam to loam soils at Mullewa and other three on a loamy sand to sandy loam soils at Katanning. The trials were laid out in a criss-cross design having six wheat varieties at each site and five rates of metribuzin (0, 75, 150, 300 and 450 g a.i. ha^{-1}) with three replications. Metribuzin at the different rates was applied as pre-emergent during 2013 and post-emergent during 2014 and 2015 at both the sites.

Metribuzin 75 g a.i. ha^{-1} was tolerated well by all the wheat varieties with a good crop safety margin. Metribuzin 150 g a.i. ha^{-1} was also tolerated well by all the wheat varieties with a good crop safety margin except Bonnie Rock at Mullewa during 2013 and Mace at Katanning during 2014 and 2015 registered a low crop safety margin. However, under shallow seeding condition at Mullewa during 2015, the wheat varieties registered significant yield loss with rates above 75 g a.i. ha^{-1}.

Keywords Wheat, metribuzin, tolerance, grain yield, weeds.

Social factors influencing landholder management of silverleaf nightshade

Vaughan Higgins[1], Karen Herbert[2], Phillip Bowden[2], Hanwen Wu[3] and Karl Behrendt[4]
[1] School of Humanities and Social Sciences, Charles Sturt University, PO Box 789, Albury, New South Wales 2640, Australia
[2] Murrumbidgee Landcare Inc, PO Box 710, Wagga Wagga, New South Wales 2650, Australia
[3] Wagga Wagga Agricultural Institute, NSW Department of Primary Industries, Pine Gully Road, Wagga Wagga, New South Wales 2650, Australia
[4] Graham Centre for Agricultural Innovation (Industry and Investment NSW and Charles Sturt University), Charles Sturt University, School of Agricultural and Wine Science, Orange, New South Wales 2800, Australia
(vhiggins@csu.edu.au)

Summary Silverleaf nightshade (*Solanum elaeagnifolium* Cav.) is recognised as one of the world's worst agricultural weeds. Yet, little is known about the broader social drivers or barriers that influence farmer adoption of particular control strategies. To better understand these drivers and barriers, individual semistructured interviews were conducted with 18 farmers involved in the management of silverleaf nightshade (SLN) across four Australian states. Findings from this research reveal that the most significant driver for landholders to control SLN is the potential implications for property value in the district if SLN is not properly controlled. Time and associated labour costs represent the main challenges in controlling SLN, but they do not act as a barrier to ongoing control due to the high importance placed by farmers on controlling SLN. Based upon these findings, the paper makes three key recommendations for future SLN extension.

Keywords Silverleaf nightshade, social drivers and barriers, best management adoption.

INTRODUCTION

Silverleaf nightshade is a deep-rooted, summer growing, perennial weed, estimated to infest over 350,000 hectares in Australia. It competes directly with spring, summer and autumn growing pastures and crops for moisture, sunlight and nutrients. To date, research has focused primarily on quantifying the farm-level impacts of SLN in terms of losses in production and control costs (e.g., McLaren *et al.* 2004, Wu *et al.* 2016). However, little is known about the social drivers and barriers for landholders to adopt particular SLN control strategies.

Social factors, ranging from farmers' personal goals, knowledge and networks, to the broader cultural, environmental and economic context in which they make decisions, have a significant impact on adoption of new practices or innovations (Vanclay 2004, Wilkinson 2011). Understanding these social factors can provide valuable insights into farmers' weed control strategies and decisions, and help in identifying opportunities to improve weed management practices (Van der Meulen *et al.* 2006). Drawing upon qualitative social research across four states, the aim of this paper is to investigate the social drivers and barriers for SLN control, and to reflect on the implications for adoption of SLN best management practices.

MATERIALS AND METHODS

The research reported in this paper forms part of a larger national project funded by Meat and Livestock Australia and Australian Wool Innovation aimed at improving awareness and adoption of best management practices for SLN control (Herbert *et al.* 2016, in these proceedings). With the aim of benefitting the largest number of growers possible, farm scale participatory on-farm validation sites of best management practice (BMP) were established in 57 locations across four states through active coordination with state and local agencies, private sector contractors and advisors and local grower groups. As part of this process, a small cross-section of 18 growers were selected, using purposive sampling, for case studies aimed at promoting the successful management of SLN to other farmers and to increase national adoption of the BMP strategies. These growers were selected based on their experiences in applying a range of successful approaches in managing SLN – including the 'Dual Action' approach, which is the current BMP recommended to control this weed.

Individual semi-structured interviews were conducted with these 18 farmers between October 2014 and September 2015. The farmers interviewed represent different states (Vic – 4 farmers, NSW – 7 farmers, SA – 3 farmers and WA – 4 farmers) and cover a range of soil and climatic conditions. The majority (15) run mixed farming operations. While the principal aim of the interviews was to construct case studies of best practice SLN control, an equally important aim was to gain an understanding of the social factors that influence how and why landholders control SLN. All interviews were analysed using open and axial coding (Miles and Huberman 1994) as a way of firstly finding common descriptors and then the relationships between those descriptors. Finally, the data were analysed using a thematic analysis to derive key themes from across the data. For example, 'impacts of SLN on farm production' and 'potential impacts on property value' were identified as two common descriptors during the analysis process. These descriptors were also considered to be closely related in that farmers in the interviews discussed potential, or actual, impacts of SLN as a strong driver for adopting control strategies. As a consequence, both of these descriptors were grouped with other similar descriptors to form part of a broader theme – key drivers to control SLN.

In total, four key key themes emerged from analysis of the landholder interviews. In summary, these themes surround the following issues: priority given to controlling SLN; key drivers to control SLN; barriers to the control of SLN; and, current control strategies as drivers and barriers to BMP adoption. Each of these themes is discussed in more detail below.

RESULTS

Priority given to controlling SLN All of the farmers interviewed believed that SLN was a high priority, if not the highest priority, weed needing to be controlled on their farm. This was due largely to the persistence of the weed and the fact that it is extremely difficult to keep under control. For example:

> *[SLN is a priority] because I know if you don't [control it] it'll just take over.* (Landholder 4)

> *It would be number one [priority]... because everything else we can spray and you kill and it's gone. Whereas you go and spray this nightshade and you go back two weeks later and there'll be another one there. You just keep coming back and it's constant.* (Landholder 18)

Key drivers to control SLN The key driver for controlling SLN mentioned most frequently by farmers was its potential impacts on property value if not sufficiently managed each year to reduce infestations. While concerned about SLN infestations on their farm, of equal concern was the effect of infestations within neighbouring properties on property prices in the local district.

> *Last week when I was in South Australia there was a farmer that's been trying to sell a block there, part of his paddock, he reckons ... it's probably wall-to-wall silverleaf because it's never been controlled. He said he can't sell it, nobody wants to buy it.* (Landholder 5)

> *There's one farm in a similar area, probably ten kilometres to the north of [property] that's not been sold and it's just riddled with silver leaf and deep sand.... Someone will buy it [eventually] ... but it's been sitting there for probably ten years.* (Landholder 12)

The prospective or actual observed impacts of SLN on farm production and viability were considered important by many farmers. For example:

> *You've got to control it because it will have an impact on your crop yields and it uses moisture, it uses nutrients, it does well. In good summers it does very well.* (Landholder 12)

> *SLN has reduced the sheep carrying capacity of the property by out-competing pastures and reducing the amount of feed available for stock. It has also depleted the soil moisture available to the cropping phase of rotations. By controlling SLN, the potential impact of the weed on the farm's carrying capacity and crop yields has been reduced.* (Landholder 16)

However, the impacts of SLN on farm production, including reports of production losses, were discussed less frequently and in far less depth than impact on property value.

Barriers to the control of SLN Time and cost were the predominant challenges for farmers in controlling SLN. In many cases the time invested by farmers to control the weed, particularly during summer, was significant ranging from two weeks up to three months full-time work (for spot spraying over widespread areas).

> *Most of my summer is spent [spraying SLN], a third of the year. Eight hour days. Seven days a week virtually.* (Landholder 5)

> *The chemicals are not expensive. It's the time involved. If the chemicals were five times more expensive than what I'm using I'd still use the five times more expensive because chemicals are not expensive. I might use two drums of Roundup,*

$200/$300 a year for chemicals. And here I am three months driving up and down the bloody paddock on a motorbike, out in the sun cooking. So you know the [main] expense is the time. (Landholder 17)

Clearly this is costly labour time that could otherwise be devoted to other farming work. The cost of chemical controls was mentioned less frequently by farmers, perhaps because spending money on chemicals to control SLN is considered a normal part of weed management. Rather than chemical costs, farmers emphasised the labour costs involved in controlling SLN.

It's certainly cost us a lot in labour.... I'm not the only who's been attending to it but it certainly costs in labour, big time. (Landholder 5)

Sometimes expenditure has been over $25,000 per annum as controlling SLN is a full-time job for a farmhand for about six weeks in January and February. (Landholder 3)

Nonetheless, chemical expenditure for controlling SLN was frequently higher than $10,000 per annum and in a few cases over $40,000. For two farmers interviewed, the relatively high expenditure meant that the price of chemicals was a major factor in determining which chemical options to use.

Glyphosate is used to control SLN because it is cheaper in comparison to other chemical options, easier to apply, and there is always some in the shed. (Landholder 4)

The price of chemicals dictates which option to use. We have increased rates of Roundup in the last ten years as the cost of this chemical has dropped. (Landholder 12)

Current control strategies as drivers and barriers to BMP adoption All of the farmers interviewed used a control program focusing predominantly on the use of chemicals. Indeed, a mix of chemicals was by far the most widely used SLN control strategy reported by farmers.

[I use] mostly just Roundup. When I spot spray with Roundup, if I've just got a hand sprayer, I'll probably wet the leaves, so it could be 10 litres to the hectare or something like that it's going on. (Landholder 4)

I was spraying to burn the top off it. Next year I would try and stop it from seeding but it used to damn well seed, you couldn't get round it all. And next year you go back, it was there, still there. You were back, back, back, effort, effort. And it was seeding and getting bigger but since we've been using a better wetter, I think the wetter's definitely the answer. We used 2,4-D for start off, we used Tordon, that was 15 years ago. And then we used Starane and then we used Roundup. But now Roundup and a good wetter, and I use a bit of, sometimes a bit of Ally.... The only thing that's keeping it under control is my spraying. And that's the only way we're going to head this. (Landholder 17)

Chemical use was, however, often combined with one or more of the recommended practices detailed in the SLN BMP guide – mostly competitive crops and pasture or mechanical control. Significantly, each of these control strategies was tailored by farmers to suit their farming system and personal values and goals.

Chemicals and competition [through vigorous groundcover] are used to keep SLN under control. We have recently adopted a dual action approach but it is too early to assess the results. (Landholder 7)

Cultivation is the main source of spread for SLN. Therefore, control strategies need to be focused primarily on machinery hygiene. In addition to this, sheep brought onto the property or from other paddocks are quarantined for two days to ensure that weed fragments are not spread. (Landholder 10)

Minimum till has meant a change from specific spraying for SLN to general summer spraying programs. One third of the farm is leased and is continually cropped. A substantial reduction in SLN infestation in these areas has been observed. With no-till, SLN is now no worse than other weeds. (Landholder 16)

Almost all the farmers interviewed recognised that SLN is one of the most persistent weeds to kill and they understood the economic consequences on the farming operation if controls were not applied early to stop the seed set and run the root reserves. Livestock producers stressed that getting the herbicides and timing right to prevent seed set was critical for success. Farmers interviewed who had cropping production systems, consistently applied controls every year in summer and autumn, over a number of years, to reduce infestations.

DISCUSSION

Control of SLN is a high priority for farmers because it is readily spread by seed and is very difficult to control. While the prospective and actual impacts of SLN on production are important drivers for controlling the weed, of greatest concern to farmers are the potential implications for property values in the district if SLN is not properly controlled. The interviews highlight a strong collective concern, and even a sense of social responsibility, among farmers to control SLN so that it does not adversely affect local property values. This shows that farmers' socio-cultural values and goals relating to the broader viability of farming in the district may be more significant drivers to control SLN than short-term economic impacts on production.

The time involved in controlling SLN, and associated labour costs, represent the most significant challenges for farmers in controlling SLN. These constraints are recognised in the adoption literature (e.g., Pannell *et al.* 2006). However, our research shows that because SLN is considered by farmers as a priority weed, such challenges do not act as a barrier to ongoing control of SLN. This applies also to the cost of chemicals. For most of the farmers interviewed, cost does not appear to be a barrier to the use of chemical controls. This is perhaps because chemicals are considered by these farmers as a normal part of effective weed management, and the chemicals that are used to control SLN can be applied also to control a range of other weeds.

Farmers tailor the use of different SLN control strategies to suit local conditions, personal goals, and existing farming practices. While multiple chemical controls are the predominant approach used in managing SLN, these are typically used in combination with one or more non-chemical controls such as the use of competitive crops and pastures, farm and machinery hygiene practices, and minimum/no-till cropping. The social and environmental context underpinning farmers' practices is crucial to take into account in understanding why farmers use specific combinations of control strategies, and their willingness to trial new control options.

CONCLUSION

The research reported in this paper has three principal implications for future SLN extension. First, since potential impact on property values is a key driver for farmers in controlling SLN, this needs to be given greater emphasis in extension in addition to the current focus on production impacts. Second, greater emphasis is needed on the time and labour cost benefits of adopting best practice SLN control strategies. Third, when developing regionally-specific best practice approaches to controlling SLN, emphasis needs to be given to the ways in which best management SLN control strategies may (a) complement existing chemical use, and (b) be integrated with other effective non-chemical control practices.

ACKNOWLEDGMENTS

The authors gratefully acknowledge the financial support of Meat & Livestock Australia and Australian Wool Innovation for funding this project.

REFERENCES

Herbert, K., Wu, H., Bowden, P., Behrendt, K. and Higgins, V. (2016). A national program to improve awareness and adoption of best management practice for silverleaf nightshade control. Proceedings of the 20th Australasian Weeds Conference, eds R. Randall, S. Lloyd and C. Borger, pp. 60-3. (Weeds Society of Western Australia, Perth).

McLaren, D., Morfe, T., Honan, I. and Holtkamp, R. (2004). Distribution, economic impact and attitudes towards silverleaf nightshade (*Solanum elaeagnifolium*) in Australia. Proceedings of the 14th Australian Weeds Conference, Wagga Wagga, p.701.

Miles, M. and Huberman, A. (1994). 'Qualitative data analysis: an expanded sourcebook'. (Sage, Thousand Oaks, CA).

Pannell, D. *et al.* (2006) Understanding and promoting adoption of conservation practices by rural landholders. *Australian Journal of Experimental Agriculture* 46, 1407-24.

Vanclay, F. (2004). Social principles for agricultural extension to assist in the promotion of natural resource management, *Australian Journal of Experimental Agriculture* 44, 213-22.

Van der Meulen, A.W., Reeve, I. and Sindel, B.M. (2006). The sociology of weeds. Motivating, building capacity, and educating graziers who fail to control weeds. Final report. Meat and Livestock Australia, North Sydney.

Wilkinson, R. (2011). The many meanings of adoption. *In* Pannell, D. and Vanclay, F. (eds), 'Changing land management: adoption of new practices by rural landholders', pp. 39-49. (CSIRO Publishing, Canberra).

Wu, H., Stanton, R. and Lemerle, D. (2016). Herbicidal control of *Solanum elaegnifolium* Cav. in Australia. *Crop Protection* 88, 58-64.

Weed management priorities of lifestyle rural landholders

Nicholas Gill[1], Natalia Adan[1], Shaun McKiernan[1] and Laurie Chisholm[2]
[1] School of Geography and Sustainable Communities and [2] School of Earth and Environmental Science, University of Wollongong, New South Wales 2522, Australia
(ngill@uow.edu.au)

Summary In areas where lifestyle-oriented rural landholders constitute a large proportion of landowners, motivations and priorities for weed control are diverse. Not only do such landowners differ from traditional farmers, but they are diverse amongst themselves as well in terms of land use, skill, resources, and land management aspirations. Based upon surveys and interviews with lifestylers we examine the weed management priorities of such landowners in south-eastern NSW along several axes. We find that, despite, the significance of weeds affecting environmental values associated with their land, lifestylers tended to focus their management effort on agricultural weeds. Among other variables, this was best explained by membership of natural resource management groups and action by weeds inspectors. We also draw on interviews to examine more generally how and why landholders make decisions about which plants to remove and which plants to keep. Such decisions are not necessarily based around whether a plant is native or exotic, although this distinction may drive overall management goals such as forest restoration. Within such overall management, however, landholders will leave or remove plants according to a range of reasons including aesthetics, landscape meaning, perceived lack of harm, plant behaviour, soil stability, food value, restoration interventions, the value of even significant weeds for productive land uses, and their value for fauna. Such prioritisations may mean that landowner weed management does not conform to ideal practice but their alignment with landowner interests and idiosyncrasies may make it more sustainable.

Keywords Invasive plants, lifestyle landowners, agricultural weeds, land use change, priorities.

A National program to improve awareness and adoption of best management practice for silverleaf nightshade control

Karen Herbert[1], Hanwen Wu[2], Phillip Bowden[1], Karl Behrendt[3] and Vaughan Higgins[4]
[1] Murrumbidgee Landcare Inc, PO Box 710, Wagga Wagga, New South Wales 2650, Australia
[2] Wagga Wagga Agricultural Institute, NSW Department of Primary Industries, Pine Gully Road, Wagga Wagga, New South Wales 2650, Australia
[3] Graham Centre for Agricultural Innovation (Industry and Investment NSW and Charles Sturt University), Charles Sturt University, School of Agricultural and Wine Science, Orange, New South Wales 2800, Australia
[4] School of Humanities and Social Sciences, Charles Sturt University, PO Box 789, Albury, New South Wales 2640, Australia
(kherbert@mli.org.au)

Summary Silverleaf nightshade (SLN) is recognised as one of the world's worst agricultural weeds that competes directly with spring, summer and autumn growing pastures and crops for moisture, sunlight and nutrients (Stanton *et al.* 2009). With the aim of benefitting the largest number of growers possible, farm scale participatory on-farm validation sites of best management practice were established with support from state and local agencies, private sector contractors and advisors, and local grower groups.

The participatory research, in conjunction with farmer case studies and a baseline survey targeting 500 growers and advisors (combined) has generated large and diverse baseline information focused on weed prioritisation, appropriateness of information, effective information delivery mechanisms and recommendations for packaging and instruction of weed management. This paper highlights key findings captured for improved adoption of integrated weed management practices.

Keywords Silverleaf nightshade, control strategies, participatory research; best management adoption.

INTRODUCTION

Silverleaf nightshade (*Solanum elaeagnifolium* Cav.) is an introduced summer active perennial from the Solanaceae family that can dominate pastures and cropping areas (Wu *et al.* 2016). In 2013 a National project funded by Meat & Livestock Australia and Australian Wool Innovation commenced to improve awareness and adoption strategies for silverleaf nightshade control across four states. It reproduces from seed and extensive root fragments, so poses a large problem for control. SLN was selected as a Weeds of National Significance (WoNS) in 2012 due to factors related its invasiveness, competitiveness, fecundity, lack of effective long-term control options, and for the fact that this weed only currently occupies a small proportion of its potential distribution.

SLN was first reported in Australia at Bingara (northern New South Wales) in 1901 (Leys and Cuthbertson 1977) and now covers over 350,000 ha in SA, NSW and Vic across a wide range of soil-types and rainfall regions. There are also isolated infestations in WA and Qld. It continues to spread within the important Australian cereal cropping production zone, primarily by seed transported by livestock, farm machinery and produce, and also by root and stem fragments.

The economic impact of SLN comprises direct control costs, production losses (crop and fodder conservation yields), reduced land value and marketability, environmental degradation and social costs. Crop yields can be impacted significantly by SLN, with cereal and summer crops reduced by up to 75 per cent (Stanton *et al.* 2009). McLaren *et al.* (2004) estimated that the average financial impact for farms infested with SLN was $1730 per year in direct control costs and $7786 in lost production.

Previous work by Stanton *et al.* (2011) and Heap (2012) outlined a number of underlying barriers influencing the adoption of recommended best management practices for control of SLN. These included farmers having a limited understanding of SLN biology and a range of integrated management packages that provide control recommendations targeting both seed set and root bank for different production scenarios

The aim of this study was to develop approaches to overcome these barriers to adoption by improving the regional relevance, effectiveness and the delivery of SLN best management practices, thereby increasing the levels of capacity, and willingness of farmers to effectively manage this weed.

This was done in two main ways through: (a) conducting a baseline survey and utilising multiple extension channels to develop a national community of interest, to raise awareness of best management practice information in an interactive and coordinated manner; and (b) encourage farmers to establish participatory trials to compare the economic benefit of different control strategies against current practice for SLN infestations, in different production systems and agro-ecological zones.

MATERIALS AND METHODS

A number of sub-components were completed to support the aims of the project.

Baseline survey A baseline survey targeting growers and advisors was completed at the start of the project to detect change in awareness, attitude, knowledge and management practice. The survey was conducted predominately online with a small number of paper surveys completed via mail outs and at workshops. The survey determined issues around the information delivery mechanisms, the prioritisation of weed management, the social, economic, environmental and management issues affecting adoption appropriateness and recommendations for packaging and instruction of weed management (including weed management plans and cost/ returns over time). The baseline survey also collected comments on the current best management practice (BMP) for further improvement to better adapt local farming scenarios.

Nationwide 'community of interest' partnerships Strong partnerships with 15 collaborators across four states were established throughout 2014–2015 in conjunction with active interactions between grower groups, through the completion of more than 40 grower group meetings, workshops and farm walks. The workshops were primarily to promote the best management practice information to advisors, farmers, contractors, Weeds Officers and NRM personnel, to raise awareness and to attract 'qualified' advocate growers to trial the BMP based on their farming scenarios. Participants were invited to share their stories at the workshops or through the news forum (blog) to cover different regions, soil types, production scenarios and infestation levels.

Promote champion growers Across four states, 18 champion growers were identified to develop individual case studies to share their stories, news with other farmers at the workshops or through the blog/web site. The purpose of the case studies was to promote the successful management of SLN to other farmers, to increase the adoption of the BMP strategies across all geographic areas and productions systems of Australia.

Participatory on-farm trials Participatory research was a key component of this program to generate a large and diverse baseline data on farming scenarios, weed pressure and failure across large geographical zones. This data will be used to refine the BMP, tailoring the need of different farming enterprises in broader geographic regions across multiple states.

Approximately 350 farm scale participatory on-farm validation sites of BMP were established across four states through active coordination of state and local collaborators. The BMP selected on each site was dependent on the individual producer's business context, farming scenario and current management practice.

In most instances, each participatory grower compared the recommended BMP with current practices at paddock scale for a two-year period, using a standardised protocol provided. Basic information on farming/infestation history, initial weed density, control application/timing and efficacy, climatic and soil conditions and stem emergence, control cost, impact of weeds and pasture/crop production general farm feedback was collected by farmers, or via the assistance of local community partners, collaborators or private contractors.

Mixed model analyses is to be conducted at the completion of the project to identify factors (regional, soil type, farming enterprise) contributing most to the variability in control. Analysis of contrasts between current and BMP by regression with explanatory covariates to look for trends in treatment effects and variability within the same treatments.

RESULTS

Baseline survey Results to the baseline survey for 436 Famers and 72 Advisors, indicated that greater than 80% of the farmers resided in NSW and SA, reflecting the SLN problem in these two states. SLN occurs commonly throughout the cereal cropping zone of these states but there are very large areas that are not yet invaded and are at risk because of their close proximity to propagule sources and exposure to invasion pathways.

Farmers were asked to rank their most preferred methods of receiving weeds advice from a range of sources supplied. The highest first preferences were directly from advisors, either independent (40%) or reseller agronomists (32%), or from attending workshops and field days (16%. Lower preferences were through social media (1%), the internet (2%) or radio (1%), irrespective of their age category. This shows that for most farmers, direct contact with an advisor either one on one or at a workshop or field day is still the preferred way of receiving technical information.

Insight into how decisions are made by farmers about adopting new practices are summarized in Figure 1. Results of this survey show that farmers would most like to see economic proof that new control measures are beneficial first.

Nationwide 'community of interest' partnerships
An interactive community of interest was established across four states (VIC, NSW, SA and WA), supported by workshops. The distribution of 1500 plus best management practice brochures, 40 media articles nationally and information sharing on the SLN blog/website and information from the network of interested growers and advisors. A comprehensive database was compiled which included affected growers, growers groups, Landcare groups, public and private partners, noxious weeds advisory groups, local councils and spray contractors in each state.

On-farm participatory research trials
Trials were established in 57 locations nationally (Figure 2), utilising six generalized land descriptions (cropping, pasture, mixed farming, unproductive farmland, roadside and reserve) and three broad treatment types incorporating either best management practice (BMP), farmer conventional management practice (CMP), or zero control.

Common feedback from participatory farmers so far is that the experience gained in implementing the BMP practices on their farm has improved their capacity, motivation and confidence to manage SLN more effectively.

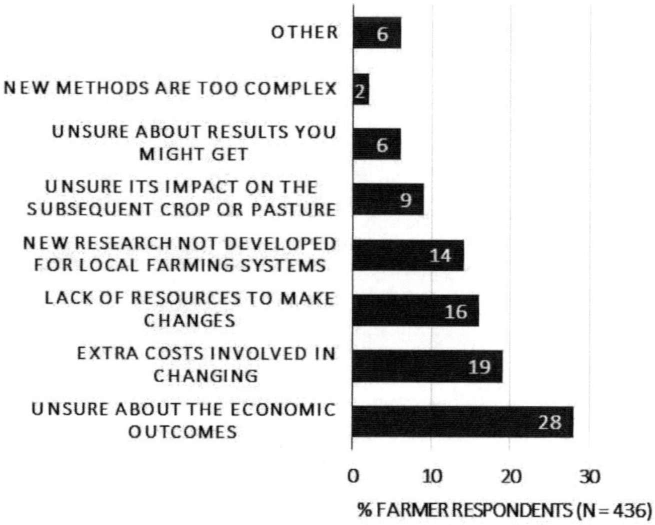

Figure 1. Results to the baseline survey question which asked farmers what prevents them from adopting new control methods on their farms.

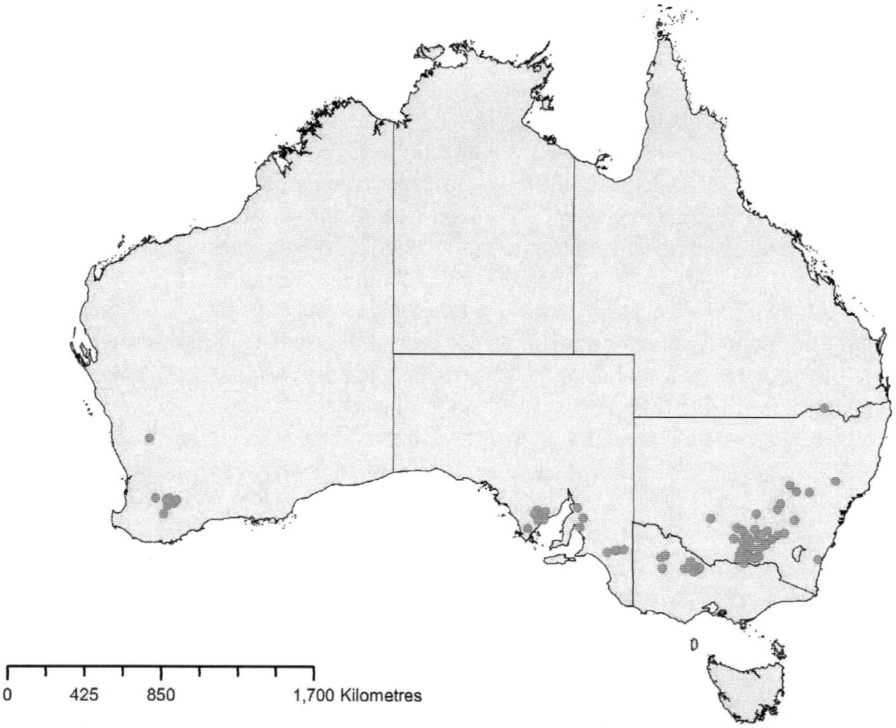

Figure 2. Overview of the 57 on-farm participatory trial locations throughout Australia (highlighted green).

DISCUSSION

The on-farm demonstrations will continue for another nine months; in that time additional data will be collected from summer and autumn applications with the dataset analysed to look for treatment effects and variability within the same treatments across similar production systems and climatic zones.

With weeds like SLN where there are no simple solutions or new chemical options, therefore adoption of BMP strategies is based on timing of a range of control measures over the long term (3–5) years (beyond the scope of this project).

Farmers need assurance using local knowledge from advisors or experiences from trials either in their district or on their own farms to reassure them that it is worthwhile making the change. SLN management is different to many other problematic weeds in that it is unlikely to be completely eradicated, so farmers need to focus on reducing plant densities to levels that do not have economic effects on production of crops and pastures.

Once present SLN becomes a perennial problem and this requires the economic analysis of management strategies to consider the inter-temporal and dynamic nature of SLN infestation, its effects on agricultural production, and the efficacy of control options (chemical, biological and cultural) in affecting SLN fecundity, survival and dispersal (Holst *et al.* 2007, Jones 2004). Simple economic threshold models based on a single season time-step do not capture the dynamic nature (carry-over effects) of controlling the SLN population between time-steps (Jones 2004), in particular a weed of perennial nature.

Near the completion of the project, data sourced from the trial sites and literature will be used to inform the parameterisation of the dynamic and stochastic simulation model of SLN infestation, and its interaction with climatic conditions, herbicide efficacy and agricultural production. At the farm level, the objective of identifying and implementing SLN best practice management is to maximise profit over the long term (King *et al.* 1998). The aim of the bioeconomic model will be to account for the dynamic and stochastic nature of SLN infestation on production, and test the economic and financial consequences of different control strategies over a 10 year planning horizon.

Effective adoption of SLN best management practices relies on the development and maintenance of partnerships between community, industry, government and recognition of the roles of each stakeholder.

CONCLUSION

Return on investment will result from reducing current and future impacts of SLN achieved through an integrated and coordinated approach that: (a) limits new incursions, (b) manages established infestations to reduce propagule spread and increase productivity; and (c) builds capacity, willingness and resilience to manage this weed into the future.

ACKNOWLEDGMENTS

The authors gratefully acknowledge the financial support of Meat & Livestock Australia and Australian Wool Innovation for funding this project.

REFERENCES

Heap, J.W. (2012). Weeds of National Significance: silverleaf nightshade (*Solanum elaeagnifolium*) Strategic Plan. Australian Weeds Committee, Canberra.

Holst, N., Rasmussen, I.A. and Bastiaans, L. (2007). Field weed population dynamics a review of model approaches and applications. *Weed Research* 47, 1-14.

Jones, R. (200). The economic benefits of IWM: the role of risk and sustainability in farming systems, Proceedings of the 14th Australian Weeds Conference, Wagga Wagga, pp. 576-9.

King, R.P., Swinton, S.M., Lybecker, D.W. and Oriade, C.A. (1998). The economics of weed control and the value of weed management information. *In* 'Integrated Weed and Soil Management', pp. 25-41.

Leys, A.R. and Cuthbertson, E.G. (1977). *Solanum elaeagnifolium* in Australia. Proceedings of the Southern Weed Science Society 20, 137-41.

McLaren, D., Morfe, T., Honan, I. and Holtkamp, R. (2004). Distribution, economic impact and attitudes towards silverleaf nightshade (*Solanum elaeagnifolium*) in Australia. Proceedings of the 14th Australian Weeds Conference, Wagga Wagga, p. 701.

Stanton, R.A., Heap, J.W, Carter, R.J. and Wu, H. (2009). *Solanum elaeagnifolium* Cav. (revised). *In* 'The biology of Australian weeds, Volume 3', ed. F.D. Panetta, pp. 274-93. (R.G. and F.J. Richardson, Melbourne, Australia).

Stanton, R.A., Wu, H. and Lemerle, D. (2011). Integrated management of silverleaf nightshade. 23rd Asian-Pacific Weed Science Society Conference, 26–29 September, pp. 544-48.

Wu, H., Stanton, R. and Lemerle, D. (2016). Herbicidal control of *Solanum elaeagnifolium* Cav. in Australia. *Crop Protection* 88, 58-64.

Silverleaf nightshade (*Solanum elaeagnifolium*) – field research on biology and management in South Australia

John W. Heap
GPO Box 1671, Adelaide, South Australia 5001, Australia
(john.heap@sa.gov.au)

Summary Silverleaf nightshade (*Solanum elaeagnifolium* Cav.) is an introduced deep-rooted perennial weed that reduces crop and pasture production in the wheat-sheep growing regions of Australia. It is a Weed of National Significance (WoNS) and is the subject of a new biological control project in Australia. Silverleaf nightshade is amongst the most difficult agricultural weeds to manage, and effective and economic strategies have not yet been found for large established infestations.

This paper reports and discusses the results of a three-year field research study funded by the South Australian Grains Industry Trust (SAGIT) and Primary Industries and Regions South Australia (PIRSA). The research was conducted in South Australia between 2014 and 2016, and addressed a range of biology and management aspects of silverleaf nightshade management.

Results are presented on:

- the effects of consecutive in-crop and non-selective spot-spraying herbicide treatments on silverleaf nightshade population density;
- the influence of soil moisture and leaf dustiness on the efficacy of glyphosate; and
- seedling emergence, establishment and survival under a range of simulated rainfall regimes.

Keywords *Solanum elaeagnifolium*, silverleaf nightshade, herbicides, seedling emergence, glyphosate efficacy.

Integration of chemical control and burning practices for the management of gamba grass (*Andropogon gayanus*) in northern Australia

Rowena J. Eastick
Department of Land Resource Management, PO Box 496, Palmerston, Northern Territory 0831, Australia
(rowena.eastick@nt.gov.au)

Summary Gamba grass (*Andropogon gayanus*) is an invasive grass species in the semi-arid tropics. There has been abundant research conducted into the impacts of gamba grass, which has identified it as a habitat transforming species, primarily due to its ability to modify intensity and seasonality of fire regimes. However, there has been little work to develop practical management options integral with gamba grass ecology. A long-term study site was initiated in 2013 evaluating gamba grass management in a native savanna environment to develop a multi-season best-practice strategy. Herbicide application and burning at different times of the year were imposed. The current practice of glyphosate application in the wet season was effectively extended into the dry season, attributed to ecophysiological traits which increase gamba grass resource use efficiency. Seed viability was also reduced with early dry season glyphosate application. The efficacy of chemical control was enhanced when applied post-burning to regenerating vegetative growth. Burning without follow-up herbicide application provides a competitive advantage to gamba grass over native species, and exacerbates the grass-fire feedback to accelerate gamba grass invasion. Land managers should incorporate both burning and chemical control, and this paper discusses advantages and disadvantages for consideration in the development of best-practice strategies.

Keywords Fire regimes, invasive grass species, glyphosate.

INTRODUCTION

Gamba grass (*Andropogon gayanus* Kunth) was introduced to northern Australia as an improved pasture species, but is now considered invasive, and was listed as a declared weed in the Northern Territory (NT) in 2008. It is acknowledged as habitat transforming, primarily due to its ability to modify intensity and seasonality of fire regimes, and numerous studies have been conducted in the NT on the ecological impacts of this species (Rossiter *et al*. 2004, Setterfield *et al*. 2014). However, there has been minimal work evaluating integration of on-ground control methods and the interaction with gamba grass ecology, especially in native habitats.

Weed management is rarely prescriptive; numerous confounding factors ranging from environmental to economic to ecological will influence control decisions and efficacy of control actions. Herbicide application and burning are two key tools for gamba grass management. Manipulating timing of both these practices within the wet season (WS) and the dry season (DS), and the corresponding integration of the two, may enhance control outcomes. The ecology of gamba grass needs to be considered when integrating control options and is essential for cost-effective and strategic gamba grass management.

A long-term reduction in cover and density of infestations needs to incorporate control options for mature perennial tussocks, as well as a reduction in the seedbank and potential for spread. It is beyond the scope of this paper to address the decision-making for prioritisation of management areas; this is addressed to some extent by Adams and Setterfield (2012).

The focus of this paper is to discuss results derived from on-ground gamba control actions over three years at a site near Darwin, and present the advantages and disadvantages of different timing of herbicide application and of burning practices. This study provides factors to be considered from a practitioner's perspective in the development of a multi-season gamba grass management strategy.

MATERIALS AND METHODS

A study was initiated in 2013–14 WS to evaluate gamba grass management in Charles Darwin National Park (CDNP) in Darwin, Northern Territory (NT). This represents a native savanna environment consistent with that discussed by Ens *et al*. (2015) and possesses environmental, conservation and recreational values threatened by gamba grass. Sites were selected within CDNP characterized by dense infestations (>10 t biomass ha^{-1}) of gamba grass and were treated at a range of times over concurrent years, categorised here as either early WS (November-January) or early DS (April-June).

The current herbicide recommendation for gamba grass was followed; foliar application of 1% solution glyphosate (360 g L^{-1}) during the active growth period, applied using a QuikSpray® unit consistent with commercial operator practice. Sprayed (2014–15 WS) and non-sprayed control plots were subject to a WS burn on the 7th Jan 2016. Post-burn glyphosate application versus a non-sprayed control was then imposed within the burnt plots two months after burning. Herbicide treatment effect was measured as percent brown-out of total cover two weeks after application, and two months after commencement of the following wet season to assess regeneration. WS burn effect was evaluated as percent regeneration of total cover, two months after fire. Photo monitoring points were established to assess long-term trends in gamba grass infestations and native species regeneration. Constraints and advantages to different on-ground control actions were also documented.

RESULTS AND DISCUSSION

Timing of herbicide application The effect of seasonal timing of glyphosate application on native plant communities was considered by Brooks *et al.* (2006). However, this was not quantified for the target species, gamba grass, itself. The majority of land managers currently finish spraying gamba grass once it has commenced seeding, coinciding with rapid bolting of reproductive tillers up to 4 m in height, usually in April-May.

Foliar glyphosate application was an effective method to control gamba grass in CDNP, with total brown-out observed within two weeks of herbicide application, even in very dense infestations (>15 t ha^{-1}), irrespective of time of application. However, the time of the season at which treatment occurred, imposed different advantages and disadvantages both operationally and ecologically (Table 1).

Gamba grass without prior burning treatment was characterised at the commencement of the WS by tall (>3 m) rank material remaining from the reproductive tillers from the previous WS. Stimulation of active vegetative growth at the start of the WS, commenced as a green collar of leaf around the base of the plant surrounding tall rank growth; this was not conducive to foliar herbicide uptake. Early WS application resulted in isolated tussock regeneration and seedling recruitment later within that WS, requiring follow-up with consequent additional mobilisation costs (spray team to access site). Collateral damage of native species was also observed, effectively limiting the opportunity for seed production and subsequent recruitment desirable for habitat restoration, although the degree of damage was gamba grass density dependent.

Early DS glyphosate application during- and post-seeding continued to be effective on gamba grass (100% brown-out). Gamba grass possesses a number of ecophysiological traits such as carbon assimilation in the dry season, for greater resource use efficiency. These traits extend its growing season and provide a competitive advantage over native species in savanna habitats (Ens *et al.* 2015). This extended growth period also provides the opportunity to extend the current practice of foliar application only in the WS to later into the DS.

Little or no off-target damage was observed at early DS, coinciding with senescence of surrounding native species, suggesting they were not receptive to herbicide uptake at that time. Brooks *et al.* (2006) concluded that glyphosate produced off-target damage when treated in April; delaying application until further into the DS may mitigate this effect, whilst also maintaining effectiveness of control on gamba grass.

Follow-up gamba grass control in the subsequent year was required irrespective of time of spraying, due either to regeneration of perennial tussocks, or seedbank recruitment.

Timing of burning Fire is intrinsic to the ecology of northern Australia, and the effect of timing of burning has been examined extensively for native savanna species. However, the effects of fire on gamba grass at different phenological stages and conducted at different times of the year, is less documented.

Manipulation of timing of burning is difficult due to the resource intense nature of conducting a controlled burn, and the difficulties in preventing wildfires. Current practices for gamba grass burning are primarily for risk mitigation, generally in the early DS, rather than for strategic weed control. Fires from August-October (late DS) are often wildfires (non-controlled).

A significant advantage of WS burning was the removal of rank gamba grass growth and promotion of vegetative production. This effectively manipulated proportion of green leaf to rank stalk and increased efficiency of subsequent glyphosate herbicide uptake. However, if follow-up spraying was not conducted, burning provided gamba grass with a competitive advantage, and the cycle repeated itself for the following season. Table 1 discusses the advantages and disadvantages of different seasonal burning times.

Irrespective of timing of burning, gamba grass regenerated to original cover by the end of the WS if follow-up herbicide application did not occur.

Integration Chemical control and fire are both recognised as control options for a range of weeds in northern Australia, where fire can increase mortality of species such as rubbervine and bellyache bush (Vitelli and Pitt 2006). However, fire provides a competitive advantage to gamba grass, initiating a grass-fire

Table 1. Considerations for spraying and burning at different times for gamba grass management.

Time of season	Advantages	Disadvantages
Early WS (Dec–Feb)	**Gamba grass is actively vegetative:** • Uses smaller volume of herbicide and less time per plant or per unit area than when plants have matured **Allows integration of spraying with previous DS or current WS burn:** • Stimulates lush green re-growth for improved herbicide efficacy • Removes rank tiller material from previous WS for decreased interception (waste) of herbicide • Increased operator safety and easier application by not working in thick tall gamba (logs, snakes, caught up hoses etc) **Allows staggering of spray and burn times:** • Prioritise spray areas previously burnt and now re-growing; follow-up areas from previous season • Less hectic than DS for Fire Officers' burning workload • Rainfall settles smoke haze	**Gamba grass if not recently burnt:** • Dormant tillers not yet emerged; often requires follow-up spraying later that WS to target regenerating tillers plus new recruited seedlings • Rank tillers remaining from previous WS not susceptible to herbicide uptake • Spraying may cause collateral damage of desirable species **Hot and humid, it rains a lot:** • Logistics more difficult; rain delays • Discomfort; poor worker efficiency **To conduct a burn at this time:** • Requires a short break in WS • May require fire exclusion in preceding DS for fuel load to carry • Removal of ground cover may lead to erosion from intense WS rain • May require additional DS burn if follow-up herbicide application NOT conducted
Early DS (Apr–June)	**Gamba grass is reproductive:** • More distinct in the environment; easier to locate and to identify for spraying. **Extends the spray window:** • Decreases work load for next season • Glyphosate may induce seed sterility • High gamba grass mortality; tussock plus new seedlings susceptible to herbicide • Resultant dead gamba grass mat reduces probability of seed soil contact at seedfall **Less collateral spray damage of native species:** • Already seeded; allows later recruitment • Dense gamba grass infestations can effectively 'shield' understory species **Logistics:** • Accessibility; decreased risk of getting bogged spraying or burning • More hours per day spraying/burning; less risk of delay from rain • Better climate condition: worker efficacy **Integrate spraying with an early cool DS burn:** • Consistent with burning practices for native species, such as *Sorghum* spp. dominated woodland • Prioritise burnt areas for early subsequent WS spraying	**Gamba grass (unless recently burnt):** • Increased herbicide volume and spray time per plant or unit area • Increased operator risk from spray drift due to taller spray height • Allowing seed set increases risk of seed spread **To conduct a burn at this time:** • Removal of ground cover coinciding with time of seedfall from gamba grass provides post-fire suitable habitat for seed soil contact • Smoke haze and associated human health issues
Late DS (Aug–Oct)	**Decreased biomass of gamba grass and subsequent regeneration of green growth provides opportunity for early WS herbicide application**	**Later curing of gamba than native species:** • Intense fires; habitat transforming due to increased fuel load and temperatures • Wildfires versus prescribed burns

feedback which accelerates gamba grass growth and invasion (Ens *et al.* 2015). Fire needs to be combined with timely herbicide application to maximise control of both established perennial tussocks, as well as reducing the seedbank for strategic management.

Gamba grass seed viability is low under natural conditions (Flores *et al.* 2005). The effect of timing of glyphosate application and of burning on mature seed viability has not been assessed, but preliminary results (this author) indicate that glyphosate can decrease gamba grass seed viability when applied at flowering and seeding.

CONCLUSION

Herbicide application (foliar glyphosate) was effective in reducing gamba grass cover in sites selected within CDNP. Early DS application on reproductive gamba grass extended the seasonal spray window compared to the current practice of WS treatment prior to flowering. This provided a number of advantages for gamba grass management including reduced collateral damage of already senesced native species. Fire was predominantly used as a risk mitigation tool in the early DS rather than for weed control. The WS burn demonstrated advantages in reducing rank biomass of gamba grass which had not been previously sprayed, to enhance subsequent herbicide efficacy (WS burn/WS spray). Plots which had been sprayed previously, then were subject to a WS burn, resulted in no regenerating gamba grass to date (2014–15 WS spray/2015–16 WS burn). In plots which were burnt (WS or DS) but herbicide not subsequently applied, gamba grass re-established to its original cover by the end of the WS. These results support the integration of burning and herbicide application for gamba grass control. Fire alone will continue to exacerbate the gamba grass problem; land managers should conduct strategic DS or WS burns, and then prioritise these areas for subsequent spraying.

Results here are based on qualitative data. Future work could better quantify relationship between gamba grass ecology and treatment outcomes, and also incorporate other factors such as grazing and the use of residual herbicides. However, current results provide a platform to demonstrate principles of integrating and rotating herbicide and fire regimes to develop a multi-season gamba grass management strategy.

ACKNOWLEDGMENTS

Thanks to Parks and Wildlife Commission of the NT and the NT Department of Land Resource Management for their provision of resources and assistance in Charles Darwin National Park.

REFERENCES

Adams, V.M. and Setterfield, S.A. (2012). Spatial prioritisation for management of gamba grass (*Andropogon gayanus*) invasions: accounting for social, economic and environmental values. Proceedings of the 18th Australasian Weeds Conference, ed. V. Eldershaw, pp. 49-52. (Weed Society of Victoria, Melbourne).

Brooks, K., Setterfield, S.A. and Douglas, M.M. (2006). Seasonal timing of glyphosate application: impacts on native plant communities in a north Australian tropical savanna. Proceedings of the 15th Australian Weeds Conference, eds C. Preston, J.H. Watts and N.D. Crossman, pp. 223-6. (Weed Management Society of South Australia, Adelaide).

Ens, E., Hutley, L.B., Rossiter-Rachor, N.A., Douglas, M.M. and Setterfield, S.A. (2015). Resource-use efficiency explains grassy weed invasion in a low-resource savanna in north Australia. *Frontiers in Plant Science*. 6:560. Dio: 10.3389/fpls.2015.00560. 10 pages.

Flores, T.A., Setterfield, S.A. and Douglas, M.M. (2005). Seedling recruitment of the exotic grass *Andropogon gayanus* (Poaceae) in northern Australia. *Australian Journal of Botany* 53 (3), 243-9.

Rossiter, N., Setterfield, S.A., Douglas, M.M., Hutley, L. and Cook, G. (2004). Exotic grass invasion in the tropical savanna of northern Australia: ecosystem consequences. Proceedings 14th Australian Weeds Conference, eds B.M. Sindel and S.B. Johnson, pp. 168-71. (Weed Society of New South Wales).

Setterfield, S.A., Rossiter-Rachor, N., Douglas, M.M., McMaster, D. Adams, V. and Ferdinands, K. (2014). The impacts of *Andropogon gayanus* (gamba grass) invasion on the fire danger index and fire management at a landscape scale. Proceedings of the 19th Australasian Weeds Conference, ed. M. Baker, pp. 125-8. (Tasmanian Weed Society, Hobart).

Vitelli, J.S. and Pitt, J.L. (2006). Assessment of current weed control methods relevant to the management of the biodiversity of Australian rangelands. *The Rangeland Journal* 28, 37-46.

Weed management of legacy mine sites in the Northern Territory

Daniel L. McIntyre
Department of Mines and Energy, Northern Territory Government, GPO Box 4550, Darwin, Northern Territory 0801, Australia
(daniel.mcintyre@nt.gov.au)

Summary The Northern Territory has a number of legacy mine sites which may pose a high risk to the natural environment and human safety. The Legacy Mines Unit (LMU) of the Northern Territory Government Department of Mines and Energy was founded in 2013 to inventory, prioritise and manage remediation works for legacy sites in the Northern Territory. Weed management has been identified as an area of interest for the LMU. Most legacy sites have ongoing weed management issues leading to an increased potential for weeds to spread off-site. Weeds at legacy sites can compromise revegetation success, spread into surrounding clean country, increase fuel loads and fire frequency/intensity and diminish visual amenity for tourism. We present two weed management case studies that highlight the work to date that has been done by the Unit: weed survey and control works conducted by an indigenous land management team at Tennant Creek and a baseline weed survey at a former gold mine in the Darwin region.

Keywords Legacy mine, weeds, Northern Territory, management, survey.

INTRODUCTION

The Northern Territory (NT) has a number of legacy mine sites which may pose a high risk to the natural environment and human safety. Most of these legacy sites were created before 2005 and pre-dated the NT Government's policy of requiring operators to lodge a 100% rehabilitation security bond. In 2013 the Mining Management Act was amended to require all operators to pay a 1% levy on their security. The objective of the levy is to generate the necessary funds to begin addressing historical mining impacts. As a result of this, the Legacy Mines Unit (LMU) of the NT Government Department of Mines and Energy was founded in 2013 to inventory, prioritise and manage remediation works for legacy sites in the NT.

The LMU has commenced weed survey and control programs at selected legacy mine sites in the NT. Weeds at legacy sites can compromise re-vegetation success, spread into surrounding clean country, increase fuel loads and fire frequency/intensity and diminish visual amenity for tourism. Weed survey and control at these sites will assist in planning (e.g. producing weed management plans) and implementing follow-up control. In this paper we present two weed management case studies that highlight the work to date that has been done by the LMU: weed survey and control works conducted by an indigenous land management team at Tennant Creek and a baseline weed survey at a former gold mine in the Darwin Region.

CASE STUDIES

1. Weed management at Tennant Creek The town of Tennant Creek is located 1000 km south of Darwin and 500 km north of Alice Springs, in the Barkly Region of the NT (Figure 1). The region has a semi-arid climate, with a cool winter from May to September and a hot summer from October to April. The region has had intermittent prospecting, mining and exploration since gold was first discovered in 1933. A gold rush to the area shortly afterwards formed the Tennant Creek Township in 1934, which remains the major population centre for the region. The peak of gold and copper mining occurred in the 1970s with the Nobles Nob, Juno, Peko and Warrego mines producing significant amounts of gold. Most operations ceased production by the 1980s with re-treatment of tailings occurring at some sites up to the mid-late 1990s. Many of the sites are within 15 km of the town and can be accessed by existing tracks.

Many of the mine sites around Tennant Creek have limited information on the location and density of weeds. Conducting detailed surveys and mapping of weeds on mining tenements will allow the information to be used to facilitate on-going monitoring and follow-up work after initial surveys; and to assist with creating weed management plans for these areas, based on the Tennant Creek Regional Weed Management Plan. The first site at Tennant Creek selected for weed survey and control was a 243 ha area of interest located approximately 4 km south of Tennant Creek and comprising of existing mining tenements and contains several inactive mines, most notably the historic Eldorado gold mine.

In early 2016 the LMU engaged the land management team from the Tennant Creek based Julalikari Council Aboriginal Corporation to conduct the work. The land management team was comprised of six

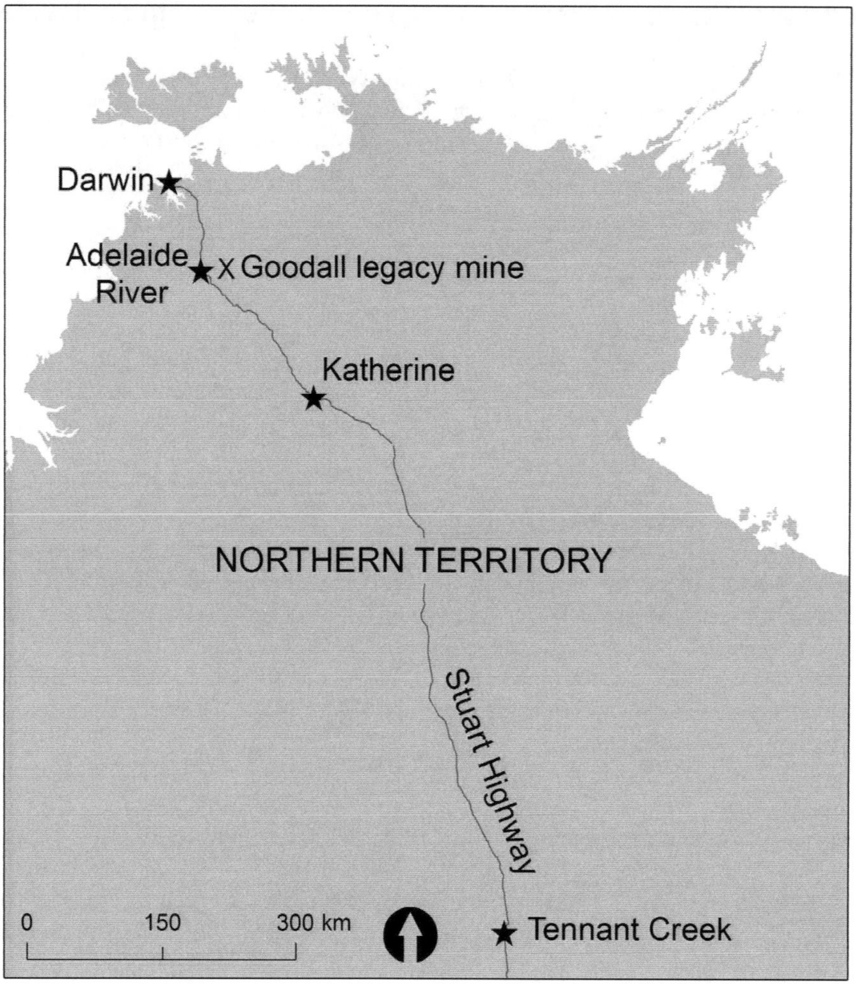

Figure 1. Location of Tennant Creek and Goodall legacy mine.

indigenous staff, with two additional indigenous 'work for the dole' participants. The team had previous experience with weed management at a number of indigenous communities in the vicinity of Tennant Creek. All members of the team had been trained in chemical handling and most had completed (or were working towards) certification in horticulture and conservation and land management.

The work was conducted during April and May 2016. The area was surveyed from four-wheel-drive vehicles, quad bikes and on foot. Data were captured with handheld Global Positioning System (GPS) units and recorded onto hard copy sheets. The format of the data was compatible with the NT Weed Data Collection Guidelines (Weed Management Branch, Northern Territory Government 2015). Overall the survey area had a very sparse distribution of weeds. Three rubber bush (*Calotropis procera* Aiton) plants were found and treated using the basal bark method with Access™ and diesel mixed at a 1:60 L ratio. Rubber bush is a Class B weed species in the NT (growth and spread to be controlled by the landowner) and is a common weed species in the Barkly Region.

2. Baseline weed survey at Goodall legacy mine

The former Goodall gold mine is located approximately 100 km south-east of Darwin and approximately 30 km east of the Adelaide River township (Figure 1). Operations at the mine commenced in 1988 and gold bearing ore was mined in four pits until the end of 1992 and then from two pits in 2002 and 2003. Some rehabilitation was conducted in 1993 and 1994, which included capping and re-vegetation of waste rock dumps.

Unlike most legacy sites, Goodall has no current lease or operator, making it a true legacy site.

Normally an operator of a site would have a current Mine Management Plan (MMP), which would include provision for environmental management of the site. In the absence of an operator and MMP, the responsibility of land management falls with the LMU on behalf of the NT Government.

The Goodall site had limited information on the location and density of weeds. In early 2016, the LMU engaged an NT-based environmental consultancy (Ecoz Pty Ltd.) to conduct a baseline weed survey at Goodall. The objective of the survey was to provide information on the location and density of weeds on the site and surrounding area, with a particular focus on declared weeds. This information would assist the LMU to plan and implement weed management at the site in the future.

The survey was conducted over a 665 ha area, which comprised: the Goodall mine footprint (300 ha) which included semi-disturbed areas of eucalypt woodland with a grassy sorghum understory and disturbed areas with an acacia shrubs; 92 ha of riparian vegetation along two creek lines; undisturbed/semi-disturbed eucalypt woodland with a grassy sorghum understory; and a disturbed area consisting of the former mine worker accommodation.

The survey was conducted from June 20–23 2016 by a botanist with extensive local experience in weed identification, with assistance from two Larrakia indigenous rangers. The area was mostly surveyed by quad bike/all-terrain vehicle. Data were recorded using Cybertracker™ software on a handheld device, with data entry fields complying with the NT Weed Data Collection Guidelines. The survey took 20 hours (60 person hours) to complete, with 84 km of transects traversed.

A total of 300 weed patches were surveyed, in which most were found in drainage areas, along tracks and areas of disturbance including waste rock dumps. The majority of the weed cover recorded within the survey area was the Class B weed hyptis (*Hyptis suaveolens* L.) with 211 patches. The survey also recorded 12 patches of Chinee apple (*Ziziphus mauritiana* Lamk.), which is a Class A weed (to be eradicated in all areas of the NT, except where classified as Class B). In total there was one Class A weed, four Class B weeds and nine environmental weeds recorded during the survey.

DISCUSSION

A small number of weeds was surveyed and controlled within the selected mining tenements at Tennant Creek. Nonetheless, surveying mining tenements regarded as being relatively weed-free remains a relevant management activity for the LMU, as it is important to record and map both the presence and absence of weeds at a given site. There are a number of tenements and associated mine sites around Tennant Creek that have larger disturbance footprints and a correspondingly larger potential for weed infestations. These sites and surrounding mining tenements will be progressively surveyed over the next two years and weed control will be conducted concurrently. The detailed weed survey and control data will allow treated infestations to be effectively monitored and timely follow-up treatment applied where necessary.

The Goodall survey provided valuable baseline data at a mine site where data were previously lacking. In the short-term, weed management at the site will be focused on preventing further weed spread by treating the weed infestations identified in the survey and conducting follow up treatment where required. In the longer term, a weed management plan will be produced.

This plan will focus on two main aspects:

1. reduce the likelihood that weeds are introduced and spread at the site, and
2. ensure that future rehabilitation of the site is not compromised by weeds.

ACKNOWLEDGMENTS

We gratefully acknowledge Mr. Daniel Steel, the Tennant Creek District Weed Officer from the Northern Territory Government Weeds Branch, for sharing valuable information and conducting an informative induction.

We thank the managers of the Bridge Creek Station (Mr. Marcus Rathmann) and Mount Ringwood Station (Mr. Don White) for their cooperation with access to the Goodall mine.

REFERENCES

Weed Management Branch, Northern Territory Government (2015). Northern Territory Weed Data Collection Manual. (Northern Territory Government of Australia, Darwin).

Are root exudation or rainfall on treated plants likely causes of pasture damage after wiper application of herbicides?

Kerry C. Harrington[1], Hossein Ghanizadeh[1], Clyton Moyo[1], Peter D. Kemp[1] and Han (J.P.J.) Eerens[2]
[1] Institute of Agriculture and Environment, Massey University, Palmerston North, New Zealand
[2] Bayer New Zealand Ltd, 30B Aspenleigh Drive, RD 3, Hamilton, New Zealand
(K.Harrington@massey.ac.nz)

Summary Wiper application of herbicides to weeds sometimes causes damage to pasture plants, especially white clover, growing immediately under the wiped weeds. Two experiments were conducted to determine the potential for either exudation of herbicide from roots or rainfall washing herbicide off treated plants to cause damage to white clover after wiper application. In Experiment 1, Californian thistle and white clover were grown in the same pots within a glasshouse, then either metsulfuron or a triclopyr/picloram mixture were applied to the Californian thistle using a Rotowiper. White clover growing in pots below treated plants was significantly affected by the metsulfuron but not the triclopyr/picloram mixture, when compared with the untreated control plants. In Experiment 2, Californian thistle, white clover and perennial ryegrass were grown in separate pots within a glasshouse. Glyphosate, clopyralid, metsulfuron and a triclopyr/picloram mixture were applied to Californian thistle plants using a Rotowiper. The treated Californian thistle plants were then exposed to artificial rain from a rainfall simulator either 1 or 5 days after herbicide treatment, with pots of white clover and perennial ryegrass positioned under each Californian thistle plant during the rainfall event. All of the herbicides except glyphosate significantly affected white clover for both rainfall events, whereas only metsulfuron significantly affected perennial ryegrass for the Day 1 washing. Thus although damage from herbicide exudation does appear possible, rainfall following wiper applications is probably more likely to cause damage to clover growing under treated plants.

Keywords Herbicide, wiper application, clover damage, root exudation, Californian thistle.

INTRODUCTION

Vegetation growing under tall weeds is often unintentionally affected by application of herbicides using weed wipers (Grekul *et al.* 2005). White clover (*Trifolium repens* L.) damage as a result of application of herbicides to Californian thistle (*Cirsium arvense* (L.) Scop.) using a weed wiper has been recorded in trial work though damage to perennial ryegrass (*Lolium perenne* L.) was found to be negligible (Moyo 2008). Herbicide movement to non-target plants from herbicide-treated plants could occur after wiper application via loss of herbicide from foliage through rainfall (Campbell and Nicol 1998). Herbicide washed off foliage during rainfall landed on pasture and soil where it could be absorbed by plants and cause damage to pasture. Triclopyr, picloram, metsulfuron and clopyralid can remain active in the soil for several months after application, can be absorbed through roots and white clover is very sensitive to all these herbicides (Tomlin 2009). Glyphosate, the other herbicides often used in wipers, would be rendered inactive within the soil (Franz *et al.* 1997), though all herbicides could also penetrate foliage of pasture plants.

Herbicides could also be released from treated weeds through root exudation (Hickman *et al.* 1990). Root exudation is defined as the release of substances from inside a plant by either secretion or passive movement into the surrounding media from plant roots (Baetz and Martinoia 2014). Several herbicides have been shown to move from treated plants and cause injury to untreated neighbouring plants through exudation (Coupland and Peabody 1981, Dinelli *et al.* 2003, Hickman *et al.* 1990). The objective of this study was to evaluate the relative importance of root exudation and rainfall washing off herbicides from treated plants in damaging white clover following wiper application to weeds.

MATERIALS AND METHODS

Experiment 1 In this experiment the occurrence and significance of root exudation of foliar-applied herbicides from Californian thistle was evaluated. Californian thistle root fragments (at least 5 cm long) were collected on 8 November 2006 from pasture near Palmerston North, New Zealand. The root fragments were planted into 2000 cm^3 PB3 planter bags with potting mix as described by Moyo (2008). The bags were kept in a glasshouse and watered with a capillary matting system. White clover seeds were sown into the same planter bags (70 seeds per bag) on 11 December 2006.

The Californian thistle plants were treated using a Rotowiper on 9 February 2007, with either a mixture of 100 g L^{-1} picloram as an amine salt plus 300 g L^{-1} triclopyr as butoxy ethyl ester (Tordon Brushkiller®) at 1 part herbicide to 39 parts water, or with 200 g a.i. kg^{-1} metsulfuron as a methyl ester (Answer®) at 3 g product per litre of water (recommended rates for both products when used in wipers). The plants, supported by concrete blocks, were arranged in a single row on asphalt pavement. The bags and the white clover growing in them were covered by a plastic skirt to prevent any contamination of the clover with herbicide from the wiper. The thistles were at the early post-flowering stage and averaged 60 cm in height.

After treatment and when the plants had dried, the plastic skirts were removed and plants were returned to the glasshouse to sub-irrigated benches. The mean daily temperature in the glasshouse during the experiment ranged from 23.2°C in February to 14.0°C in May. The experiment was a randomised complete block design with three treatments, i.e. the two herbicide treatments and an untreated control. Plants were allocated to blocks based on the amount of white clover in each bag. Each treatment was replicated ten times. Visual assessments of foliar injury symptoms were conducted on average every three weeks until 14 May when the foliage of both the thistles and white clover were separately harvested, dried at 80°C for 24 hours and weighed.

Experiment 2 The objective of this experiment was to investigate whether rainfall might wash off sufficient herbicide from treated plants to damage pasture underneath. Californian thistle plants were established as described for Experiment 1, except root fragments were collected on 6 October 2006 and no clover was grown in the same pots. White clover was instead sown in separate 200 cm^3 pots on 5 January 2007 with 35 seeds per pot. Perennial ryegrass was also established in other 200 cm^3 pots on 10 January 2007 with 32 seeds per pot. All plants were grown in a glasshouse and irrigated as described for Experiment 1.

The herbicides used were Tordon Brushkiller® and Answer® at the same rates as in Experiment 1, clopyralid (Versatill®, 300 g a.e. L^{-1} as amine salt) at 1 part herbicide to 39 parts water, and glyphosate (Roundup Renew®, 360 g a.e. L^{-1} as isopropylamine salt) at 1 part herbicide to 19 parts water. The thistles were treated on asphalt pavement as in Experiment 1 on 9 February 2007 using a Rotowiper, and a different roller was used for each herbicide. Using a rainfall simulator, treated plants later received rainfall 1 or 5 days after the herbicides were applied. The thistles were placed under the simulator for 30 seconds and received the equivalent of 2.0 mm of rainfall. Four pots of white clover and four pots of perennial ryegrass were placed around each Californian thistle plant during rainfall simulation to intercept any herbicide that washed off the treated thistle

The experiment had a randomised complete block design with four herbicide treatments, two untreated controls for each species, and two simulated rainfall periods (1 day and 5 days) after treatment. Treatments were blocked on the size of the thistle plants. Each treatment was replicated six times, and as each of the six thistle plants had four pots of white clover and four pots of perennial ryegrass, each treatment had 24 pots of both clover and ryegrass. Clover and ryegrass were analysed separately. Visual assessments of foliar injury symptoms were recorded by scoring the white clover and perennial ryegrass plants as described for Experiment 1. The scoring was done until the foliage of the clover and ryegrass was harvested in mid-April by cutting at ground level, drying at 80°C for 24 hours then weighed.

Statistical analysis The visual scores and dry weights were subjected to an analysis of variance using the statistical software package SAS (SAS 2004) and means were separated using LSD tests at the 5% level of probability.

RESULTS

Experiment 1 White clover plants growing under Californian thistle plants treated with metsulfuron had significantly lower dry weight than both the untreated control and the triclopyr/picloram treatment (Table 1). Visual injury scores also showed a similar result (data not shown).

The dry weight of the thistle stems treated with either metsulfuron or triclopyr/picloram was similarly reduced relative to the untreated control (Table 1). Injury scores on thistle stems showed a similar result to the dry weight (data not shown). Not all of the thistles died as a result of the treatments with only 60% and 70% of plants dying completely for metsulfuron

Table 1. The dry weight (g pot^{-1}) of white clover and Californian thistle grown together in pots, measured 13 weeks after treatment of the thistle with metsulfuron and triclopyr/picloram mixture.

	White clover	Californian thistle
Metsulfuron	1.65 b*	11.0 b
Triclopyr/picloram	6.64 a	10.3 b
Untreated	7.10 a	16.8 a

* Means with the same letter within a column are not significantly different at P ≤0.05.

and triclopyr/picloram, respectively, probably due to insufficient herbicide being applied.

Experiment 2 There was extensive damage to white clover 20 days after simulated rainfall events at 1 and 5 days after treatment (DAT) had washed off the herbicide from the Californian thistle plants. All herbicides except glyphosate caused significant levels of damage to white clover plants under the Californian thistle when artificial rain was applied (Table 2). Clopyralid and metsulfuron washed off the thistle plants 1 DAT were particularly damaging. Herbicide injury symptoms ranged from minor leaf discolouration to complete necrosis.

Perennial ryegrass showed few signs of being affected 20 days after the simulated rainfall event, with the only significant difference in scores occurring for the metsulfuron washed 1 DAT.

Nine weeks after treatment (WAT) when the plants were harvested, severe herbicide damage to white clover was measured following metsulfuron and clopyralid washing off the Californian thistle plants on both Day 1 and Day 5 (Table 3). There was a smaller amount of damage to clover caused by the triclopyr/picloram mixture, which was significant only for the Day 1 washing. Effects from glyphosate were not significant.

As with the scoring 20 days after the rainfall event, the only significant herbicide damage to perennial ryegrass 9 WAT was a reduction in growth of plants under metsulfuron-treated thistles for the Day 1 washing (Table 4). The ryegrass exposed to metsulfuron at Day 5 was not significantly lower in dry weight than untreated plants, though it was significantly less than ryegrass plants exposed to glyphosate or the triclopyr/picloram mixture. Although white clover exposed to metsulfuron washings at Day 1 had a dry weight only 15% that of untreated plants, the perennial ryegrass plants exposed to metsulfuron at Day 1 were 71% the weight of untreated plants.

Table 2. The effect after 20 days on white clover (using scores from 0 = severe damage to 10 = no damage) of herbicides washed off Californian thistle 1 or 5 days after treatment (DAT).

	1 DAT	5 DAT
Clopyralid	5.1 c*	6.1 c
Glyphosate	9.1 a	9.1 ab
Metsulfuron	4.9 c	6.1 c
Triclopyr/picloram	8.0 b	8.7 b
Untreated	10.0 a	10.0 a

* Means with the same letter within a column are not significantly different at $P \leq 0.05$.

Table 3. The dry weight (g pot^{-1}) of white clover plants harvested 9 weeks after being placed under Californian thistle plants from which herbicides were washed off 1 or 5 days after treatment (DAT).

	1 DAT	5 DAT
Clopyralid	2.36 c*	2.37 b
Glyphosate	4.83 a	4.96 a
Metsulfuron	0.78 d	0.98 c
Triclopyr/picloram	3.63 b	4.74 a
Untreated	5.13 a	5.27 a

* Means with the same letter within a column are not significantly different at $P \leq 0.05$.

Table 4. The dry weight (g pot^{-1}) of perennial ryegrass plants harvested 9 weeks after being placed under Californian thistle plants from which herbicides were washed off 1 or 5 days after treatment (DAT).

	1 DAT	5 DAT
Clopyralid	1.23 a*	1.17 ab
Glyphosate	1.19 a	1.24 a
Metsulfuron	0.79 b	0.97 b
Triclopyr/picloram	1.07 a	1.23 a
Untreated	1.12 a	1.15 ab

* Means with the same letter within a column are not significantly different at $P \leq 0.05$.

DISCUSSION

There was some evidence that root exudation of metsulfuron contributes to pasture damage caused by wiper application of herbicides. Damage caused by exudation in Experiment 1 may have been accentuated as roots were restricted to pots in which they were growing. Other studies have concluded that unless a large amount of the root system of a companion crop comes into contact with roots of treated plants, herbicide release from treated plants appears unlikely to be important (Penn and Lynch 1982).

There is no obvious reason to explain why metsulfuron damaged the white clover in Experiment 1, yet triclopyr/picloram did not, unless metsulfuron is more easily exuded than triclopyr/picloram. Field results from wiper application of herbicides showed that metsulfuron and triclopyr/picloram both caused damage to pasture (Moyo 2008). If exudation of herbicides was the primary cause, then both herbicides in

Experiment 1 should have caused some damage to the white clover growing with the thistles. Unfortunately no soil tests were conducted of herbicide residues within these pots to assist with interpretation of results.

The widespread damage to clover under treated Californian thistles when rain was simulated showed that washing off of herbicides is a more likely cause of pasture damage. The results of this experiment confirm results from earlier weed-wiping studies in which damage was caused to plants by rainfall washing off herbicides from the treated plants (Campbell and Nicol 1998).

Metsulfuron and clopyralid were highly damaging to white clover while the effect of glyphosate was minimal, and the triclopyr/picloram mixture caused moderate damage. The lack of damage to white clover by glyphosate following wiping is in agreement with field studies (Grekul *et al.* 2005, Moyo 2008, Thompson 1983). Low rates of glyphosate tend to be ineffective against both white clover and perennial ryegrass (Casey *et al.* 2000) which could explain the results. The ester formulation of the triclopyr/picloram mixture may have allowed it to enter the Californian thistle rapidly enough to prevent much being washed off by rain. The partial damage to perennial ryegrass by metsulfuron has been noted in the field following wiping (Moyo 2008).

Another possible mechanism for clover under Californian thistle to be damaged by herbicides applied by wipers is through splattering of herbicide at the time of impact by the wiper. This was investigated by Moyo (2008) and appears to be less important than wash off from rainfall.

Although there can be some damage to pasture following rainfall, Moyo (2008) showed that this only occurs with some applications, and no damage appeared if rain did not occur for a few days after application, or if either a small or very large amount of rainfall occurred. These observations involved work with glyphosate, metsulfuron and clopyralid. Thus further research is required about this interaction between rainfall and damage to pasture.

In conclusion, although it appears that herbicide damage can occur due to herbicide washing off pasture plants, observations in the field by Moyo (2008) and results from Experiment 2 suggest that glyphosate causes the least damage to both white clover and perennial ryegrass of the herbicides investigated.

ACKNOWLEDGMENTS

The authors would like to thank the Foundation for Research, Science and Technology for funding this study, and also the staff of Massey University's Plant Growth Unit.

REFERENCES

Baetz, U. and Martinoia, E. (2014). Root exudates: the hidden part of plant defense. *Trends in Plant Science* 19, 90-8.

Campbell, M.H. and Nicol, H.I. (1998). Effects of wiping herbicides on serrated tussock (*Nassella trichotoma* (Nees) Arech.) and African lovegrass (*Eragrostis curvula* (Shrad.) Nees). *Plant Protection Quarterly* 13, 36-8.

Casey, M.J., Brown, C. and Stevens, D.R. (2000). A summary of research into the use of low rates of glyphosate as a pasture management tool. Proceedings of the NZ Grasslands Association 62, 123-8.

Coupland, D. and Peabody, D.W. (1981). Absorption, translocation and exudation of glyphosate, fosamine and amitrole in field horsetail (*Equisetum arvense*). *Weed Science* 29, 556-60.

Dinelli, G., Bonetti, A., Minelli, M., Marotti, I. and Catizone, P. (2003). Root exudation of diclofop-methyl from treated ryegrass (*Lolium* spp.) and durum wheat (*Triticum durum* Desf.). Proceedings of the XII Symposium Pesticide Chemistry, Piacenza, Italy, 4-6 June 2003.

Franz, J.E., Mao, M.K. and Sikorski, J.A. (1997). Glyphosate: a unique global herbicide. American Chemical Society.

Grekul, C.W., Cole, D.E. and Bork, E.W. (2005). Canada thistle (*Cirsium arvense*) and pasture forage responses to wiping with various herbicides. *Weed Technology* 19, 298-306.

Hickman, M.V., Messersmith, C.G. and Lym, R.G. (1990). Picloram release from leafy spurge roots. *Journal of Range Management* 43, 442-5.

Moyo, C. (2008). Improving the efficiency of herbicide application to pasture weeds by weed-wiping and spot-spraying. PhD thesis, Massey University, Palmerston North, New Zealand, 206 pp. http://hdl.handle.net/10179/779

Penn, D.J. and Lynch, J.M. (1982). Toxicity of glyphosate applied to roots of barley seedlings. *New Phytologist* 90, 51-5.

SAS (2004). SAS/STAT User's Guide Version 9.1. Statistical Analysis Systems Institute, Cary, 1030 pp.

Thompson, A. (1983). Pasture weed control by rope wick applicator. Proceedings of New Zealand weed and pest control conference 36, 96-98.

Tomlin, C.D.S. (2009). 'The pesticide manual: a world compendium'. British Crop Production Council, Alton, UK.

Management of emerging weeds in the western region: fleabane (*Conyza bonariensis* (L.) Cronquist) and sowthistle (*Sonchus oleraceus* L.)

Mohammad Amjad and Abul Hashem
Department of Agriculture and Food Western Australia, Northam, Western Australia 6401, Australia
(mohammad.amjad@agric.wa.gov.au)

Summary Fleabane and sowthistle (also commonly known as flaxleaf fleabane (*Conyza bonariensis* (L.) Cronquist) and milk thistle (*Sonchus oleraceus* L.) are emerging summer weeds in Western Australia. Changes in farming systems, farm management practice and climate are resulting in changes in the summer weeds spectrum throughout the cropping area. Trials were conducted to examine the efficacy of herbicides to control fleabane and sowthistle during summer and the winter seasons of 2015. After the 2014 harvest of winter crops, two experimental sites with infestation of fleabane were selected across a wheat paddock on sandy loam soils at Grass valley (31° 38' 20" S and 116° 48' 22" E) and Geraldton (28° 78' 83" S and 114° 69' 47" E). A winter-fallow site with sowthistle infestation was also selected at Geraldton. During summer of 2014/2015, eight herbicide treatment combinations alone, as tank mix or in sequence were applied at different growth stages of fleabane at Grass Valley and only at flowering stage of fleabane at Geraldton. Similarly the sowthistle infestation at Geraldton was treated in winter-fallow with 12 herbicides combinations in winter 2015.

At Grass valley, all treatments were found effective in controlling fleabane although the efficacy of herbicides to control fleabane during summer varied with plant growth stages. In some treatment, surviving fleabane plants were affected to variable degrees by herbicides but plants regrew and developed when assessed 3 weeks after the second spraying. Glyphosate or glyphosate + 2,4-D amine applied as a single knock also provides good but incomplete fleabane control. However, double knockdowns with a mixture of 2,4-D ester + glyphosate followed by Spray.Seed® provided 98% control while double knockdowns with a mixture of 2,4-D ester + glufosinate followed either alone, as a tank mix or in sequence at by Spray.Seed® provided 94% control of fleabane when applied at rosette stage. Most herbicide treatments provided 80% to 94% control of fleabane when applied at flowering stage. The most effective treatment identified was a double knock application of glyphosate or glyphosate + 2,4-D amine followed by paraquat seven days later kills 100% of mature fleabane.

At Geraldton, all treatments worked very well and almost 100% control was observed in fleabane. Herbicides were found more effective on controlling fleabane when applied at seedling or flowering stage than rosette stage. Application of a mixture of glyphosate and 2,4-D followed by the application of Spray.Seed® was highly effective on fleabane whether applied in December 2014 at Grass Valley or in March 2015 at Geraldton.

Single application of glyphosate as well as double knockdown of glyphosate followed by Para-Trooper® or Spray.Seed® provided up to 100% control of sowthistle at Geraldton. However, double knockdown is preferred to single knockdown to minimise risks of herbicide resistance development. Double knockdowns of glyphosate followed by Para-Trooper® or Spray.Seed® were also effective on most of the other weed species such as capeweed, ryegrass, blue lupin, wild turnip, doublegee, fleabane, flatweed and couch.

Keywords Flaxleaf fleabane, sowthistle, couch grass, double knockdowns, tank mixes, summer, winter, weed control.

Unpacking the 'invasion curve': a discussion and workshop session

Hillary Cherry
Pest and Weeds Team, National Parks and Wildlife Service, PO Box 1967, Hurstville, New South Wales 1481, Australia
(hillary.cherry@environment.nsw.gov.au)

Summary Over the last ten years, there has been widespread adoption/recognition of the invasion curve (or biosecurity continuum) across Australasia. Given its importance in driving policy and legislation, this facilitated discussion forum and workshop will provide an opportunity to debate the validity and usefulness of the curve. Questions may include: How do we accurately portray the cost: benefits of management actions? For example, the current cost-benefit of asset-protection is said to be 1:1-5 however this may not reflect the biodiversity benefits gained by targeted weed management for biodiversity protection. Should the curve 'taper down' over time as we develop better ways to manage widespread weeds, such as a robust biocontrol agent? The benefits of eradication are said to outweigh the costs 25:1, however feasibility of eradication is often low, and few successful eradications exist on which to base this information.

- Would investment be better applied between the containment and asset-protection stages, (i.e. in the exponential growth stage) to stop a weed becoming widespread?
- How does scale factor in when using the curve (e.g. can local, regional, state or national decisions 'feed up' into overall management effectiveness)?
- Over 60% of Australia's weeds are garden escapes, so by default may already be widespread: what is the relevance of the curve to these species?
- How does feasibility of control factor in when using the curve to determine actions?

Participants will first discuss the principles underlying the curve and determine key issues to be debated, then split into smaller groups to workshop key ideas to improve future application of the invasion curve.

Keywords Biosecurity continuum.

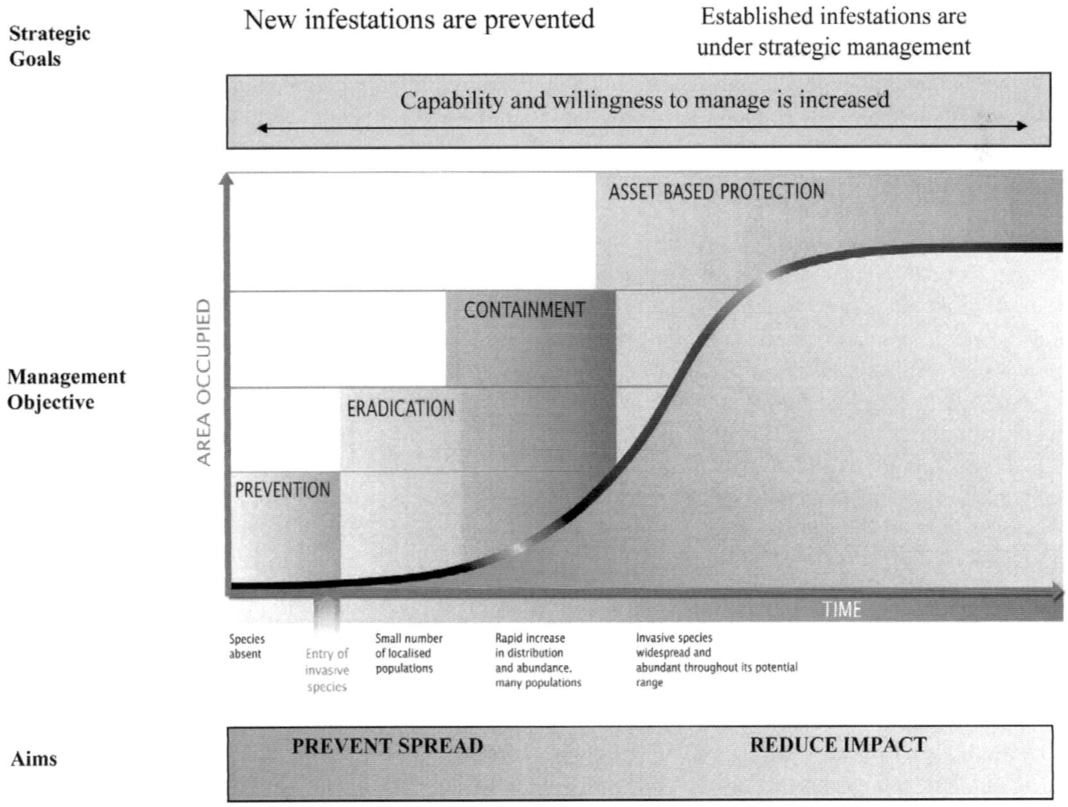

Impact reduction and asset protection

Bruce A. Auld[1,2] and Stephen B. Johnson[1,2]
[1] Weed Research Unit, New South Wales Department of Primary Industries, Locked Bag 6006, Orange, New South Wales 2800, Australia
[2] Graham Centre for Agricultural Innovation, Charles Sturt University, PO Box 833 Orange, New South Wales 2800, Australia
(stephen.johnson@dpi.nsw.gov.au)

Summary Management of weed impacts occurs across a spectrum of activities from: quarantine, to exclude species entirely; eradication or containment, where feasible; and asset protection strategies to reduce the impact of established weed species. The transition between containment and asset protection strategies occurs after containment of a weed species is no longer economically rational. Weed management efforts are then directed at reducing or maintaining a weed's rate of spread and/or its population below a notional threshold level.

Management activities aligned with a putative weed invasion curve have 'containment' followed by 'asset protection'. We propose that the weed management activities previously commonly termed 'asset protection' be split into at least two categories: 'impact reduction' and a new 'asset protection' category. The transition between the two proposed categories occurs when management activities are no longer directed at the management of a weed species, that is species-led.

'Impact reduction' would involve species-led management activities, generally aimed at managing the recovery of some asset/s. The long-term program to manage the recovery of threatened species populations and ecological communities from bitou bush (*Chrysanthemoides monilifera* (L.) Norlindh subsp. *rotundata* (DC.) Norl.) in eastern Australia is an excellent example. In contrast, the new more restricted 'asset protection' category occurs after this point, when a decision to protect areas/assets from multiple species weed invasions is made. The New South Wales state framework evaluating and prioritising biodiversity priorities from widespread weeds is an example of this. Both new categories will help better tailor weed investment, and management activity parameters.

Keywords Weed invasion curve, containment.

INTRODUCTION

All weed management activities can be placed on a generalised weed invasion curve, for example Charlton *et al.* (2009). This paper focuses on the latter part of the curve where asset (based) protection (AP) is the primary motivation for weed control, and the transition from containment to asset protection.

The initial phases of most plant invasions can be described by a constant exponential *population* growth curve (Cousens and Mortimer 1995). However linear models may well fit *area occupied* vs. time data (Lacey 1957, Auld and Coote 1980, Cousens and Mortimer 1995) for long periods. This will be somewhat dependent on the size of the unit area used: the larger the unit area, the slower the rate of total area occupied will be, and thus a linear fit may be suitable in spite of exponential population growth.

While the initial shape of the curve may be difficult to characterise as an invasion proceeds, at some point saturation of an area must be approached and the invasion curve asymptotes to a maximum value. This is the basis for the generalised weed invasion curve (Figure 1, Auld and Johnson 2014).

Often, the first management response to a high risk invasion is eradication. Experience has shown that eradication is only feasible in the very early stages of an invasion (Panetta 2015). Should this fail to be achieved, the next weed management tactic employed is containment although there is no identifiable turning point in the invasion curve at which to switch tactics from eradication to containment.

CONTAINMENT TO ASSET PROTECTION VIA IMPACT REDUCTION

Many weeds have invaded to such an extent that containment (C, see Figure 1), across their entire occupied range is no longer feasible, for example the Weeds of National Significance (WoNS, AWC 2015). While containment activities proceed, the weed population may still be notionally increasing exponentially and if the invaded area continues to grow at a constant (exponential or linear) rate there is, likewise, no identifiable point indicating where containment is no longer feasible. Having said this, a signal may occur when an increasing number of satellite populations occurs (Auld *et al.* 1979) and there are insufficient resources to contain the dispersed populations.

Overall, implementation of a containment strategy does not mean that Quarantine (Q)/exclusion and prevention, and eradication activities (E, see Figure 1) should not occur, particularly in areas where the weed is not found but could still invade. Such activities often aim to protect environmental assets from damage (Cherry 2008, 2010, Cherry *et al.* 2012, Grice *et al.* 2013, Auld and Johnson 2014). These efforts are usually focused on the expanding margins and satellite infestations of existing weed invasions, even though there has been little research examining how spatial characteristics of an invasion or landscape characteristics affect optimal control policies and resourcing: for example, see Epanchin-Neill and Hastings (2010) and Auld and Johnson (2014), but note recent exceptions include Steel *et al.* (2014) and Sydes and Murphy (2014).

Policy advisors commonly tend to aggregate a range of weed management activities at the asset protection end of the weed invasion curve. Auld and Johnson (2014) suggested splitting what has been termed the asset protection stage into two stages, these being impact reduction (IR) and asset protection (AP, see Figure 1).

In contrast to the earlier part of the invasion curve, there is an identifiable region in the latter part of the curve where the invasion rate slows and asymptotes towards a maximum area. Beyond this region, impact reduction (in the broad sense) is no longer logical and asset protection remains as the final tactic. The effect of 'impact reduction' may not necessarily reduce the area finally occupied but may increase the time taken to reach that point, potentially providing added net economic benefits (dotted line, Figure 1). Delaying the spread of a weed may also buy time for finding a more effective means of controlling it; for example introducing a biological control agent.

Local eradication and containment may form part of 'impact reduction'. An excellent example is the long-term program to manage the recovery of threatened species populations and ecological communities from bitou bush in eastern Australia. Impact reduction will involve management activities located closest to (to the immediate right of) the containment side of the invasion curve.

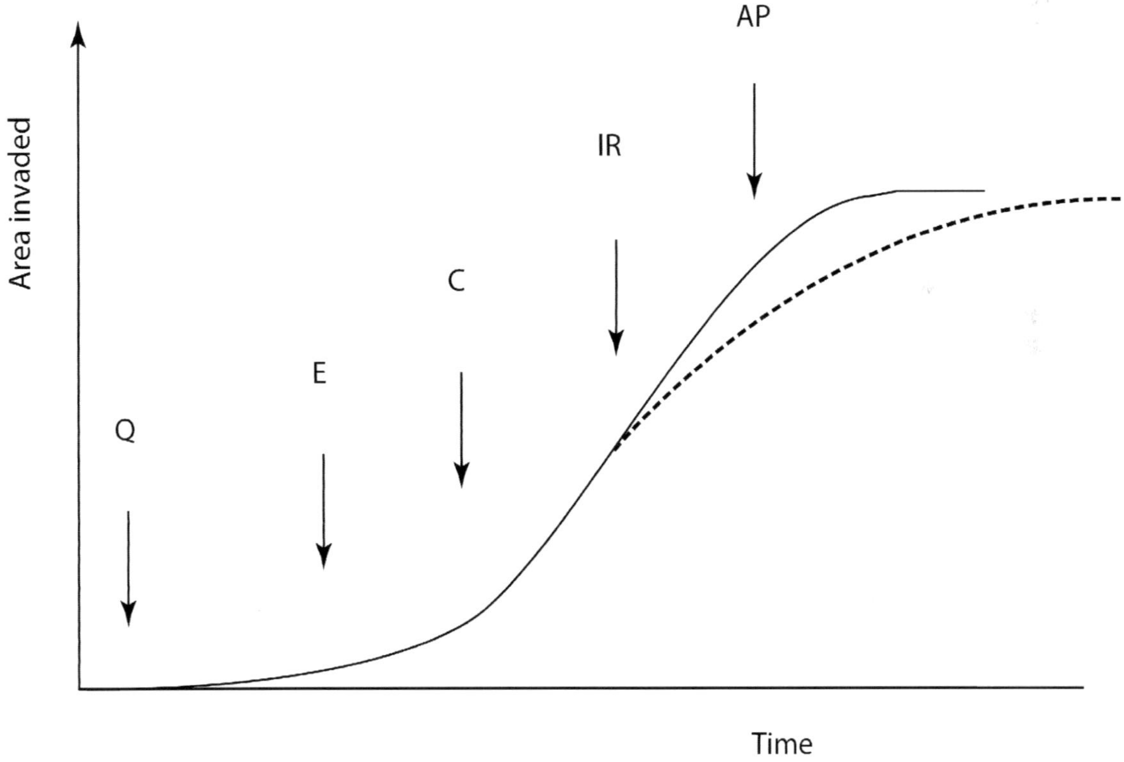

Figure 1. Notional weed management strategies/activities at various stages of the weed invasion curve with the following stages highlighted: Q, Quarantine/prevention; E, Eradication; C, Containment; IR, Impact Reduction; and AP, Asset (based) Protection (from Auld and Johnson 2014 c.f. Charlton *et al.* 2009, originally based on Chippendale 1991). The addition of stages to this curve also originated in Chippendale (1991), was expanded in Hobbs and Humphries (1995) and further refined in Environmental Weeds Working Group (2007). The dotted line indicates the possible effect of impact reduction activities.

In contrast, traditional asset (based) protection activities focus on assets, rather than the invading weed species. The transition point between impact reduction and the refined asset protection activities is, necessarily, when it is no longer economically feasible to focus on the species across its invaded range (species-led management) and a decision to protect areas/assets (asset-led protection) from weed invasion occurs (Auld and Johnson 2014). This acknowledges more recent work within, for example, biodiversity conservation in New South Wales, that environmental assets are often overlooked in natural resource management (Williams *et al.* 2009, Auld and Johnson 2014). To redress this imbalance, a state-wide framework was prepared across former Catchment Management Authorities (CMAs) in NSW:

> '*prioritizing areas for protection from widespread weeds based on the conservation status of ecosystems (e.g. the number of endangered species present) under threat from widespread weeds*' (Auld and Johnson 2014, also see Whiffen *et al.* 2011).

The scope of asset protection may vary with asset type and invading species. For instance, some exotic species are tolerated in native pastures and rangeland but not in annual crops. In nature conservation areas, ideally, no exotic species are allowed. Hence some assets are not species-specific in their 'asset protection' strategy and therefore actually have a local quarantine policy. While excluding all exotic species may not be feasible in many cases, those plant types that could transform ecosystems such as vines, legumes, shrubs and trees may be the target of species-specific asset protection.

ACKNOWLEDGMENTS

We thank Professor Clem Tisdell for his comments. Annie Johnson kindly assisted with drawing up Figure 1 and Philip Blackmore did some of the initial research in highlighting the Chippendale (1991) reference to us.

REFERENCES

Auld, B.A. and Coote, B.G. (1980). A model of a spreading plant population. *Oikos* 34, 287-92.

Auld, B.A. and Johnson, S.B. (2014). Invasive alien plant management. CAB Reviews: Perspectives in Agriculture, Veterinary Science, Nutrition and Natural Resources 9, No. 37. 12 pp.

Auld, B.A., Menz, K.M. and Monahan, N.M. (1979). Dynamics of weed spread: implications for policies of public control. *Protection Ecology* 1, 141-8.

AWC, Australian Weeds Committee (2015). Weeds of National Significance. http://www.weeds.org.au/WoNS/ (accessed 10 June 2016).

Charlton, S.A., Downey, P.O. and Johnson, S.B. (2009). Weed categorisation – applying the 'PECA code' in NSW. Proceedings of the 15th Biennial New South Wales Weeds Conference, Narrabri, pp. 19-28. (Industry and Investment New South Wales, Orange). http://www.nswweedsoc.org.au/conferences/past_proceedings

Cherry, H. (2008). Eradication versus long-term surveillance and removal: Contrasting approaches to *Chrysanthemoides monilifera* (L.) Norlindh management in Australia. Proceedings of the 16th Australasian Weeds Conference, eds R.D. van Klinken, V.A. Osten, F.D. Panetta and J.C. Scanlan, pp. 427-9. (Weed Society of Queensland, Cairns).

Cherry, H. (2010). Boneseed in Australia: Research, management and coordinated action. Proceedings of the 17th Australasian Weeds Conference, ed. S.M. Zydenbos. (New Zealand Plant Protection Society, Christchurch).

Cherry, H., Turner, P.J., Strehling, N. and English, D. (2012). Meeting objectives of the Australian Weed Strategy through collaborative management of icon weed species: From the national to the local level. Proceedings of the 18th Australasian Weeds Conference, ed. V. Eldershaw, pp. 187-90. (Weed Society of Victoria, Melbourne).

Chippendale, J.F. (1991). Potential returns to research on rubber vine (*Cryptostegia grandiflora*). Master of Agricultural Studies Thesis, University of Queensland, Brisbane.

Cousens, R. and Mortimer, M. (1995). 'Dynamics of weed populations'. (Cambridge University Press, England).

Environmental Weeds Working Group (2007). Guidelines and procedures for managing the environmental impacts of weeds on public lands in Victoria 2007. Department of Sustainability and Environment, Melbourne.

Epanchin-Niell, R.S. and Hastings, A. (2010). Controlling established invaders: integrating economics and spread dynamics to determine optimal management. *Ecology Letters* 13, 528-41.

Grice, A.C., Clarkson, J.R., Murphy, H.T., Fletcher, C.S. and Westcott, D.A. (2013). Containment as a strategic option for managing plant invasion. *Plant Protection Quarterly* 28, 62-5.

Hobbs, R.J. and Humphries, S.E. (1995). An integrated approach to the ecology and management of plant invasions. *Conservation Biology* 9, 761-70.

Lacey, W.S. (1957). A comparison of the spread of *Galinsoga parviflora* and *G. ciliata* in Britian. *In* 'Progress in the study of the British flora', ed. J.E. Lousley, pp. 109-15. (Botanical Society of the British Isles, London).

Panetta, F.D. (2015). Weed eradication feasibility: lessons of the 21st century. *Weed Research* 55, 226-38.

Steel, J., Weiss, J. and Morfe, T. (2014). To weed or not to weed? The application of an agent-based model to determine the costs and benefits of different management strategies. *Plant Protection Quarterly* 29, 101-10.

Sydes, T.A. and Murphy, H.T. (2014). Pest adaptation response planning: A practical application of species distribution science in forecasting strategic planning for weed managers. Proceedings of the 19th Australasian Weeds Conference, ed. M. Baker, pp. 104-7. (Weed Society of Tasmania, Hobart).

Whiffen, L.K., Williams, M.C., Izquierdo, N., Downey, P.O., Turner, P.J., Auld, B.A. and Johnson, S.B. (2011). Biodiversity priorities for widespread weeds. Report prepared for the 13 Catchment Management Authorities (CMAs) by New South Wales Department of Primary Industries and Office of Environment and Heritage, Orange.

Williams, M.C., Auld, B.A., Whiffen, L.K. and Downey, P.O. (2009). Elephants in the room: widespread weeds and biodiversity. *Plant Protection Quarterly* 24, 120-2.

An integrated view of the invasive *Acacia longifolia* (Andrews) Willd. in Portugal: from impacts to biocontrol

Elizabete Marchante[1], Fco. Alejandro López-Núñez[1], Ruben Heleno[1], Helena Freitas[1], John Hoffmann[2] and Hélia Marchante[1,3]
[1] Centre for Functional Ecology, Department of Life Sciences, University of Coimbra, Portugal
[2] University of Cape Town, South Africa
[3] Department of Environment, Agrarian School, Polytechnic Institute of Coimbra, Portugal
(emarchante@uc.pt)

Summary More than ten species of Australian wattles are invasive in Portugal. Amongst them, Sydney golden wattle (*Acacia longifolia* (Andrews) Willd.) is one of the worst invasive plants in coastal areas. In these areas, *A. longifolia* drastically transforms dune ecosystems. Mechanical and chemical control methods in use are very expensive and frequently unsuccessful, due to the persistent, long-lived seed-bank accumulated in the soil. The biocontrol agent *Trichilogaster acaciaelongifoliae* Froggatt targets reduction of *A. longifolia* seeds and has been successfully used in South Africa. This agent presents a potentially good alternative option for the control of *A. longifolia* in Portugal. Specificity tests, using 40 non-target plant species, indicated that *T. acaciaelongifoliae* would be a safe biocontrol agent. Oviposition in non-choice tests was detected in three non-target species but subsequent trials involving potted plants and surveys in the field (in Australia and South Africa) showed that galls only developed on *A. longifolia*. These promising results encouraged an application for its release in the wild in Portugal, back in 2011. Afterwards, the permit to release this agent was granted in July 2015 and in November 2015 the first insects were released in the wild. Impacts measured showed changes in native plant communities, alteration of soil ecology and impoverishment of interaction networks of plants and associated gall communities. The first results of biocontrol process will be also presented.

Keywords *Acacia longifolia*, impacts, biocontrol, Portugal, dune ecosystems, *Trichilogaster acaciaelongifoliae.*

A matter of perspective: associations between *Acacia* and rhizobia across three spatial scales

Elizabeth M. Wandrag[1], Christina Birnbaum[2] and Metha M. Klock[3]
[1] Institute for Applied Ecology, University of Canberra, Australian Capital Territory 2617, Australia
[2] School of Veterinary and Life Sciences, Murdoch University, Western Australia 6150, Australia
[3] Department of Bioscience, Louisiana State University, United States of America
(lizzie.wandrag@canberra.edu.au)

Summary A disproportionate number of *Acacia* species have become invasive outside their native range, which is partly attributed to their ability to form symbiotic associations with nitrogen fixing bacteria in the soil (rhizobia). The genus has thus become a model system to study the role of plant-soil interactions in driving invasion success. However, this role remains equivocal, not least because of the now global distribution of many *Acacia* species and potential idiosyncrasies associated with the location of introductions.

We review several recent studies that have examined interactions between *Acacia* and rhizobia in the native and introduced range and cover three spatial scales (within Australia, between Australia and the geographically close New Zealand, and between Australia and the USA). Species examined span a spectrum of invasive success from those that are not invasive anywhere in the world to some of the world's worst weeds. While there were some idiosyncratic responses among species and scales, two generalisations emerge.

First, all *Acacia* species were able to form symbiotic associations with rhizobia in the introduced range, regardless of spatial scale and invasive status. Second, all species showed a similar growth response to soils in each range, again regardless of spatial scale and invasive status. Moreover, where examined, the rhizobial communities associated with species in their native and introduced ranges were often very similar. Our findings highlight the generalised nature of the symbiosis between *Acacia* and rhizobia, which may explain the global success of *Acacia*. There is no evidence that these interactions explain variable invasion outcomes.

Keywords Casual, naturalised, nodulation, plant-soil feedbacks.

Do soil microbes drive *Acacia* species invasion in non-native ranges in Australia?

Christina Birnbaum
Murdoch University, School of Veterinary and Life Sciences, 90 South Street, Murdoch, Western Australia 6150, Australia
(C.Birnbaum@murdoch.edu.au)

Summary Across Australian states, acacias became invasive or even naturalised after being introduced to ecosystems outside their natural distribution range. The relative importance of soil biota in their invasion success remains unknown, particularly that of rhizobial and fungal communities. We examined the role of soil biota on the invasion success of four *Acacia* spp. (*A. cyclops* A.Cunn. ex G.Don, *A. longifolia* (Andrews) Willd., *A. melanoxylon* R.Br. and *A. saligna* (Labill.) H.L.Wendl.) and the closely-related *Paraserianthes lophantha* (Willd.) I.C.Nielsen in Australia. Samples were collected from five native and five non-native populations of each species in four states (i.e. New South Wales, Victoria, South Australia and Western Australia).

We used a plant-soil feedback experiment to measure the net effect of beneficial and detrimental soil microbiota on plant performance. In addition, we used 454 sequencing to identify the nitrogen-fixing bacterial and fungal communities in nodules and soil. The plant-soil feedback experiment showed that soil origin had no effect on the performance of these five host species in their non-native range soils. However, seed origin influenced the performance of two species, i.e. *A. cyclops* and *A. saligna*. Overall, 454 sequencing results revealed that geographic location had an effect on fungal, but not on rhizobial composition. Rhizobial and mainly fungal composition of *A. cyclops* were significantly different from other four host species suggesting that this species encounters and plausibly associates with different soil microbiota compared to other acacias.

In conclusion, our results suggest that soil biota is unlikely to have impacted the invasion success of these five species in Australia.

Keywords Australia, invasion, fungi, legumes, plant-soil feedback, rhizobia.

The ecology and determinants of successful invasion by a native Australian tree *Acacia longifolia* subsp. *longifolia* (sallow wattle) in south-eastern Australia

Samantha Barron[1], Singarayer Florentine[1], Nick Schultz[1], Per Milberg[2], Michael Tausz[3], Marie Keatley[4] and Ryan Duffy[5]
[1] Faculty of Science and Technology, Centre for Environmental Management, Federation University, Mount Helen, Victoria 3353, Australia
[2] IFM Biology, Conservation Ecology Group, Linköping University, Sweden
[3] School of Ecosystem and Forest Sciences, Faculty of Veterinary and Agricultural Sciences, University of Melbourne, Parkville, Victoria 3052, Australia
[4] Parks Victoria, Science and Management Effectiveness Branch, Bourke Street, Melbourne, Victoria 3000, Australia
[5] Parks Victoria, Halls Gap, Victoria 3381, Australia
(Samanthabarron88@hotmail.com)

Summary *Acacias* are some of the most successful and detrimental global invaders. Their impacts on ecosystems are often substantial, long-lasting and irreversible. The Australian native tree *Acacia longifolia* (Andrews) Willd. subsp. *longifolia* is currently recognised as a significant invasive species both within Australia and abroad. While *A. longifolia* has been extensively studied overseas, within Australia there has been little work undertaken to determine its impact. In 2014, we investigated the impacts of *A. longifolia* on above and belowground vegetation in Grampians National Park (GNP) in south-eastern Australia, and found that *A. longifolia* is altering vegetation structure and composition to the detriment of native species. More specifically, above ground, increasing cover of *A. longifolia* is gradually transforming diverse open herb-rich woodland to densely populated closed shrubby woodland with limited diversity. Below ground, by contrast, the soil seed bank indicated that although there were some minor changes from the original state, overall the seed bank consisted largely of ruderal annual and short-lived perennials, particularly graminoids and forbs. Nevertheless, we still have very little understanding of the factors which drive invasion success. There is unlikely a single mechanism facilitating the invasion process. We are undertaking research to identify the mechanisms which enhance the invasiveness of this species in order to improve long-term management in the GNP. We aim to (1) determine which environmental factors are facilitating the species, (2) compare plant functional traits and genetic differences between *A. longifolia* within the home and invaded ranges, and (3) determining the competitive ability of the species with functionally similar native species under future climate change scenarios.

Keywords Acacia, biological invasion, competition, climate change, drought, phenotypic plasticity, genetic variation, physiology.

Sydney golden wattle (*Acacia longifolia* (Andrews) Willd.) on the south coast

John Moore
Department of Agriculture and Food Western Australia, 444 Albany Highway, Albany, Western Australia 6330, Australia
(john.moore@agric.wa.gov.au)

Summary There was one Sydney golden wattle for every 23 native trees on average in untreated bushland in the area surveyed around Albany. At both the local level and the landscape level it would appear that the area of Sydney golden wattle is doubling every five years. At this rate 18% of the local native bushland will be replaced by Sydney golden wattle by 2020. This makes it the most serious environmental weed in the area.

The optimal window for good levels of detection is about three weeks of which only six days are likely to be suitable for aerial surveillance, one third to one half of the infestation could be mapped each year. The cost of Sydney golden wattle surveillance was 22c per hectare and mapping of the Albany region is estimated to cost around $54,000 for an in-house project. This should be coupled to a control program of around $75,000 to arrest the spread and allow strategic control and evaluation to occur simultaneously. From the preliminary survey, control programs in the Torndirrup National Park appear to be preventing the establishment of Sydney Golden Wattle in the park, in spite of significant adjacent infestations.

Control and education programs are run by the Green skills, South Coast NRM, City of Albany, Department of Parks and Wildlife, Department of Main Roads, Water corporation, Denmark Weed Action Group, Friends of the Porongurups, Plantation managers, Noongar Community, Bushcarers, Department of Agriculture, the Wellstead CRC, Great Southern Development Corporation and many individuals or local groups.

Keywords Sydney golden wattle, *Acacia longifolia*, South Coast, Albany, Western Australia, invasive, weeds.

Temperature-induced variations in gas exchange and chlorophyll-a fluorescence of invasive *Acacia mangium* and two co-occurring tropical heath forest species (*Dillenia suffruticosa* and *Buchanania arborescens*) in Brunei Darussalam

Kushan U. Tennakoon[1,2], Quang-Vuong Le[2], M.H. Ibrahim[1], Faizah Metali[1] and Rahayu S. Sukri[1]
[1] Environmental and Life Sciences Program, Faculty of Science, Universiti Brunei Darussalam, Jln Tungku Link, Gadong, BE 1410, Brunei Darussalam
[2] Institute for Biodiversity and Environmental Research (IBER), Universiti Brunei Darussalam, Jln Tungku Link, Gadong, BE 1410, Brunei Darussalam
(kushan.tennakoon@ubd.edu.bn)

Summary The better performance of exotic alien plants compared to co-occurring native flora, which facilitates their successful invasion to new habitats, often depends on microhabitat conditions. We investigated photosynthetic characteristics of alien invasive tree *Acacia mangium* Willd. and two co-occurring tropical heath forest species (*Dillenia suffruticosa* (Griff. ex Hook.f. & Thomson) Martelli and *Buchanania arborescens* (Blume) Blume) in response to varying temperatures by measuring their gas exchange at 25°C and 30°C, and fast chlorophyll *a* fluorescence kinetics at 25°C, 30°C and 42°C respectively. Invasive *A. mangium* exhibited higher maximum photosynthesis than *D. suffruticosa* and *B. arborescens* at 30°C but not at 25°C. At comparable temperatures (25°C, 30°C or 42°C), *A. mangium* showed higher maximum quantum yield of primary photochemistry, higher quantum yield of electron transport and lower quantum yield of energy dissipation. Elevated temperature scenarios (exposure to 42°C for 1 h) significantly influenced the chlorophyll *a* fluorescence kinetics of *A. mangium*, but did not affect those of the native plants *D. suffruticosa* and *B. arborescens*. The contrasting differences shown in relation to plasticity, capacity and responses of photosynthetic apparatus to heat stress by *A. mangium* and two cohabiting native plant species investigated here provide insights into the efficacious spread of exotic Acacias into Bornean tropical heath forests.

Keywords Competitive advantage, alien invasive plants, OJIP test, light harvesting efficiency.

Resistance to glyphosate found in feathertop Rhodes grass populations (*Chloris virgata* Sw.) in Australia

The D. Ngo, Peter Boutsalis, Christopher Preston and Gurjeet Gill
School of Agriculture, Food and Wine, University of Adelaide, PMB 1, Glen Osmond, South Australia 5064, Australia
(ducthe.ngo@adelaide.edu.au)

Summary Feathertop Rhodes grass (*Chloris virgata* Sw.) is annual, warm season, C4 grass. It has become a significant weed in agricultural systems in northern Australia. This weed species has spread to southern Australia, where it is found on roadsides and vineyards in South Australia. *C. virgata* is viewed as difficult to control with glyphosate. It was reported that *C. virgata* had a high risk of evolving glyphosate resistance, particularly in summer fallows. Dose response experiments were conducted on one susceptible (CV7) and two suspected resistant (V15 and V14.2) populations of *C. virgata* collected in Victoria and South Australia, Australia. The V15 and V14.2 populations exhibited LD50 values 5.7 and 16.9-fold greater than the sensitive biotype CV7, respectively. Sequencing of the target-site gene (EPSPS) of the three populations confirmed the presence of mutations at position 106 leading to a change from proline to serine in V15 and from proline to leucine in V14.2 populations. This is the first report to show that these substitutions confer resistance to glyphosate in *C. virgata*.

In vivo and *in vitro* application of a mutagen to induce herbicide resistance in Sri Lankan rice (*Oryza sativa* L.) varieties

Sachini I. Ekanayaka[1,2], Shyama R. Weerakoon[1], S. Somarathne[1], O.V.D.S.J. Weerasena[2] and Tara D. Silva[3]
[1] Department of Botany, The Open University Sri Lanka, Nawala, Nugegoda, Sri Lanka
[2] Institute of Biochemistry, Molecular Biology and Biotechnology, University of Colombo, Sri Lanka
[3] Department of Plant Sciences, University of Colombo, Sri Lanka
(shyamaweerakoon@gmail.com)

Summary Inducing herbicide resistance (HR) in cultivated rice is a novel approach to enhance selectivity and crop safety. Studies on induced HR in Sri Lankan rice varieties are limited and further research are required to include HR rice in a cropping program. Ethyl methyl sulfonate (EMS) is a chemical mutagen used for functional mutations in plants. The present study is an attempt to develop HR rice lines through conventional breeding methods using the chemical mutagen EMS. A detailed AFLP analysis was made to identify molecular markers for HR induced varieties. *In vivo* seed mutation induced HR in 14 varieties and *in vitro* mutated calli also exhibited resistance to glyphosate. E12M32 was found to be a specific AFLP marker for induced HR varieties.

Keywords Herbicide resistance, glyphosate, ethyl methyl sulfonate (EMS), seed culture, AFLP analysis.

INTRODUCTION

Effective weed control in any crop production system is a prerequisite if high yield and good quality are to be achieved. Introduction of herbicide resistant (HR) rice would improve efficiency of weed management thus rice growers throughout the world could benefit from HR cultivars. There is evidence that HR crops can bring significant benefits to farmers, consumers and the environment (Duke *et al*. 2002). Over the last two decades, mutational techniques have become one of the most important tools available to progressive rice-breeding programs (Sandhu *et al*. 2002) and most of the herbicide-tolerant mutants were developed through chemical mutagenesis followed by herbicide selection (Tan *et al*. 2005). Among the chemical mutagens, ethyl methyl Sulfonate (EMS) is a strong chemical mutagen which can make the chromosome structure different (Barro *et al*. 2001). Plant tissue culture represents the simplest of the biotechnologies available at present for crop improvement (Sudhakar *et al*. 2009) and mutagenesis techniques in tissue culture are commonly used in developing herbicide resistant (HR) rice varieties. Studies on induced herbicide resistance in Sri Lankan rice varieties are limited and the present study focused on inducing HR in rice by *in vivo* and *in vitro* EMS mutation.

MATERIALS AND METHODS

Materials Bg94-1, Bg250, Bg300, Bg304, Bg305, Bg352, Bg357, Bg359, Bg360, At362, Bw364, Ld365, Bg366, Bg369, Bg379-2, H4, Bg403, Bg450, Bg 454 inbred-developed rice varieties and five traditional varieties (KaluHeenati, Kurulu thuda, Suwadal, Rathhal, Madel, and Pachchaperumal) were collected from Rice Research Development Institute at Bathalagoda, Ambalanthota, Bombuwela and Labuduwa, Sri Lanka for the study.

Method 1 – Seed mutation Seeds from each variety were pre-soaked in distilled water for 24 h at room temperature and exposed to 4.5mmoll^{-1} EMS for 12 hours. The seeds were then washed with running tap water and allowed to leach the residual chemicals and let it to germinate. The germinated seedlings were transferred into mud pots and glyphosate (0.5 g a.e. L^{-1}) was applied at two weeks after sawing (S_0 generation). Following the application of glyphosate, the dead plants were considered as susceptible to the herbicide and surviving plants with a substantial growth were considered as resistant to herbicide. For each rice variety, the number of resistant plants and percentage resistance was calculated. Plants with ≥50% resistance to glyphosate (0.5 g a.e. L^{-1}) were arbitrarily considered as resistant varieties. The morphological and yield data were collected from the resistant varieties and seeds of self-pollinated S_0 generation of mutated plants were designated as S_1 seeds. Three panicles per S_1 plants were harvested from mutated resistant surviving plants and their resistance to glyphosate was evaluated using the same procedure (S_2).

Descriptive statistics were performed on the dataset. The mean and standard deviation was computed and ANOVA test were used to compare the mean. One-way-analysis of variance (ANOVA) was performed on agro-morphological characters. All statistical analyses were carried out using SAS Version 9.2 (SAS 2008).

Method 2 – AFLP molecular study AFLP analysis was carried out for the mutated seeds using 16 AFLP primer combinations. Genomic DNA was extracted from seven day old leaves using a PhytoSpin™ plant genomic DNA extraction kit (Ceygen Biotech). The protocol followed the traditional procedure described by Vos *et al.* (1995) with several modifications.

Method 3 – Callus mutation A glyphosate susceptible rice variety, Bg250, was selected for the study (Weerakoon *et al.* 2013) and glyphosate resistant variety, Pachchaperumal, was taken as the reference/control. Mature rice seeds were de-husked and surface disinfected after soaking for 1 minute in 70% ethanol and subsequently seeds were washed with distilled water and sodium dodecyl benzene sulphonate (Teepol) for 1 minute. Bleach-sterilization was carried out for 20 minutes in 50% commercial bleach. Following three rinses with sterilized water, seeds were cultured (10 seeds per Petri dish) in solid MS medium for callus induction (Murashige and Skoog 1962). Cultures were incubated in a culture room maintaining at 25°C ± 2°C in complete dark. For the mutation process three weeks old callus masses were transferred into Falcon™ 50 mL tubes which containing 30 mL, liquid MS media without 2, 4-D.

Subsequently, the tubes were immediately mutagenised by adding different concentrations of EMS (0.0, 0.1%, 0.2%, 0.3%and 0.4% v/v) (Sigma-Aldrich). Each treatment consists of 10 calli and there were three replicates for each treatment. Culture tubes were covered with aluminium foil and placed on an orbital shaker (150 rpm) for two hours. The incubated calli were rinsed 10 times with MS liquid medium and transferred to new tubes containing MS liquid media with 2,4-D and shaken (120 rpm) for two more days (28°C ± 2°C, in darkness). The treated calli were subcultured on to a basal media of MS containing NAA and kinetin. Initial screening of viable/non-viable calli was carried out based on the changing from creamy white colour to brown. The browning calli were considered as susceptible to EMS and white calli with substantial growth in the subculture media were selected as resistant to EMS. To confirm viability, a tetrazolium chloride (1% TTC) test was conducted (Towill *et al.* 1975). Data were arbitrarily collected on callus and they were expressed as percentage response against total treated callus. Sub-cultured media was treated with 0.02 g a.e. L^{-1} glyphosate solution after ten days and observed the characteristics of the callus compared to untreated callus. Regenerated plantlets were also exposed to glyphosate and evaluated for their HR.

RESULTS AND DISCUSSION

Results indicated that fourteen varieties increased their resistance to glyphosate (Bg94-1, Bg352, Bg359, Bg360, At362, Bw364, Ld365, Bg366, Bg379-2, Bg403, Bg454, Kalu heenati, Pachcha perumal and Madel) in the first generation-S_0 (Figure 1). The pair-wise statistical analysis of variance of agro-morphological characters clearly showed significant differences ($P \leq 0.05$) between control and treated plants. The percentage of resistance in S_1 was almost similar in all the varieties except Bg454 and 'Madel'. In contrast, there were statistically significant differences ($P \leq 0.05$) between the EMS-mediated-mutated rice plants with non-mutated rice plants related to agro-morphological and yield characters such as plant height, number of panicles per plant, seeds per panicle and 1000 grain weight. However, differences in number of leaves per plant and number of tillers per plant were not statistically significant when comparing mutated rice plants with non-mutated rice plants.

AFLP analysis of EMS mutated rice lines indicated variations in several fragments. Out of sixteen primer combinations, five primer sets (E10M31, E10M33, E11M32, E12M32, and E12M33) indicated the possibility to be used in differentiating HR-varieties. Among those the 78 bp fragment of E11+M13 primer (TGT AAA ACG ACG GCC AGT GAC TGC GTA CCA ATT CAC and M32 primer (GAT GAG TCC TGA GTA AAA) was identified as a specific marker for the resistant lines. That marker was common to all the HR induced varieties after mutation. Cluster analysis based on UPGMA for the mutated rice varieties indicated that control and mutated lines are genetically different to each other. The varieties which increased their resistance after the mutation appeared more or less in the same cluster. The varieties with induced HR after mutation clustered in separate groups (Figure 2).

In vivo application of EMS revealed that the viability of seed-derived callus decreased when the EMS concentration was increased. At 0.2% v/v EMS concentration, 60% of the calli survived and the same concentration was selected for further studies. One week after EMS treatment, no noticeable differences were detected in calli between control and treatments. During 7–10 days after mutation, 0.3 and 0.4 of treated calli showed some degree of browning. Subsequent to the application of glyphosate, EMS-mutated calli indicate a positive response for TTC test. Sixty percent (60%) of the calli turned into red colour following the staining process. The control treatment which was not mutated with EMS signified negative results and the colour of the calli were not changed in staining. It indicated that glyphosate resistance has developed

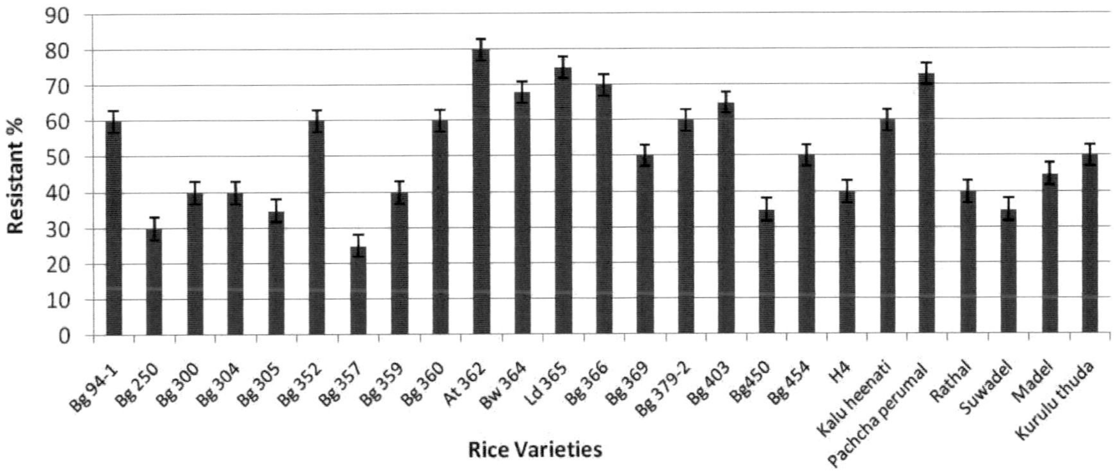

Figure 1. Resistant percentage (No. of survived plants) of S_0 rice varieties 0.5 g a.e. L^{-1} glyphosate concentration.

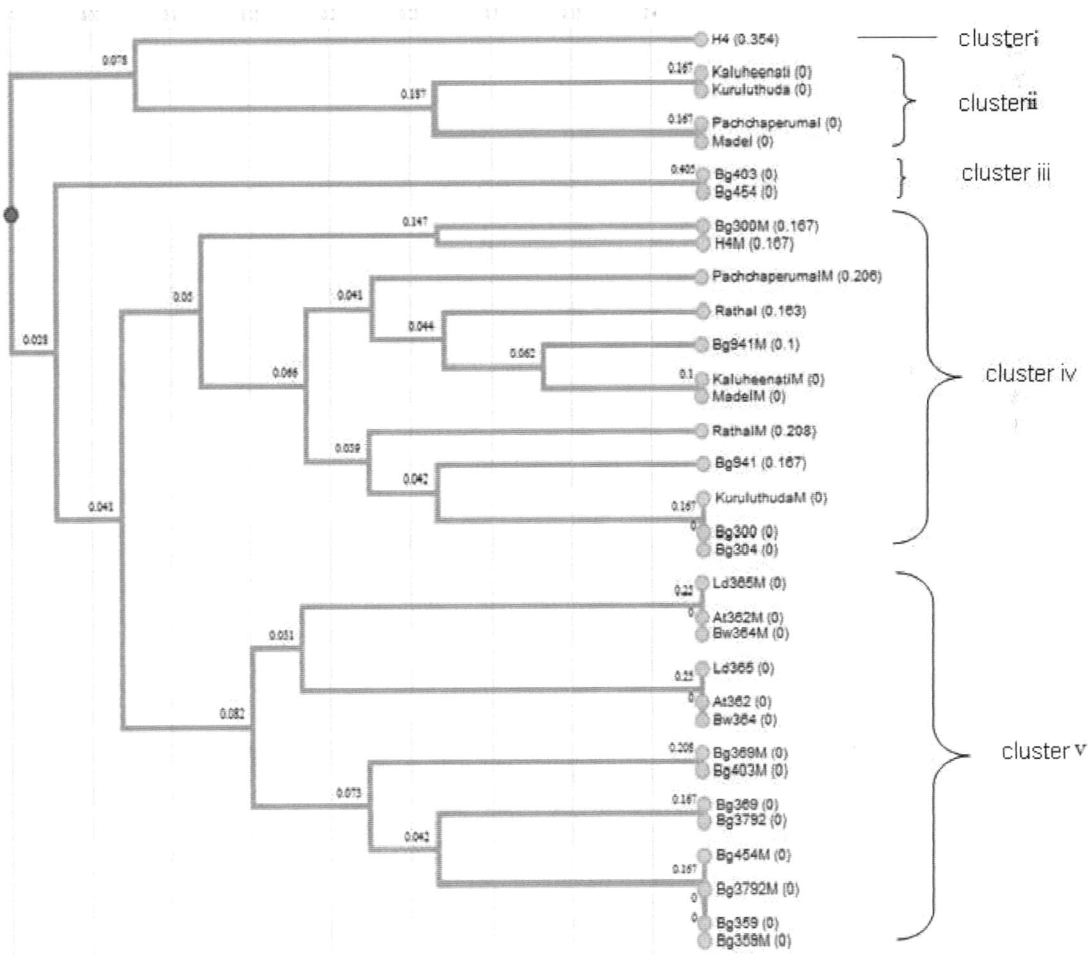

Figure 2. The dendrogram showing genetic diversity among rice varieties (M denote – S_0 varieties) (based on Jaccard's similarity coefficient).

in the EMS-mutated callus compared to the control. Callus derived from Pachcaperumal seeds also showed positive results to TTC test without EMS mutation. The regenerated plants of mutated calli also showed resistance to glyphosate.

The technique of induced mutations has over the past 50 years played a major role in the development of superior crop varieties. The present study revealed, EMS application *in vivo* and *in vitro* was successful in inducing HR in certain Sri Lankan rice varieties and induced HR rice varieties have a higher potential in rice breeding programs. Furthermore, the identified AFLP markers may be potentially useful for differentiating HR rice varieties in future.

ACKNOWLEDGMENTS

The research grant provided by the National Research Council, Sri Lanka (Grant No. NRC 12-037) is greatly appreciated and the proper guidance of Mrs A.S.K. Abeysekara, Senior Research Officer, RRDI, and Bathalagoda, Sri Lanka is also greatly acknowledged.

REFERENCES

Barrio, F., Fernandez-Escobar, J., De La Vega M. and Martin, A. (2001). Doubled haploid lines of *Brassica carinata* with modified erucic acid content through mutagenesis by EMS treatment of isolated microspores. *Plant Breeding* 120 (3), 262-4.

Duke, S.O., Scheffler, B.E., Dayan, F.E. and Dyer, W.E. (2002). Genetic engineering crops for improved weed management traits. Proceedings of ACS Symposium Series No. 829, pp. 52-66. American Chemical Society, Washington, DC.

Gealy, D.R., Mitten, D.H. and Rutger, J.N. (2003). Gene flow between red rice (*Oryza sativa*) and herbicide-resistant rice (*O. sativa*): implications for weed management. *Weed Technology* 17, 627-45.

Murashige, T. and Skoog, F. (1962). A revised medium for rapid growth and bioassay with tobacco tissue cultures. *Physiologic Plantarum* 15, 473-97

Sandhu, S.S., Bastos, C.R., Azini, L.E, Neto, A.T and Colombo, C. (2002). RAPD analysis of herbicide-resistant Brazilian rice lines produced via Mutagenesis. *Genetics and Molecular Research* 1, 359-70.

Sudhakar, P., SaradaMani, N. and Ramana, T. (2009). Long-term maintenance of callus cultures from immature embryo of *Sorghum bicolor*. *World Journal of Agricultural Sciences* 5 (4), 415-21.

Smith, J.H., White, W.T. and Brown, B.R. (1994). Yield response of wheat to herbicides. Proceedings of the First International Control Congress, p. 310. Weed Science Society of Victoria, Melbourne.

Tan, S., Evans, R.R., Dahmer, M.L., Singh, B.K. and Shaner, D.L. (2005). Imidazolinone-tolerant crops: history, current status and future. *Pest Management Science* 61, 246-57.

Towill, L.E. and Mazur, P. (1975). Studies on the reduction of 2,3,5-triphenyltetrazolium chloride as a viability assay for plant tissue cultures. *Canadian Journal of Botany* 53 (11), 1097-102.

Vos, P., Hogers, R., Bleeker, M., Vandelee, T., Hornes, M., Fritjers, A., Pot, J., Peleman, J., Kuiper, M. and Zabeau, M. (2001). AFLP: a new technique for DNA finger printing. *Nucleic Acid Research* 23, 4407-14.

Investigation of herbicide resistance in Indian hedge mustard (*Sisymbrium orientale* L.)

Hue Thi Dang, Christopher Preston, Jenna Malone, Peter Boutsalis and Gurjeet Gill
School of Agriculture, Food and Wine, the University of Adelaide PMB1, Glen Osmond, South Australia 5064, Australia
(huethi.dang@adelaide.edu.au)

Summary Indian hedge mustard (*Sisymbrium orientale* L.) (IHM) is a problematic broadleaf weed in southern Australia, which has become more difficult to control in field crops due to the evolution of herbicide resistance. Despite several previous studies, there is little information available on resistance in IHM to some herbicide groups.

This study investigated the extent of resistance and multiple resistance to different herbicide groups. Fifty five populations were screened for herbicide resistance in 2014 and 2015. Resistance was confirmed to group B, C, F and I herbicides. Resistance was more common to group B (n = 24, 44%) followed by resistance to group F (n = 11, 20%); seven populations were resistant to group C and four populations were resistant to group I herbicides. Several populations exhibited multiple resistance to more than one herbicide mode of action with four populations resistant to group B, C and F; four populations resistant to group B and I and one population resistant to group F and I herbicides. Representative resistant populations for each herbicide group and one known susceptible population were used in dose response experiments to quantify the level of herbicide resistance.

The resistance level of all populations was generally high and varied depending on the herbicide from 20 to 1940-fold compared to the susceptible control. The most resistant populations were selected for the study of genetics of herbicide resistance to group I, C and F.

Keywords Herbicide resistance, Indian hedge mustard, diflufenican, atrazine, 2,4-D.

Herbicide resistance management and tillage reforms in the rice–wheat cropping systems of South Asia

R.K. Malik[1], Virender Kumar[2] and Andrew McDonald[1]
[1] International Maize and Wheat Improvement Center (CIMMYT), India
[2] International Rice Research Institute (IRRI), Los Banõs, Philippines
(RK.Malik@cgiar.org)

Summary The shift in weed flora in favour of wild canary grass (*Phalaris minor* Retz) from among wide spectrum weed flora in 1970s and towards resistant populations within *P. minor* during 1990s is the classical example of weed shifts happened in wheat in North-West India. Herbicide resistance crisis during 1994–2003 which resulted in significant yield losses in wheat in NW India demanded development of long-term sustainable solutions. To develop integrated solution, a paradigm shift is needed in the ways we innovate, develop, test, and disseminate technologies. Zero tillage (ZT) research which was at the dead end in 1996 was taken up using farmer's participatory approach in a project funded by Australian Centre for International Agricultural Research (ACIAR) and supported by International Maize and Wheat Improvement Center (CIMMYT). By 2002 more than 0.3 M ha were brought under ZT. Long-term studies indicate that early wheat planting combined with ZT+ residue as much can overcome the problem of herbicide resistant *P. minor* and herbicides can be skipped once in 3–4 years. These past projects and current one 'Cereal Systems Initiative in South Asia (CSISA)' conveyed a powerful message in this regard. The work explains how crisis like herbicide resistance led to opportunities that are being extended to whole Indo-Gangetic Plains. The sustainable intensification with tillage reforms and better bet agronomy need to be aligned with research, teaching and extension system in national system. While implementing various projects (CIMMYT, ACIAR, National Agricultural Technology Project and CSISA), we saw an opportunity to design a more practical and pragmatic template for development and dissemination of technologies at the same time.

What drives practice change? Influence, not information

Peter Newman, Lisa Mayer and Brogan Micallef
Australian Herbicide Resistance Initiative, School of Plant Biology, The University of Western Australia, Crawley, Western Australia 6009, Australia
(petern@planfarm.com.au)

Summary Our job, as communicators, is to filter information, and then present it to farmers. We don't do away with the data, we filter it and present it in a way that makes sense, utilising the power of influence. Data communication is best done between scientists. Practice change is not about scientists giving the answer to farmers, it is about working with farmers to come up with the solutions. Do with, not for.

We are all in a world of information overload. There is more information than ever, and there are more platforms to communicate this information than ever. In the world of agriculture, the latest science is communicated to farmers to encourage practice change to improve their profitability. In this saturated world of information, how do we best communicate with farmers to drive practice change in the right direction? The answer is to think influence, not information.

There is an extensive world of scientific research going on that is aimed at improving agriculture. If all of the data from this science is presented to farmers as information, it is overwhelming and difficult for farmers to adopt. We need to we filter this data and present it to farmers in a way that makes sense.

AHRI is a research group based at The University of WA and is focused on sustainable cropping and herbicide resistance in broad-acre farming. Significant AHRI funds are directed to the communication program, and AHRI's communication activities are enhanced by WeedSmart, which is managed by AHRI on behalf of the WeedSmart stakeholders. WeedSmart is an industry funded herbicide stewardship initiative

Both AHRI and Weedsmart communication are in the world of influence, however there are differences.

As can be seen in the table below, a good communication program has many facets, all of which are linked.

In the same way that researchers are the best people to do research, communicators are the best people communicate research. Many communication programs of research organisations are left up to researchers. We at AHRI believe that we are reaping the rewards of investing in a dedicated communication team and we encourage other research organisations to consider doing the same. Influence, not just information.

Keywords Communication, resistance, crops, weeds, AHRI, WeedSmart.

	AHRI	Weedsmart
Primary target audience	Agronomists/Industry	Farmers
Level of information	Often high level science simplified as well as practical solutions	Awareness/practical solutions – herbicide resistant weeds
Main activities	AHRI insight newsletter, Twitter, media, science articles, workshops, website, conferences	Diversity era online learning, media, case studies, champion farmer presentations/ video, workshops, website, social media

First case of glyphosate resistance in *Tridax procumbens*

Jingbo Li[1,2], Qin Yu[1] and Stephen Powles[1]
[1] Australian Herbicide Resistance Initiative (AHRI), School of Plant Biology, The University of Western Australia, Crawley, Western Australia 6009, Australia
[2] Collaborative Innovation Center for Field Weed Control, Hunan University of Humanities, Science and Technology, Loudi 417000, China
(jingbo.li@uwa.edu.au)

Summary *Tridax procumbens* L. (coat buttons or tridax daisy) is a dicot weed species native to the tropical Americas and can be either cross-or self-pollinated. Globally, *T. procumbens* is a major weed of 31 crops in 60 countries. In Australia, it is mostly found in Northern Territory, Queensland and north Western Australia (WA), infesting vegetables, pasture, irrigated and dry-land crops. Glyphosate, 2,4-D, atrazine and diuron are usually used for *T. procumbens* control. However, a *T. procumbens* population from Kununurra, WA is suspected for glyphosate resistance. This population has been managed by repeated glyphosate treatments over several years, with poor control only evident recently. Seeds of the putative glyphosate resistant *T. procumbens* population were collected from Kununurra in November 2014, and preliminary pot experiments conducted in February 2016 at the University of Western Australia. The Kununurra *T. procumbens* showed a low level resistance to glyphosate, with 100% survivorship at the field glyphosate rate of 540 g ha^{-1} and 74% at 1080 g ha^{-1}. In contrast, the susceptible *T. procumbens* population (collected from Queensland in February 2016, with no history of glyphosate exposure) displayed 100% mortality at glyphosate 540 g ha^{-1}. Glyphosate dose response experiments were conducted and reveled a three-four fold resistance. Target-site and non target-site resistance mechanisms are being investigated.

Keywords *Tridax procumbens*, glyphosate resistance.

Herbicide metabolic resistance gene discovery in *Lolium rigidum* Gaudin using RNA-sequencing

Heping Han[1], Qin Yu[1], Susana Gonzalez[2], Roland Beffa[2] and Stephen B. Powles[1]
[1] Australian Herbicide Resistance Initiative, 35 Stirling Highway, Nedlands, Western Australia 6009, Australia
[2] Bayer CropScience, Industripark Höchst, Frankfurt/Main 65926, Germany
(heping.han@uwa.edu.au)

Summary Evolution of multiple herbicide resistance is widespread in *Lolium rigidum* Gaudin in Australia. This resistance is often endowed by enhanced rates of herbicide metabolism involving cytochrome P450, GST or other genes. To identify the herbicide-metabolizing genes in *L. rigidum*, the subset populations with metabolic resistance mechanism to diclofop and chlorsulfuron were purified from two field-evolved resistant populations SLR31 and VLR69. No herbicide resistance endowing mutations in ALS and ACCase were detected in individual plants by DNA sequencing. The increased diclofop metabolism in purified subsets was further confirmed by HPLC compared to the susceptible population (VLR1). The F_2 populations were generated by crossing the susceptible plants (VLR1) and two purified resistant subsets. Genetic studies showed the resistance to diclofop and chlorsulfuron in the SLR31 subset is linked but segregated in the VLR69 subset, indicating different genes conferring resistance to chlorsulfuron and diclofop in the two subsets. The most susceptible and resistant plants were individually phenotyped from the F_2 and parent subsets. RNA was extracted and RNA-sequencing performed. A few P450s and GSTs were selected as candidate genes for further validation and functional characterization by yeast and plant transformation. Herbicide-metabolizing and resistance-endowing gene discovery will greatly extend our currently limited knowledge and understanding of metabolic herbicide resistance.

Keywords Chlorsulfluron, diclofop, herbicide metabolic resistance, *Lolium rigidum*, RNA-Seq.

Pyroxasulfone (Sakura) resistance mechanism

R. Busi[1], A. Porri[2] and S.B. Powles[1]
[1] Australian Herbicide Resistance Initiative, School of Plant Biology, University of Western Australia, Crawley, Western Australia 6009, Australia
[2] Plant Energy Biology, University of Western Australia, Crawley, Western Australia 6009, Australia
(roberto.busi@uwa.edu.au)

Summary Repeated herbicide field selection led to the evolution of a ryegrass (*Lolium rigidum* Gaudin) multiple-resistant (MR) population with complex pattern of resistance to the ACCase-inhibitor diclofop-methyl, the ALS inhibitor chlorsulfuron, the micro-tubule assembly-inhibitor trifluralin, and the very long chain fatty acid elongase (VLCFAE)-inhibiting herbicides S-metolachlor and triallate. Subsequent experimental evolutionary studies have selected, through recurrent pyroxasulfone selection, high-level pyroxasulfone resistance and cross-resistance to prosulfocarb + S-metolachlor and triallate. Recent RNA-Seq transcriptome analysis identified several putative genes (contigs) strongly associated with the resistance phenotype in ryegrass.

In this study we have attempted to explain and correlate the observed phenotypic resistance to pyroxasulfone evolved in the population MR against five individual annotated putative cytochrome P450 (P450s) resistance genes and five glutathione-S-transferase (GSTs) putative resistance genes.

We present the phenotypic characterization of pyroxasulfone resistance observed in the parental MR population versus the evolved pyroxasulfone resistance in the progeny P6, the relative increased expression of putative resistance GST genes assessed via q-PCR and thin-layer chromatography experimental data indicating that the pyroxasulfone metabolites formed in the pyroxasulfone-resistant ryegrass population P6 population are characteristic of the glutathione conjugation pathway.

This research emphasizes there are complexities associated to multiple-resistance due to enhanced herbicide metabolism. However, greater knowledge of resistance mechanism(s) can lead to improved management of metabolic herbicide resistance in the major weed ryegrass.

Keywords GST, herbicide detoxification, herbicide metabolism, herbicide resistance, metabolism, ryegrass, weed control, weed management.

Trifluralin resistance in *Lolium rigidum* Gaudin

Jinyi Chen, Qin Yu, Heping Han and Stephen Powles
Australian Herbicide Resistance Initiative (AHRI), School of Plant Biology, The University of Western Australia, 35 Stirling Highway, Crawley, Western Australia 6009, Australia
(jinyi.chen@research.uwa.edu.au)

Summary *Lolium rigidum* Gaudin is Australia's number one damaging crop weed, and has evolved resistance to many herbicides with different sites of action (e.g. ACCase, ALS inhibitors). Trifluralin, a pre-emergence herbicide, was first introduced for *L. rigidum* control in Australia in the 1960s and is vital for Australian minimum/zero tillage agriculture as it remains effective in controlling multiple resistant *L. rigidum*. Trifluralin resistance was early reported in 1995, and now the resistance has reached a worrisome level.

Pot experiments were conducted with three putative resistant *L. rigidum* populations, and trifluralin resistance confirmed in these populations. In addition, a Petri dish seedling-growth system has been established to rapidly quantify trifluralin resistance in terms of seedling root and shoot growth. This system has been adjusted to allow for determination of cross-resistance patterns to other microtubule-inhibiting herbicides.

Resistance mechanisms are being investigated. Preliminary α-tubulin gene sequencing in one population revealed three novel mutations. Rice genetic transformation is underway to confirm these putative resistance-endowing mutations. In contrast, non-target-site based resistance is most likely to be involved in two resistant populations with no mutation detection, and these populations will be used for trifluralin uptake, translocation and metabolism studies. Work is currently underway to characterise the mutations using purified sub populations with plants homozygous for a specific tubulin mutation. Genetic inheritance and fitness of trifluralin resistance will also be investigated.

Keywords Trifluralin resistance, tubulin gene mutation, *Lolium rigidum*.

Controlling patches of glyphosate-resistant weeds: is victory possible?

Tony Cook[1] and Bhagirath S. Chauhan[2]
[1] NSW Department of Primary Industries, 4 Marsden Park Road, Calala, New South Wales 2340, Australia
[2] The University of Queensland, Queensland Alliance for Agriculture and Food Innovation, 13 Holberton Street, Toowoomba, Queensland 4350, Australia
(tony.cook@dpi.nsw.gov.au)

Summary Management of weeds is frequently undertaken on a whole of paddock perspective. This is more often a result of weeds occupying a large proportion of the paddock. However, earlier intervention could prevent this from happening.

Four case studies investigating four different weed species has demonstrated a variety of ways farmers have managed patchy glyphosate resistant weeds. A common link between all case studies is the well apportioned use of chemical and non-chemical tactics. In addition, a common theme is the use of patch specific treatments superimposed by whole of paddock treatments and an understanding that weed seed production / prevention contributes to population development / decline.

Gauging the level of success for these case studies is a subjective measurement. The opinion of effective control can be highly variable; as some farmers consider total eradication the end point whereas others may consider maintaining or gradually reducing patch size a form of success.

These case studies have shown that management of patchy glyphosate-resistant (GR) weeds can result in positive outcomes. Some improvements have been identified to further reduce weed patchiness, namely fence line weed control. Despite this, an overriding feature of these case studies is the early action taken and the persistent objective of farmers to push for victory in the battle against patchy weeds.

Keywords Glyphosate resistance, patch management, eradication, integrated weed management, annual ryegrass, *Lolium rigidum,* awnless barnyard grass, *Echinochloa colona,* feather-top Rhodes grass, *Chloris virgata,* liverseed grass, *Urochloa panicoides*.

INTRODUCTION

Literature reporting broad-acre glyphosate-resistant weed control occasionally provides research or advice on the management of small patches. Good examples of such research have either used computer modelling (Thornby *et al.* 2014) or interim field results that required more time to validate success (Werth *et al.* 2014). Despite the lack of field based evidence, there are many farmers trying to manage these patchy situations. Therefore, it would be more prudent to have information that could prevent new glyphosate-resistant weed patches spreading beyond the early phases of establishment, using examples from case studies.

Most of the emerging glyphosate-resistant issues are emerging from the northern grains region of New South Wales and Queensland. This is predominantly due to the overuse of glyphosate in the fallow period and the diversity of weed species in this region. It is therefore common to find examples where glyphosate-resistant weeds are affecting a small proportion of a paddock. In some cases, if little or no intervention is practiced, the problem will spread beyond the patch stage. Developing patch management advice to growers in the north is therefore a high priority.

There is a good variety of weed management tactics that can be used for patch management (McGillon and Storrie 2006). Combinations of chemical and non-chemical methods are being used by grain growers with varying degrees of success. Although there are few examples of eradication, success can also be seen as containing or reducing the size of patches, preventing them from affecting large proportions of the farm.

In this paper, a total of four case studies will be examined. All of these are based within the northern grains region. Each will focus on a different glyphosate-resistant weed. The species investigated are annual ryegrass (*Lolium rigidum* Gaudin), awnless barnyard grass (*Echinochloa colona* (L.) Link), feather-top Rhodes grass (*Chloris virgata* Sw.), and liverseed grass (*Urochloa panicoides* P.Beauv.).

MATERIALS AND METHODS

All case studies were completed as face to face interviews of farm managers in spring and summer of 2015/16. A set list of questions were asked and explanation of some responses were required if the interviewer deemed it was required, particularly if the response was not specific enough.

The structure of the questionnaire and the some of the questions about critical aspects of management are listed in Table 1. Background information of the property was first investigated; giving a snap-shot of the property operations at the time of interview. The next section is characterized by questions that investigated what happened prior to the confirmation of resistance. Accordingly the next group of questions dealt with the management decisions and tactics required to deal with the patchy weeds. Finally the last section of responses was directed at evaluating the result of their changed management tactics.

Case study 1 investigated GR annual ryegrass within Central West slopes of NSW, near the town of Coolah. The second case study, at Pallamallawa within North West plains of NSW focused on GR liverseed grass. Case studies 3 (GR awnless barnyard grass) and 4 (feather-top Rhodes grass) were located in the western (Tara) and eastern (Dalby) Darling Downs region of southern Queensland, respectively.

RESULTS

Case study 1 (annual ryegrass) This case study highlights the attempts to manage GR annual ryegrass in a mixed farming system. The 18000 hectare property, located near Coolah, has 5000 ha devoted to cropping and 3000 ha for dual purpose cropping and pastures. There is opportunity to rotate from winter to summer crops (sorghum) for management flexibility and herbicide choice.

Glyphosate resistance was suspected in 2009 and tested soon afterwards. It was thought to have originated on the property due to over reliance on glyphosate based products. The rotation in the past had three seasons of sorghum followed by wheat then canola. Patches can be seen in the middle of paddocks and along fence lines, a consequence of glyphosate resistance. Approximately 30 to 40% of cropping paddocks are affected by GR annual ryegrass with 10% being termed 'a real problem'. Some of the annual ryegrass (ARG) movement can be attributed to movement down a slope following heavy rainfall. Sheep and the occasional cultivation have been used as non-chemical methods for control, although the manager admits the cultivation was used mainly for other purposes.

What is the approach to manage the GR ARG? The cropping rotation has changed to a canola/sorghum/chickpeas/barley and dual purpose wheat cycle. This allows annual ryegrass control with a robust range of modes-of-action chemistry. The ARG populations on this property were still susceptible to Group A herbicides and most pre-emergence options. Paddocks that are not infested by GR ARG, are treated as though they have GR ARG, with similar cropping system changes to ensure herbicide choice flexibility. The farm manager has also considered dryland cotton to add more summer options.

Table 1. Structure of the questionnaire and type of information sought.

BACKGROUND INFORMATION
Farm location (town, region and state), size of property (area devoted to cropping and grazing), types of enterprises (specific crops grown, stock type), a description of the climate (annual rainfall and whether it is summer or winter dominant), use of any conservation farming techniques – if yes what types are used and the reason for implementing them, number of years the owner has managed the property, information about current resistance issues, resistance patch sizes and where on paddocks are they located – this can be supported by photographic evidence of some resistance patches
HISTORY
What year was glyphosate first suspected of losing efficacy, what year was glyphosate resistance confirmed, likely cause of resistance, any other weed species of concern (what modes of action are they resistant to), cropping rotation and details about the integrated weed management used over that cropping cycle
DEALING WITH THE PROBLEM
Who is involved in developing the weed management plan and how are they involved, where is weed management information sourced from (e.g. agronomist/government advisor/other framers/grower advisory days/written material etc.), describe the changes in crop rotations, describe changes to whole of paddock treatments, describe any specific patch management treatments, are any of the stages of the weeds' life cycle targeted (e.g. actions taken to stop weed seed set), what measures are taken to prevent weed movement (e.g. weed control in non-crop areas, wash down vehicles after travelling through weed patches etc.), monitoring (e.g. frequency of inspections and areas on property more frequently inspected than others) ,duration of patch management activities/plan or is the management of patches more reactive
RESULTS AND REFLECTIONS
Describe how the patch size has changed over time, are outlier patches detected, what are barriers to adopting other techniques, anything challenging with new strategies and how were they overcome, comment on any failures with some treatments, recommendations to other farmers, profitability of changed weed management – have benefits been seen in other farm enterprises, how many years will it take to get problem under control, improvements – could things have been managed better?

Some difficulties encountered were the availability and timeliness for spraying operations. Machinery is a limiting resource and the advent of double knocking has restricted the availability of spray rigs.

Fence line weed management was a reasonable concern. The manager did not know of alternative options for this situation. However, whilst the case study interview was being conducted, NSW Department of Primary Industries advised him of alternative options. Furthermore, there were some fallow pre-emergence options that were discussed that would find a good fit into the cropping rotation.

A positive aspect of this case study was the understanding that the manager knew that stopping weed seed set was the key to success. In summary, the manager has minimised the spread of this weed but wants to take it one step further and reduce weed seed banks significantly.

Case study 2 (liverseed grass) A 600 ha farm, located at Pallamallawa was acquired by the current owner in 1998.

Approximately 10 paddocks are used in a barley/chickpea/wheat/long fallow and summer crop rotation. Resistance was suspected and tested in the summer of 2007/08 and confirmed glyphosate resistant. The resistance is believed to have been the consequence of many years of consecutive applications of glyphosate with poor cultivation thereafter (not full disturbance).

The patch of liverseed grass is relatively small, about one hectare. The farmer now has managed the property for 18 years and has recently said in the case study interview, "I'm sure I have won this battle."

What is the approach to manage the GR liverseed grass? For this case, the patch was contained and very small, thus would allow the farmer to "throw the kitchen sink at it". The objective of the farmer was to eradicate the patch; most of the tactics mentioned further on were specifically directed at the patch. Thereafter, in 2009 the patch was treated with herbicides using alternatives modes-of-action to glyphosate. Cultivation was used frequently and changed to full disturbance (wide sweeps) so that nothing could survive. Very importantly the patch was monitored closely to check that the job resulted in 100% control. Double knocking was used to good effect along with pre-emergence herbicides.

One year in the rotation a winter cereal was brown manured and the biomass was used as mulch in the period that liverseed grass would normally emerge.

The farmer stated one noteworthy comment in relation to managing his resistance, "Get onto it", confirming his intention to avoid procrastination and to take control. The result of this case study is what could be seen as eradication. However, since being affected by resistance this farmer is always on the lookout for other potential small patches of resistance or hard to control tolerant weeds.

Case study 3 (awnless barnyard grass) A mixed farming property located near Tara, Queensland, consisting of 1750 ha cultivation/farming paddocks and 500 ha pasture has infestations of GR awnless barnyard grass and feather-top Rhodes grass (FTR). The glyphosate resistance status of the FTR was unknown. Glyphosate resistance in awnless barnyard grass was confirmed in 2012 and again in 2015, reportedly surviving glyphosate (450 g a.i L^{-1}) at 4.5 L ha^{-1}. FTR was first seen in 2012 and has progressively got worse. Glyphosate was the only herbicide used and often it may be applied 3 to 4 times per fallow. The owner and agronomist blame the lack of chemical group rotation for their problems.

The cropping rotation is wheat/wheat/wheat/long fallow/sorghum/sorghum and long fallow (duration of long fallow approximately 16 months).

What is the approach to manage the GR awnless barnyard grass? Having major changes to chemical and non-chemical options has led to improved results. By having an occasional long fallow it has allowed the option to use cultivation if weed 'blowouts' occur. However, pre-emergence herbicides are now frequently used, namely imazapic (prior to wheat) and metolachlor (in preparation for sorghum). WeedSeeker® technology (Brownhill 2006) is used for isolated patches but requires high rates of glyphosate to manage the problem.

The results are mixed. It appears the FTR can be managed due to residual herbicides, cultivation and WeedSeeker®. The awnless barnyard grass still remains the issue. The owners treat the affected paddocks differently to the rest of the farm. Currently the practice of extremely high rates of glyphosate via a WeedSeeker® are working well but they struggle to justify the cost of the more expensive Group A options and understand most of them are not registered for use. Adjustments to some activities have been implemented since resistance. Now sheep are used as a second knock after the initial spray whereas previously they were used as the first knock, making plants less susceptible to herbicides when used as the second knock.

Attempting to patch spray many isolated patches on a large property was a difficult task and many outlier plants were missed and allowed to set seed. Therefore over time some patches have slowly expanded. The management rule applied within this property is to make overall changes to paddock-wide treatments so that the time consuming effort of spot treating can be avoided. Fence line weed management is very active on this farm, the choice is to either allow native grasses to colonise these areas to provide some competition to the weedy grasses or they use an array of herbicides with no consistent pattern around the farm.

In summary, the owners of this property are still reasonably happy with their outcome as they know that if they continue with the glyphosate based fallow strategy, the property would be fully infested with awnless barnyard grass and FTR. They surmised the main treatment for control in this situation would be to resort to cultivation in every paddock several times per year.

Case study 4 (feather-top Rhodes grass – FTR) A 1350 ha farmer-owned property near Dalby was used for this report. The split between winter and summer cropping is about 75:25. Prior to the discovery of FTR the cropping rotation was sorghum/sorghum/sorghum/chickpea and long fallow. The first sighting of the weed commenced soon after the floods of 2011. Rapid change was made in response to this new weed incursion as FTR was widely known as a troublesome weed around the Darling Downs. The largest patch on the property was 5 ha.

What is the approach to manage the GR FTR? A change in cropping rotation was the first action. One sorghum crop was substituted for wheat and additional barley followed, then mungbeans. As such the new rotation is now sorghum/sorghum/chickpeas/wheat/barley and mungbeans. By incorporating additional winter crops, more summer fallows can allow better surveillance of the summer growing weed and permit the use of some pre-emergence herbicides. Some paddock-wide treatments are employed in the summer fallow to gain control. The pre-emergence herbicide imazapic is sometimes used depending on plant-back restrictions (ideal prior to winter cereals). Thicker patches are also treated with haloxyfop followed by paraquat as a double knock. However, some of the critical management tactics used are hand chipping the occasional isolated plant or using fire to burn isolated mature plants that have a good seed load. If patch size warrants the need, the farmer patch cultivates.

Delayed sowing of summer crops is a tactic this farmer uses so that cultivation can be used to control early season emergence of weeds. Other summer crop options are the use of pendimethalin and imazethapyr in mungbeans and metolachlor and atrazine in the sorghum phase.

Fence line management is either the use of haloxyfop tank mixed with glyphosate or promoting non-weedy grasses as a form of competition against FTR. The property seemed nearly 100% clean of FTR along the fence lines so internal farm hygiene must be a high priority.

The five hectare infestation that was the largest is now isolated plants due to sacrificing part of a chickpea crop by using glyphosate then burning the dry remains.

In conclusion, this farmer is very strict about his FTR patch management. He used combinations of non-chemical and chemical tactics and sets rules for various scenarios. For example, if the FTR is between 5 and 15% of the paddock he would use blanket treatments, if between 1 and 5% the WeedSeeker® would be used and infestations less than 1% he would consider hand removal.

This farmer had shown that monitoring, determination and setting plans are crucial factors to success.

DISCUSSION

There is no set recipe for successful patch management of glyphosate resistant weeds. However common themes appear within many of these case studies.

It appears some keys to winning the battle include:

- Maintain the pressure on the weeds by using effective treatments regularly.
- Always undertake chemical and non-chemical options.
- Get the weed patches early, if possible.
- Consider using pre-emergence herbicides and cultivation in fallows.
- Optical Spot Spray technology is useful.
- Fence line management is critical.
- Consider the battle may go on for at least 5 to 10 years.
- Farm hygiene practices must be used.
- Monitor every paddock after every treatment.
- Patch treatments are complementary to paddock-wide treatments.
- Stop weed seed set.

ACKNOWLEDGMENTS

The authors gratefully acknowledge the support of the Grains Research and Development Corporation for funding this work and the farm managers that cooperated in the case studies.

REFERENCES

Brownhill, D. (2006). Using 'Weedseeker'spot spraying technology in cropping systems. Proceedings of the10th Annual Symposium on Precision Agriculture in Australasia.

McGillion, T. and Storrie, A. (2006). Integrated weed management in Australian cropping systems – a training manual for advisors. CRC for Australian Weed Management, Adelaide, Australia.

Thornby, D., Werth , J. and Walker, S. (2014). Patch management solves early infestations of glyphosate resistant awnless barnyard grass. Proceedings of the 19th Australasian Weeds Conference, Hobart, ed. M. Baker, pp. 214-7. (Tasmanian Weed Society, Hobart).

Werth, J., Thornby, D., Keenan, M. and Walker, S. (2014). Managing patches of glyphosate resistant *Echinochloa colona*: can they be eradicated? Proceedings of the 19th Australasian Weeds Conference, Hobart, ed. M. Baker, pp. 3-6. (Tasmanian Weed Society, Hobart).

Antagonistic and synergistic effect of 2,4-D on herbicides of different sites of action in weed control

Jingbo Li[1,2], Heping Han[1], Qin Yu[1] and Stephen Powles[1]
[1] Australian Herbicide Resistance Initiative (AHRI), School of Plant Biology, The University of Western Australia, Crawley, Western Australia 6009, Australia
[2] Collaborative Innovation Center for Field Weeds Control, Hunan University of Humanities, Science and Technology, Loudi 417000, China
(jingbo.li@uwa.edu.au)

Summary Tank mixing herbicides of different sites of action is practiced for broader spectrum weed control, reducing herbicide application cost and delaying resistance evolution. For example, the auxinic herbicide 2,4-D is often tan mixed with herbicides of ACCase, PSII or EPSPS inhibitors for grass and broadleaf weed control. However, sometimes tank mix can have synergistic or antagonistic effect on grass weed control. We have found that tank mix of 2,4-D amine and diclofop-methyl or clodinafop greatly reduces grass weed control (e.g. ryegrass and wild oats). In contrast, tank mixing 2,4-D with atrazine or metribuzin enhances grass weed control (e.g. ryegrass, wild oat, goose grass). Our previous studies demonstrated that antagonism of diclofop-methyl by 2,4-D is due to 2,4-D induced enhanced diclofop metabolism via gene over expression of herbicide metabolizing enzymes (P450, GST, GT etc). Our current studies reveal that synergism of metribuzin by 2,4-D is associated with enhanced metribuzin translocation. In addition, reduced efficacy of glyphosate grass weed control when tank mixing with 2,4-D have been experienced by WA farmers and reported elsewhere. We also found that tank mixing 2,4-D (amine) with glyphosate (potassium salt) compromises barnyard grass control, especially for glyphosate-resistant populations. Based on our observation, we hypothesize that 2,4-D antagonism to glyphosate is likely due to reduced glyphosate uptake/translocation, as has been reported in Johnson grass (*Sorghum halepense* (L.) Pers.). Experiments are under way to testify this. As 2,4-D has multiple, unpredictable effects on other herbicides, hence care must be taken when tank mixing herbicides with 2,4-D.

Keywords 2,4-D, tank mix, herbicide translocation.

It's time for new thinking about herbicide resistance

Andrew Storrie[1,2], Peter Boutsalis[3] and John Broster[4]
[1] AGRONOMO, PO Box 5944, Albany, Western Australia 6330, Australia
[2] Australian Glyphosate Sustainability Working Group
[3] Plant Science Consulting, 22 Linley Avenue, Prospect South Australia 5082, Australia
[4] Graham Centre for Agricultural Innovation (Charles Sturt University and NSW Department of Primary Industries), Charles Sturt University, Locked Bag 588, Wagga Wagga, New South Wales 2678, Australia
(andrew@agronomo.com.au)

Summary The increasing number of species and populations of weeds being selected for resistance to herbicides appears to continue unabated. Globally there are currently 249 species resistant to one or more herbicide modes of action. Australia now has 45 species resistant to one or more of 13 of the 19 known modes of action.

Despite herbicide resistance being first confirmed in Australia 34 years ago the majority of farmers still wait for a spray failure, and sometimes repeated spray failures, before reacting to a resistance problem. Surveys of growers have shown that formal testing for resistance is very low and this is corroborated by the demographic data from testing services.

Decisions to manage herbicide resistance are often made by the adviser or farmer with little hard data. Spray failures can be the result of poor spray practice combined with stressed and or high weed densities.

So what are the reasons for the low use of testing by growers? Previous farmer surveys have shown that many growers think there will be a new herbicide to solve the problem and or resistance is imported from somewhere else. They think it is largely out of their control and therefore fail to act. Also have weed researchers and extensionists promoted a largely negative message that promotes inertia in farmers? Is the collection of samples for testing too difficult and or the testing too expensive?

This presentation investigates these questions and suggests turning the herbicide resistance message on its head and promoting the maintenance of herbicide susceptibility as a positive message that gives farmers hope and see herbicide resistance as a problem to be managed, not ignored.

Keywords Spray failure decisions.

Investigation of glufosinate resistance mechanisms in *Eleusine indica* (L.) Gaertn.

Adam Jalaludin[1,3], Qin Yu[1], Roland Beffa[3] and Stephen B. Powles[1]
[1] Australian Herbicide Resistance Initiative, School of Plant Biology, University of Western Australia, Western Australia 6009, Australia
[2] Weed Resistance Research Lab, Bayer AG, Industriepark Hoechst, Building H872, 65926 Frankfurt, Germany
[3] Queensland Department of Agriculture and Fisheries, Toowoomba, Queensland 4350, Australia
(adam.jalaludin@daf.qld.gov.au)

Summary An *Eleusine indica* (L.) Gaertn. population from Malaysia has evolved resistance to glufosinate, an important post-emergence herbicide. This population was analysed for target-site (glutamine synthetase) and non-target-site (glufosinate uptake, translocation and metabolism) resistance mechanisms. Glutamine synthetase (GS) activity extracted from susceptible (S) and resistant (R) plants was equally sensitive to glufosinate inhibition, with IC_{50} values of 0.85 mM and 0.99 mM, respectively. The extractable GS activity was also similar in S and R samples. Foliar uptake of [^{14}C]-glufosinate did not differ in S and R plants, nor did glufosinate net uptake in leaf discs. Translocation of [^{14}C]-glufosinate into untreated shoots and roots was also similar in both populations, with 44% to 47% of the herbicide translocated out from the treated leaf 24 h after treatment. The HPLC and LC-MS analysis of glufosinate metabolism revealed no major metabolites in S or R leaf tissue.

Therefore, glufosinate resistance in this resistant population is not due to an insensitive GS, GS over production or altered glufosinate uptake and translocation, nor enhanced glufosinate metabolism. The exact resistance mechanism(s) remain to be determined.

Keywords Glufosinate, herbicide resistance, mechanisms, *Eleusine indica*.

Survey of glyphosate resistance in common sowthistle (*Sonchus oleraceus*) across the Australian Northern Grains Region

Annie W. van der Meulen[1], Michael Widderick[1], Tony Cook[2], Bhagirath S. Chauhan[3] and Kerry Bell[1]
[1] PO Box 2282, Toowoomba, Queensland 4350, Australia
[2] Tamworth Agricultural Institute, NSW Department of Primary Industries, Tamworth, New South Wales 2340, Australia
[3] Queensland Alliance for Agriculture and Food Innovation (QAAFI), University of Queensland, Queensland, Australia
(annemieke.vandermeulen@daf.qld.gov.au)

Summary Common sowthistle (*Sonchus oleraceus* L.) is a major weed of conservation tillage systems in Australia. The major problem associated with the weed is that it reduces stored soil moisture in fallow, impacting growth of the subsequent crop. Glyphosate has been a reliable control option, but resistance to this herbicide has now been confirmed in populations growing in the northern cropping region of Australia.

To benchmark glyphosate resistance in common sowthistle across the study region, a survey was conducted (2014–2016). Common sowthistle seeds were collected from cropping properties as part of a non-targeted sampling approach. Plants were grown to a two-to-four leaf stage, and then treated with a commercial glyphosate formulation at a discriminating rate (field rate), to distinguish between resistant and susceptible populations.

At the time of writing, 20% of the populations tested have exhibited resistance to glyphosate (i.e. less than 80% of seedlings were killed by treatment with glyphosate at the discriminating dose). Most of the glyphosate resistant populations identified through this testing were clustered in the Liverpool Plains region of NSW, where long term use of conservation tillage practices has resulted in reliance on glyphosate for sowthistle control. By determining the geographic extent of glyphosate resistant common sowthistle in the northern region, the results of this survey can assist industry stakeholders to respond with suitable management solutions for the control of this problematic weed.

Keywords Herbicide resistance, survey, common sowthistle, *Sonchus oleraceus*.

Understanding the mechanisms of triallate resistance in annual ryegrass

Mahima Krishnan, Peter Boutsalis and Christopher Preston
The University of Adelaide, Adelaide, South Australia 5005, Australia
(mahima.krishnan@adelaide.edu.au)

Summary Annual ryegrass (*Lolium rigidum* Gaudin.) is a major weed in Australian cropping systems and has developed resistance to all post-emergent herbicides registered for use in cereals. This has increased the importance of pre-emergent herbicides for control of annual ryegrass in cereal cropping systems. In 2014, two populations from South Australia and one from New South Wales were reported as having survived triallate application (a Group J mode of action herbicide).

Seed samples from these three populations were confirmed as having triallate resistance and also cross-resistance to prosulfocarb, another Group J herbicide. These cross-resistant populations were, however, mostly controlled by the addition of S-metalochlor (Group K) to prosulfocarb. Pyroxasulfone, an alternative pre-emergent herbicide from Group K also controlled the resistant populations.

The failure of Group J chemistry to control annual ryegrass in a cereal cropping system places pressure on Group K chemistry, which is the only alternative pre-emergent herbicide providing effective control in the field. Understanding the basis of resistance to group J herbicides and the number of genes involved will help with developing alternative strategies that growers can use to effectively control ryegrass in their fields and manage existing resistant populations.

Keywords Triallate, prosulfocarb, pyroxasulfone, Group J, Group K, annual ryegrass, resistant, cross-resistant, pre-emergent, gene.

Mechanism and inheritance of resistance to clethodim in *Lolium rigidum* populations from Australia

Rupinder K. Saini, Jenna M. Malone, Gurjeet S. Gill and Christopher Preston
University of Adelaide, PMB1, Glen Osmond, South Australia 5064, Australia
(jenna.malone@adelaide.edu.au)

Summary *Lolium rigidum* Gaudin. is a major weed of agriculture in Australia and has developed resistance to many herbicides. Clethodim was the last in-crop herbicide available for its control, however recently there has been a rapid evolution of clethodim resistance in *L. rigidum* across Australia. The mechanism and inheritance of clethodim resistance was studied in a number of resistant populations. Several different patterns of inheritance of clethodim resistance were observed and included a single gene, partially dominant, nuclear encoded trait, two different patterns of two-gene inheritance, as well as maternal inheritance of the resistance trait. Amino acid substitutions at five (1781, 2041, 2078, 2088, and 2096) of the seven known positions in the *ACCase* gene documented to confer resistance to *ACCase*-inhibiting herbicides in grass weed species were identified in clethodim-resistant populations. In addition to these previously document substitutions, many resistant individuals also contained one or more amino acid substitutions at other positions in the *ACCase* gene. Studies into whether these substitutions also confer clethodim resistance are being undertaken.

Keywords Clethodim, *Lolium rigidum,* resistance, inheritance.

Pathogens for weed biological control: a solid past towards a rosy future?

Louise Morin
CSIRO Health and Biosecurity, GPO Box 1700, Canberra, Australian Capital Territory 2601, Australia
(louise.morin@csiro.au)

Summary Biological control (biocontrol) using plant pathogens has delivered exciting solutions for weed management in Australia. Interest in pathogens has steadily increased since the deliberate introduction of the rust fungus *Puccinia chondrillina* Bubak & Syd. in 1971, which has had a spectacular impact on the widespread form of skeleton weed (*Chondrilla juncea* L.). Another highlight has been the impact of the rust fungus *Puccinia myrsiphylli*, introduced from South Africa in 2000, on coastal populations of bridal creeper (*Asparagus asparagoides* (L.) Druce) in temperate Australia. More recent observations of the impact of the white-smut fungus *Entyloma ageratinae* R.W.Barreto & H.C.Evans on mistflower (*Ageratina riparia* (Regel) R.King & H.Robinson) have further increased interest in pathogens as biocontrol agents. Current projects using exotic pathogens are expected to deliver similar levels of impact on targeted weeds. The rust fungus *Baeodromus eupatorii* (Arthur) Arthur was released in 2014 for the biocontrol of Crofton weed (*Ageratina adenophora* (Spreng.) R.King & H.Robinson), and all signs so far are that this agent will provide a sustainable tool to reduce populations of this weed. An application to release the leaf-smut fungus *Kordyana brasiliensis* D.M.Macedo, O.L.Pereira & R.W.Barreto for the biocontrol of wandering trad (*Tradescantia fluminensis* Vell.) has just been submitted following completion of host-specificity testing. A new project is underway to explore prospects of developing as a biocontrol tool the key organism (*Phytophthora bilorbang* S.Aghighi, G.E.S.J.Hardy, J.K.Scott & T.I.Burgess) identified as a contributor to the extensive decline of invasive blackberry (*Rubus anglocandicans* A.Newton) observed in riparian areas in south-east Western Australia. Whilst not all pathogens investigated for weed biocontrol lead to novel tools for land managers, past successes are sufficiently significant to promote continued investments in this area of research.

Keywords Biological control, pathogens, impact, research investments.

Why has resistance evolved more slowly to pre-emergent herbicides than to post-emergent herbicides?

Gayle J. Somerville[1], Stephen B. Powles[1], Michael Walsh[2] and Michael Renton[1]
[1] Australian Herbicide Resistance Initiative, MO 86, University of Western Australia, Crawley, Western Australia 6009, Australia
[2] Plant Breeding Institute, University of Sydney, 12656 Newel Highway, Narrabri, New South Wales 2390, Australia
(gayle.somerville@research.uwa.edu.au)

Summary Populations of *Lolium rigidum* Gaudin (annual ryegrass) have generally evolved resistance to pre-emergent herbicides more slowly than to post-emergent herbicides. One example is the pre-emergent herbicide trifluralin, which was in steady use from the 1960s until the 1980s with few indications of resistance evolution. In contrast, diclofop, a post-emergent herbicide released in the 1980s, lost effectiveness due to resistance evolution within just a few years of its first use. It has been speculated that the slower evolution of resistance to pre-emergent herbicides may be due to a number of factors including lower initial frequencies of resistance genes, differences in herbicide use patterns or reduced herbicide efficacy.

Individual based simulation modelling was used to investigate the effect of two herbicide application timings, (pre and post emergent), herbicide efficacy and resistance allele frequency on the evolutionary rate of a single dominant gene for herbicide resistance within a population of annual ryegrass in a wheat field.

This study determined that the observed slower evolution of resistance to pre-emergent herbicides (e.g. trifluralin) in annual ryegrass populations is most likely due primarily to reduced efficacy (70%) of trifluralin with a low initial gene frequency (1×10^{-8}) contributing secondarily. Differences in herbicide use patterns (pre- versus post-emergent) had minimal effect on rates of resistance evolution when similar weed cohorts were affected.

Keywords Resistance, modelling, herbicide, trifluralin.

Improving IWM practice of emerging weeds in the Western Region

Abul Hashem, Mohammad Amjad and Catherine Borger
Department of Agriculture and Food Western Australia, 75 York Road, Northam,
Western Australia 6401, Australia
(abul.hashem@agric.wa.gov.au)

Summary Changes in farming systems, management practices and climate are resulting in a change to the weed spectrum that is causing increased cost and yield loss for growers. Emerging summer weed species such as fleabane (*Conyza* spp.), tar vine (*Boerhavia* spp.), sowthistle (*Sonchus* spp.) and button grass (*Dactyloctenium radulans* (R.Br.) P.Beauv.), and emerging winter weeds such as brome grass (*Bromus* spp.), Indian hedge mustard (*Sisymbrium orientale* L.) and wild radish (*Raphanus raphanistrum* L.) are becoming increasingly resistant to a broad number of herbicide modes of action. As a consequence these species are now increasing in crop, resulting in increased control costs and crop losses. This project aimed at delivering new knowledge on the biology and management of emerging summer weeds and winter weeds in the western region. Research results showed that emergence of button grass, fleabane, tar vine and sowthistle were not restricted to summer. Seed germination of exhumed fleabane and sowthistle was low and declined over time, while exhumed tar vine and button grass seed germination showed a cyclic pattern of dormancy release. Double knockdowns (mixture of 2,4-D + glyphosate or 2,4-D + glufosinate followed by paraquat-based mixture controlled 94% to 98% fleabane, sowthistle and tar vine. Narrow row spacing and high density of a wheat crop with a high level of soil disturbance resulted in negligible emergence of summer weeds. Glyphosate alone was also effective on button grass and tar vine over summer. High herbicide rates were required to kill summer weeds when temperatures were high, but lower rates achieved the same result on cooler days. Low harvest height and windrowing of wheat crop collected 70% of brome grass seed on windrows. Subsequent burning of the windrows reduced the seedling emergence of brome grass by 93% to 99%. A roadside weed survey across the Western Australian wheatbelt during late summer and early autumn of 2015 showed that the most common summer weed species were African love grass, fleabane, windmill grass and wild radish. Sowthistle and fleabane seed samples collected during the survey was tested for resistance to glyphosate and three sowthistle populations were found resistant to 1080 g a.i. ha^{-1} glyphosate. The results have been communicated in four media releases, two factsheets, eight field walk/field days, and 11 GRDC/Grower Groups Research Update articles.

Keywords Emerging summer and winter weeds, seed dormancy, emergence pattern, seed exhume, windrow burning, herbicide resistance, summer weed control.

Non-chemical tactics for improved control of key northern region weeds

Michael J. Widderick[1], Andrew R. McLean[1], Michelle D. Keenan[1] and Bhagirath S. Chauhan[2]
[1] Queensland Department of Agriculture and Fisheries, PO Box 2282, Toowoomba, Queensland 4350, Australia
[2] The University of Queensland, Queensland Alliance for Agriculture and Food Innovation, PO Box 2282, Toowoomba, Queensland 4350, Australia
(michael.widderick@daf.qld.gov.au)

Summary Under existing conservation farming systems of the northern cropping region of Australia, herbicides are heavily relied upon for fallow and in-crop weed control. Subsequently, herbicide resistant and difficult to control weeds are common throughout the region. To improve management of key weeds and to retain useful herbicides, there is a need to identify alternative non-chemical tactics for weed management. This paper will discuss the effects of targeted tillage, strategic burning and cover crops on the management of key weed species including feathertop Rhodes grass (*Chloris virgata* Sw.), awnless barnyard grass (*Echinochloa colona* (L.) Link.), windmill grass (*Chloris truncata* R.Br.), flaxleaf fleabane (*Conyza bonariensis* (L.) Cronquist) and common sowthistle (*Sonchus oleraceus* L.). Targeted tillage has been evaluated across four seasons and sites and has included tillage of different intensity and multiple passes. All forms of tillage have generally reduced the emergence of target weeds. However, the results are not consistent across seasons, likely due to interacting factors of rainfall, temperature and soil disturbance. Strategic burning of *C. virgata* patches reduced surface-lying seeds by 93% and emergence within the patch from 785 plants m^{-2} to 450 plants m^{-2}. Different cover crops, cover crop combinations and termination times were compared for efficacy in suppressing *E. colona* and *C. virgata*. All cover crop treatments with millet reduced *E. colona* emergence compared with cowpea and lablab on their own. The integration of these alternative non-chemical tactics will improve weed management and further reduce the impact of weeds on crop production systems.

Keywords Herbicide resistance, non-chemical, tillage, cover crops, burning.

AgPest

Katherine N. Tozer[1], Colin Ferguson[2] and Trevor K. James[1]
[1] AgResearch, Ruakura Research Centre, Hamilton, 3240, New Zealand
[2] AgResearch, Invermay Agricultural Centre, Mosgiel, 9053, New Zealand
(katherine.tozer@agresearch.co.nz)

Summary AgPest (www.agpest.co.nz) is a website that was conceived and designed by AgResearch weed scientists and entomologists in collaboration with farmers and farm consultants. The website provides a decision support system, with relevant, up-to-date and independent information for endusers on the identification and control of pasture weeds and pests in New Zealand. There are currently 101 pages on AgPest, comprising 33 pest and 68 weed pages, with information on the biology, control and management for each weed and pest, and pictures that assist in their identification. A tool, based on descriptions of plant characteristics, insect pest characteristics and insect pest damage, enables the weeds and insect pests on AgPest to be identified. Additionally, a free subscription service disseminates region-specific email and text alerts on weed and pest threats that reduce the need for 'fire-fighting' decisions and enable pre-emptive action to minimise weed and pest damage. In this paper we summarise why and how AgPest was developed, highlight issues encountered when developing AgPest, and outline plans for its future.

WHY WAS AGPEST DEVELOPED?

Weeds and insect pests severely limit pasture and livestock production on New Zealand farms (Bourdot *et al.* 2007, Zydenbos *et al.* 2011) and information on their control is often fragmented and difficult for farmers to access. Fact sheets on weed and pest management are available but they become out of date, or lost. In addition, while there is significant knowledge on the control of key pasture weeds and pests within AgResearch, the Crown Research Institute with a mandate to improve productivity of New Zealand's pastoral sector, there was concern that valuable scientific knowledge would be lost as key scientists retired.

These issues provided a catalyst for scientists, farmers and farm consultants to collaborate and apply for funding to develop a website for the benefit of the pastoral industry that provided independent information on the control of pasture insect pests and weeds. The hosting group for the application and a driver of the project was the Otago Beef+LambNZ Monitor Farm Group, which was concerned about this industry-critical issue.

HOW WAS AGPEST DEVELOPED?

Government and industry funding was granted to build a pilot website over two years. The steering group, comprising farmers, scientists and rural professionals from throughout New Zealand determined the scope of the site. There was unanimous agreement that the website should:

- provide independent and farmer-friendly information on the biology, control, impact and identification of pasture weeds and pests,
- provide information that assisted farmers and consultants in making pre-emptive pest management decisions, leading to better control and increased pasture productivity,
- enable pest and weed control solutions to be more readily available to farmers and industry professionals, and
- be a supplementary, easily accessible source of information for farmers to use alongside other existing sources (such as consultants and rural retailer employees).

Based on these aims, a pilot website was developed, which contained a small number of weeds and pests. The steering group, and in particular the farmers, tested the pilot website and provided feedback on its ease of use. The site was fine-tuned to ensure it was farmer-friendly when it went live. The website, now known as 'AgPest' (www.agpest.co.nz), contained six major weeds and pests when it was launched at the National Agricultural Field Days at Mystery Creek, Hamilton, in the North Island of New Zealand in June 2011. The Field Days provided an excellent setting for its launch, as thousands of farmers from throughout New Zealand attend these field days each year. Feedback was positive.

Additional funding was granted, and this allowed the addition of 20 weed and pest species over a further three years. Species were chosen in consultation with key farming, industry and research personnel. The information provided in the pages was based on peer-reviewed scientific literature and contextualised for regional conditions by the AgResearch weed scientists and entomologists. The pages were written and reviewed by scientists. An alert function, available by free subscription, was developed. This enabled farmers to receive free region-specific emails or text alerts about outbreaks or potential outbreaks of insect pests and to a lesser extent weeds, along with recommendations for their management. This service has been used by the New Zealand government to alert farmers and rural professionals about black grass (*Alopecurus myosuroides*) and velvet leaf (*Abutilon theophrasti*), which have been recent biosecurity incursions. In addition, social media such as Twitter and Facebook have been launched, posting information about weeds and pests.

A search function was developed and expanded with each new weed and pest page, enabling users to search AgPest by the weed or pest name or easily identifiable plant and insect characteristics. Once the weed or pest is identified, links enable the user to access the information on its biology, impact and management. The series of pictures of each weed and pest assist in its accurate identification.

The website is interactive and allows users to text, email, or use the direct web enquiry tool to ask for information on identification and control of pasture weeds and pests.

ISSUES ENCOUNTERED

A number of issues were encountered while developing AgPest. The major issues and solutions developed are discussed below.

Choice of the hosting platform A difficulty faced by all websites is ensuring their maintenance and longevity. Originally, AgPest was hosted by an industry body. To ensure that the website could be easily upgraded and that changes could be easily made, the website was transferred to a domain under the jurisdiction of AgResearch. This has allowed scientists to rapidly update the site and has harnessed the skills of personnel in the AgResearch Communications Team. These personnel are critical for ensuring that enduser's queries are rapidly directed to the relevant weed scientists and entomologists who can answer the queries.

Intellectual property To overcome intellectual property issues, trademarking was necessary. This ensured that the name 'AgPest' was not accessible to another company which could potentially stop us from using the 'AgPest' brand.

Commercial sensitivities AgPest provides information on novel endophytes of perennial ryegrass, which differ in the protection that they confer against insect attack (Anon. 2012). As there are commercial sensitivities regarding the ownership of particular endophytes and the protection that endophytes provide, industry-agreed on protection levels are used in any AgPest endophyte communications.

Another commercially sensitive issue pertains to chemical control options. Given that tradenames can frequently change, AgPest provides information on the generic active ingredient rather than the tradename. The enduser can then take this information to their agchem technical representative and rural retailer who can suggest the appropriate commercial product.

Fielding questions Some of the queries received regarding weeds and pests are outside the scope of AgPest, which is focused on serving the New Zealand pastoral industry. There are a growing number of questions being referred via AgPest and so if the question does not relate to agricultural weeds and pests the enquirer is directed elsewhere, such as to the National Herbarium. It is also important to determine the country of origin of the query to prevent scientists' time being spent on answering questions from other countries with no relevance to the New Zealand pastoral sector.

Funding There has been strong industry and local council support for AgPest, which has been critical for its success. From its inception, the sheep and beef industry, represented by Beef+LambNZ, have been strong advocates of AgPest. Beef+LambNZ have also had a critical role in disseminating text alerts, email alerts and tweets. Industry collaboration is imperative for AgPest maintenance and future development.

The role of advertising to provide funding for AgPest has been discussed at length by the steering group, in response to requests from agri-chemical companies for advertising space. The steering group have taken the approach of no advertising, to maintain the independence of the site and the information presented. We believe this is to the advantage of the whole pastoral industry.

Continued promotion There can be a large turnover in industry technical staff, including beef, sheep and dairy industry agronomists, who are key users of AgPest and who inform farmers of how to use AgPest. Therefore continual promotion of AgPest is needed to inform industry professionals and farmers of its presence and of new weeds and pests that have been added. Promotion has included speaking about AgPest at field days and rural professional training days, and featuring AgPest in articles disseminated via rural media.

AgPest is now recognised across the New Zealand pastoral sector as the premiere online weed and pest information source for New Zealand farmers. This is demonstrated by AgPest being cited as an authoritative information source by major New Zealand seed companies and rural retailers, being used by the Ministry of Primary Industries for alerting farmers to biosecurity threats, and promotion by key industry bodies, such as Beef+LambNZ and the Pasture Improvement Leadership Group.

FUTURE DEVELOPMENT

While the most damaging pasture insect pests are now on AgPest, we aim to add five new weeds to AgPest each year over the next five years. Updating of the site will be on-going, given changes in chemical registrations for weed and pest control, changes in weed and pest prevalence and the sporadic nature of weed and pest outbreaks, new biosecurity incursions (from outside and inside New Zealand) and the development of new weed and pest management strategies.

Further work is required to simplify the identification key. As part of this initiative, more photos have recently been added to assist endusers in accurately identifying weeds and pests. A mobile app. is being discussed, which will enable farmers to use AgPest when out on the farm. This will assist in weed and pest identification and pre-emptive action to achieve more rapid control, reducing the impact of weeds and pests.

From its inception eight years ago, AgPest has grown to comprise information on over 100 weeds and pests and is successfully serving the pastoral sector.

ACKNOWLEDGMENTS

Thanks to Beef+LambNZ and AgResearch for funding AgPest. We are also grateful to Ian Popay for his input into the weed pages and the many AgResearch scientists and staff who have assisted with weed and pest page development and reviewing.

REFERENCES

Anon. (2012). DairyNZ pasture renewal guide, p. 32. DairyNZ, Hamilton, New Zealand.

Bourdot, G.W., Fowler, S.V., Edwards, G.R., Kriticos, D.J., Kean, J.M., Rahman, A. and Parsons, A.J. (2007). Pastoral weeds in New Zealand: status and potential solutions. *New Zealand Journal of Agricultural Research* 50, 139-61.

Zydenbos, S.M., Barratt B.I.P., Bell, N.L., Ferguson, C.M., Gerard, P.J., Mcneill, M.R., Phillips, C.B., Townsend, R.J. and Jackson, T.A. (2011). The impact of insect pests on pasture persistence and their interrelationship with biotic and abiotic factors. *Pasture Persistence – Grassland Research and Practice Series* 15, 109-18.

Search and destroy: a new tool in the war on weeds

Clayson J. Howell and Stu Cockburn
Department of Conservation. PO Box 10-420, Wellington New Zealand
(chowell@doc.govt.nz)

Summary The New Zealand Department of Conservation (DOC) has developed a new tool to record search paths and herbicide usage. The unit combines a flow meter with an accurate GPS and records daily search paths and the volume of herbicide dispensed at each location in a single .gpx file. Once switched on, all data recording is completely hands-free, and is being widely used to record operations. Once operations are complete, the data are easily uploaded into DOC's Weeds application, an on-line GIS-based tool that allows DOC staff to record all weed control in a consistent way. Once entered, data are available to all staff, which enables more effective planning over very large tracts of land administered for conservation.

Keywords Search, destroy, herbicide, GPS.

INTRODUCTION

Tracking herbicide use is of increased importance. Typically staff record spray information in diaries and transfer these data when reporting on operations. It is vital that this information is recorded to monitor effectiveness, respond to public queries, and pass on historic information to new staff. Our staff were increasingly being asked to improve the detail of the information they recorded about spraying operations and this reduced the amount of work they were able to deliver.

PRODUCT DEVELOPMENT

With DOC's electronics workshop we developed a small unit to improve field data capture of weed spraying information. The unit comprises a turbine-type flow meter integrated with a GPS. The flow-meter can be connected to any spray unit (Figure 1) and records herbicide volume and location every time herbicide is dispensed. The unit also records a track that clearly illustrates the search path taken (Figure 2).

All information is stored in .gpx file format and upload of data and battery recharging is completed with a standard USB cable. All software was developed internally and can be customized. The data from the unit can easily be uploaded to standard GIS tools where information on the herbicide formulation used and weeds targeted can be appended.

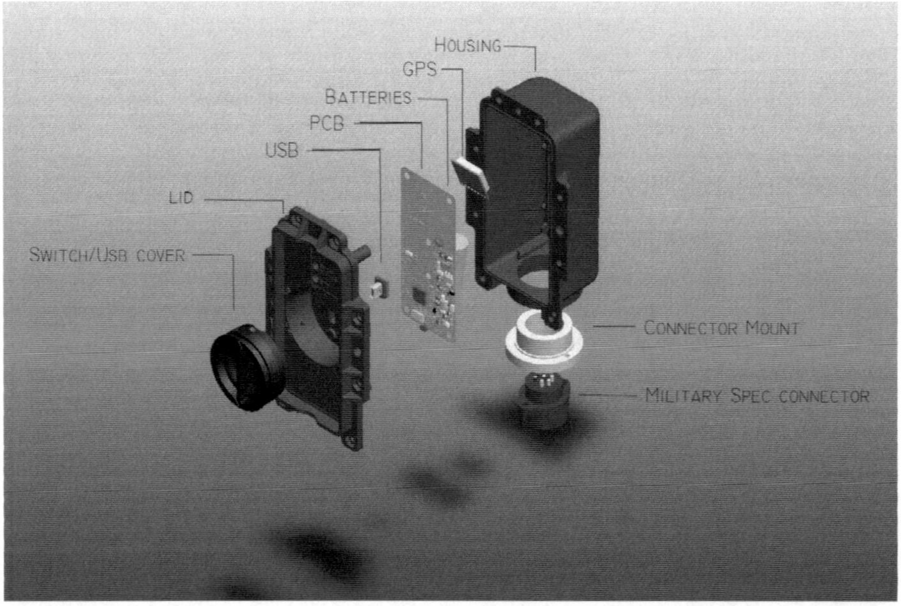

Figure 1. The search and destroy unit is small, hands-free and can be attached to a knapsack, gun and hose or helicopter spray wand.

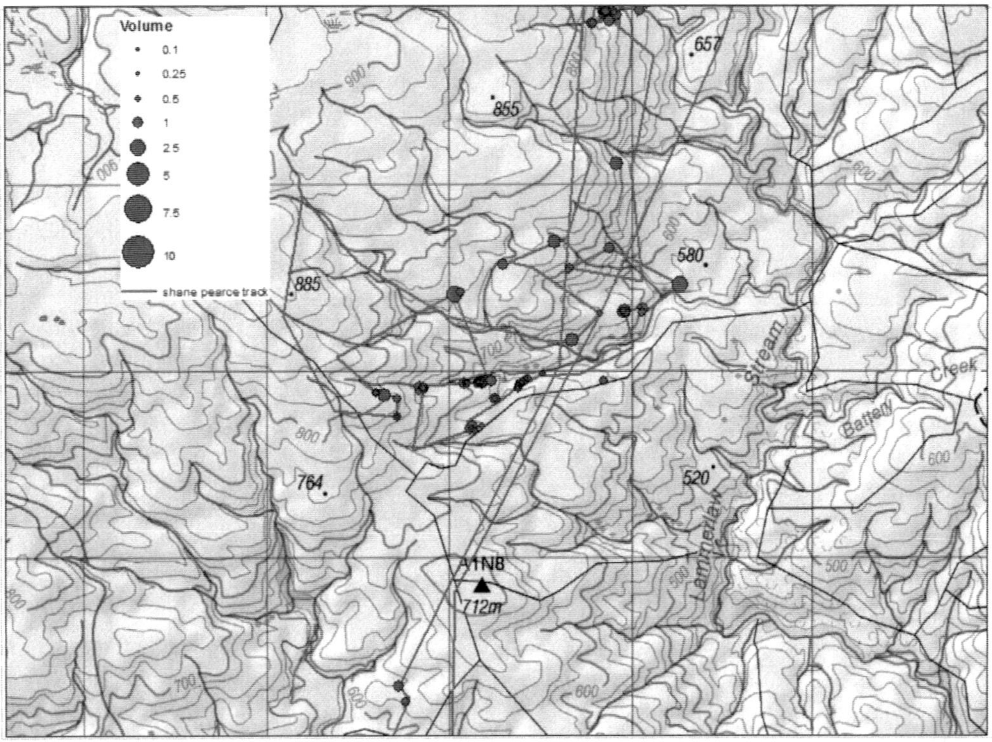

Figure 2. Example of search path and volume data overlaid on topographic map.

APPLICATIONS

After the final prototype was built feedback on the unit has been positive. Staff no longer have to carry a GPS that is separate from the spray equipment. While not mandatory, it is hoped that uptake of the unit will be widespread.

Because operation is very simple, potential applications can range from very small scale spraying around sensitive plants in a wetland to spot treatments of wilding conifers over thousands of hectares.

The inclusion of the search path allows simple checks that surveillance plans have been followed, and can highlight points of interest for follow-up surveys.

The unit has also proven useful in operational experiments to test efficacy of spot spraying. When mortality is variable, the dose that individual plants received can explain some of the variation in results without an expensive controlled experiment.

Within DOC, the information is easily integrated with our web-based GIS tool for recording weed work. But the standard format can be viewed in many applications.

DISCUSSION

Sophisticated systems are available for large agricultural weed spraying, but this simple unit has great potential for recording control actions for environmental weeds. The unit can be connected to a variety of tools and the cost is comparable to hand-held GPS units so most budget managers can afford multiple units.

Herbicide usage is likely to come under increasing scrutiny so recording information as accurately as possible is seen as a priority. This system makes recording easier and allows operators more time to execute safe and effective weed control.

ACKNOWLEDGMENTS

Nick Pederson and Steve McKevitt from the DOC electronics lab developed the hardware and electronics. Prototype units were field-tested by DOC Operations staff Shane Pearce, Dave Walford and Chris Woolmore.

ARM software for weed research management

Steven R. Gylling
405 Martin Boulevard, Brookings, SD 57006, United States of America
(steve@gdmdata.com)

Summary ARM is commercial software specifically designed to manage crop protection and crop production efficacy experiments, including weed research. Available tools to plan, randomize, establish, manage, analyse, graph, and report information for field and greenhouse experiments will be briefly reviewed. The software provides a structured way to enter treatments, assessments, site and application information, tasks, and attachments. The software can also automatically randomize and statistically analyse results for the 11 single and multi-factor experimental designs, review randomization quality, calculate product amounts to achieve desired application rates, find statistical outliers, perform data quality checks, identify potential site (non-treatment) effects, automatically detect and correct common violations of AOV assumptions. Reports are also included for data collection and verification, labels, plot signs, spray/seeding plans, site information, maps, field tours, and several assessment data reports that include statistical analysis. All reports can be printed to Word, Excel, and Adobe .pdf files. An integrated trial and protocol overview database automatically indexes information to conveniently search current and historical trials based on more than 75 different types of information. Similar trials can be readily combined across locations and years using an optional multi-trial summary analysis module. The software has been available for more than 30 years, includes several language translations, and is currently used by more than 6000 clients in approximately 80 countries.

Keywords Software, research data management.

MyWeedWatcher: mapping weeds in Western Australia

Tamrika Lanoiselet[1], Miriam Sanchez[1], Neil Steventon[1], Steve Collins[1], Scott Carmichael[1], Alex Bergin[1], Sandy G. Lloyd[1] and David Hill[2]
[1] Department of Agriculture and Food Western Australia, 3 Baron-Hay Court, South Perth, Western Australia 6151, Australia
[2] Department of Agriculture and Food Western Australia, 4703 Jalbarragup Road, Vasse, Western Australia 6280, Australia
(tamrika.lanoiselet@agric.wa.gov.au)

Summary The general community and community groups can now use smartphone and tablet devices to identify, survey and report weeds and view results online. MyWeedWatcher enables users to identify weeds, conduct surveys on weeds of interest and report on the presence of declared weeds.

The identification guide allows users to quickly search for a weed according to plant characteristics such as flower colour, leaf shape, and plant type. The survey/reporting feature enables users to map weeds, add images and record survey data such as weed density, weed counts, confidence of identification, and notes on control activities done.

The information gathered through MyWeedWatcher will be used to help Community Groups, Local Biosecurity Groups, and DAFWA make more informed management decisions on weeds of interest in the State. In addition, users help to maintain Australia's pest-free status, support farm biosecurity and become participants in a new monitoring and surveillance community.

Keywords Weed surveillance, apps, app, report, map, identify.

Current perspectives on the impact of weed seed contamination in sheep

Jane E. Kelly[1,2], Jane C. Quinn[1,2], Panayiotis Loukopoulos[1,2], Sharon G. Nielsen[3], Paul Weston[1,2], John C. Broster[1,2] and Leslie A. Weston[2]
[1] Charles Sturt University, Locked Bag 588, Wagga Wagga, New South Wales 2678, Australia
[2] Graham Centre for Agricultural Innovation, Charles Sturt University, Locked Bag 588, Wagga Wagga, New South Wales 2678, Australia
[3] QCU, Research Office, Charles Sturt University, Locked Bag 588, Wagga Wagga, New South Wales 2678, Australia
(janekelly@csu.edu.au)

Summary Grass seed contamination of sheep carcasses and skins results in significant losses across the Australian lamb and mutton value chains. Seven grass species found across Victoria (VIC) and New South Wales (NSW) are known to significantly impact carcass quality, with barley grass and brome grass as key perpetrators. To evaluate current trends associated with carcass weed seed contamination, extensive abattoir datasets (2009–2014) provided by Animal Health Australia (AHA) were analysed by state and biogeographic region. Significant differences in regional infestation were noted, with reduced contamination observed in Tasmania (TAS) and VIC in contrast to other states. No significant differences were observed in contamination between NSW and Queensland (QLD). Highest contamination occurred throughout the mixed cropping and pastoral zones, with notable events also in the high rainfall zone. Sex and age of animal were identified to have impacted seed contamination. These findings have implications for integrated weed management (IWM) strategies for weed species of importance to the livestock production chain.

Keywords Sheep, seed contamination, barley grass, brome grass, carcass, invasive weed.

INTRODUCTION

Physical damage from grass seed contamination in sheep results in reductions in wool and carcass quality in Australian flocks, through lodgement of seed within fleece, skin, carcass, and organs of grazing sheep. Direct and indirect costs are incurred by both producers and processors due to contamination (Collins 2013). Barley grass (*Hordeum* spp. Link.) and brome grass (*Bromus* spp. Roth.) are two species recently listed in the top 20 summer fallow weeds of crops in Australia (Llewellyn *et al.* 2016) and are also associated with carcass damage in Australia. The prevalence of both species is increasing across southern Australia, with some populations displaying herbicide resistance and variable seed dormancy (Gill and Blacklow 1985, Fleet and Gill 2012, Owen *et al.* 2015). Little information on the regional severity of seed contamination is currently available, limiting capacity to regionally tailor research efforts to match changing weed dynamics. To address this issue, statistical analyses were performed on abattoir datasets for seed-contaminated sheep located across Australian states provided by Animal Health Australia as part of the National Sheep Health Monitoring Project. Weed survey data was also spatially joined to identify associations between the distribution of barley grass and brome grass seed in sheep and existing weed infestations.

MATERIALS AND METHODS

Datasets All available abattoir datasets detailed the numbers of animals with significant grass seed lesions across all Australian agro-ecological zones during the period of 2009–2014. Barley grass and brome grass population density and distribution data from 2007 to 2015 were obtained using the methods described by Llewellyn *et al.* (2009) and were combined with abattoir datasets for New South Wales (NSW), Victoria (VIC), South Australia (SA) and Tasmania (TAS) using ArcGIS Desktop (ESRI, 2015) for spatial representation. Natural breaks (Jenks) classification was used to create classes of infestation levels within each map.

Analysis To explore the relationship between the percentage of carcass infestation (PIN) and the predictor variables (state, region and age) linear mixed models were used. The predictor variable was fitted as a fixed factor and the abattoir was fitted as a random variable in the model. The validity of conclusions from statistical analysis is dependent on the validity of the assumptions associated with the analysis; in this case the linear mixed model assumptions are: 1) that the residuals are normally distributed, 2) they have a constant variance and 3) they are independent. It is assumed that the factor level variances are equal for the predictor variable. It was necessary to use a natural logarithm to transform the response data and a weighted least squares analysis to correct for heterogeneity of variance. A significance level of 5% was used for reporting results.

RESULTS

Seed contamination in Australian sheep, averaged over states and territories, was 23.3% of lambs, ewes and male animals. Significant differences in the incidence of weed seed carcass contamination was noted between states ($P < 0.001$), and also within states for NSW, VIC, South Australia (SA) and Western Australia (WA) only.

The predicted means for individual states show that contamination levels in WA (19.0% ± 1.3) were not significantly different to NSW (14.9% ± 1.2), Queensland (QLD) (14.1% ± 1.2), VIC (11.2% ± 1.2) or SA (18.1% ± 1.2), but were significantly higher than levels in TAS (9.2% ± 1.2). No significant difference was observed in seed contamination between sheep surveyed from TAS and VIC or between QLD and NSW. Predicted means for seed contamination by region showed the majority of contamination to be distributed across regions designated as mixed farming and/or pastoral zones. Greatest contamination was observed in WA regions, while southern SA, central western NSW and central QLD regions also showed high incidence of contamination (Figure 1). Regions in southeastern Australia showed high incidence of barley grass and brome grass infestation, and were positively correlated with regions displaying high carcass contamination (Figure 2).

Significant effects of sex and age of animal were also noted. Carcass contamination levels were highest in animals greater than two years of age and also in animals of 'unknown' age. Levels were lowest for animals under two years of age and for groups of mixed

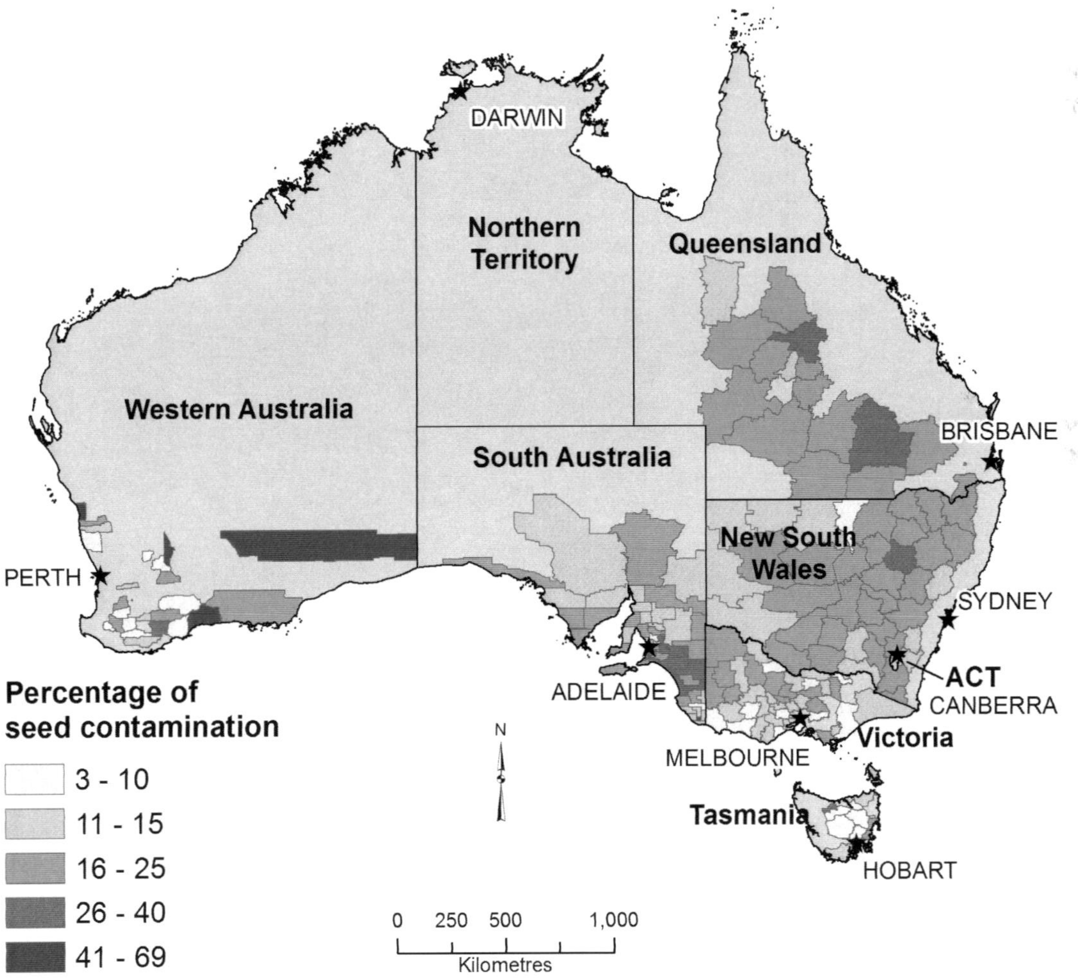

Figure 1. Distribution and the total density of sheep carcasses showing weed seed contamination between 2008 and 2014.

age (Table 1). Contamination levels were higher for all entire males and lower for castrated males than all other sex groups (Table 2). Contamination levels in females were not different to those of mixed or unknown sex groups (Table 2) but were significantly higher than in castrated males and lower than in entire males.

Table 1. Mean percentage of animals contaminated with weed seed by age group.

Age group	Mean percentage carcass contamination by weed seed (%) ± SE
Mixed age	11.3 ± 1.202 a
<2 years	12.4 ± 1.655 a
Unknown	13.6 ± 1.169 ab
>2 years	15.2 ± 1.168 b

Values with the same letter within each group are not significantly different (P <0.05).

Table 2. Mean percentage of animals contaminated with weed seed by sex group.

Sex group	Mean percentage carcass contamination by weed seed (%) ± SE
Castrated male	12.5 ± 1.161 a
Mixed sex	13.9 ± 1.159 b
Female	14.3 ± 1.160 bc
Unknown	16.4 ± 1.165 c
Entire male	24.0 ± 1.170 d

Values with the same letter within each group are not significantly different (P <0.05).

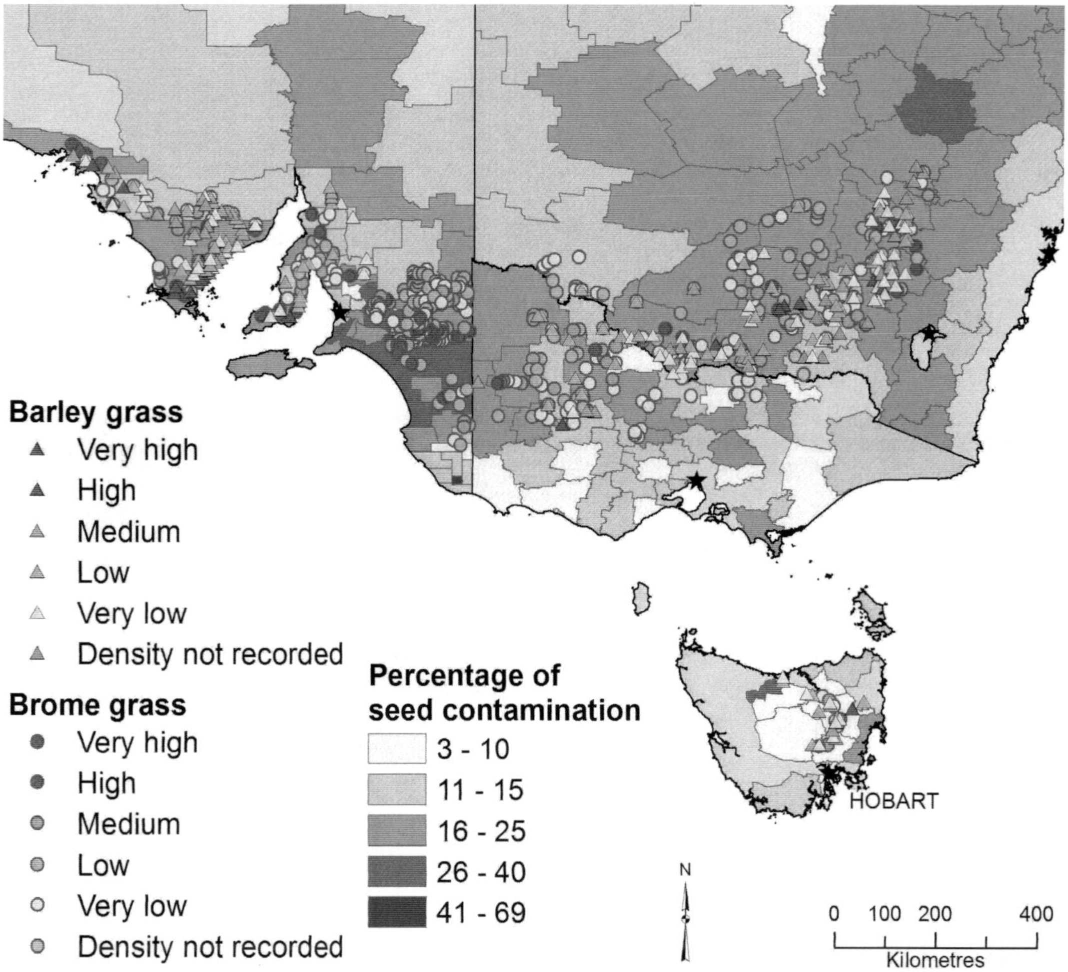

Figure 2. Regional distribution within NSW, VIC, TAS and SA of total density of carcass contamination combined with distribution of barley grass and brome grass populations across southeastern Australia between 2007 and 2015.

DISCUSSION

Key regional differences noted in WA, SA, VIC and NSW revealed that the highest incidence of contamination occurred within discrete regions within each agro-ecological zone. Widespread seed contamination occurred throughout Australia's pastoral zone and mixed farming zone where grain cropping and sheep production are typically concentrated (Puckridge and French, 1983). Barley grass and brome grass are annual weeds associated with carcass damage and are also common in pastures and grain crops in southern mixed farming regions (Broster *et al.* 2012). The distribution patterns of weed infestation and carcass contamination observed in this study are also in agreement with the findings of Llewellyn *et al.* (2016), who identified barley grass and brome grass as two of the most problematic weeds of grain crops in southern and western Australia. Given the strong integration of grain cropping with sheep production in the mixed farming zone, the potential for carcass contamination by contact with barley grass and brome grass is high. Consequently, it is not surprising that the distribution of weed infestation was spatially associated with regions displaying higher weed seed contamination, highlighting the significance of barley grass and brome grass in causing carcass damage in Australia.

The lack of differences in seed contamination between NSW and QLD and between TAS and VIC may be associated with early seed contamination of QLD and TAS re-stocker animals before they are transported (East and Foreman 2011) and identified as NSW and VIC sheep upon slaughter. Additional experimentation to determine specifically when and where highest rates of contamination occur in young stock is required to draw meaningful conclusions about seed source populations.

High contamination levels associated with sex and age were likely due to the length of exposure of sheep to infested sites producing ample weed seed. Males and females over two years of age and entire males sold for slaughter (normally cast for age rams), typically experience longer periods of seed exposure. Animals under two years of age commonly comprise castrated males and females sold for slaughter into the lamb market before considerable seed exposure. Variable contamination in all other groups was likely associated with variable age of sheep comprising each group and the associated variation in the length of exposure to mature weeds producing seed.

Increased weed seed contamination in sheep by barley grass and brome grass is potentially associated with changes in rainfall patterns and farming practices in recent years. Increased variation in rainfall, as evidenced by drier autumns and prolonged periods of drought, coupled with warmer temperatures and higher soil moisture evaporation rates are experienced across southern Australia (Murphy and Timbal 2008) and may limit growth of desirable pasture species. Weed populations displaying variable dormancy patterns, drought tolerance and rapid germination, are likely to be highly competitive in such conditions. Barley grass and brome grass also potentially utilise such mechanisms to enhance survival under the climatic stresses of southern Australia (Rumball 1971, Gill and Blacklow 1985).

In addition, the adoption of reduced tillage, strong reliance on herbicides and earlier sowing of grain crops have also likely selected for barley grass and brome grass biotypes now displaying herbicide resistance and variable seed dormancy (Gill and Blacklow 1985, Fleet and Gill 2012, Owen *et al.* 2015). High incidence of carcass contamination in the pastoral and high rainfall regions is also possibly due to seed dispersal from mixed farming regions via attachment of seed to fleece during animal movement and transport.

Anatomically, barley grass and brome grass seed possess long awns, enabling them to easily adhere to fleece and later penetrate dermal tissues. Specialty lamb producers and traders in the high rainfall zone and properties within the pastoral zone also often introduce large numbers of sheep from adjacent mixed farming regions (East and Foreman 2011) that are possibly already contaminated with seed upon arrival. Both weed species display significant phenotypic plasticity across Australia, and readily germinate and establish across all regions. Higher infestation rates in the high rainfall zones may be assisted by overgrazing, reduced soil fertility and the use of less competitive pasture legumes or legume mixtures.

Focused chemical and rotational strategies to manage these weeds could be employed in areas of high weed infestation to result in reduced infestation to carcasses over time. Future research into the development of focused IWM strategies will be essential for regional control of these weeds and their biotypes, permitting a more targeted approach in reducing seed contamination in Australia.

ACKNOWLEDGMENTS

We acknowledge the support of the AW Howard Memorial Trust, Meat and Livestock Australia (Project B.WEE 0416), the Grains Research and Development Corporation (Project UCS 00020) and the Graham Centre for Agricultural Innovation (Charles Sturt University and NSW Department of Primary Industries). We are also grateful to Wool Producers Australia and the Sheepmeat Council of Australia through Animal Health Australia and Dr Peter Boutsalis of the

University of Adelaide for data provision and to the Spatial Data Analysis Network at Charles Sturt University for map development.

REFERENCES

Broster, J., Koetz, E. and Wu, H. (2012). Weed species present in cereal crops in southern New South Wales, Proceedings of 18th Australasian Weeds Conference, ed. V. Eldershaw, pp. 241-2. (Weed Society of Victoria, Melbourne).

Collins, A. (2013). 'Winning against seeds: resources review and update'. pp. 1-23. (Meat and Livestock Australia Ltd, North Sydney).

East, I. and Foreman, I. (2011). The structure, dynamics and movement patterns of the Australian sheep industry. *Australian Veterinary Journal,* 89, 477-89.

ESRI. (2015). ArcGIS 10.4 for Desktop. (Environmental Systems Research Institute, Redlands, CA).

Fleet, B. and Gill, G. (2012). Seed dormancy and seedling recruitment in smooth barley (*Hordeum murinum* ssp. *glaucum*) populations in southern Australia. *Weed Science,* 60, 394-400.

Gill, G. and Blacklow, W. (1985). Variations in seed dormancy and rates of development of great brome, *Bromus diandrus* Roth., as adaptations to the climates of southern Australia and implications for weed control. *Australian Journal of Agricultural Research,* 36, 295-304.

Llewellyn, R., D'Emden, F., Owen, M. and Powles, S. (2009). Herbicide resistance in rigid ryegrass (*Lolium rigidum*) has not led to higher weed densities in Western Australian cropping fields. *Weed Science,* 57, 61-5.

Llewellyn, R., Ronning, M., Ouzman, J., Walker, S., Mayfield, A. and Clarke, M. (2016). 'Impact of weeds on Australian grain production – the cost of weeds to Australian grain growers and the adoption of weed management and tillage practices. Report for GRDC'. (CSIRO, Australia).

Murphy, B.F. and Timbal, B. (2008). A review of recent climate variability and climate change in southeastern Australia. *International Journal of Climatology,* 28, 859-79.

Owen, M.J., Martinez, N.J. and Powles, S.B. (2015). Herbicide resistance in *Bromus* and *Hordeum* spp. in the Western Australian grain belt. *Crop and Pasture Science* 66, 466-73.

Puckridge, D.W. and French, R.J. (1983). The annual legume pasture in cereal—ley farming systems of southern Australia: a review. *Agriculture, Ecosystems & Environment* 9, 229-67.

Rumball, P.J. (1971). The ecology and control of barley grass: a review. Proceedings of the 24th New Zealand Weed and Pest Control Conference, ed. L.J. Matthews, pp. 63-9. (The New Zealand Weed and Pest Control Society Inc.).

Wind assisted dispersal of seeds not adapted for flight

Jillian K. Hetherington
Department of Geography, University of Otago, PO Box 56, Dunedin, New Zealand
(jillian.hetherington@otago.ac.nz)

Summary Fortuitous dispersal of invasive plant seeds over a longer than 'normal' distance can be catastrophic for management programmes. It is often assumed that only seeds with morphological adaptations for flight are capable of distance dispersal and those not adapted for such will disperse in close proximity to the parent plant. However, the mapped distribution of an invasive legume, *Lupinus arboreus* Sims, on a dune system in New Zealand suggested the prevailing wind was facilitating sporadic distance dispersal. The role of wind in phase I and II dispersal of these non-wind adapted seeds was examined. Phase I was examined through the placement of seed traps around three mature individual plants; wind speed and direction was recorded on the dunes during the period of seed expulsion. Phase II, rolling, was examined in a wind tunnel where seeds of known weight and sphericity were placed on one of five surfaces, and the wind speed that initiated rolling was recorded.

More than 80% of the seed dispersed at phase I accumulated within half a metre of the canopy edge, less than 30% of the seed produced was expelled via phase I with no strong evidence of wind assisted dispersal. Transportation of seed via rolling occurred on all surfaces, but the wind speed to initiate rolling increased with increasing surface roughness, and the presence of obstacles prevented rolling. Perpendicular orientation of the seed to the wind increased the likelihood of phase II dispersal, and sphericity appeared to be more important than seed weight. Under the right conditions seed not adapted for flight can travel longer distances.

Keywords Wind dispersal, non-wind adapted seeds, sphericity, surface roughness.

Can herbicides affect seed dormancy and viability of flaxleaf fleabane (*Conyza bonariensis* (L.) Cronquist)

Hanwen Wu[1], Adam Shepherd[1], Md Asaduzzaman[2] and John Broster[2]
[1] Graham Centre for Agricultural Innovation (NSW Department of Primary Industries and Charles Sturt University), Invasive Plants and Animals, Wagga Wagga, New South Wales 2650, Australia
[2] Graham Centre for Agricultural Innovation (NSW Department of Primary Industries and Charles Sturt University), Wagga Wagga, New South Wales 2678, Australia
(hanwen.wu@dpi.nsw.gov.au)

Summary Flaxleaf fleabane (*Conyza bonariensis* (L.) Cronquist) has been considered a major weed in dryland cropping systems in Australia. Resistance in populations of *C. bonariensis* to glyphosate has also been confirmed. The weed is very difficult to control with any single herbicide application especially when herbicides are applied after the bolting stage. Anecdotal evidence suggests that some chemicals might affect fleabane seed viability and/or dormancy based on the observations that seeds collected from certain paddocks did not germinate. Glasshouse experiments were conducted to investigate whether applications of glyphosate and paraquat could sterilise the weed seeds and affect seed dormancy of fleabane. Results showed that both herbicides affected the seed viability and dormancy, depending on the application timing and the herbicide. The herbicide-induced dormancy decreased during storage under laboratory conditions.

Keywords Seed dormancy, viability, herbicides, germination.

INTRODUCTION

Conyza bonariensis (flaxleaf or hairy fleabane) is a cosmopolitan plant of temperate and sub-tropical climates, native to South America and a member of the Asteraceae (Everett 1990, Prieur-Richard *et al.* 2000). It invades a range of habitats including roadsides, wastelands, fencelines, fallows, cropping and pasture paddocks.

Research efforts have been concentrated on improved herbicide control efficiency particularly in cereal crops and fallows (Wu *et al.* 2008, 2010, Werth *et al.* 2010). It has been found that chemical control is more effective at the early rosette stage, and that control efficacy decreases with increasing plant age, especially after the bolting stage (Wu *et al.* 2007, Walker *et al.* 2012). A 'double-knock' technique, glyphosate + 2,4-D followed by paraquat ± diquat, has been widely adopted by growers to effectively control mature fleabane plants.

Flaxleaf fleabane has evolved resistance to glyphosate, with more than 60 resistant populations of fleabane confirmed (Walker *et al.* 2011, Preston 2016). When evaluating fleabane seed samples collected from the field for their glyphosate resistance status, it was found that some of these seed samples have little to no germination under ideal laboratory conditions (Wu, unpublished data), although Wu *et al.* (2007) has previously reported that fleabane seeds have no dormancy requirement for germination. It is suspected that heavy reliance on herbicides to control fleabane in the summer might have resulted in the loss of seed viability and/or induced dormancy. Herbicidal effects on seed viability have been previously documented in many weeds, including fleabane (Keenan *et al.* 2014). However, little information is available on the herbicidal impact on seed dormancy.

Seed dormancy refers to a physical and/or physiological block to the germination of viable seeds when they are imbibed under ideal germination conditions (Finch-Savage *et al.* 2006). It is a survival mechanism which contributes to the persistence of certain weed species, such as *Lolium rigidum* Gaudin (Owen *et al.* 2015). Seed dormancy is a highly complex trait, regulated by both genetic and environmental factors. Many environmental factors have strong influences on dormancy, especially during seed development and maturation, such as temperature, soil moisture, light quality and photoperiod (Gutterman *et al.* 2000, Swain *et al.* 2006). Farming practices such as nitrogen input could also induce dormancy (Luzuriaga *et al.* 2006). It has been documented that intensive cropping systems with heavy reliance of herbicides have altered weed seed biological traits, selecting for greater seed dormancy in such species as smooth barley (*Hordeum murinum* L. subsp. *glaucum* (Steud.) Tzvelev), ripgut brome (*Bromus diandrus* Roth) and *L. rigidum* (Fleet and Gill 2012, Kleemann and Gurjeet 2013, Owen *et al.* 2015). However, there is little information available about the direct linkages of herbicide applications with dormancy induction.

Glasshouse trials were established in 2014 to evaluate the impact of two most commonly used herbicides, glyphosate and paraquat, on fleabane seed viability and dormancy. Flaxleaf fleabane plants were sprayed at two growth stages (early budding and late budding). Mature seeds were collected from main stems as well as subsequent branches for determination of germination, dormancy and seed viability.

MATERIAL AND METHODS

Fleabane seed collection and sowing Mature flax-leaf fleabane seeds were collected in summer 2013/14 near Wagga Wagga Agricultural Institute (WWAI), New South Wales, Australia. Plastic pots (20 cm in diameter and 18 cm in depth) were filled with field soil. The pots were irrigated to field capacity and two hundred mature seeds of fleabane were sown in the pots on 10 June 2014 in a glasshouse. After sowing, seeds were slightly harrowed to mix with the surface soil. The pots were then covered with a piece of tissue towel to maintain moist conditions for four days to encourage emergence. After the emergence, the pots were maintained with automatic irrigation on daily basis. Seedlings were manually thinned to two plants per pot at rosette stage.

Herbicide applications Herbicides were applied at reduced rates in an attempt not to kill the plants, with glyphosate applied at 360 g a.i ha^{-1} and paraquat at 100 g a.i ha^{-1}. The two herbicides were applied at two different growth stages of fleabane plants: at early budding (small visible buds just formed on the main stem) and late budding (large buds just about to open or slightly opened on the main stem) on 17 September and 11 October 2014, respectively. For both growth stages, the herbicides were applied using an automated laboratory sized cabinet sprayer with a moving boom applying a water volume of 77 L ha^{-1} equivalent from a flat fan nozzle at 300 kPa pressure. Pots without herbicide applications were included as controls.

Seed collection after herbicide treatments After the herbicide applications, mature seeds from the treated fleabane plants were collected between November 2014 and February 2015. Two sources of seeds were collected, seeds from the main stem (young buds directly exposed to herbicide application) and seeds from healthy branches (indirect effect from translocated herbicides). At the time of herbicide applications, the branches had not formed the buds yet. After the seed collections from the main stems and branches, the mature plants were then cut at the base (10 cm above the soil surface), on 20 January 2015 for the early budding (EB) treatments, and on 2 February 2015 for the late budding (LB) treatments, to assess residual herbicidal impacts on seeds produced on re-grown stems. The seeds produced from the re-grown stems were collected in March and April 2015, respectively. The collected seeds were tested for viability by both germination and tetrazolium tests.

Germination and viability tests For germination bioassay, one hundred of seeds from each treatment were placed onto 9 cm Petri dishes lined with one layer of filter paper (No. 1 Whatman). Five mL of distilled water was delivered to each Petri dish. Each Petri dish with its cover was sealed with a piece of Parafilm to reduce evaporation and was placed in growth incubators at 15°C/25°C on a 12 hour day/night cycle. After seven days of incubation, the germinated seeds were counted based on the emergence of cotyledons. Seeds that did not germinate were further assessed for viability by a tetrazolium test (Stanton *et al.* 2012). The tetrazolium [2,3-5-triphenyltetrazolium chloride (TTC) solution was added into a new Petri dish and 20 non-germinated seeds were left in contact with the solution for two days at room temperature. After this time, seeds which turned red in colour were classed as viable but dormant. The embryos that did not stain red were considered unviable. Total viable seeds include the germinated seeds and the 'viable but dormant' seeds.

Design and statistical analysis The experiments were designed as a randomised complete block with three replications. ANOVA was conducted using Genstat.

RESULTS AND DISCUSSION

Effects of herbicides on seed viability and dormancy of the maternal fleabane plants Data analysis confirmed significant ($P < 0.001$) differences in seed viability of fleabane in response to herbicides applied at the EB and the LB stages. Application of glyphosate at 360 g a.i ha^{-1} at the EB stage completely stopped the seedset on the main stem, while paraquat at 100 g a.i ha^{-1} did not affect fleabane seed production and viability on the main stem (Table 1). Glyphosate application at the EB stage desiccated the premature buds.

Glyphosate significantly affected the germination and dormancy of seeds collected from the earliest branches on the 10 December 2014, although the total seed viability was not affected as compared to the untreated control. The germination and dormancy was 9.3% and 69.8% for the glyphosate treatment, compared to 6.1% and 77.7% for the untreated control, respectively. The impact of glyphosate on seed germination and dormancy decreased over time when fleabane seeds were collected from later branches on or after 24 December 2014.

Table 1. Impact of herbicides on seed production and dormancy in *Conyza bonariensis*.

Treatment	Date of collection	Seed source	Germinable	Viable but dormant	Total viable
EARLY budding					
Glyphoste	20/11/14	Main stem	NA*	NA	NA
Glyphoste	10/12/14	Branch	9.3	69.8	79.1
Glyphoste	24/12/14	Branch	73.0	20.2	93.2
Glyphoste	2/01/15	Branch	73.3	17.0	90.3
Glyphoste	20/01/15	Branch	78.3	14.2	92.5
Paraquat	20/11/14	Main stem	61.0	23.7	84.7
Paraquat	2/01/15	Branch	49.0	14.1	63.1
Paraquat	20/01/15	Branch	65.0	9.3	74.3
Control	20/11/14	Main stem	66.7	8.6	75.3
Control	10/12/14	Branch	77.7	6.1	83.7
Control	24/12/14	Branch	73.3	6.7	80.0
Control	2/01/15	Branch	82.3	3.6	85.9
Control	20/01/15	Branch	80.7	3.7	84.4
$LSD_{0.05}$			13.5	5.6	15.5
LATE budding					
Glyphosate	19/11/14	Main stem	34.0	26.8	60.8
Glyphosate	23/01/15	Branch	30.0	18.6	48.6
Glyphosate	2/02/15	Branch	68.0	10.7	78.7
Paraquat	19/11/14	Main stem	28.0	30.7	58.7
Paraquat	23/01/15	Branch	56.3	19.7	76.0
Paraquat	2/02/15	Branch	45.3	10.7	56.1
Control	19/11/14	Main stem	77.7	4.6	82.2
Control	23/01/15	Branch	80.7	3.7	84.4
Control	2/02/15	Branch	76.3	5.2	81.5
$LSD_{0.05}$			14.4	6.7	13.2

*Glyphosate application at the early budding stage completely killed the premature buds.

The contact herbicide paraquat at 100 g a.i. ha^{-1} applied at the EB stage had little desiccation effect on the premature buds and it only desiccated some leaves. The plant recovered completely and the young buds continued to progress and set seeds, resulting in no differences in seed viability compared to the untreated control. However, paraquat-treated seeds on the main stem were more dormant (23.7%) than the untreated control (8.6%). The impact of paraquat on dormancy also declined as the seed collection time progressed from 20 November 2014 to January 2015. When collected on 20 January 2015, its impact on dormancy dropped to 9.3%., which is not significantly different to the untreated control (3.7%) for the same period.

Herbicide applications at LB stage did not stop the growth of fleabane plants which continued to develop, flower and set seeds. However, the LB application of glyphosate and paraquat also had significant effects on seed viability and dormancy. Both herbicides reduced seed viability on the main stem by 26–29% but increased the dormancy by 5–6 times when compared with the untreated control (Table 1). The herbicides also induced the dormancy of seeds collected from branches. However, the herbicidal impact on dormancy decreased as the seed collection progressed. These results suggest that fleabane seed dormancy could be induced by herbicides, depending on the herbicide and the application timing. This finding is in contrast with the previous report that seeds did not possess innate dormancy (Wu *et al.* 2007). The discrepancy could be partly due to the seed used by Wu *et al.* (2007) was not from fleabane plants after herbicide application. In addition, the seed has been stored for a few months prior to the experiments. Any innate dormancy might have been lost during storage as shown in the current study.

Dormancy release after ripening The induced dormancy due to herbicides underwent a steady decline during storage (Figure 1). The dormant seeds (%) decreased significantly from the initial 26.8% in the glyphosate treatment to 2.3% after seven months of storage under room temperature. The significant dormancy release in the paraquat treatment also ranged from 30.7% to 0.3%, while there was no difference in the dormancy status in the control treatments, with an average of 3.6%.

Carry-over effects of herbicides on seed viability and dormancy of the re-grown fleabane plants after cutting Glyphosate and paraquat did not have any significant effects at either EB or LB stage on viable-but-dormant seeds collected from re-grown plants after cutting, when compared with the untreated control (Table 2). However, glyphosate applied at the EB stage and paraquat applied at the LB stage significantly reduced the germinable seeds and the total viable seeds.

The impact of both herbicides on seed viability is in general accordance with previous research. Carrithers *et al.* (2004) found that picloram applied at bud stage reduced seed production of yellow starthistle (*Centaurea solstitialis* L.) by 42 to 86% and viability by 80 to 99%. Application of picloram, which caused both bud abortion and lower seed germination, was more effective at the bud stage than the flowering stage in reducing viable seed. Jha and Norsworthy (2012) found that a range of herbicides applied at the first visible sign of inflorescence reduced seed production of glyphosate resistant Palmer amaranth (*Amaranthus palmeri*) by 75 to 95% depending on the biotypes. The herbicides also reduced viability, with seeds produced by treated plants being 45 to 61% viable, compared with 97% seed viability in non-treated plants. Keenan *et al.* (2014) also reported that a range of herbicides can reduce seed viability of fleabane by up to 100%, depending on the application timing and the herbicide.

These studies demonstrate that herbicides could potentially reduce seed production and viability of many agricultural weeds, thereby reducing seedbank replenishment and subsequent impact in the following season. However, on the other hand, herbicide applications could also induce seed dormancy, which could prolong the seed persistence in the field, adding another layer of complexity to seedbank management. The herbicide-induced dormancy reported in the present study could contribute to the altered seed dormancy found in *H. murinum*, *B. diandrus* and *L. rigidum* in intensive cropping systems which relied heavily on herbicide weed control (Fleet and Gill 2012, Kleemann and Gurjeet 2013, Owen *et al.* 2015).

More research is needed to determine if herbicides applied at seed maturation will have broad effects on seed dormancy in other weeds. Further studies on the physiological and molecular responses to herbicidal applications might unravel the mechanisms for this dormancy induction.

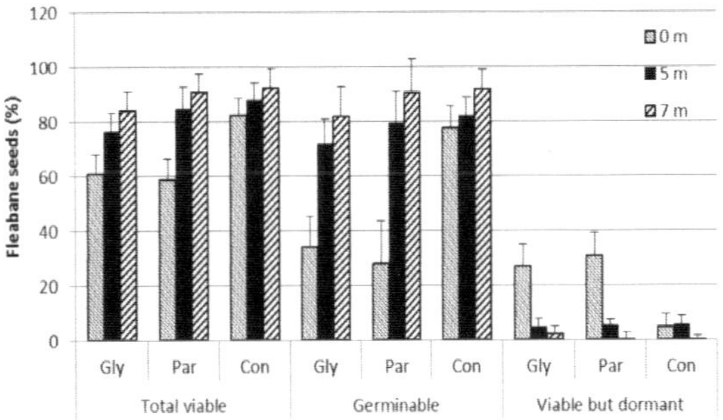

Figure 1. Reduction in fleabane dormancy after ripening.

Table 2. Herbicide carry-over impacts on viability and dormancy of seeds collected on 13 April 2015 from the re-grown fleabane plants after cutting.

Herbicide	Germinable seeds (%)	Viable but dormant seeds (%)	Total viable seeds (%)
EARLY budding			
Glyphosate	25.3a	8.8a	34.1a
Paraquat	95.0b	1.3a	96.3b
Control	69.3c	11.2a	80.5b
LATE budding			
Glyphosate	77.3c	8.9a	86.2b
Paraquat	54.7d	14.4a	69.0c
Control	74.0c	11.6a	85.6b

ACKNOWLEDGMENTS

The authors gratefully acknowledge the financial support of Grains Research and Development Corporation for funding this research.

REFERENCES

Carrithers, V.F., Roché, C.T., Gaiser, D.R., Horton, D., Duncan, C.L. and Scherer, P.N. (2004). Herbicides reduce seed production in reproductive-stage yellow Starthistle (*Centaurea solstitialis*). *Weed Technology* 18, 1065-71.

Everett, J. (1992). *Conyza*. *In* 'Flora of New South Wal*es'*. Ed. G.J. Harden, Volume 3, pp. 197-200. (New South Wales University Press, Sydney).

Finch-Savage, W.E. and Leubner-Metzger, G. (2006). Seed dormancy and the control of germination. *New Phytologist* 171, 501-23.

Fleet, B. and Gill, G. (2012). Seed dormancy and seedling recruitment in smooth barley (*Hordeum murinum* ssp. *glaucum*) populations in southern Australia. *Weed Science* 60, 394-400.

Jha, P. and Norsworthy, J.K. (2012). Influence of late-season herbicide applications on control, fecundity, and progeny fitness of glyphosate-resistant Palmer Amaranth (*Amaranthus palmeri*) biotypes from Arkansas. *Weed Technology* 26, 807-12.

Keenan, M.D., Werth, J., Thornby, D. and Walker, S. (2014). Reducing seed viability of flaxleaf fleabane, feathertop Rhodes grass and awnless barnyard grass. Proceeding of the Nineteenth Australasian Weeds Conference, pp. 15-19. 31 August to 3 September 2014, Hobart, Tasmania.

Kleemann, S.G. and Gurjeet, G. (2013). Seed dormancy and seedling emergence in ripgut brome (*Bromus diandrus*) populations in Southern Australia. *Weed Science* 61, 222-9.

Owen, M.J., Goggin, D.E. and Powles, S.B. (2015). Intensive cropping systems select for greater seed dormancy and increased herbicide resistance levels in *Lolium rigidum*. *Pest Management Science* 71, 966-71.

Preston, C. (2016). The Australian Glyphosate Sustainability Working Group. Online. Internet. Monday 13 Jun 2016. Available http://www.glyphosateresistance.org.au/

Prieur-Richard, A.H., Lavorel, S., Grigulis, K. and Dos Santos, A. (2000). Plant community diversity and invasibility by exotics, invasion of Mediterranean old fields by *Conyza bonariensis* and *Conyza canadensis*. *Ecology Letters* 3, 412-22.

Stanton, R., Wu, H. and Lemerle, D. (2012). Factors affecting silverleaf nightshade (*Solanum elaeagnifolium*) germination. *Weed Science* 60, 42-7.

Swain, A.J., Huges, Z.S., Cook, S.K. and Moss, S.R. (2006). Quantifying the dormancy of *Alopecurus myosuroides* seeds produced by plants exposed to different soil moisture and temperature regimes. *Weed Research* 46, 470-9.

Walker, S., Bell, K., Robinson, G. and Widderick, M. (2011). Flaxleaf fleabane (*Conyza bonariensis*) populations have developed glyphosate resistance in north-east Australian cropping fields. *Crop Protection* 30, 311-17.

Walker, S., Boucher, L., Cook, T., Davidson, B., McLean, A. and Widderick, M. (2012). Weed age affects chemical control of *Conyza bonariensis* in fallows. *Crop Protection* 38, 15-20.

Werth, J.A., Walker, S.R., Boucher, L.R. and Robinson, G.R. (2010). Applying the double knock technique to control flaxleaf fleabane (*Conyza bonariensis*). *Weed Biology and Management* 10, 1-8.

Wu, H., Walker, S., Rollin, M.J., Tan, D.K.Y. and Werth, G. (2007). Germination, persistence and emergence of flaxleaf fleabane (*Conyza bonariensis* L. Cronq.). *Weed Biology and Management* 7, 192-9.

Wu, H., Walker, S. and Robinson, G. (2008). Control of flaxleaf fleabane (*Conyza Bonariensis* L. Cronq.) in winter fallows. *Plant Protection Quarterly* 23, 162-5.

Wu, H., Walker, S. and Robinson, G. (2010). Control of flaxleaf fleabane (*Conyza Bonariensis* L. Cronq.) in wheat and sorghum. *Weed Technology* 24, 102-7.

A stinking problem: novel control solutions and threat mitigation for invasive *Passiflora foetida* in the Kimberley and Pilbara

Bruce L. Webber[1,2], Paul B. Yeoh[1], Jana Z. Coletti[3], Raphael K. Didham[1,4], Vicki Long[5], David Pozzari[6], John Tomich[7], John K. Scott[1,4], Ruchira Somaweera[1], Tracy Sonneman[8] and Stephen J. van Leeuwen[9]
[1] CSIRO Land and Water, Private Bag 5, PO Wembley, Western Australia 6913, Australia
[2] School of Plant Biology, The University of Western Australia, 35 Stirling Highway, Crawley, Western Australia 6009, Australia
[3] School of Earth and Environment, The University of Western Australia, 35 Stirling Highway, Crawley, Western Australia 6009, Australia
[4] School of Animal Biology, The University of Western Australia, 35 Stirling Highway, Crawley, Western Australia 6009, Australia
[5] Vicki Long and Associates PO Box 713, Karratha, Western Australia 6714, Australia
[6] Pilbara Ports Authority, Mof Road, Burrup Peninsula, Dampier, Western Australia 6713, Australia
[7] Mount Gibson Iron, 2 Kings Park Road, West Perth, Western Australia 6005, Australia
[8] Western Australia Department of Parks and Wildlife, 111 Herbert Street, Broome, Western Australia 6725, Australia
[9] Western Australia Department of Parks and Wildlife, Dick Perry Avenue, Kensington, Western Australia 6151, Australia
(bruce.webber@csiro.au)

Summary Stinking passionflower (*Passiflora foetida* L.) is a vine from South and Central America that has been widely introduced into many tropical regions of the world. Across the Kimberley and Pilbara regions of north western Australia, it is considered to be one of the most significant weed problems in landscapes of high biodiversity, cultural and historical value.

Current control methods (spraying, hand-pulling) are costly (particularly in remote areas), labour intensive, and have minimal impact on long-term control. A major impediment for developing effective control strategies for stinking passionflower is that very little is known about its biology and life history. In the long term, however, the only practical solution will be implementing a biological control solution.

Our research program represents the beginning of that process. Here we document our findings to this point, providing examples of (i) the population dynamics of the weed across the Kimberley and Pilbara, (ii) its impact on high value landscapes from a biodiversity, cultural and historical perspective, and (iii) the progress towards a biological control solution to improve the efficiency and effectiveness of management.

The insights obtained support the idea that stinking passionflower represents a significant threat to iconic ecosystem values in the Kimberley and Pilbara and that this plant remains a high priority target for management via biological control.

Keywords Stinking passionflower, weed, management, impact, ecosystem values, biodiversity, biological control.

Recovery of the endangered rainforest herb *Isoglossa eranthemoides* following weed management in northeast New South Wales, Australia

Barbara Stewart and Annette McKinley
Landmark Ecological Services Pty. Ltd., PO Box 100, Suffolk Park, New South Wales 2481, Australia
(barbara.stewart@landmarkonline.com.au)

Summary Weeds are recognised as major threats to native flora species and vegetation communities, but instances of extinction resulting from weed competition are rarely documented. We described weed management history for Booyong Flora and Fauna Reserve, a remnant of subtropical rainforest in the northeast New South Wales Big Scrub region. We then estimated population size for the rainforest herb Isoglossa (*Isoglossa eranthemoides* (F.Müell.) R.M.Barker) (Endangered, NSW and Commonwealth) using randomly located plot samples. Although much history and baseline data is informal, we present convincing evidence of recovery of Isoglossa following weed management, in particular the control of the ground storey weed *Tradescantia fluminensis* Vell. Weed management has been staged since 1997, with primary control now almost complete and maintenance ongoing.

In 13 ha of habitat, the Isoglossa population was once considered to number about 50 individuals (1993–94), increasing to more than 3000 by 2000. In January 2016 we estimated the population at more than 150,000, approximately double our 2014 estimate.

As is the case for all small remnants, weed reinfestation is inevitable and, in this floodplain location, weed propagules will periodically be water-dispersed throughout the remnant. The recovery of Isoglossa is likely to quickly reverse unless weed management is maintained.

From this example, we conclude that weed management over about 20 years in Big Scrub rainforest remnants is likely to have prevented local extinctions of Isoglossa. Recovery actions for Isoglossa will also benefit the regeneration of other threatened rainforest species and ecological communities.

Keywords *Isoglossa eranthemoides, Tradescantia fluminensis*, recovery, weed, competition, extinction.

The landscape ecology of para grass (*Urochloa mutica*) on a monsoonal wetland of Kakadu National Park, NT Australia

James M. Boyden[1], Penny Wurm[1], Guy Boggs[1,2], Karen E. Joyce[3] and Peter Bayliss[4]
[1] Charles Darwin University, School of Environment, PO Box 40146, Casuarina, Northern Territory 0811, Australia
[2] Wheatbelt Natural Resource Management Northam, Western Australia 6401, Australia
[3] James Cook University, College of Science, Technology and Engineering, Cairns, Queensland 4870, Australia
[4] CSIRO Oceans and Atmosphere Flagship, 306 Carmody Road, St Lucia, Brisbane, Queensland 4072, Australia
(jamesdwn@internode.on.net)

Summary This study applied remote sensing, ground data and historical maps to monitor and understand the dynamics of para grass (*Urochloa mutica* (Forssk.) T.Q.Nguyen) spread in relation to key habitat features (water depth, fire history and native vegetation) on the 225 km^2 Magela Creek Floodplain within Kakadu National Park. Maps of these features were produced from Medium- and High-Spatial-Resolution (HSR) satellite images (Landsat 5 TM; and Ikonos, Quickbird or Worldview 2).

A trend of increasing para grass cover and patch connectivity was apparent from maps produced from HSR images for a selected 42 km^2 area, captured biennially from 2001 to 2010. Yet spatial variability, between-years, was high. An uncharacteristic net decline in para grass was also measured for 2004. The average annual increase in para grass was estimated by linear regression to be 55.4 ha when data were included with historical para grass maps for 1992 and 1998, with the exception of the 2004 outlier (R^2 0.96, $p = 0.0003$, $n = 6$). Para grass occupied 1308 ha of the total floodplain area as mapped by Landsat in 2006 and, compared to a map of the floodplain for 1986, had primarily displaced native *Oryza* and *Hymenachne* grassland.

As determined by two-way analysis of variance, the spatial dynamics of para grass was strongly associated with water-depth (i.e. shallow, moderate and deep categories: ≥1.15 m, >1.15 to <1.45 m, and ≥1.45 m) and fire in the previous dry season ($F_2 = 370$, $p < 0.0001$). For example, in flood-years following fire, para grass cover declined in the 'shallow' and 'moderate' depth habitats, but increased in 'deep' habitat compared to unburnt areas.

Medium- and HSR remote sensing provide essential information for applied research and adaptive management of site- and time-specific management strategies for para grass. This information is also likely to improve the sensitivity of spread prediction models for para grass as defined by vegetation type, ecohydrology and fire; and be useful for development of weed management strategies on monsoonal wetlands, generally.

Keywords *Urochloa mutica*, para grass, remote sensing, monsoonal wetlands, ecohydrology, habitat mapping, conservation management.

Pairing biological control species: cochineal insects *Dactylopius* (Hemiptera: Dactylopiidae) with prickly pears *Opuntia* (Cactaceae) within the South Australian Adelaide and Mount Lofty Ranges

Henry C. Rutherford
South Australian Department for Environment, Water and Natural Resources, Wittunga House, 328 Shepherds Hill Road, Blackwood, South Australia 5051, Australia
(henry.rutherford3@sa.gov.au)

Summary South Australia successfully led the 2010 nomination of the opuntioid group of cacti as Weeds of National Significance (WoNS). The National Co-ordinator Program that followed identified biological control programs that had been applied early last century in Queensland and New South Wales as key examples of how this weed could again be managed. In 2013 funding for the WoNS National Coordinator Program ceased. The South Australian Adelaide and Mount Lofty Ranges (AMLR) Natural Resources Region continued the work, conducting a series of rudimentary oviposition preference trials. Naturalised populations of cochineal insects, *Dactylopius opuntiae* Cockerell (Hemiptera: Dactylopiidae) and *Dactylopius ceylonicus* Green (Hemiptera: Dactylopiidae), were tested as biological control agents on a range of *Opuntia* spp. The susceptibility of nine species of *Opuntia* was observed following inoculation with two discrete populations of *D. opuntiae,* and one of *D. ceylonicus*. The observations were used to design the establishment of 15 nursery sites on the nine *Opuntia* targets throughout the AMLR Region, using both *D. opuntiae* and *D. ceylonicus*. Nursery sites were strategically located in easily accessible public locations to form the basis for a future expansion of the distribution program. The next step in the program is to develop a community-driven mechanism to access and translocate the agents across the wider landscape.

Keywords Weeds of National Significance (WoNS), cochineal insects, *Dactylopius*, *Opuntia*.

Improving the effectiveness of the release strategy for the array of *Dactylopius tomentosus* biotypes for the biocontrol of *Cylindropuntia* species by using DNA analysis

Peter K. Jones[1], Jane Oakey[2], Michael D. Day[1] and Joseph S. Vitelli[1]
[1] Queensland Department of Agriculture and Fisheries, Ecosciences Precinct GPO Box 267, Brisbane, Queensland 4001, Australia
[2] Queensland Department of Agriculture and Fisheries, Health and Food Sciences Precinct, Box 156, Archerfield BC, Queensland 4108, Australia
(peter.jones@daff.qld.gov.au)

Summary Six biotypes of *Dactylopius tomentosus* (Lamarck) (Hemiptera: Dactylopiidae) have been imported into Australia as biological control agents to combat the eight naturalised invasive *Cylindropuntia* species. Each *D. tomentosus* biotype tested exhibited significant differences in host specificity, host preference and performance across the *Cylindropuntia* species. Biocontrol success is dependent on the correct identification of the targeted plant species and the correct matching of an insect or pathogen to that plant. An optimal release strategy is to match the most effective biotype to the targeted *Cylindropuntia* species. Biotype matching to *Cylindropuntia* species was enhanced through the use of molecular phylogenetic tools and more thorough morphological observations of the cacti species. Molecular studies demonstrated that using two plastid markers (Rpl36-Rps8 spacer region and TrnH-PsbA spacer) and a nuclear region (phosphoenolpyruvate carboxylase) allowed for differentiation of the eight *Cylindropuntia* species. Molecular analyses using three markers enable the identification of unknown *Cylindropuntia* species when species identification is not possible using traditional phenological and morphological tools. The genetic identification of the *Cylindropuntia* provides a time efficient match of target host species and biotype within days rather than the traditional long term host trials (three month life cycle) previously required. Utilising DNA analyses will markedly improve the chance of containment of *Cylindropuntia* infestations in Australia through better matching of biocontrol agents with their most preferred host.

Keywords Host specificity, biotypes, molecular analysis, biocontrol.

INTRODUCTION

The genus *Cylindropuntia* is native to southern USA and Mexico and comprises 33 species, eight of which have naturalised in Australia. These eight species: *C. fulgida* var. *mamillata* (DC.) Backeb., *C. imbricata* (Haw.) F.M.Knuth, *C. kleiniae* (DC.) F.M.Knuth, *C. leptocaulis* (DC.) F.M.Knuth, *C. prolifera* (Engelm.) F.M.Knuth, *C. rosea* (DC.) Backeb., *C. spinosior* (Engelm.) F.M.Knuth and *C. tunicata* (Lehm.) F.M.Knuth are shrubs ranging in height from 60 cm to 3 m (Holtkamp 2012b, Botanic Gardens Trust 2013). These plants have succulent segments growing end-to-end that are armoured with large barbed spines and tiny bristles known as glochids. Collectively these weeds pose serious environmental, agricultural and recreational problems through restricting access to recreational areas, reduced productivity of agricultural land and injury to livestock and native animals.

Variation within species has been a major obstacle for correct identification and the wide variability in morphological traits is increased by environmental conditions that ultimately influence the growth habit of the plants – leading to confusion in correct identification of some species.

Two cochineal insects, *Dactylopius tomentosus* 'imbricata' biotype and *D. tomentosus* 'cholla' biotype have been released to control *C. imbricata* and *C. fulgida* respectively. There are another four *D. tomentosus* biotypes in quarantine awaiting approval for field release. These six biotypes collectively attack the majority of *Cylindropuntia* species found in Australia. However, they all show a significant difference in performance and preference between the eight *Cylindropuntia* species.

Successful biological control of these *Cylindropuntia* species is dependent on releasing the most effective biotype to control each individual *Cylindropuntia* species. Ultimately, the success of the release strategy is thus dependent on the correct identification of the infestation species.

This paper describes the molecular techniques on how to distinguish the eight naturalized *Cylindropuntia* species in Australia and the impact this has on the future success of the *Cylindropuntia* biological control program.

MATERIALS AND METHODS

Plant specimens All *Cylindropuntia* species were identified by recognized taxonomists and Table 1 details the species and localities of plants used in the trial. Identified plant specimens from the US, Mexico and Spain were also used to validate our results.

Table 1. *Cylindropuntia* specimens and the locality collected.

Species	Locality
C. fulgida var. *mamillata*	Longreach, Qld
C. fulgida var. *mamillata*	Burakin, WA
C. fulgida var. *mamillata*	Booligar, Qld
C. fulgida var. *mamillata*	Charleville, Qld
C. fulgida var. *mamillata*	Augathella, Qld
C. fulgida var. *mamillata*	Moama, Qld
C. imbricata	Boggabri, NSW
C. kleiniae	Lightning Ridge, NSW
C. leptocaulis	Curlewis, NSW
C. prolifera	Lightning Ridge, NSW
C. prolifera	Longreach, Qld
C. rosea	Grawin, NSW
C. rosea	Lorne Station, NSW
C. rosea	Willows, Qld
C. rosea	Quilberry Creek, Qld
C. rosea	Charleville, Qld
C. spinosior	Bexley, Qld
C. spinosior	Grawin, NSW
C. tunicata	Glengarry, NSW
C. tunicata	Grawin, NSW
C. tunicata	Cracow, Qld
C. tunicata	Mundubbera, Qld
C. tunicata	Kalkada, Qld

DNA extraction, amplification and sequencing Two plastid markers (Rpl36-Rps8 spacer region and TrnH-PsbA spacer) and one partial nuclear region (phosphoenolpyruvate carboxylase, ppc) were evaluated to differentiate the eight *Cylindropuntia* species. DNA was extracted using a commercial kit and three loci were amplified by PCR. Nucleotide sequences were determined for each amplicon. When needed, further discrimination was achieved using High Resolution Melt PCR of a partial region of the ppc locus.

RESULTS

The eight naturalised *Cylindropuntia* species could be differentiated from each other by amplicon sequence (Figure 1). A High Resolution Melt (HRM) PCR using partial phosphoenolpyruvate carboxylase amplicon was required to distinguish *C. rosea* from *C. tunicata* (Figure 2), as differentiation using sequence alone was not always conclusive due to heterozygosity in the *C. rosea* ppc gene (Figure 1). The HRM curve characteristics were consistent for all individuals tested of these two species.

DISCUSSION

The current biocontrol project has six *Dactylopius tomentosus* biotypes targeting the *Cylindropuntia* species. Two cochineal insects, *Dactylopius tomentosus* 'imbricata' biotype and *D. tomentosus* 'cholla' biotype have been released to control *C. imbricata* and *C. fulgida* respectively and another four *D. tomentosus* have biotypes undergone host specificity testing in quarantine and are awaiting approval for release. A recent survey in the US and Mexico collected another 16 *Dactylopius tomentosus* biotypes to target the two *Cylindropuntia* species, *C. spinosior* and *C. prolifera*, that the current suite of biotypes do not attack. Mathenge *et al.* (2009), Mathenge *et al.* (2010) and Jones *et al.* (2015, 2016) showed that populations of *D. tomentosus* displayed distinct host preferences when collected from different *Cylindropuntia* species within their native ranges. These populations (biotypes) display significant differences in feeding and development preferences and ultimately their impact on different *Cylindropuntia* species. To date, a biotype most suited to controlling each *Cylindropuntia* species has been identified through a series of host specificity, efficacy and fitness index trails.

The use of the three molecular markers can reliably identify the eight naturalised *Cylindropuntia* species and can be confirmed within a couple of working days. As a result, samples of an undetermined species recently discovered in the field can be identified and the most suitable biotype to target this species can be confirmed and released promptly at the infestation site.

Previously, without a confirmed identification, a series of host tests and feeding impact trials would be required to determine the most suitable biotype. These trials would take up to three months to complete. There are multiple benefits in adopting this molecular technique for species identification. First and most important, correct identification will lead to matching the most effective biotype. Second, time efficient confirmation allows rapid and appropriate response to new infestations. Third, these *Dactylopius tomentosus* biotypes can interbreed and the progeny display a

	7	28	64	137	149	152	155	178	181	182	189	204	208	228	283	287	288	305	307	313	325	330	339	355	360	372	376	381	392	399	402	403	404	413	423	430	435	438	455	460	461	476	477
imbricata	G	C	C	G	T	A	C	A	C	C	A	G	A	G	T	T	G	T	A	T	C	G	A	C	G	C	C	C	T	A	T	A	G	A	T	G	C	A	G	A	A	G	G
californica	G	C	C	G	T	A	C	A	C	C	A	G	A	G	T	T	G	T	A	T	C	G	A	C	G	C	C	C	T	A	T	A	G	A	T	G	C	A	G	A	A	G	G
calvenensis	G	C	C	G	T	A	C	A	C	C	A	G	A	G	T	T	G	T	A	T	C	G	A	C	G	C	C	C	T	A	T	A	G	A	T	G	C	A	G	A	A	G	G
prolifera	G	C	C	G	T	A	C	A	C	C	A	G	A	G	T	T	G	T	A	T	C	G	A	C	G	C	C	C	T	A	T	A	G	A	T	G	C	A	G	A	A	G	G
acanthocarpa	G	C	C	G	T	A	C	A	C	C	A	G	A	G	T	T	G	T	A	T	C	G	A	C	G	C	C	C	T	A	T	A	G	A	T	G	C	A	G	A	A	G	G
ramisissima	G	C	C	G	T	A	C	A	C	C	A	G	A	G	T	T	G	T	A	T	C	G	A	C	G	C	C	C	T	A	T	A	G	A	T	G	C	G	G	A	A	G	G
arbuscula	G	C	C	G	T	A	C	A	C	C	A	G	A	G	T	T	G	T	A	T	C	G	A	C	G	C	C	C	T	A	T	A	G	A	T	G	C	G	G	A	A	G	G
leptocaulis	G	C	G	A	T	A	C	A	C	C	A	G	A	G	T	T	G	T	A	T	C	G	A	C	G	C	C	C	T	A	T	A	G	A	T	G	C	A	G	A	A	G	G
kleiniae	G	C	G	A	T	A	C	A	C	C	A	G	A	G	T	T	G	T	A	T	C	G	A	C	G	C	A	C	T	A	T	A	G	A	T	G	C	A	G	A	A	G	G
rosea	G	C	C	G	T	C	T	G	C	C	A	T	A	G	T	T	G	T	G	T	C	G	A	C	G	C	C	C	T	T	T	G	G	G	C	G	C	A	G	A	A	G	G
tunicata	G	C	C	G	T	C	T	G	C	C	A	T	A	G	T	T	G	T	G	T	C	G	A	C	G	C	C	C	T	T	T	G	G	G	C	G	C	A	G	A	A	G	G
subulata	A	T	G	A	C	C	C	G	T	T	A	T	C	T	C	G	C	C	G	A	T	C	C	T	C	T	T	T	C	A	A	A	A	A	T	A	T	A	A	T	T	A	G
bigelowi	G	C	C	G	T	C	C	G	C	C	A	T	A	G	T	T	G	T	G	T	C	G	A	C	G	C	C	C	T	T	T	G	G	G	C	G	C	A	G	A	A	G	G
spinosior	G	C	C	G	T	A	C	A	C	C	A	G	A	G	T	T	G	T	G	T	C	G	A	C	G	C	C	C	T	A	T	A	G	A	T	G	C	A	G	A	A	G	G
fulgida	G	C	C	G	T	A	C	G	C	C	C	G	A	G	T	T	G	T	A	T	C	G	A	C	G	C	C	C	T	A	T	A	G	A	T	G	C	A	G	A	A	G	.

Rpl36-Rps8 spacer region

	7	24	35	39	48	49	65	67	70	90	120	133	237	291	HVR*
imbricata	C	G	A	C	A	G	A	T	A	T	T	A	T	A	4
californica	C	G	A	C	A	G	A	T	A	T	T	A	T	A	4
prolifera	C	G	A	C	A	G	A	T	A	T	T	A	T	A	4
calvenensis	C	G	A	C	A	G	A	T	A	T	T	A	T	A	3
spinosior	C	G	A	C	A	G	A	T	A	T	T	A	T	A	3
fulgida	C	G	A	C	A	G	A	T	A	A	T	A	T	A	5
rosea	C	G	A	C	A	G	G	T	.	A	T	A	T	A	2
tunicata	C	G	A	C	A	G	G	T	.	A	T	A	T	A	2
ramisissima	C	G	A	C	C	G	A	T	A	T	T	A	T	A	3
arbuscula	C	G	A	C	C	G	A	T	A	T	T	A	T	A	3
leptocaulis	C	G	A	C	C	G	A	T	A	A	T	A	T	A	6
bigelowi	C	G	A	C	A	G	G	T	.	A	T	A	T	A	2
acanthocarpa	C	G	G	C	A	G	A	T	A	T	T	A	T	A	6
kleiniae	C	T	A	A	A	G	A	T	A	A	T	C	T	A	6
subulata	T	G	A	C	A	T	G	C	.	A	C	A	G	G	1

TrnH-PsbA spacer region

	21	63	90	98	109	118	130	199	217	240	264	283	301	325
imbricata	T	C	A	A	T	A	T	G	T	G	T	T	C	T
spinosior	T	C	A	A	T	A	T	G	T	*S*	T	T	C	*K*
kleiniae	T	C	A	A	T	A	T	G	T	*S*	T	T	C	*K*
californica	T	C	*M*	*R*	T	A	T	G	T	G	T	T	C	G
calvenensis	T	C	*M*	*R*	T	A	T	G	T	G	T	T	C	G
prolifera	T	C	*M*	A	T	A	T	G	T	G	T	T	C	G
fulgida	T	C	C	G	T	G	T	G	T	G	T	T	C	G
rosea	T	C	A	A	*Y*	A	T	*R*	T	G	T	T	C	*K*
tunicata	T	C	A	A	T	A	T	A	T	G	T	T	C	G
ramisissima	T	C	A	A	T	A	T	G	T	G	T	T	C	G
arbuscula	T	C	A	A	T	A	T	G	T	G	T	T	C	G
bigelowi	T	C	A	A	T	A	T	G	T	G	T	T	C	G
alcahes	T	C	A	A	T	A	T	G	T	G	T	T	C	G
leptocaulis	T	C	A	A	T	A	T	G	T	C	T	T	C	G
acanthocarpa	T	C	A	A	T	A	T	G	T	C	T	T	C	G
subulata	C	T	A	A	T	A	*Y*	G	*Y*	C	*Y*	*Y*	T	G

phosphoenolpyruvate carboxylase, ppc

Figure 1. Differentiation sites for the three sets of primers to distinguish the eight naturalized *Cylindropuntia* species in Australia, US, Mexico and Spain. *hypervariable region between sites 140-225 for the TrnH-PsbA spacer appears to have specific patterns that may be recognisable.

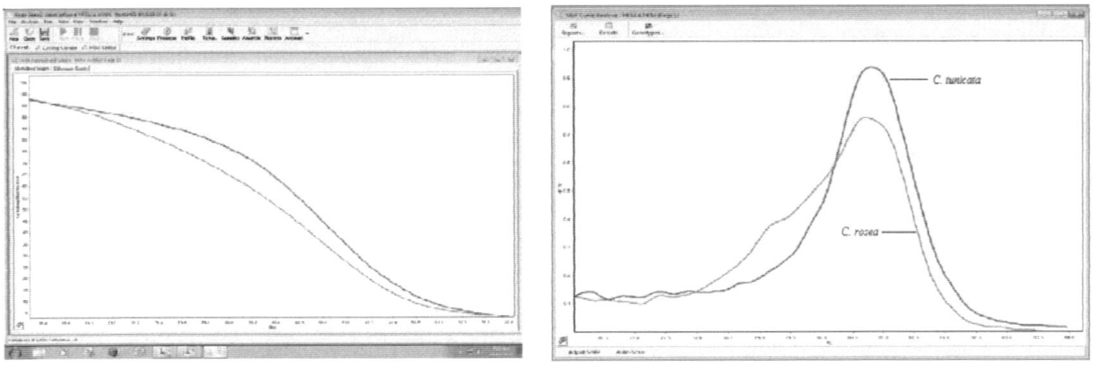

HRM Curve

HRM dF/dT Peak

Figure 2. High Resolution Melt to distinguish *C. rosea* (green) seen as heterozygous for a selected partial region of ppc gene and *C. tunicata* (red) which is homozygous at the corresponding site.

difference in host range and host performance, which could result in a reduction of the virulence and impact on the target species. Correct identification will assist in reducing cross contaminating the biotypes at an infestation site.

ACKNOWLEDGMENTS

This project was funded by Qld. Department of Agriculture Fisheries, NSW Department of Primary Industries, Western Catchment Management Authority and Rural Industries Research and Development Corporation. Authors thank Tony Pople for their advice and comments on the manuscript.

REFERENCES

Anderson, E.F. (2001). 'The cactus family'. (Oregon Timber Press, Portland pp. 776).

Botanic Gardens Trust (2013). PlantNet – The plant information network system of Botanic Gardens Trust, Sydney, Australia. Retrieved 22 November 2013, from http://www.plantnet.rbgsyd.nsw.gov.au

Jones, P.K., Holtkamp, R.H., Palmer, W.A., and Day, M.D. (2015). The host range of three biotypes of *Dactylopius tomentosus* (Lamarck) (Hemiptera: Dactylopiidae) and their potential as biological control agents of *Cylindropuntia* spp. (Cactaceae) in Australia. *Biocontrol Science and Technology* 25 (6), 613-28.

Jones P.K., Holtkamp R.H. and Day, M.D. (2016). The host range of four new biotypes of *Dactylopius tomentosus* (Hemiptera: Dactylopiidae) from southern USA and their potential as biological control agents of *Cylindropuntia* spp. (Cactaceae) in Australia: Part II. *Biocontrol Science and Technology* 26 (8), 1033-47.

Mathenge, C.W., Holford, P., Hoffmann, J.H., Spooner-Hart, R., Beattie, G.A.C. and Zimmermann, H.G. (2009). The biology of *Dactylopius tomentosus* (Hemiptera: Dactylopiidae). *Bulletin of Entomological Research* 99 (6), 551-9.

Mathenge, C.W., Holford, P., Hoffmann, J.H., Zimmermann, H.G., Spooner-Hart, H. and Beattie, G.A.C. (2010). Determination of biotypes of *Dactylopius tomentosus* (Hemiptera: Dactylopiidae) and insights into taxonomic relationships of their hosts, *Cylindropuntia* spp. *Bulletin of Entomological Research* 100 (3), 347-58.

Eradication of wheel cactus from WA: setting priorities to achieve outcomes

Kay Bailey
Department of Agriculture and Food Western Australia, PO Box 19, Kununurra, Western Australia 6743, Australia
(kay.bailey@agric.wa.gov.au)

Summary In order to provide a rigorous and objective method for prioritising government actions associated with declared plants in WA, a decision tool was developed to demonstrate the likely agricultural impacts of those declared plants. This decision tool utilized declared pest ranking assessment by an expert panel and an economic benefit cost analysis. The resulting list was used to determine a list of 15 High Priority species that potentially have the greatest impact on agriculture. Five of the High Priority species were cacti.

As a consequence of the naming of opuntioid cacti as Weeds of National Significance, 27 cacti species were declared under the *Biosecurity and Agriculture Management Act 2007* in December 2014. Control of a number of cactus infestations using various methods including spraying and biological control had taken place in Western Australia for a number of decades. Declaration and delineation as High Priority species increased the focus on these weeds for both government and community weed management activity.

With limited and reducing resources, a subsequent process was undertaken to prioritise cacti species for action by the Invasive Species program within the Department of Agriculture and Food. The prioritization matrix that was developed scored each species according to declaration status, its spread method, availability of effective control methods, number and size of recorded sites in WA, the potential for biological control and, most importantly, the presence and willingness of a biosecurity group to commit to control. Wheel cactus (*Opuntia robusta* H.L.Wendl. ex Pfeiff.) scored highest through this process. This presentation describes the cactus prioritization matrix and the subsequent action to eradicate wheel cactus from WA.

Keywords Prioritisation, matrix, cactus, wheel cactus, biosecurity group.

Progress towards eradication of bitou bush, *Chrysanthemoides monilifera* subsp. *rotundata* in Western Australia

John K. Scott[1,2], Kathryn L. Batchelor[1] and Bruce L. Webber[1,3]
[1] CSIRO Land & Water, Private Bag 5, PO Wembley, Western Australia 6913, Australia
[2] School of Animal Biology, The University of Western Australia, 35 Stirling Highway, Crawley, Western Australia 6009, Australia
[3] School of Plant Biology, The University of Western Australia, 35 Stirling Highway, Crawley, Western Australia 6009, Australia
(john.k.scott@csiro.au)

Summary Bitou bush (*Chrysanthemoides monilifera* (L.) Norlindh subsp. *rotundata* (DC.) T.Norl.) was first detected in Western Australia in 2012 at Kwinana, an industrial area south of Perth. Since 2012 we have undertaken annual surveys of the population, including mature plants, seedlings and the seed bank, as part of assessing the potential for bitou bush to spread or be eradicated. With the collaboration of landholders, plants found each year were eliminated by weeding, herbicide or excavation with earth moving equipment; leading to a marked decline in plant numbers. Very few mature plants are now found on annual surveys and most new discoveries are of plants of up to one year old and seedlings. The seed bank is also declining, but a residue of viable seeds is still present four years after the removal of adult plants. Any new plant discoveries have been within the delimited area of infestation, as defined by the original surveys made in 2012–2013. However the ongoing inaccessibility of some adjoining areas with dense vegetation prevent the achievement of absolute population delimitation. We discuss the plant characteristics that influence the potential for eradication, and the future prospects for management in Western Australia of this weed of national significance.

Keywords *Chrysanthemoides monilifera* subspecies *rotundata*, delimitation, eradication, invasion ecology, seed bank, WoNS, Western Australia.

INTRODUCTION

In 2012 bitou bush (*Chrysanthemoides monilifera* (L.) Norlindh subsp. *rotundata* (DC.) T.Norl. (Asteraceae)) was discovered in the Kwinana region, an industrial and port area south of Perth, Western Australia (Scott and Batchelor 2014). Bitou bush is a Weed of National Significance and a major threat to coastal dune ecosystems (Weiss *et al.* 2008). This subspecies is listed under state legislation in the C1 control category, which includes organisms that are prohibited and should be excluded from part or all of WA. We have been measuring the plant population since its discovery. The aim of this paper is to describe the progress we have made towards assessing the feasibility to eradicate bitou bush from the only known infestation in Western Australia, at Kwinana.

MATERIALS AND METHODS

Survey methods The majority of the area surveyed was prohibited public access and was either a classified mine site, construction site, petroleum refinery or secure maritime access zone. Time was needed to negotiate access and to complete the specific site inductions and orientation. Some sites required us to have drug and alcohol screen tests, a driving test, and the wearing of fire retardant suits with meters for detecting hydrogen sulphide. At least 20 hours per person has been spent in specific site inductions since 2012. Further details of the survey methods are given in Scott and Batchelor (2014). By 2016 over 500 person hours have been spent between three people actively undertaking visual surveillance ground surveys searching the Kwinana estate for plants.

RESULTS AND DISCUSSION

Annual survey for plants Since 2012, we found 651 plants and 1078 seedlings at Kwinana. All but 5 plants were within 500 m of a dilapidated jetty; a possible entry point for the plant invasion.

The majority of plants were removed between October and December 2012 (Table 1). Most were small, <1 m in diameter, but over 130 plants were large (implying capable of reproduction), averaging 3.5 m in diameter with the largest over 11 m in diameter. The number of new plants found in subsequent years has decreased considerably (Figure 1), and by 2016 we found mostly small plants and seedlings (Table 1, Figure 1). Some large plants were found every year, confounding the authors as to how they were missed previously (Figure 1).

After 2012, the annual survey took place between March and May. Bitou bush at Kwinana has its main flowering period at this time and the green leaves and stem, and flowers if present, contrasted well against the dead grass. Plants were often found under trees or fence lines. This is not surprising given that birds would eat the ripe fruits and disperse the seed to suitable perches near the plant of origin. Seed germination occurred throughout the months that soil was moist (usually June to November).

Remote sensing There are frequent aerial photography surveys of the Kwinana area, and available online on a bimonthly basis for Nearmap (http://www.nearmap.com.au/) or on an annual basis by Landgate (https://www.landgate.wa.gov.au/). The photos viewed online are sufficiently detailed to identify bitou plants over 1 m in diameter, particularly if the plant was isolated from other vegetation. However, it was not possible to separate bitou bush from the form and colour of the invasive tree *Schinus terebinthifolius* Raddi (Anacardiaceae) (Brazilian pepper) in these images. Brazilian pepper is abundant in the Kwinana area, and caused many false positives during the search for bitou bush.

The fleshy fruits of bitou bush means that birds and mammals are major means of dispersal (Gosper 2004, Meek 1998). This also results in many of the new plants being found under established vegetation (e.g. tobacco bush, Eucalypts) or other perches (fence lines, buildings, light posts). This distribution pattern where bitou is located under other vegetation makes detection difficult by remote means.

Table 1. Number of individuals and average diameter of plant canopy, excluding seedlings at Kwinana.

Year	Number of plants measured	Diameter (cm) (average ± SE)
2012	487	101.1 ± 7.4
2013	26	62.2 ± 14.9
2014	40	27.9 ± 4.8
2015	56	42.2 ± 6.4
2016	42	26.4 ± 5.2

Controlling plants Initially we thought that most seedlings would succumb over summer. However, rainfall events over summer, and perhaps the coastal environment, meant sufficient moisture was available for some seedling survival. There appeared to be multiple germination events following rain which resulted in early seedlings being sufficiently large to survive summer.

Plants up to ~3 m wide were readily hand weeded ensuring the entire stem and upper roots were removed. Plants are shallow rooted, which facilitates their removal by hand. No regeneration was observed from roots remaining in the soil after the aboveground and small section of below ground stem had been removed.

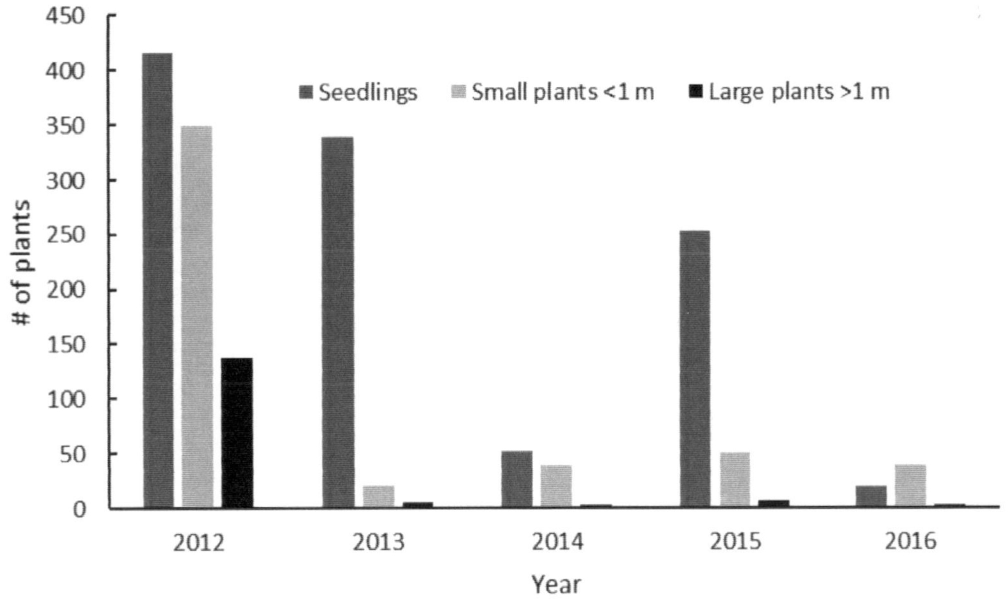

Figure 1. Progress towards eradication of *Chysanthemoides monilifera* subsp. *rotundata* at Kwinana, Western Australia. Plants at different stages (seedling, small (<1.0 m canopy diameter), large (>1.0 m canopy diameter) found since the start of surveys (October 2012).

We did however find one or two examples each year of plants that had regrown from the stem broken off at ground level in the previous year.

Larger plants required considerable physical strength to remove and in some instances needed earth moving machinery. Herbicide (this was organized by the landholder) was also used to kill large plants. However, hand removal was the quickest option, particularly where plants were growing next to buildings or other infrastructure or over potential buried services.

Large plants in areas with shifting sand (such as sand dunes near the beach) can be partially covered by sand such that the main stem is underground. Side branches in this situation will develop roots and become independent of the original plant, a process called 'layering'. Our experience was that it was difficult to find and extract all the layered stems, especially if the plant was large and in situations of considerable sand movement.

Mechanical plant removal was sometimes counterproductive. Land owners helped with digging machinery of a range of sizes from bobcats to large front end loaders. The main issue was the risk of burying pieces of stem from which future plants could arise. Revisiting the area at yearly intervals for up to three years enabled us to find and remove any surviving layered stems.

We left the stems on the soil surface next to where they grew. Dead plants were still visible four years after removal which helped in the relocation of sites with high potential for seedlings.

Herbicide spraying was very effective, and the contractor used was skilful to recognize bitou bush from the very similar bush *Scaevola crassifolia* Labill. (Goodeniaceae). Bitou bush has yellow daisy-like flowers and the growth apex and small new leaves covered with white fur (called an indumentum), whereas *S. crassifolia* has blue, zygomorphic flowers (petals on one side) and no indumentum. When neither is flowering, they look remarkably similar.

The complex land use and infrastructure at Kwinana meant that some plants were growing in locations too dangerous to access. This included access points for conveyor belts, steep-sided storm water drainage points and similar confined spaces. Binoculars at these locations were useful to determine if a plant was bitou or *S. crassifolia*. Confirmed bitou plants could then be killed by herbicide sprayed from a safe access point, as usually that did not require entry into the confined space.

Annual assessment of seed bank The most suitable time to assess the seed bank is late summer, while the soil is dry and easily sieved. Soil was first passed through a course sieve (10 mm) to remove leaves and larger material then a 2 mm sieve. Bitou bush seeds have a hard putamen and do not break easily, thus the soil can be gently forced through the sieve using a gloved hand to removed most of the material – leaving the seeds and seed fragments In 2013, cores were collected under the canopy area of 15 plants. By 2016, only nine of these areas could be sampled due to substrate relocation by heavy machinery as part of port activities. From the nine plants sampled throughout, the viable under-canopy seed bank averaged 39.3 ± 35.9 seeds m^{-2} in 2012, decreasing to 7.1 ± 12.5 in 2016.

It has now been four years since reproduction has occurred at the study site and we continue to find abundant seed parts under plants targeted for seed bank sampling. The finding of putamen fragments is a useful check that sampling is in the right place where seeds could be potentially found.

Is not known for how long viable seeds can persist in the soil. Noble and Weiss (1989) found the viability of seeds decreased over two years, and viability was greater for seeds buried up to 6 cm than on the surface. Soil cores collected in 2016 presented a low number of viable seed, indicating seeds can survive at least four years.

There is circumstantial evidence that bitou bush requires outcrossing pollen to produce seed. Isolated plants (500 m from other plants) were either non-reproductive or had very few seeds. This was verified by sampling the soil under four of the isolated plants in 2013 to find no seeds or seed parts from previous years. Another observation has been that plants in heavily shaded environments have long branches and no flowers, or if a few flowers they were on branches that have grown out from the shade.

Smoke water trial Fire can be used to control bitou bush and stimulate germination of the seedbank (Weiss *et al*. 2008), but this was not a practical option given that the infestation's proximity to an oil refinery, gas plants and other explosive infrastructure.

Smoke water is known to increase germination in dormant Boneseed, *Chrysanthemoides monilifera* subsp. *monilifera* (Reynolds *et al.* 2013). However, an experimental treatment on the under canopy area of five former plants with smoke water (Regen 2000 Smokemaster© at 10%) in 2014 known to have a viable seed did not produce adequate germination for assessment.

Delimitation It is highly unlikely that delimitation of bitou bush can be achieved for Western Australia. By delimitation we mean that we have established the limits or boundaries of the entire distribution including potential areas of distribution based on species distribution modelling.

There are numerous means by which seed, and plants, could have left the Kwinana bitou infestation and potentially started another population. Frugivorous birds (e.g. ringneck parrot, *Barnardius zonarius* were observed at the Kwinana estate) and mammals (e.g. foxes (Meek 1998)) are the usual means of seed dispersal in eastern Australia. There has been major disturbances of the area during the time bitou has been present, including: levelling and construction of a laydown area on Mason Road (southern edge of bitou distribution; 40 ha); decommissioning and transport to China of an iron smelting plant (eastern edge of distribution; 13 ha); construction of facilities related to cement production (northern edge of distribution; 10 ha). Potential seed dispersal from Kwinana also include the following vectors; ships (to and from eastern Australia, China), boats (fishing and recreational), trains (transporting iron ore, coal), trucks (carrying clinker, bauxite), horse floats (the adjacent public beach was a horse exercise area where the parking was surrounded by bitou plants), suspected drug dealers (a large plant was found at a meeting point in the parking area mentioned previously), other vehicles (port staff parking in the centre of the bitou distribution), road sweepers (plants overlapped road edges that were cleaned), and gardeners (removal of plant material for composting). What can be seen from this extensive list is that despite the general area having controlled access, there are more than enough means for seed to leave the site, and consequently we must assume that this has happened. Indeed, it is surprising that other bitou infestations have not been detected.

Conclusions It is likely that the threat of invasion is largely tempered by the unsuitability of climate and habitat across most of Western Australia. However, the abundance of seedlings and small plants without apparent morality strongly indicates that the Kwinana area, at least, is suitable for plant naturalisation. Bitou bush is largely found on the coastline in eastern Australia and this implies that the coastal region in Western Australia such as Cockburn Sound could be suitable. Further work on species distribution modelling is needed to determine if climate is limiting for bitou bush.

The next surveys for plants will be made in autumn (March–April) 2017. Seed densities are now so low that it is becoming marginally feasible to continue the seed bank survey. A proxy for the seed bank counts can be obtained by counts of seedlings within the canopy area sampled previously. It is still too early to state that eradication is possible because of the delimitation issue and because of the continued presence of a seed bank.

ACKNOWLEDGMENTS

We thank Crystal Jones for help in the field and Ruchira Somaweera for comments on the manuscript. Support of the landholders of the Kwinana and Cockburn regions is gratefully acknowledged.

REFERENCES

Gosper, C.R. (2004). Fruit characteristics of invasive bitou bush, *Chrysanthemoides monilifera* (Asteraceae), and a comparison with co-occurring native plant species. *Australian Journal of Botany* 52, 223-30.

Meek, P.D. (1998). Weed seeds and whoopsie daisies: viability of bitou bush *Chrysanthemoides monilifera* seeds in fox (*Vulpes vulpes*) scats. *Plant Protection Quarterly* 13, 21-4.

Noble, I.R., and Weiss, P.W. (1989). Movement and modelling of buried seed of the invasive perennial *Chrysanthemoides monilifera* in coastal dunes and biological control. *Australian Journal of Ecology* 14, 55-64.

Reynolds, C.J., Long, R.L., Flematti G.R., Cherry, H. and Turner, S.R. (2013). Karrikins promote germination of physiologically dormant seeds of *Chrysanthemoides monilifera* ssp. *monilifera* (boneseed). *Weed Research* 54, 48-57.

Scott, J.K. and Batchelor, K.L., (2014). Management of *Chrysanthemoides monilifera* subsp. *rotundata* in Western Australia. *Invasive Plant Science and Management* 7, 190-96.

Weiss, P.W., Adair, R.J., Edwards, P.B., Winkler, M.A. and Downey, P.O. (2008). *Chrysanthemoides monilifera* subsp. *monilifera* (L.) T.Norl. and subsp. rotundata (DC.) T.Norl. *Plant Protection Quarterly* 23, 3-14.

Landscape transformations following blackberry decline in the south-west of Western Australia: successful restoration or back to blackberry?

Paul B. Yeoh[1], Lee Fontanini[3], Jenny A. Carley[3], Andy T. Russell[3], Kathy E. Dawson[3], John K. Scott[1,4], Caroline Y.M. Delaisse[1,2] and Bruce L. Webber[1,2]

[1] CSIRO Land & Water, Private Bag 5, PO Wembley, Western Australia 6913, Australia

[2] School of Plant Biology, University of Western Australia, 35 Stirling Highway, Crawley, Western Australia 6009, Australia

[3] Warren Catchments Council, 52 Bath Street, Manjimup, Western Australia 6258, Australia

[4] School of Animal Biology, University of Western Australia, 35 Stirling Highway, Crawley, Western Australia 6009, Australia

(paul.yeoh@csiro.au)

Summary In recent years there has been a rapid decline in blackberry (*Rubus anglocandicans* A.Newton) populations along waterways within the Warren Catchments region located in the south-west of Western Australia. The soil pathogen, *Phytophthora bilorbang* S.Aghighi, G.E.S.J.Hardy, J.K.Scott & T.I.Burgess, is thought to be instrumental in the decline of these populations. For most waterways in the region, blackberry has dominated the forest understorey for decades. Consequently, the weed's rapid decline typically results in areas of bare soil or grassed landscapes lacking structure.

The blackberry decline created the opportunity to access and restore what were previously often inaccessible but, with the exception of blackberry, fairly pristine landscapes. However, the post-decline absence of blackberry plants within impacted populations is not absolute; scattered individuals persist. Moreover, the decline is not known to impact on what could be a significant blackberry seedbank. Without intervention to restore the landscape, therefore, new blackberry individuals or other weed species could just replace the previous blackberry infestations.

Consequently, a follow-up restoration project is underway to increase the resilience of the post-decline community to future weed invasion, including possible blackberry re-invasion. Here we outline the efforts being undertaken to understand these threats and to encourage the successful restoration of these areas within the Warren Catchments region.

Keywords Biological control, blackberry, community resilience, decline, *Phytophthora bilorbang*, restoration, *Rubus anglocandicans,* seed ecology, weed management.

INTRODUCTION

'Blackberry' comprises numerous species of which *Rubus anglocandicans* A.Newton (Rosaceae) is the most widespread and weedy in Australia, being found across all of temperate southern parts (Evans and Weber 2003). It is also the dominant weedy species of riparian areas in the Donnelly and Warren Catchments of south-west Western Australia, an area targeted for ecosystem restoration.

Blackberry can normally be effectively suppressed within accessible agriculture grazing systems especially if grazing can be integrated with herbicide and/or physical control methods. However, control is usually difficult and expensive within native ecosystems as the risk of off-target negative impacts on native flora or fauna limits control options, access for chemical/physical control is usually difficult and ongoing control is usually required (Adair and Bruzzese 2006). As such, very large blackberry infestations in native vegetation and rugged terrain remain untreated in Australia. In addition, blackberry is a riparian species and its presence on water courses makes chemical control more difficult. Riparian ecosystems are also known to favour alien plant invasions due to naturally higher levels of disturbance (which is associated with decreased competition and increased light and nutrient availability) and because water flow provides an effective method of rapidly spreading weed propagules (Richardson *et al.* 2007). For all of these reasons, biological control is considered to be the preferred strategy for controlling blackberry in Australia (Adair and Bruzzese 2006, Aghighi *et al.* 2012, Morin *et al.* 2006).

In 2004, eight European strains of the leaf-rust fungus, *Phragmidium violaceum* (Schultz) Winter were approved for release in Australia to improve prospects of biological control over a wider range of blackberry taxa over a broader range of habitats, relative to the two rust fungal strains introduced in the 1980–90s, which were effective but only in localized areas (Adair and Bruzzese 2006, Morin *et al.* 2006). Whilst conducting impact assessment and integrated control trials as part of the *P. violaceum* biological control release program, patches of 'blackberry decline' were discovered, where populations of mature

R. anglocandicans, hereafter referred to as 'blackberry', suddenly collapsed at a rate far greater than expected from the leaf-rust alone (Yeoh and Fontanini, unpublished). The areas of decline were subsequently mapped and the phenomenon considered to be a result of a combinations of stresses but in which the oomycetes *Phytophthora bilorbang* and '*P. cryptogea*' (but which, following DNA sequencing, is now believed to be *P. pseudocryptogea* T. Burgess pers. comm.) were consistently found associated with affected plants (Aghighi *et al.* 2014).

The blackberry decline appears to be highly specific with no other plant species recorded as impacted by the decline in any of the locations examined (Aghighi *et al.* 2014). The rapid demise of the thick blackberry infestations (previously the dominant forest understorey species) has, however, significantly altered the vegetation structure of the riparian zone. This situation is known to potentially also negatively affect ecosystem function, including riverbank stability, water quality and animal assemblages (Richardson *et al.* 2007). The newly created space, therefore, is vulnerable to future weed dominance once again (Cherry *et al.* 2006).

The aim of our project was to increase the resilience of riparian ecosystems within the Donnelly and Warren catchments, Western Australia. Approximately 25% of the area being actively restored by a large restoration program in the catchments are those where blackberry decline has occurred. These areas (Figure 1) will be the focus of this paper, while the remaining areas have been degraded primarily by grazing and other farming practices and will be discussed elsewhere.

Richardson *et al.* (2007) lists five main management actions that relate to biotic components of riparian environments and that are potentially available for facilitating the restoration of these plant communities affected by alien plant invasions

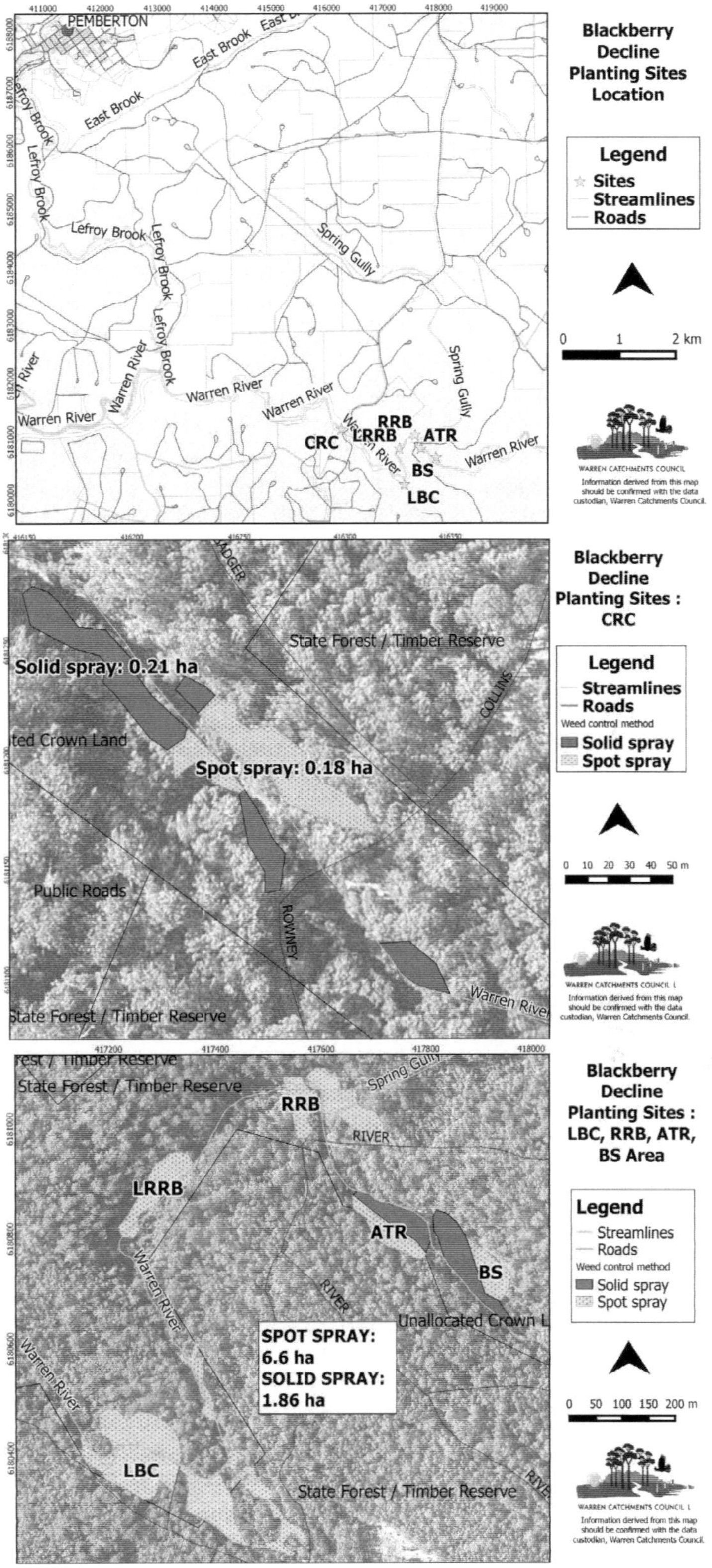

Figure 1. Location of decline sites under restoration in south-west Western Australia and their spray regimes. Collins Road Crossing (CRC), Log Bridge Crossing (LBC), River Road Bridge, (RRB), Andy T. Russell's site (ATR) and Billabong Site (BS).

(numbered i to v in the following discussion section). We will discuss our strategies addressing these and similar actions.

RESULTS AND DISCUSSION

i) Clearing the alien stands In regard to blackberry; only one of the sites where native vegetation is currently being restored was surveyed prior to and then after their decline (See Aghighi *et al.* 2014, Figure 1). General characteristics were, however, measured for heavy infestations of blackberry (*R. anglocandicans*) over a variety of habitats within the Donnelly and Warren catchments region prior to decline being discovered (Yeoh, Fontanini and Scott, unpublished data). As such, these metrics can be used as a general indication for blackberry population dynamics typical of the region. In general (based on site means ± SE, n = 8), densities were 5.4 ± 0.54 plants m^{-2} with 2.2 ± 0.17 canes per plant. Primocanes typically extended 2.2 ± 0.23 m from the crown but with the largest individuals exceeding 5 m reach. Basal diameters of the stems averaged 7.8 ± 0.48 mm and seed production was estimated at approximately 10,000 ± 3200 seeds per plant or 8600 ± 2000 seeds m^{-2}. One plant was estimated to have produced over 30,000 seeds within a season.

The measurement of the size of individual plants after the population had declined is currently ongoing. Most of the areas of declined blackberry thickets would now be considered to be successfully controlled if we used Adair and Bruzzese (2006) definition for a 'successful biocontrol program' (i.e: <1 plant m^{-1}, canes extending <1 m from crown and <2 canes per plant). Prior to populations showing decline symptoms, heavy infestations of blackberry typically extended 40 to 50 m from the river's edge. After populations of blackberry had declined, individuals of the size/density observed in pre-decline infestations now appear restricted to a 3 to 4 m wide zone growing on the steep banks where the plant's foliage is overhanging the river.

Within the restoration project, these individuals are being sprayed using metsulfuron-methyl ('Brush-off' at a rate 10 g 100 L^{-1}) with a vegetable oil based spreader ('Endorse' at a rate 300 mL 100 L^{-1}). This mixture has proven to have less off-target effects and to be more effective at killing the blackberry compared to mixtures containing glyphosate (see Anon 2009, case study p. 40).

Other alien species such as *Atriplex prostrata* DC. and *Cyperus eragrostis* Lam. or *C. congestus* Vahl were also common in some locations and spray trials (either cover spraying of whole areas of dense weeds or spot spraying of isolated individuals using backpack spray units) occurred at the sites (Figure 1, Table 1). The spraying of the exotic *A. prostrata* and *Cyperus* spp. was of dubious benefit as there was rapid re-establishment of seedlings (likely from soil seed bank), regardless of whether the areas were sprayed or not (Carley and Fontanini, unpublished data). A combination of spraying followed by the dense planting of suitable native species may prove to be the best option to establish a resilient native flora community to out-compete these alien species. Survival of transplanted native plants, however, did not seem to be effected by the presence of *Atriplex* as it appeared to provide physical protection that reduced detection and subsequent impacts from mammalian grazing.

ii) Control regenerating aliens by (iii) re-introducing biological control Adair and Bruzzese (2006) observed the greatest impact of *P. violaceum* upon

Table 1. Herbicide (metsulfuron-methyl) application at the various post-blackberry decline restoration sites. (Spot spraying = only target weed sprayed; cover spraying = general area containing predominantly the target weed sprayed).

	Spot (S) or Cover (C) sprayed				
Species	CRC	LBC	RRB	ATR	BS
Atriplex prostrata DC.	–	–	–	S/C	S/C
Cyperus congestus Vahl	S/C	–	–	–	–
Cyperus eragrostis Lam.	S/C	–	–	–	–
Dysphania ambrosioides (L.) Mosyakin & Clemants	–	–	–	S	–
Rubus anglocandicans	S	S	S	S	S
Rumex conglomeratus Murray	S/C	S	S	C	S/C
Rumex crispus L.	S/C	S	S	C	S/C
Setaria parviflora (Poir.) Kerguelen	S/C	–	–	–	–

plants growing in large infestations and receiving full-sun, whilst blackberry in shaded habitats and those experiencing temperature or moisture stress during summer appear to be only weakly affected. Our observations on impacted blackberry populations suggest the reduction in biomass due to the decline complex complements the control expected by the introduced leaf rust biocontrol. That is, plants showing symptoms in the declining populations were usually growing in the shade whereas healthier plants were usually overhanging the water and growing in the sun, where the leaf rust usually thrives. More recently, some of the blackberry populations impacted by the decline in 2008-13 have begun to show signs of the blackberry population recovering (Yeoh, unpublished data).

iv) Manipulate disturbance regime Motion detecting cameras have been used to monitor for the presence of exotic mammals in the study areas, with rabbits, foxes, cats, pigs, and sheep being observed. Based on observations, only rabbits were in sufficient numbers to be providing grazing pressure and they have been subjected to control measures. It was not feasible to fence restoration areas due to the presence of native herbivores such as quokkas, possums and kangaroos that also graze the same areas.

v) Seeding or planting: is it needed? Blackberry infestations are known to negatively impact upon native plant species with a 46% reduction in the richness of native plant species in areas where blackberry had invaded when comparisons were made between areas ahead and behind of invading blackberry fronts at five bushland sites within the south-west of Western Australia (Yeoh *et al.* 2006). Subsequent surveys from 10 sites found the reduction in native plant diversity remained at 46% and was associated with an average 79% decrease in native plant cover. The blackberry populations averaged 59% cover in the survey areas classed as 'post-blackberry invasion' and 1% cover in the 'pre-blackberry invasion' areas (Yeoh *et al.* 2006, unpublished data). The locations surveyed in this study were at the leading edge of the blackberry population, where the weed had probably been present for the shortest period of time (relative to other blackberry at the site but further from the edge). It could be anticipated that the weed's impact upon native plants would be even greater in parts of the infestations further in from the invading fronts.

It was hypothesized that soil seed banks of both native species and blackberry could be stimulated to germinate by cultivation and the addition of smoke water. This would deplete the weed's seed bank whilst possibly removing the need to plant/sow native species. A pilot trial (16 plots each 2.5 × 2.5 m) was set up within a blackberry decline area to test this with half the plots cultivated to a depth of approximately 10 cm (Ryobi 2 stroke cultivator) and either 0, 10, 20 or 100% strength Regen 2000 smoked vermiculite (7.5 g m^{-2}) applied to each plot.

Based on careful consideration of positive and negative outcomes for the community overall, the method was considered unsuitable because 1) the surface roots of the existing overstore plants could be excessively damaged, 2) any existing ground cover (mainly native grasses) could be disturbed/potentially injured and, therefore, the risk of erosion increased, 3) the cultivator posed a fire risk and was slow due to roots/sticks jamming the blades, 4) a significant amount of Regen 2000 smoked vermiculite would be needed to treat the whole restoration area creating logistical challenges, and most importantly, 5) there did not appear to be any significant increase in the germination of either blackberry seedlings (six total seen) or native tree/shrub seedlings. Numerous annual plants (especially chenopods) did germinate but it was not clear if this was beneficial or not to the restoration efforts (see spraying section).

The results (or lack thereof) from the smoke water pilot study supports the alternative hypothesis that the long term dominance of blackberry in understorey has suppressed native species, depleting their seed banks. The potentially low benefits from physically disturbing the soil surface does not seem to warrant the associated high costs/risks and so the idea was not pursued further. It was therefore decided that planting/sowing of native species into the areas of decline is essential.

vi) Selection process for species to plant Our aim was to maximize the resilience of the restored ecosystems by replanting multiple species in multiple height and vegetation type stratifications.

Our logistical restrictions for the restoration included 1) a general lack of knowledge of historical pre-blackberry species composition, 2) only local provenance seed required to be used for plantings, 3) unknown methods of propagation/germination for some species, 4) unavailable seed/propagules for some species ranked as high restoration priorities, 5) the susceptibility of some species to *Phytophthora cinnamomi,* which was known to be in some restoration areas, and 6) the cost of acquiring plants of suitable size/condition to confer resilient to transplant shock.

The definition of local provenance is highly variable (see Carr 2008). We relied on advice from project collaborators Drs. M. Byrne, T. Hopley and T. Macfarlane (Department of Parks and Wildlife, WA) who used the principle of planting within areas where gene flow from natural dispersal was probable. This meant plantings could occur within the same vegetation complex for several kilometres downstream of collection sites but it may be more restricted in an upstream direction. Plantings at our sites utilized 15 native species (three over-storey and three mid-storey trees/shrubs, five sedges/rushes, two grasses and two herbs) and we are currently 85% towards the anticipated final project totals (Table 2).

vii) Lessons learned a) Access problems to sites (requiring the use of a flying fox, canoes, waders, bush treks) slows planting rates and makes it difficult to use volunteers. This can significantly increase planting costs. b) Planting out required at least a season's pre-preparation (to allow for seed collection, after-ripening, germination and growing to transplant size). Local disasters (storms/disease etc) at the nursery stage can be catastrophic. We mitigated against such disasters by using several suppliers. c) Herbivores can also be disastrous. We had to trim the upper foliage off grasses just prior to planting to prevent them all being pulled out of the ground by kangaroos! d) Similarly abiotic factors (heat waves/floods) caused problems. Long range weather forecasts potentially can help planning when and where to plant so as to avoid plants being washed away or dehydrating before they have established adequate root systems, but in reality the long range forecasts are often unreliable. Again, we recommend spreading the risks by planting cohorts across years/sites.

CONCLUSIONS

When a weed species dominates habitats to such an extent that it is beyond chemical/physical control then normally biological control is seen as the most obvious viable option of control.

What is not so obvious is that within natural environments, the same issues that made the weed control difficult with chemicals/physical methods are likely to also make restoration as difficult. Furthermore, the severity of previous infestations make the need for restoration more likely. Forward planning is therefore needed to protect affected susceptible/sensitive habitats.

Table 2. Current plantings at the various post-blackberry decline restoration sites.

		# Individuals planted per sites					
Species	**Life form**	**CRC**	**LBC**	**RRB**	**ATR**	**BS**	**Total**
Eucalyptus rudis Endl.	*Tree (5–20 m)	100	2,900	3,630	240	1,040	**7,910**
Agonis flexuosa (Willd.) Sweet	Tree/shrub (to 10 m)	600	2,800	1,240	2,660	3,660	**10,960**
Paraserianthes lophantha (Willd.) I.C.Nielsen	Tree/shrub (1–10 m)	8,300	4,700	1,050	1,690	3,750	**19,490**
Astartea leptophylla Schauer	Tree/shrub (to 5 m)	–	–	–	–	320	**320**
Taxandria linearifolia (DC.) J.R.Wheeler & N.G.Marchant	Tree/shrub (to 5 m)	1,250	1,720	–	–	120	**3,090**
Callistachys lanceolata Vent.	Tree/shrub (1.5–7 m)	–	380	–	–	–	**380**
Carex appressa R.Br.	Sedge (0.5–2 m)	1,130	46,970	6,540	10,650	17,200	**82,490**
Carex tereticaulis F.Muell.	Sedge (0.7 m)	1,370	1,920	610	570	2,670	**7,140**
Lepidosperma persecans S.T.Blake	Sedge (0.5–2 m)	1,040	14,080	–	4,290	6,160	**25,570**
Juncus amabilis Edgar	Rush (0.3–1 m)	920	6,580	5,750	1,700	2,360	**17,310**
Juncus pauciflorus R.Br.	Rush (0.45–1 m)	6,200	23,690	3,550	7,860	10,030	**51,330**
Microlaena stipoides (Labill.) R.Br.	Grass (0.3–0.75 m)	770	3,260	–	1,860	4,950	**10,840**
Poa porphyroclados Nees	Grass (0.4–1.0 m)	–	1,450	1,360	–	–	**2,810**
Alternanthera nodiflora R.Br.	Herb (0.02–1 m)	60	250	1,390	–	1,130	**2,830**
Persicaria prostrata (R.Br.) Sojak	Herb (0.05–0.1 m)	70	–	–	–	2,100	**2,170**
Total		**21,810**	**110,700**	**25,120**	**31,520**	**55,490**	**244,640**

*Life Forms are as described Florabase https://florabase.dpaw.wa.gov.au

ACKNOWLEDGMENTS

The Australian Government's Clean Energy Future Biodiversity Fund, Warren Catchments Council and CSIRO financially supported this work. Jody Quinn and many volunteers have also helped.

REFERENCES

Anon. (2009). Blackberry control manual: management and control options for blackberry (*Rubus* spp.) in Australia. Department of Primary Industries, Victoria. Accessed 25/6/16 via http://www.weeds.org.au/WoNS/blackberry/.

Adair, R.J. and Bruzzese, E. (2006). Blackberry: treading a prickly path to effective biological control in Australia. Proceedings of the 15th Australian Weeds Conference, eds C. Preston, J.H. Watts, and N.D. Crossman, pp. 557-60. (Weed Management Society of South Australia, Adelaide).

Aghighi, S., Fontanini, L., Yeoh, P.B., Hardy, G.E.S., Burgess, T.I. and Scott, J.K. (2014). A conceptual model to describe the decline of European blackberry (*Rubus anglocandicans*), a weed of national significance in Australia. *Plant Disease* 98 (5), 580-9.

Carr, D. (2008). There's more to seed than local provenance. Thinking Bush (Land and Water Australia) Issue 7, pp. 5-8. Accessed 25/6/16 via http://www.florabank.org.au/files/Carr%20 2008%20Thinking%20Bush%207%20More%20 to%20seed_low.pdf

Cherry, H., Turner, P.J. and French, K. (2010). Restoration and resilience in weed-invaded ecosystems. Proceedings of the 17th Australasian Weeds Conference, ed. S.M. Zydenbos, pp. 437-40. (New Zealand Plant Protection Society, Christchurch).

Evans, K.J. and Weber, H.E. (2003). *Rubus anglocandicans* (Rosaceae) is the most widespread taxon of European blackberry in Australia. *Australian Systematic Botany* 16, 527-37

Morin, L., Aveyard, R., Batchelor, K.L., Evans, K.J., Hartley, D. and Jourdan, M. (2006). Additional strains of *Phragmidium violaceum* released for the biological control of blackberry. Proceedings of the 15th Australian Weeds Conference, eds C. Preston, J.H. Watts, and N.D. Crossman, pp. 565-8. (Weed Management Society of South Australia, Adelaide).

Richardson, D.M., Holmes, P.M., Esler, K.J., Galatowitsch, S.M., Stromberg, J.C., Kirkman, S.P., Pyšek, P. and Hobbs, R.J. (2007). Riparian vegetation: degradation, alien plant invasions, and restoration prospects. *Diversity and Distributions* 13, 126-39.

Yeoh, P.B., Fontanini, L., Cranfield, R.J. and Scott, J.K. (2006). How bad is blackberry in south west Australia. Proceedings of the 15th Australian Weeds Conference, eds C. Preston, J.H. Watts, and N.D. Crossman, p. 293. (Weed Management Society of South Australia, Adelaide).

Chloroplast genome of serrated tussock (*Nassella trichotoma*): structure and evolution

Aisuo Wang[1,2], Hanwen Wu[1,2] and David Gopurenko[1,2]
[1] NSW Department of Primary Industries, Wagga Wagga Agricultural Institute, PMB, Wagga Wagga, New South Wales 2650, Australia
[2] Graham Centre for Agricultural Innovation, Locked Bag 588, Wagga Wagga, New South Wales 2678, Australia
(aisuo.wang@dpi.nsw.gov.au)

Summary The genus *Nassella* contains many economically and environmentally important weed species in Australia, such as *Nassella trichotoma* (Nees) Hack. ex Arechav. (serrated tussock) and *Nassella neesiana* (Trin. & Rupr.) Barkworth (Chilean needle grass), which are both Weeds of National Significance (WoNS). Availability of *Nassella* chloroplast genome sequences can provide a versatile tool for identifying novel gene targets for DNA barcoding of these weeds species. We report here the first chloroplast genome of *Nassella* (*N. trichotoma*) obtained through de novo assembly of Illumina paired-end reads produced by total DNA sequencing. The *N. trichotoma* chloroplast genome is 112,102 bp in size, encodes 140 genes including 99 protein-coding genes, 37 tRNA genes and 4 ribosofmal RNA genes. The total size of intergenic regions within the genome is up to 48,764 bp. The total GC content of the genome is 37.88%, which is relatively lower than that of the reported genomes within the *Stipeae* tribe. A comparison of this chloroplast genome with other *Stipeae* chloroplast genomes (*Oryzopsis asperifolia* Michx., *Piptochaetium avenaceum* (L.) Parodi, *Achnatherum hymenoides* Roem. & Schult., *Stipa lipskyi* Roshev. and *Stipa purpurea* Griseb.) provides new insights into their chloroplast genome evolution. Our future work will report genomes of other *Nassella* species as a means to identify novel DNA barcode regions useful for distinguishing these species.

Keywords DNA barcoding, invasive weeds, *Stipeae*, Next Generation Sequencing, Illumina.

INTRODUCTION

The genus *Nassella* contains many economically and environmentally important weeds species in Australia, such as *Nassella trichotoma* (Nees) Hack. ex Arechav. (Serrated tussock) and *Nassella neesiana* (Trin. & Rupr.) Barkworth (Chilean needle grass). Effective management of these invasive weeds relies on correct identification of them at all growth stages. Traditional morphological identification plays a crucial role in identifying these weeds species, but relies on the availability of flowering materials of the weeds and expert taxonomic knowledge. A genetic approach such as DNA barcoding provides a hope for identifying these weeds at all growth stages (Syme *et al.* 2013, Wang *et al.* 2014). Nevertheless, a lack of universal and robust markers has been hindering the development of DNA barcoding technology. Screening ideal DNA barcoding markers to identify *Nassella* species and other invasive weeds thus becomes a necessity.

The chloroplast genome is a major source for markers applied in plant DNA barcoding technology. Most of the well-known DNA barcoding markers for plants such as *rbcL, matK, psbK–psbI, trnH–psbA, atpF–atpH, rpoB* and *rpoC1* (CPW Group 2009) were selected from sequences of chloroplast genome of plants. With the fast development of next-generation sequencing technology (NGS) in recent years, it becomes increasingly practical and cheaper to sequence a whole grass chloroplast genome. The availability of whole chloroplast genome sequences of invasive weeds will undoubtedly facilitate the marker selection process that has been a bottleneck for weeds DNA barcoding technology.

Here we report the chloroplast genome sequences of *N. trichotoma*, the first chloroplast genome for the *Nassella* genus, obtained through de novo assembly of Illumina paired-end reads. Information on this *Nassella* chloroplast genome sequence will be instrumental for DNA barcoding, phylogeny analysis and other related research for weeds of the *Nassella* genus.

MATERIALS AND METHODS

Sample collection Fresh *N. trichotoma* leaves were collected from Wagga Wagga NSW in Australia (34° 58' 40.7", 147° 26' 19.5"). The voucher samples have been deposited into Wagga Wagga Agricultural Institute (WWAI) (voucher number: ww19856).

Genomic DNA extraction and sequencing Total genomic DNA was extracted from fresh leaves of *N. trichotoma* using traditional phenol chloroform protocol (Sambrook and Russell 2006) with modifications.

Basically the plant tissue was digested in CTAB extraction buffer (100mM Tris-HCl (pH 7.5), 25mM EDTA, 1.5 M NaCl, 2% (w/v) CTAB) at 55°C overnight before being extracted twice with chloroform: isoamyl alcohol 24:1. About 10% volume of 5M NaCl was added into the supernatant before being precipitated with 100% ethanol. DNA quality was measured by running gel electrophoresis (1% agarose, Bioline) and using NanoDrop 2000 Spectrophotometer (Thermo Scientific, Australia). Only DNA samples that meet the quality check criteria (sample concentration: ≥30 ng μL^{-1}; total DNA quantity ≥3 µg; no or partial degradation; DNA absorbance ratios OD260/280:1.8-2.0 and OD260/230: 2.0-2.2) were sent to Beijing Genomics Institute (BGI) in Hong Kong for NGS sequencing. At BGI, high quality DNA samples were applied in library construction (insert size 500 bp) before being sequenced in an Illumina HiSeq2000 platform.

Chloroplast genome assembly and annotation Illumina sequencing data were de novo assembled using SOAPdenovo-Trans (Xie *et al.* 2014). Annotation of the *N. trichotoma* chloroplast genome was performed using DOGMA with default settings (Wyman *et al.* 2004). In addition, all tRNA genes were further verified online using tRNAscan-SE search server (Lowe and Eddy 1997) (http://lowelab.ucsc.edu/tRNAscan-SE/). The circular *N. trichotoma* chloroplast genome map was drawn using OGDraw v1.2 (Lohse *et al.* 2007).

Genome analysis The chloroplast genome of *N. trichotoma* was comparatively studied with the chloroplast genomes of other species of the *Stipeae* tribe, including *Stipa purpurea* (NC_029390.1), *Stipa lipskyi* (NC_028444.1), *Piptochaetium avenaceum* (NC_027483.1), *Oryzopsis asperifolia* (NC_027479.1) and *Achnatherum hymenoides* (NC_027464.1) (originally defined as *Stipa hymenoides*). All of these chloroplast genomes were re-analysed in DOGMA for better comparison with the results of *N. trichotoma*. GMATo (Wang *et al.* 2013) was applied to search Simple Sequence Repeats (SSR) existing in the genomes. Whole genomes across these six species were aligned using progressive MAUVE implemented by MAUVE v2.3.1 software (Darling *et al.* 2004).

RESULTS AND DISCUSSION

General features of the *N. trichotoma* chloroplast genome The obtained *N. trichotoma* chloroplast genome in present study is 112,102 bp in size, encodes 118 unique genes including 85 protein-coding genes, 29 tRNA genes and 4 ribosomal RNA genes. Some genes were duplicated in the genome, including *atpF*, *ndhA* , *ndhB* , *ndhH* , *orf56* , *rpl2* , *rpl23* , *ycf1* , *ycf2*, *ycf3*, *trnA-UGC* , *trnI-GAU* , *trnL-UAA* , *trnM-CAU* , *trnT-GGU* and *trnV-UAC*. This increases the total gene number of *N. trichotoma* chloroplast genome to 140. The total GC content of the genome is 37.88%, which is relatively lower than that of the reported genomes within the *Stipeae* tribe (Table 1).

AT-rich regions in the *N. trichotoma* chloroplast genome are intergenic (66.17%) and protein-coding (60.24%), while rRNAs (45.3%) and tRNAs (48.53%) have a much lower AT content. This pattern is similar to other five *Stipeae* species (Table 1). AT content in genomic regions were reported to be associated with the dynamics of repeats, the codon bias of chloroplast protein-coding genes and the regulation of gene expression (Rouwendal *et al.* 1997, Morton 2003). Further analyses are thus needed to investigate the links between AT content and its significance to the functions of *N. trichotoma*.

The annotated chloroplast genome of *N. trichotoma* is shown in (Figure 1). The genome is similar to that of other *Stipeae* species but shorter in size (25,752 bp shorter than the largest chloroplast genome of *Stipa lipskyi*). Only one inverted repeat (IRA) was identified to separate the long single copy section (LSC) and the short single copy section (SSC) in the genome, which is different from the remaining *Stipeae* species.

Genome comparison The *N. trichotoma* chloroplast genome was aligned with the chloroplast genomes of other *Stipeae* species to compare the organization of these genomes. Three major locally collinear blocks (LCBs) across these genomes were identified (Figure 2). These blocks suggest a high level of similarity in genome organization of *S. purpurea, A. hymenoides, P. avenaceum* and *O. asperifolia*. By comparison, the gene arrangements of *S. lipskyi* and *N. trichotoma* are different from these four chloroplast genomes as several inversions and translocations events occurred through these two chloroplast genomes when compared with that of the other four *Stipeae* species. The alignment appears to suggest a close evolutionary relationship between *N. trichotoma* and *S. lipskyi*, and a more distant relationship between *N. trichotoma* and the remaining *Stipeae* species. Nevertheless, further research is required to verify this conclusion.

It was noted that the gene content of the five *Stipeae* species chloroplast genomes are different from those previously published in NCBI (Table 1), which are probably due to differences in annotation methodology. There were three major differences between *N. trichotoma* and the remaining *Stipeae* species. Firstly, the copies of duplicated genes in *N. trichotoma* were only half of the corresponding gene

Table 1. Genome features of the chloroplast genome of six species from the *Stipeae* tribe (gene numbers shown in parentheses are the results of DOGMA re-analyses).

Characteristics	*Nassella trichotoma*	*Oryzopsis asperifolia*	*Piptochaetium avenaceum*	*Achnatherum hymenoides*	*Stipa lipskyi*	*Stipa purpurea*
GenBank Accession No.	TBA	NC_027479	NC_027483	NC_027464	NC_028444	NC_029390.1
Size (bp)	112,102	134,281	137,701	137,742	137,854	137,370
GC content (%)	37.88	38.8	38.8	38.8	38.8	38.8
Total number of genes	140	130 (175)	129 (179)	130 (181)	129 (181)	130 (181)
Total number of unique genes	118	119	119	119	119	119
Protein-coding genes	99	84 (118)	83 (122)	84 (124)	81 (124)	84 (124)
Ribosomal RNAs	4	8	8	8	8	8
Transfer RNAs	37	38 (49)	38 (49)	38 (49)	40 (49)	38 (49)
Protein-coding genes (bp)	58974	63717	66447	66510	66201	66231
Ribosomal RNAs (bp)	4596	9192	9192	7972	9192	9192
Transfer RNAs (bp)	2442	3110	3110	3112	3110	3110
Intergenetic regions (bp)	48764	59333	60234	60240	60198	59942
AT content (%)						
Genome	62.04	61.23	61.23	61.18	61.20	61.21
Protein-coding genes	60.24	60.32	60.35	60.22	60.25	60.30
Ribosomal RNAs	45.3	45.3	45.28	45.48	45.28	45.28
Transfer RNAs	48.53	48.04	48.07	48.04	47.91	47.91
Intergenic regions	66.17	65.3	65.25	65.26	65.23	65.26

copies in the five *Stipeae* species. These include 11 protein-coding genes (*ndhB, orf42, orf56, rpl2, rps12, rps12_3end, rps15, rps19, rps7, ycf1, ycf15)*, eight tRNA genes *(trnA-UGC, trnH-GUG, trnI-CAU, trnI-GAU, trnL-CAA, trnN-GUU, trnR-ACG, trnV-GAC*) and all rRNA genes (*rrn16, rrn23, rrn4.5, rrn5*). Secondly, two genes were not present in the five *Stipeae* species but were found in the chloroplast genome of *N. trichotoma* (*psbA* and *psbG*). Thirdly, another two genes (*matK* and *trnK-UUU*) were not found in *N. trichotoma* but were present in the five chloroplast genomes of *Stipeae* species.

Simple Sequence Repeats Three Simple Sequence Repeats (SSR) were identified in the chloroplast genome of *N. trichotoma*, including a 'TC' 5× repetitions between the range of 12636 and 12645, a 'TA' 5× repetitions between the range of 30178 and 30187 and a 'AT' 5× repetitions between the range of 89616 and 89625. The SSR numbers in *N. trichotoma* was less than that of the five *Stipeae* species [*A. hymenoides* (4), *O. asperifolia* (7), *P. avenaceum* (7), *S. lipskyi* (7), and *S. purpurea* (7)].

Intergenic regions *Nassella trichotoma* contains 125 intergenic regions, which is less than other *Stipeae* species (160 in *O. asperifolia*; 162 in *P. avenaceum* and *A. hymenoides*; 163 in *S. lipskyi* and 164 in *S. purpurea*). Among them, seven intergenic regions are greater than 1000 bp in size, whist 29 other intergenic regions have sequences greater than 500 bp in size. This provides a valuable resource for selecting robust DNA barcodes for differentiation of *N. trichotoma* from other grasses because intergenic regions have been widely used as plastid barcodes for species differentiation (Dong *et al.* 2012, Suzuki *et al.* 2014), particularly for those with highly variable sequence areas and in good size.

In summary, the present study represents the first attempt to sequence the chloroplast genome of *Nassella* using NGS technology. In comparison to chloroplast genomes of five published *Stipeae* species, the available data indicates that gene re-arrangements and inversions have likely occurred in the evolution of *N. trichotoma* chloroplast genome. Our data provided valuable resources for selecting robust markers for DNA barcoding of *Nassella* weeds species, and made it easier to sequence the chloroplast genomes of other *Nassella* species (as the current data could be used as a reference genome for genome assembly). Nevertheless, more research is required for the chloroplast genome of *N. trichotoma* as the evidences from the present study (e.g. shorter genome size and reduced gene numbers of the genome) suggested that the obtained chloroplast genome of *N. trichotoma* is likely to be incomplete.

ACKNOWLEDGMENTS

The authors would like to thank Dr Shanlin Liu and Dr Chentao Yang at Beijing Genomics Institute (BGI) for technical support of the NGS sequencing; the authors would also like to acknowledge Mr Jeremy Crocker at Wagga Wagga City Council for assisting to collect *N. trichotoma* samples.

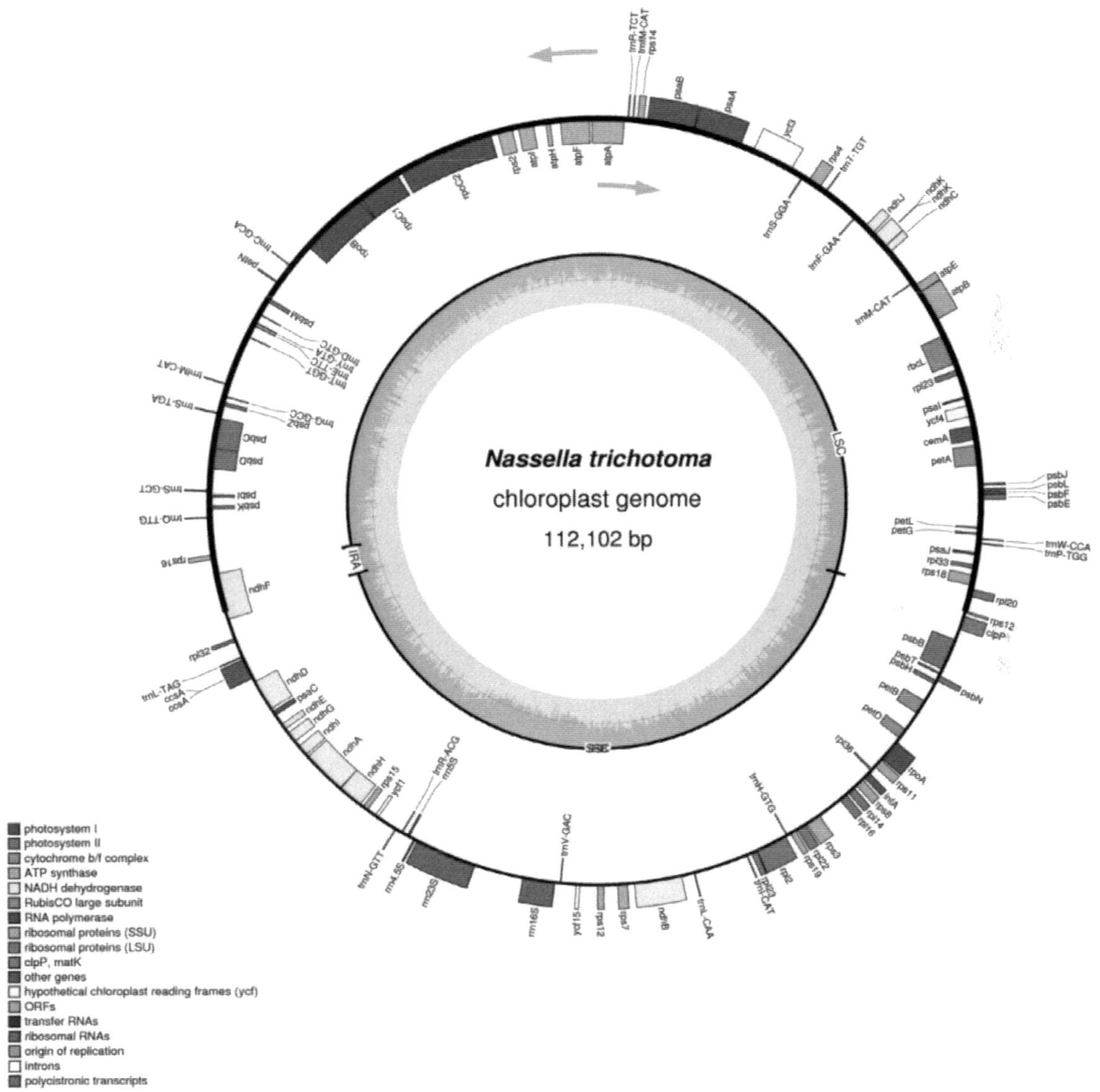

Figure 1. Sequence map of the *Nassella trichotoma* chloroplast genome. Genes drawn inside of the circle are transcribed clockwise, while genes shown outside of the circle are transcribed counter-clockwise. Genes belonging to different functional groups are colour-coded. The darker grey in the inner circle indicates GC content, while the lighter grey corresponds to AT content.

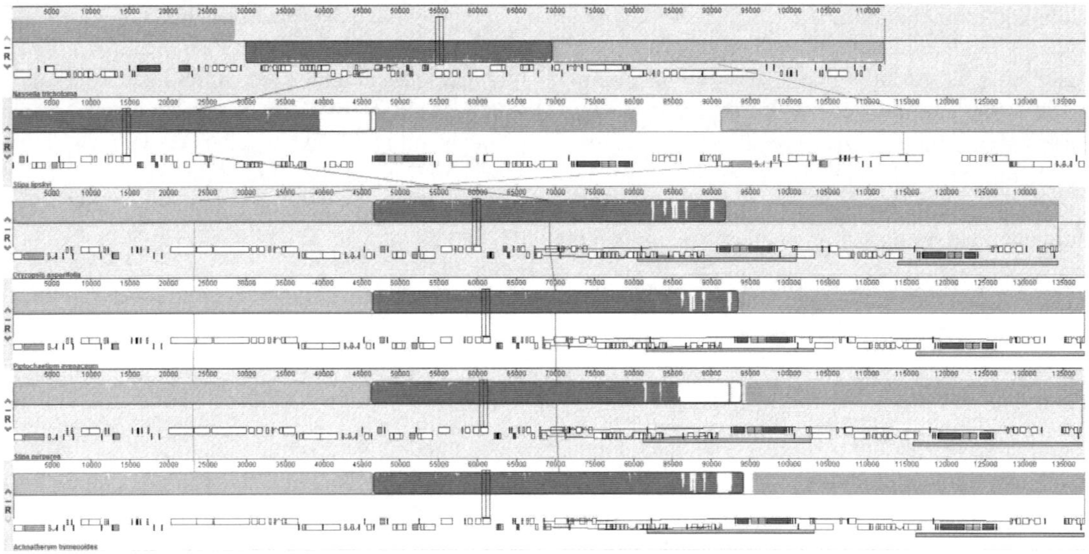

Figure 2. Alignment of *N. trichotoma* chloroplast genome with the chloroplast genomes of five *Stipeae* species (*Stipa purpurea, Stipa lipskyi, Piptochaetium avenaceum, Oryzopsis asperifolia* and *Achnatherum hymenoides*). Three major locally collinear blocks (LCBs) were labelled with three different colours (orange, red and green).

REFERENCES

Darling, A.C., Mau, B., Blattner, F.R. and Perna, N.T. (2004). Mauve: multiple alignment of conserved genomic sequence with rearrangements. *Genome Research* 14, 1394-1403.

Dong, W.P, Liu, J., Yu, J., Wang, L. and Zhou, S.L. (2012). Highly variable chloroplast markers for evaluating plant phylogeny at low taxonomic levels and for DNA barcoding. *PLoS One* 7 (4): e35071. doi: 10.1371/journal.pone.0035071

Group CPW. (2009). A DNA barcode for land plants. *Proceedings of the National Academy of Sciences U.S.A.* 106, 12794-97.

Lohse, M., Drechsel, O. and Bock, R. (2007). OrganellarGenomeDRAW (OGDRAW): a tool for the easy generation of high-quality custom graphical maps of plastid and mitochondrial genomes. *Current Genetics* 52, 267-74.

Lowe, T.M. and Eddy, S.R. (1997). tRNAscan-SE: a program for improved detection of transfer RNA genes in genomic sequence. *Nucleic Acids Research* 25, 955-64.

Morton, B.R. (2003). The role of context-dependent mutations in generating compositional and codon usage bias in grass chloroplast DNA. *Journal of Molecular Evolution* 56, 616-29.

Rouwendal, G.J., Mendes, O., Wolbert, E.J. and Douwe de Boer, A. (1997). Enhanced expression in tobacco of the gene encoding green fluorescent protein by modification of its codon usage. *Plant Molecular Biology* 33, 989-99.

Sambrook, J. and Russell, D.W. (2006). Purification of nucleic acids by extraction with phenol: chloroform. *Cold Spring Harbor Protocols* 2006.

Suzuki, J.Y., Matsumoto, T.K., Keith, L.M. and Myers, R.Y. (2014). The chloroplast psbK-psbI intergenic region, a potential genetic marker for broad sectional relationships in *Anthurium*. *Hortscience* 49, 1244-52.

Syme, A.E., Udovicic, F., Stajsic, V. and Murphy, D.J. (2013). A test of sequence-matching algorithms for a DNA barcode database of invasive grasses. *DNA Barcodes* 1, 19-26.

Wang, A, Gopurenko, D., Wu, H., Stanton, R. and Lepschi, B. (2014). DNA barcoding for identification of exotic grass species present in eastern Australia. Proceedings of the 19th Australasian Weeds Conference – Science, Community and Food Security: the Weed Challenge, 1–4 September: Hobart, Tasmania, pp. 444-7.

Wang, X., Lu, P. and Luo, Z. (2013). GMATo: a novel tool for the identification and analysis of microsatellites in large genomes. *Bioinformation* 9, 541-4.

Wyman, S.K., Jansen, R.K. and Boore, J.L. (2004). Automatic annotation of organellar genomes with DOGMA. *Bioinformatics* 20, 3252-5.

Xie, Y., Wu G., Tang, J., Luo, R., Patterson, J., Liu, S., Huang, W., He, G., Gu, S., Li, S., *et al.* (2014). SOAPdenovo-Trans: de novo transcriptome assembly with short RNA-Seq reads. *Bioinformatics* 30, 1660-6.

Introduction of Paterson's curse (*Echium plantagineum*) to Australia – unravelling the story by DNA sequence analysis

Xiaocheng Zhu[1], David Gopurenko[2], Dominik Skoneczny[1], Brendan J. Lepschi[3], Mark A. Spencer[4], Geoff M. Gurr[1], Ragan M. Callaway[5], Miguel Serrano[6] and Leslie A. Weston[1]

[1] Graham Centre for Agricultural Innovation (Charles Sturt University and NSW Department of Primary Industries), Charles Sturt University, Wagga Wagga, New South Wales 2678, Australia

[2] NSW Department of Primary Industries, Wagga Wagga Agricultural Institute, Wagga Wagga, New South Wales 2650, Australia

[3] Australian National Herbarium, Centre for Australian National Biodiversity Research, Canberra, Australian Capital Territory 2601, Australia

[4] Department of Life Sciences, Natural History Museum, London, SW7 5BD, United Kingdom

[5] Division of Biological Sciences, University of Montana, Missoula, 59812, United States of America

[6] Department of Botany, Faculty of Pharmacy, University of Santiago de Compostela, 15782 Santiago de Compostela, Spain

Correspondence and requests for materials should be addressed to L.A.Weston (leweston@csu.edu.au)

Summary Paterson's curse (*Echium plantagineum* L.) is an annual/biennial species in the family Boraginaceae. Native to N.W. Africa, the Iberian Peninsula and Atlantic Western Europe, it is a noxious weed in Australia and has spread over 30 million ha resulting in >A$250 million annual losses to agricultural industries (Anon. 2009). Determination of the pathway of introduction is critical to gain further insight into the invasion process including the sources of propagules.

Two hypotheses regarding the origin of Australian populations were formulated: 1) introduction as an ornamental via seed importation from England; 2) multiple introductions via South Africa from the Iberian Peninsula and UK.

In this study, 131 Australian *E. plantagineum* samples were analysed using one nuclear gene region (ITS) and three chloroplast gene regions (*trn*L intron, *trn*L-*trn*F spacer and *trn*H-*psb*A) and compared with successfully sequenced samples from the Iberian Peninsula (N = 43), UK (N = 15), South Africa (N = 24) and the USA (N = 4). A high level of genetic diversity was found in Australia, which suggests a complex history of introduction of Australian *E. plantagineum*. The most abundant haplotype in Australia was shared with the Iberian Peninsula and UK populations, which suggests that they are both critical source populations for Australian *E. plantagineum*.

Importantly, however, 10 of 12 haplotypes found in Australia were represented in South Africa, indicating its potentially important historical role as an intermediate source population for *E. plantagineum* introductions. The Australian *E. plantagineum* population was genetically most similar to the UK population and significantly different from other populations, which supports UK *E. plantagineum* as potential founding population and the Iberian Peninsula and South African populations as additional sources of genetic diversity.

Keywords Genetic diversity, invasive weed, introduction history, DNA barcoding.

INTRODUCTION

Paterson's curse (*Echium plantagineum* L.) is a noxious weed in Australia. It produces toxic pyrrolizidine alkaloids in the shoots (Skoneczny *et al.* 2015, Weston *et al.* 2013) and bioactive shikonins in the root (Weston *et al.* 2011, Zhu *et al.* 2016a). It competes with pasture grasses and is often toxic to grazing livestock. *Echium plantagineum* now infests more than 30 million ha and causes over A$250 million in annual losses (Anon. 2009).

Determining the path of introduction of an invasive species is critical in comprehending the plant invasion process, managing current infestations, and preventing future introductions. *Echium plantagineum* is native to N.W. Africa, the Iberian Peninsula and Atlantic Western and was first recorded in 1843 in the Camden Gardens, Sydney as an ornamental obtained from the UK (Piggin 1977).

This record, for decades, has been considered the first introduction event (Piggin 1982). However, considering the introduction of Merino sheep from the native range of *E. plantagineum* in the late 1800s, Zhu *et al.* (2016b) hypothesized that *E. plantagineum* was introduced to Australia much earlier than 1843, as an accidental by-product of livestock shipments to the country.

Molecular approaches such as DNA sequence analysis have been widely used for genetic structure and introduction history studies (Shaik *et al.* 2015). However, information on the genetic background of *E. plantagineum* is very limited. Burdon and Brown (1986) reported a similar level of genetic diversity of Australian *E. plantagineum* compared to the species in the native ranges using isozyme markers. A nuclear gene region (ITS) and three chloroplast regions (*trn*H-*psb*A spacer, *trn*L intron and *trn*L-*trn*F spacer) were recently investigated to separate two closely related species: *E. plantagineum* and *E. vulgare* (Zhu *et al.* 2014). These gene regions were further used to investigate genetic diversity and structure of both species in Australia (Zhu *et al.* 2016b). Region-specific haplotypes were detected, which supports the theory of multiple introduction events of *E. plantagineum* in Australia (Zhu *et al.* 2016b).

In this study, four previously explored gene regions (Zhu *et al.* 2014) were used to investigate the introduction history of Australian *E. plantagineum* by comparing the genetic diversity of populations from the native Iberian Peninsula population to Australia, South Africa and the UK. We hypothesized that, *E. plantagineum* was historically sourced directly from the UK and the Iberian Peninsula, as well as indirectly through South Africa.

MATERIALS AND METHODS

Leaf tissue was sampled from both the field and preserved herbarium specimens, provided by the Australian National Herbarium and the Natural History Museum, London UK. Geographic distribution of sampled plants comprised: the Iberian Peninsula (N = 53), Australia (N = 131), South Africa (N = 25), the UK (N = 40) and the USA (N = 4).

DNA extraction, PCR, sequencing, alignment and network analysis were performed as described by Zhu *et al.* (2013). Samples were sequenced for one nuclear region (ITS) and three chloroplastic regions (*trn*H-*psb*A spacer, *trn*L intron and *trn*L-*trn*F spacer).

RESULTS

Only two alleles were found in the ITS region. The level of polymorphism in the ITS region was not sufficient for genotype investigation comparing the native and invasive ranges and was disregarded for further analysis.

By contrast, the chloroplastic regions are more polymorphic than the nuclear region. Only samples with all three chloroplastic regions successfully sequenced were used in further analysis, including; 43, 131, 24, 15 and 4 samples from the Iberian Peninsula, Australia, South Africa, the UK and the USA, respectively (Table 1). Low PCR success in UK samples probably because of DNA degradation due to long term preservation of herbarium specimens. Analysed regions were represented by 8, 12, 11, 4 and 2 haplotypes, respectively (Table 1). Levels of nucleotide (π) and haplotype (h) diversity at the cpDNA regions were similar among the Iberian Peninsula ($\pi = 0.0016$; $h = 0.7409$), Australia ($\pi = 0.0020$; $h = 0.7657$) and South Africa ($\pi = 0.0023$; $h = 0.8913$) (Table 2). In contrast, *E. plantagineum* from the UK was genetically much less diverse compared to other populations ($\pi = 0.0010$; $h = 0.3714$).

Ten out of 12 haplotypes in Australia were recovered in South Africa despite a relatively small sample size (N = 24, in South Africa; Table 1 and Figure 1). However, the most common haplotype (Hap 4) in Australia was absent in South Africa, but present in the UK as 80% of haplotypes encountered, despite a small sample size in the UK. No significant difference was found between UK and Australian populations of *E. plantagineum* using a population pairwise F_{ST} test. In contrast, significant pairwise differences were found between all other *E. plantagineum* populations (Table 3).

Table 1. *Echium plantagineum* haplotypes found in the Iberian Peninsula (native range), Australia (Aus), South Africa (SAf), UK and USA. *n*: number of samples successfully sequenced at the three chloroplastic regions. * indicates the presence of particular haplotype.

Region	*n*	H1	H2	H3	H4	H5	H6	H7	H8	H9	H10	H11	H12	H13	H14
Native	43		*	*	*		*	*	*			*		*	
Aus	131	*	*	*	*	*	*	*	*	*	*	*	*		
SAf	24	*	*	*		*	*	*	*	*		*	*		*
UK	15		*		*	*	*								
USA	4		*		*										

DISCUSSION

High levels of chloroplast genetic diversity evident in *E. plantagineum* sampled from its native range in the Iberian Peninsula, and from invasive populations in South Africa and Australia, suggest a complex history of introduction of this species. Interestingly, most haplotypes detected in Australia were also present in the invasive South Africa population as opposed to the native population in the Iberian Peninsula (Table 1 and Figure 1), indicating either un-sampled diversity is present in the Iberian Peninsula, or was lost from that region after introduction of the species into South Africa. Seven and 10 out of 12 Australian haplotypes were found in the Iberian Peninsula and South Africa, respectively, which suggests that the Iberian Peninsula and South Africa were likely important sources of *E. plantagineum* introduction to Australia.

Table 2. Nucleotide (π) and haplotype (h) diversity of *E. plantagineum* populations from the Iberian Peninsula (native range), Australia, South Africa, the UK and the USA.

Populations	π	h
Native	0.0016 ± 0.0011	0.7409 ± 0.0565
Australia	0.0020 ± 0.0012	0.7657 ± 0.0297
South Africa	0.0023 ± 0.0014	0.8913 ± 0.0395
UK	0.0010 ± 0.0008	0.3714 ± 0.1532
USA	0.0006 ± 0.0007	0.6667 ± 0.2041

Table 3. Pairwise comparison (shown as F_{ST} value) of genetic structure between *E. plantagineum* populations.

Populations	Australia	UK	South Africa
Australia			
UK	0.03		
South Africa	0.16***	0.32***	
Native	0.21***	0.38***	0.06*

Statistical significance indicated as *** P <0.001 and * P <0.05. USA samples were excluded in this figure due to limited sample size.

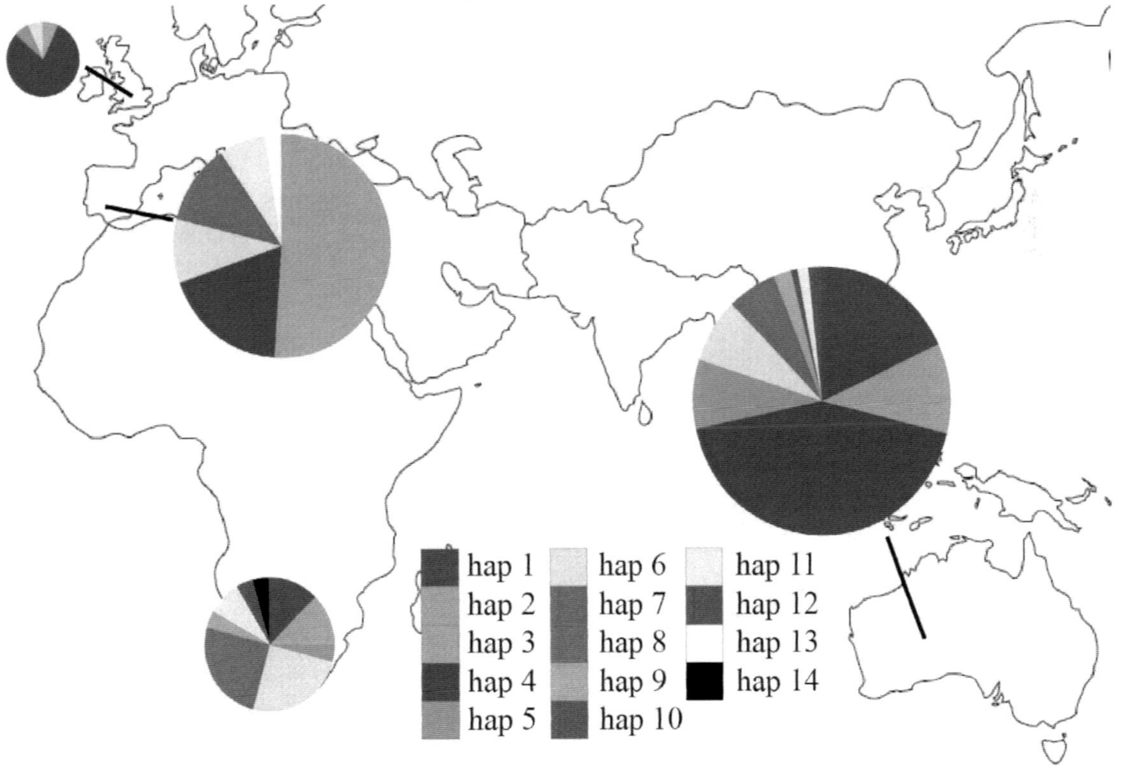

Figure 1. Distribution of haplotypes of *E. plantagineum* in its native range (Iberian Peninsula), Australia, South Africa and the UK. The USA samples were excluded in this figure due to limited sample size.

Interestingly, although *E. plantagineum* has been cultivated since 1658 in the UK (Kloot 1982), it showed a very limited sequence variation. Four haplotypes were found in the UK with one haplotype predominating (Hap 4, Figure 1), while the other three haplotypes were represented by only a single specimen. The predominant Hap 4 in the UK is also the most frequent and widely distributed haplotype in Australia (Figure 1), indicating a portion of the *E. plantagineum* diversity in Australia was most likely sourced either directly as an ornamental import from the UK, and/or indirectly from native populations in the Iberian Peninsula where the haplotype is also evident. Historically, climate conditions in the UK have generally been unsuitable for the establishment of *E. plantagineum* (Dr. Mark Spencer, pers. comm.).

The limited diversity evident in UK samples suggests adaptive selection for a restricted range of haplotypes tolerant of UK climate. Alternatively, population bottleneck processes associated with invasion history may also have limited the diversity present in the UK. All herbarium specimens collected from the UK evaluated in this study were relatively old (from 1845–1959). Established populations of this species in the UK are now generally restricted to the Channel Islands and Cornwall. It would be useful to monitor current genetic composition in the UK.

The high level of genetic diversity in South Africa suggests either multiple introductions of *E. plantagineum* or single introduction of multiple genotypes to this region. However, introduction of 11 haplotypes at one point in time is unlikely, especially considering the presence of several rare haplotypes from the native range (Hap 2, 9, 11, 12 and 14). The South African samples investigated in this study were collected from localities near Cape Town, which was the most important port connecting Europe to the Far East, before the opening of the Suez Canal in 1869. Since 1488, shipments of agricultural products from the UK and the Iberian Peninsula travelling to Asia and Australia (onwards from 1788) travelled to Cape Town to replenish their supplies before crossing the Indian Ocean. Centuries of trade likely enabled the introduction of less common haplotypes to South Africa from Europe with seed contamination of agricultural products leading to the high level of genetic diversity observed in South Africa. Similarly, introductions via South Africa to Australia were also noted in other invasive species such as capeweed (Scott 1990) and redlegged earth mite (Ridsdill-Smith 1997). Surprisingly, the most common Hap 4 in the UK and Australia was not observed in South Africa. This may be associated with the limited sampling size or lack of this genotype in South Africa.

Colonial British settlement of Australia after 1788 allowed the export of produce to Australia from Europe via South Africa. Serial accidental introductions of *E. plantagineum* from South Africa to Australia would therefore likely have increased the probability of export of *E. plantagineum* genetic diversity in South Africa to Australia.

In conclusion, the genetic diversity of invasive *E. plantagineum* in Australia was likely derived from deliberate introduction of plants as ornamental subjects from the UK in 1840s or the accidental introduction via agricultural products from the UK since 1788. In addition, Australian introductions of *E. plantagineum* were reinforced by trade shipping and accidental importations of seeds from genetically diverse, invasive *E. plantagineum* populations in South Africa, which supplements additional genetic diversity in Australia. Considering the large number of ships arriving in Australia via South Africa from UK, the first introduction event of *E. plantagineum* to Australia could also potentially have occurred much earlier than that first recorded in 1843.

ACKNOWLEDGMENTS

The authors acknowledge the Australian Research Council (Discovery Project: DP130104346) for financial support, the Australian National Herbarium and the Natural History Museum, London for access to herbarium specimens and associated data. We also thank Drs P.J. Pieterse and Charlie Reinhardt for providing South Africa samples.

REFERENCES

Anon. (2009). Paterson's curse. Weeds of Southern Tasmania. NRM South and the Southern Tasmanian Councils Authority. Retrieved 25th January 2016, from http://www.nrmsouth.org.au/wp-content/uploads/2014/10/patersons_curse.pdf

Burdon, J.J. and Brown, A.H.D. (1986). Population genetics of *Echium plantagineum* L. – target weed for biological control. *Australian Journal of Biological Sciences* 39, 369-78.

Kloot, P.M. (1982). The naturalization of *Echium plantagineum* L. in Australia. *Australian Weeds* 1, 29-31.

Piggin, C.M. (1977). The herbaceous species of *Echium* (Boraginaceae) naturalized in Australia. *Muelleria* 3, 215-44.

Piggin, C.M. (1982). The biology of Australian weeds. 8. *Echium plantagineum* L. *Journal of the Australian Institute of Agricultural Science* 48, 3-16.

Ridsdill-Smith T.J. (1997). Biology and control of *Halotydeus destructor* (Tucker) (Acarina:

Penthaleidae): a review. *Experimental and Applied Acarology* 21, 193-223.

Scott, J.K. amd Way, M.J. (1990). A survey in South Africa for potential biological control agents against capeweed, *Arctotheca calendula* (L.) Levyns (Asteraceae). *Plant Protection Quarterly* 5, 31-4.

Shaik, R., Gopurenko, D., Urwin, N.R. *et al.* (2015). Population genetics of invasive *Citrullus lanatus*, *Citrullus colocynthis* and *Cucumis myriocarpus* (Cucurbitaceae) in Australia: inferences based on chloroplast and nuclear gene sequencing. *Biological Invasions* 17, 2475-90.

Skoneczny, D., Weston, P.A., Zhu, X. *et al.* (2015). Metabolomic profiling of pyrrolizidine alkaloids in foliar of two *Echium* spp. invaders in Australia – a case of novel weapons? *International Journal of Molecular Sciences* 16, 26721-37.

Weston, L.A., Weston, P.A. and McCully, M. (2011). Production of bioactive naphthoquinones by roots of Paterson's Curse (*Echium plantagineum* L.) – implications for invasion success? 23rd Asian-Pacific Weed Science Society Conference, eds S. Adkins and R. McFadden, Cairns, Australia, pp. 576-84.

Weston, P., Weston, L. and Hildebrand, S. (2013). Metabolic profiling in *Echium plantagineum*: presence of bioactive pyrrolizidine alkaloids and napthoquinones from accessions across southeastern Australia. *Phytochemistry Reviews* 12, 831-37.

Zhu, X., Skoneczny, D. Weidenhamer, J.D. *et al.* (2016a). Identification and localization of bioactive naphthoquinones in the roots and rhizosphere of Paterson's curse (*Echium plantagineum*), a noxious invader. *Journal of Experimental Botany* 67, 3777-88.

Zhu, X., Meyer, L., Gopurenko, D. *et al.* (2014). Selection of DNA barcoding regions for identification and genetic analysis of two *Echium* invaders in Australia: *E. plantagineum* and *E. vulgare*. Proceedings of 19th Australasian Weeds Conference, ed. M. Baker, 396-400. (Tasmanian Weed Society, Hobart, TAS, Australia).

Zhu, X., Weston, P., Skoneczny, D. *et al.* (2016b). A tale of two plant invaders: comparison of the ecology and genetics of *Echium plantagineum* and *E. vulgare* in southern Australia. *Scientific Reports* under review.

Herbicide effectiveness under elevated CO_2

Paul O. Downey[1], Tanja I. Lenz[2], Pawel Waryszak[3] and Michelle R. Leishman[4]
[1] Institute for Applied Ecology, University of Canberra, Australian Capital Territory 2601, Australia
[2] School of Agriculture, Food and Wine, Adelaide University, South Australia 5005, Australia
[3] Terrestrial Ecology Research Group, Murdoch University, Western Australia 6150, Australia
[4] Department of Biological Sciences, Macquarie University, New South Wales 2109, Australia
(pawel.waryszak@gmail.com)

Summary Chemical control is the most ubiquitous management technique for invasive plant species worldwide. Although the increase in atmospheric CO_2 is thought to affect the efficacy of herbicides, there have been few studies that have investigated the effect of elevated CO_2 concentration on herbicide efficiency in invasive plants and these have focused on CO_2 effect on efficacy of the widely used herbicide glyphosate.

In this study we assessed the effect of two herbicides on four functional types of exotic invasive plant species: grasses, herbs, shrubs and trees. In total, we grew 15 common Australian environmental weed species at ambient and elevated CO_2 concentrations and applied recommended and double-recommended concentrations of glyphosate and metsulfuron methyl herbicides. Increased CO_2 showed a positive fertilising effect on plant biomass but the effect of the herbicides on plant survival was very variable.

We found no clear relationships between species' responses to herbicide under elevated CO_2 and a range of functional and allocation traits. Our results suggest that plant responses and herbicide efficacy under future CO_2 conditions may change but that the differences in response will be species-specific. The potential implications for weed management in the future will be discussed.

Keywords Herbicide, chemical control, invasive species, climate change, elevated carbon dioxide.

Hawkweed (*Hieracium* spp.) surveillance: development of a targeted and robust plan for the Victorian Alps

Angela Constantine[1], Cindy E. Hauser[2], Keith Primrose[3] and Neil Smith[1]
[1] Department of Economic Development, Jobs, Transport and Resources, 3/2 Codrington Street, Cranbourne, Victoria 3977, Australia
[2] School of BioSciences, The University of Melbourne, Parkville, Victoria 3010, Australia
[3] Parks Victoria, PO Box 422, Mount Beauty, Victoria 3699, Australia
(angela.constantine@ecodev.vic.gov.au)

Summary In Victoria, hawkweeds (*Hieracium* spp.) are declared as State prohibited weeds, the highest category of noxious weeds, which are managed with the aim of eradication from the state due to their significant threat to Victoria's environmental, agricultural and social values. Surveillance is a vital step in the eradication process to ensure that all plants are detected and controlled.

There are a number of small hawkweed infestations in Victoria, the largest of which is in the Falls Creek and Bogong High Plains area of the Victorian Alps. It is not possible to survey the entirety of the landscape each year, so creating a surveillance plan is necessary, but complex. The landscape must be closely examined to prioritise areas where surveillance will be most beneficial.

To assist in surveillance planning, a model developed by the University of Melbourne was used to prioritise areas for surveillance. The model requires inputs of information for each potential survey location: the likelihood of hawkweed occurrence; the detectability of hawkweed; and the value of finding a new hawkweed infestation, relative to other areas. The model uses this information, plus the hours of surveillance effort available, to inform the priority grids for survey.

The model was used to develop a three year surveillance plan for hawkweed at Falls Creek and the surrounding area. The results and learnings from the first year of the plan will also be discussed.

Keywords Hawkweed, *Hieracium*, surveillance.

It takes a village: detection dogs, partnerships and volunteers aid hawkweed eradication in mainland Australia

Hillary Cherry[1], Angela Constantine[2], Keith Primrose[3], Cindy Hauser[4] and Kirsten Tasker[5]
[1] National Parks and Wildlife Service, PO Box 1967, Hurstville, New South Wales 1481, Australia
[2] Department of Economic Development, Jobs, Transport and Resources, 3/2 Codrington Street, Cranbourne, Victoria 3977, Australia
[3] Parks Victoria, 225 Kiewa Valley Highway, Tawonga South, Victoria 3698, Australia
[4] School of BioSciences, University of Melbourne, Parkville, Victoria 3010, Australia
[5] ACT Parks and Conservation Services, 500 Cotter Road, Weston, Australian Capital Territory 2611, Australia
(hillary.cherry@environment.nsw.gov.au)

Summary Weed eradication is costly and challenging, but the benefits of successful eradication are considerable. Hawkweed (*Hieracium* L.) species pose a significant threat to south eastern Australia, where several small incursions are documented. Four jurisdictions are partnering to develop best practice management techniques and work towards eradication of hawkweeds from the Australian mainland. Here we discuss the application of three key tools: 1) extensive partnerships necessary to maintain the programs, 2) the sustained participation of large numbers of volunteers in surveillance, and 3) the training of Weed Eradication Detector Dogs to 'verify' hawkweed freedom and assist in delimitation. We highlight the benefits of these tools, as additions to conventional surveillance and control, and how they can aid success of eradication programs.

Keywords Detection dog, weed eradication, surveillance, partnerships, hawkweed, *Hieracium*.

INTRODUCTION

Eradication of invasive plants (weeds) is considered the most cost-effective management action once an incursion is documented in a previously un-invaded area, yet there are few examples of weed eradications globally (Panetta 2014). In part, this may be because weed eradication is costly and challenging, and requires continued commitment from those affected or potentially affected by the weed, as well as social and political commitment to resourcing the program to the end. Eradication involves the elimination of every individual and its propagules from an area where recolonisation is unlikely to occur (Myers *et al.* 1998), hence eradication programs must continue until all plants are found and controlled, and the weed seedbank is exhausted (Panetta 2014). Depending on the seed persistence of the weed, this may require ten or more years, which is longer than most governmental political and budget cycles. There are many other criteria to address when considering the feasibility of an eradication attempt (e.g. ability to delimit and control entire infestations, etc.: c.f. Wittenberg and Cock 2001, Dodd *et al.* 2014, Panetta 2014), many of which are costly or difficult to meet. Thus, weed eradication is not often attempted, despite the high cost-benefit of significant early investment in eradication versus continued costs of indefinite weed control (Wittenberg and Cock 2001).

In contrast to on-going weed control, eradication has an 'end point' where weed management costs and potential weed impacts are eliminated: These significant benefits make eradication worthy of consideration. This is especially true when the known weed risks are high, and when the weed has shown considerable negative impacts in similar environments elsewhere in the world. This is the case for hawkweed (*Hieracium* L.) species. Weed risk assessments, including potential distribution modelling, show that hawkweeds pose a significant threat to south eastern Australia, where there are similar habitats and climates to already invaded areas (Brinkley and Bomford 2002, Beaumont *et al.* 2009), such as in New Zealand (NZ) and North America, where hawkweeds cause extensive damage, impacting agriculture and the environment (Espie 2001, Williams and Holland 2007). Several small hawkweed incursions are documented from mainland Australia, primarily in the bordering alpine regions of New South Wales (NSW), Australian Capital Territory (ACT) and Victoria, and these are the subject of an eradication attempt.

Hawkweeds can be suitable eradication targets as they have short-lived soil seedbank (Bear *et al.* 2012), effective control methods, and can be detected in the landscape, especially when in flower. However, detectability decreases when flowers are not present, and thus a key challenge in hawkweed eradication is detecting small, non-flowering plants, especially those growing amongst dense vegetation. To overcome this and other challenges, a group of partners working towards hawkweed eradication on mainland Australia are developing or refining tools to complement existing best practice weed surveillance and control methods. This paper discusses the application of three key 'tools': 1) the extensive partnership arrangements between multiple agencies, research institutions, governments and the community necessary to maintain the programs, 2) the sustained participation of large numbers of volunteers in surveillance and delimitation, and 3) the training and implementation of Weed Eradication Detector Dogs to 'verify' hawkweed freedom at individual sites and to assist in delimitation of infestations. We discuss the interactions between these components, including a recent evaluation comparing detector dog and human surveillance, and the benefits and challenges of incorporating these components into eradication programs. Background and further detail on the jurisdictional programs, including the use of other novel approaches, as well as conventional surveillance and control, are available in other publications (Hamilton *et al.* 2015, Primrose *et al.* 2016, Constantine *et al.* 2016).

Photo 1. Orange hawkweed flowers are readily detectable, but the flowering season is restricted to summer months, and non-flowering rosettes are easily camouflaged in other vegetation. Photo: Hillary Cherry.

PARTNERSHIPS

Three high-risk hawkweed species, orange hawkweed (*Hieracium aurantiacum* L.), mouse-ear hawkweed (*H. pilosella* L.), and king devil hawkweed (*H. praealtum* Vill. ex Gochnat) are documented from four jurisdictions in Australia: ACT, NSW, Tasmania and Victoria. These are either 1) not currently considered feasible for eradication (orange hawkweed in Tasmania), 2) presumed eradicated (mouse-ear hawkweed in ACT) or 3) under eradication (in NSW: orange hawkweed since 2002 and mouse-ear hawkweed since 2015; in Victoria: orange hawkweed since 1999, king devil hawkweed since 2002 and mouse-ear hawkweed since 2011). In NSW and Victoria, the known infestations are contained mainly to publicly-managed land and distributed over a relatively limited extent, which improves the feasibility of eradication. While each jurisdiction manages a collaborative hawkweed program within their region, agency staff from the four jurisdictions, as well as other experts, partner together through a National Hawkweed Working Group (NHWG) to develop and share best practice management techniques and coordinate eradication.

The NHWG comprises conservation and primary industries agency representatives involved in on-ground aspects of the program, as well as in state-level weed strategic planning, coordination and management. This broad membership base allows networking and information sharing with international and national eradication experts and weed managers, as well as with a range of researchers to improve eradication methods, including understanding seed dispersal patterns to inform search effort (Cousens *et al.* 2012), novel detection methods (Hung and Sukkarieh 2015), and improving delimitation ability (Hauser *et al.* 2016).

While collaboration between the lead agencies and experts is critical, so too are comprehensive partnerships across government and the community. NHWG members work to ensure all levels of government are aware and supportive of the program (e.g. members liaise with and update relevant Australian Government agencies on eradication progress, and regional and local representatives are regularly invited to attend NHWG meetings). Partners also interact with community members (often via local government contacts and events such as field days), as well as the media and politicians, to encourage community support (see sections below). Given the scope of stakeholders potentially affected by hawkweeds, partners must 'throw the net widely' to seek support from all levels. The NHWG is a valuable tool that enables coordination, collaboration and improvement in the eradication program.

VOLUNTEERS

Community volunteers are an integral part of the hawkweed eradication program, and complement the work of contractors and agency staff. Volunteers assist with surveillance (actively in the eradication program and passively, e.g. when bushwalking) and improve delimitation, they promote good weed hygiene practices, and they encourage political and community support for the program. Consequently, the NSW and Victorian programs have invested time and resources to establish and support a large group of volunteers. Because activities primarily occur on high-value public land (e.g. alpine national parks), the hawkweed eradication program has strong opportunities to recruit volunteers who are passionate about protecting these landscapes, or who may already be involved in other park activities. This is also advantageous in retaining volunteers because they establish ownership of the eradication program and value their role in protecting public land, which means many volunteers return year after year.

Photo 2. NSW Hawkweed Team members (2015/16) – volunteers assist with surveillance and are critical to program success. Photo: Steve Austin.

Victoria and NSW agency staff have developed extensive volunteer programs over the last eight years. Each year, volunteer information packages are developed, and opportunities to participate are advertised on agency websites and in other appropriate media. Volunteers are sourced from a range of entities such as community and 'friends' groups, associations (e.g. National Parks Associations), societies (e.g. Weed Society of Victoria), clubs such as bushwalking and field naturalist clubs (e.g. Canberra and NSW Bushwalking Clubs), schools, universities and other areas. They are trained by agency staff in plant identification, surveillance, safety, hygiene and, in some cases, chemical control (hand spot-spraying) of hawkweed in endangered plant communities. Lead agencies also monitor and report on volunteer participation and outcomes, and provide updates and coverage of volunteer and program activities via media and websites, (e.g. flickr pages dedicated to the program or other social media that can be shared back through the volunteer community).

Together the programs have engaged over 750 volunteers, many of which have returned for four years or more. During any given season, approximately 70% of volunteers are past participants. Most participate for one or more week-long blocks, and are provided with food, accommodation and all necessary training and tools. Volunteers have made major contributions to hawkweed surveillance: to date, volunteers have contributed over 10,000 hours of surveillance and control activities. Examples from the Victorian program illustrate the value of volunteer participation: since 2008, volunteers have surveyed over 825 ha for hawkweed (32% of the total 2574 ha searched), and approximately 21% of all known hawkweed 'sites' were found by volunteers (site = approximately 2m radius around centre point of incursion).

There are multiple benefits in training volunteers in plant identification and surveillance: Important hawkweed discoveries have been detected outside of targeted programs, assisting with delimitation. For example, volunteers discovered mouse-ear hawkweed for the first time in the Victorian Alps during recreational bushwalking. Also in Victoria, volunteers found a hawkweed infestation in a private, metropolitan park, and a backyard infestation was discovered when a NSW hawkweed volunteer recognised orange hawkweed in a photo of her Victorian friend's yard.

In January 2015, the first occurrence in NSW of mouse-ear hawkweed was detected by a bushwalker in a remote area of Kosciusko National Park, where it may have been introduced via camping equipment. A sample was rapidly identified and reported to relevant agencies by herbarium staff in Victoria, who are also involved in the Victorian eradication network. The infestation was treated within two weeks of detection and within one month, agency staff and community volunteers surveyed the surrounding landscape to ensure no other detectable infestations (Hamilton *et al.* 2015). Due to early detection and reporting, the small infestation size, and the rapid management response, it is highly feasible that this infestation can be eradicated. This was all made possible via volunteer and community networks established by the eradication programs, again illustrating the importance of these types of partnerships to support eradication attempts.

WEED ERADICATION DETECTION DOGS

Hawkweed surveillance is undertaken by trained staff, contractors and volunteers. Detection experiments have demonstrated that searchers are unable to detect hawkweeds with 100% confidence (Moore *et al.* 2011, Hauser unpublished data). Delimitation is the greatest hurdle in the initial stages of the eradication program, because all plants must be found and controlled, and reproduction prevented, before eradication can proceed. Ideally, plants should be found prior to flowering, and controlled before setting seed. But as mentioned above, non-flowering hawkweed rosettes can be difficult to detect by humans. In addition, as management units ('sites') are cleared of hawkweed, a method is needed to quickly and effectively declare them 'free' of hawkweed so they can be retired from further surveillance effort, allowing those resources to be allocated elsewhere in the program. To address these issues, program partners worked with a professional dog trainer to train three spaniels, of varied age and experience, to distinguish hawkweed from other plant species and detect hawkweed in field situations.

Dogs have long been used for detecting illegal substances and, more recently, in conservation to find endangered or pest animals (Wollett *et al.* 2013). A trial has also shown that dogs can detect a target weed in an agricultural field setting (Goodwin *et al.* 2010). Hawkweed does not have a distinctive odour to humans, and it was unknown whether it could be distinguished by a dog. The first and most vital step

Photo 3. Orange hawkweed rosettes are difficult to detect when not flowering, and are visually similar to many other species in the invaded areas, thus an ideal target for scent detection dogs. Photo: Pete Turner.

was to determine if dogs could detect hawkweed and distinguish it from other plant species. Victorian partners conducted an initial feasibility trial to see whether a detection dog, that was already trained to detect feral animals, could detect hawkweed. The dogs are trained on a reward system, where they receive a 'ball play' period when they correctly indicate a hawkweed plant to the handler. The trials were conducted with potted hawkweed plants in controlled environments (e.g., potted plants were hidden among other foliage at the trainer's home or nearby parks). These demonstrated that the dog could reliably detect potted hawkweed.

The next stage of the feasibility trial was to conduct a detection experiment in Victoria at Falls Creek, in a field environment similar to that of the largest Victorian hawkweed infestations (c.f. Hauser, these proceedings for further details). The experiment provided promising results, but importantly identified several factors in the dog training and handling protocols that could limit the dog's effectiveness in detecting hawkweed. Feral animals, or their scents, were present in the field trial area, and this appeared to distract the dog from hawkweed detection at times. It was also evident that training the dog on potted plants may have confounded the 'scent picture', making it unclear to the dog if it was searching for hawkweed, the pots and/or the potting mix.

Learnings from this initial trial were incorporated into a larger Weed Eradication Detector Dog (WEDD) project undertaken by NSW partners, who engaged the same dog trainer. Two new dogs were purchased and trained from pups on hawkweed only (i.e. no prior training on pest animals or other detection targets). They were trained using only hawkweed plant fragments rather than potted plants, meaning it was only hawkweed that the dogs were learning to detect, and not a combination of scents. Much of the training occurred in situ in the hawkweed infestation at Kosciuszko National Park on naturalised hawkweed plants. Thus the dogs were trained on the exact target desired, and in same terrain and environmental conditions where detection will occur. Training began when the dogs were approximately 6 months old, and dogs were proficient at detecting hawkweed in field situations within four months. All training followed strict biosecurity protocols to ensure no spread of hawkweed.

During the 2015/16 field season, the dogs were trained and tested at hawkweed sites in the alpine region. They consistently detected hawkweeds plants, often among dense vegetation and, several times, where plants had been undiscovered by humans. Thus far, they have only indicated on hawkweed (i.e. not on any non-target species), which illustrates their ability

Photo 4. Sally, a working cocker spaniel, is the first dog trained specifically to detect hawkweeds (seen here searching alpine habitats of Kosciusko National Park, NSW). Photo: Hillary Cherry.

to discern the target. There were also instances where the dogs did not detect plants that were known to the handler: This was likely due to weather conditions (e.g. high winds) rather than a lack of ability to discern hawkweed. Many factors contribute to detection ability, including weather and terrain, as well as handler skills and ability.

In March 2016, partners conducted field trials to monitor dog detection capabilities under different environmental conditions, and evaluate dog detection abilities compared to those of humans in a field situation (see Hauser, these proceedings). While results are still being analysed, the dogs successfully found hawkweed plants that were overlooked by human searchers, thus indicating that they can significantly contribute to the eradication program. An important finding reinforced in the trial is that dogs are most effective when the target is at a very low density, as too much target scent can be overwhelming and may confuse the dogs' efforts. For this reason, dogs may be best used to 'clear' areas where all plants are assumed to be extirpated, and are not likely be useful for detecting 'widespread' or common weeds.

Photo 5. Connor, an English springer spaniel, began successfully detecting hawkweed plants in the field at 8 months old. Photo: Glen Sanders

While we now know that dogs can effectively detect hawkweed in field situations, further work remains to determine where they can most cost-effectively and efficiently be integrated into eradication efforts. The cost of purchasing and training dogs is relatively affordable, but there is a significant cost to either hiring a skilled handler or training a staff member and resourcing their time to handle the dog. There are other considerations regarding the housing and ownership of the dog (e.g. contractor vs. staff member), as well as upkeep and on-going training. As part of the WEDD project, partners are developing best practice protocols for weed detection dogs, which may assist further development of dogs as a 'tool' in weed eradication programs.

An unexpected consequence of the detection dog project is the extensive coverage by news media and other entities. The dogs generate great community interest that has in turn provided a platform for partners to highlight the hawkweed eradication program, and the need for continued support, including to politicians and government agencies. There are significant positive opportunities to engage the community and land managers when using detector dogs. Everyone, including the media, love dogs, thus they can be used as weed 'ambassadors' or 'champions', allowing eradication managers to interact with the public, politicians and funding bodies more successfully. For example, one of the dogs, 'Sally', and her handler were awarded a travel grant to attend the 20th Australasian Weeds Conference and demonstrate the dogs' weed detection skills. This allowed the WEDD project and the hawkweed program, to be presented to over 200 weed professionals from across Australasia.

DISCUSSION

The above tools were developed for the hawkweed eradication program, however they are likely to have applications to other eradication projects. The benefits of formal partnerships extend beyond ensuring strategic, best practice control and coordinated programs. The NHWG partnership has facilitated the development of research that is being actively applied and is making the eradication program more cost-efficient. The strong partnerships also support a robust volunteer program that is a necessary part of maintaining the eradication effort.

The contribution of volunteers has greater benefits than targeted surveillance during formal programs. Incorporating volunteers into the eradication program and acknowledging the value of their contribution inspires a sense of personal investment and ownership of the eradication effort. This in turn has prompted volunteers to be vocal advocates for the program and further raise awareness in the wider community. In addition, there are limited resources to search for hawkweed outside of known areas, and large numbers of volunteers with the ability to detect and appropriately respond to hawkweed increases overall surveillance effort and improves eradication feasibility. Many volunteers recreate in environments that are at risk from hawkweed, but that may not be targeted for active surveillance: passive surveillance by volunteers in these areas extends the reach of the program.

Photo 6. The dogs are effective in a variety of habitats, including dense grassland. Photo: Jo Caldwell.

The need for improved hawkweed detection has prompted trialling of weed detection dogs, which to our knowledge has not been done in Australia previously. Early signs indicate that detector dogs are likely to improve our chances of finding hawkweed, therefore increasing the likelihood of eradication. In the future, dogs may also assist with other eradication programs, as it appears they can be easily trained on other weeds. Partners are also investigating the use of drones and remote imaging to improve surveillance (Hamilton *et al.* 2015), and ultimately hope to incorporate these into the program to complement the suite of tools necessary for eradication. But despite these new technologies, volunteers are still the best resource in the hawkweed eradication program to ensure necessary surveillance and increase delimitation area: we simply couldn't do it without them.

No surveillance method – human teams, dogs, drones, or other – is ever likely to find hawkweeds with 100% accuracy. Therefore it's important that we understand the different strengths and weaknesses of each approach, and develop a complementary set of methods that gives us the best capacity to detect hawkweeds across the variety of conditions in which they occur.

The hawkweed eradication program on mainland Australia, like most eradication attempts, is complex and requires multiple resource inputs and continued support. Formal partnership and volunteer programs are not traditionally 'costed' when scoping eradication feasibility, however given their benefits to the hawkweed program, they should be considered where relevant. Depending on the weed and the eradication environment, new or improved technologies for better detection and delimitation are also likely to be needed in novel eradication attempts, and should also be costed or considered. Many of the issues described here may apply to other high-risk weeds being considered for eradication, thus when scoping the feasibility of an eradication program, don't underestimate the costs or benefits of developing partnerships, volunteer programs and better detection methods.

ACKNOWLEDGMENTS

Many thanks to the dedicated hawkweed volunteers, and to Steve Austin, the hawkweed dog trainer. The ACT, NSW and Victorian Governments provide support and funding for hawkweed eradication.

REFERENCES

Bear, J.L., Giljohann, K.M., Cousens, R.D. and Williams, N.S.G. (2012). The seed ecology of two invasive *Hieracium* (Asteraceae) species. *Australian Journal of Botany* 60, 615-24.

Beaumont, L.J., Gallagher, R.V., Downey, P.O., Thuiller, W., Leishman, M.R. and Hughes, L. (2009). Modelling the impact of *Hieracium* spp. on protected areas in Australia under future climates. *Ecography* 32, 757-64.

Brinkley, T.R. and Bomford, M. (2002). Agricultural sleeper weeds in Australia: what is the potential threat? Bureau of Rural Sciences, Canberra.

Constantine, A., Hauser, C.E., Primrose, K. and Smith, N. (2016). Hawkweed (*Hieracium* spp.) surveillance: development of a targeted and robust plan for the Victorian Alps. Proceedings of the 6th biennial conference of the Weed Society of Victoria, ed. S.B. Johnson, Creswick, Victoria. Weed Society of Victoria, Melbourne.

Cousens, R., Heydel, F., Giljohann, K., Tackenberg, O., Mesgaran, M. and Williams, N. (2012). Predicting the dispersal of hawkweeds (*Hieracium aurantiacum* and *H. praealtum*) in the Australian Alps. Proceedings of the 18th Australasian Weeds Conference, October 2012, ed. V. Eldershaw, pp. 5-8. Weed Society of Victoria, Melbourne.

Dodd, A., Ainsworth, N., Burgman, M.A. and McCarthy, M.A. (2014). Plant extirpation at the site scale: implications for eradication programmes. *Diversity and Distributions*, 21, 151-62.

Espie, P. (2001). '*Hieracium* in New Zealand: ecology and management'. (AgResearch Ltd, Mosgiel, New Zealand). 66 pp.

Goodwin, K.M., Engle, R.E. and Weaver, D.K. (2010). Trained dogs outperform human surveyors in the detection of rare spotted knapweed (*Centaurea stoebe*). *Invasive Plant Science and Management* 3, 113-21.

Hamilton, M.A., Cherry, H. and Turner, P. (2015). Hawkweed eradication from NSW: could this be 'the first'? *Plant Protection Quarterly* 30, 110-15.

Hauser, C.E., Weiss, J., Guillera-Arroita, G., McCarthy, M.A., Giljohann, K.M. and Moore, J.L. (2016). Designing detection experiments: three more case studies. Proceedings of the 20th Australasian Weeds Conference, eds R. Randall, S. Lloyd and C. Borger, pp. 171-8. (Weeds Society of Western Australia, Perth).

Hauser, C.E., Giljohann, K.M., Rigby, M., Herbert, K., Curran, I., Pascoe, C., Williams, N.S.G., Cousens, R.D. and Moore, J.L. (2016). Practicable methods for delimiting a plant invasion. *Diversity and Distributions* 22, 136-47.

Hung, C. and Sukkarieh, S. (2015). Using robotic aircraft and intelligent surveillance system for orange hawkweed detection. *Plant Protection Quarterly* 30, 100-2.

Moore J.L., Hauser C.E., Bear J.L., Williams N.S.G. and McCarthy M.A. (2011). Estimating detection-effort curves for plants using search experiments. *Ecological Applications* 21, 601-7.

Myers, J.H., Savoie, A. and Van Randen E. (1998). Eradication and pest management. *Annual Review of Entomology* 43, 471-91.

Panetta, F.D. (2014). Weed eradication feasibility: lessons of the 21st century. *Weed Research* 55, 226-38.

Primrose, K., Constantine A., Smith, N. and Pascoe, C. (2016) Eradicating hawkweeds (*Hieracium* spp.) from the Victorian Alps: improving the efficiency and effectiveness of control whilst mitigating off-target impacts. Proceedings of the 6th biennial conference of the Weed Society of Victoria, ed. S.B. Johnson, Creswick, Victoria. Weed Society of Victoria, Melbourne.

Williams, N.S.G. and Holland, K.D. (2007). The ecology and invasion history of hawkweeds (*Hieracium* species) in Australia. *Plant Protection Quarterly* 22, 76-80.

Wittenberg, R. and Cock, M. (2001). Invasive alien species: a toolkit of best prevention and management practices. GISP/CAB International, Wallingford.

Woollett, D.A, Hurt, A. and Richards, N.L. (2013). The current and future roles of free-ranging detection dogs in conservation efforts. *In* Free-ranging dogs and wildlife conservation, ed. M.E. Gompper, pp. 239-64, Oxford University Press, Oxford, UK.

Designing detection experiments: three more case studies

Cindy E. Hauser[1,2], John Weiss[3,4], Gurutzeta Guillera-Arroita[1], Michael A. McCarthy[1], Katherine M. Giljohann[1] and Joslin L. Moore[1,2]
[1] School of BioSciences, University of Melbourne, Parkville, Victoria 3010, Australia
[2] School of Biological Sciences, Monash University, Clayton, Victoria 3800, Australia
[3] BioSciences Research, Department of Economic Development, Jobs, Transport and Resources, Bundoora, Victoria 3083, Australia
[4] Plant Biosecurity Cooperative Research Centre, Bruce, Australian Capital Territory 2617, Australia
(chauser@unimelb.edu.au)

Summary Detection experiments are designed to reveal which kinds of targets are likely to be detected and left undetected during environmental surveys. These experiments can help plan surveys and reliably interpret survey data. In this paper we outline the designs of three detection experiments. The first evaluates UAVs and human inspectors for detecting a simulated disease in a vineyard. The second and third experiments are two phases in the evaluation of dogs trained to detect the scent of hawkweeds. The experiments illustrate a variety of approaches, each selected to serve the management objectives, priorities, knowledge and resources available in that program at the time.

Keywords Surveys, experimental design, plant disease, weed management, detection dog, unmanned aerial vehicle (UAV), *Hieracium* species.

INTRODUCTION

Imperfect detection of invasive individuals and populations underpins all surveillance. It has implications for early detection of new incursions (Mehta *et al.* 2007), and monitoring established populations (Epanchin-Neill *et al.* 2012), through to finding every last infestation before declaring eradication (Regan *et al.* 2006).

It follows that survey designs and analyses of survey data should account for detection rate. A detection rate describes the expected frequency of detection events for a species, population or individual under a given set of survey conditions (Hauser *et al.* 2015).

Scientific experiments can be designed specifically to estimate detection rate (Bulman *et al.* 1999, Bulman 2008, Christy *et al.* 2010, Britton *et al.* 2011, Moore *et al.* 2011, Stringer *et al.* 2011). We have previously discussed the tricks and trade-offs involved in designing such a detection experiment (Hauser *et al.* 2012). In this paper we show how we have applied these principles in three more field experiments.

1. HUMANS INSPECT VINEYARDS TO DETECT SIMULATED DISEASE

The Victorian Department of Economic Development, Jobs, Transport and Resources (DEDJTR), in conjunction with the Plant BioSecurity Cooperative Research Centre, are exploring the feasibility of using unmanned aerial vehicles (UAVs) to detect plant diseases. We compared the performance of UAVs and human inspectors when detecting a simulated disease in a Victorian vineyard.

Influencing variables We expected that the severity and prevalence of disease would affect detection rate.

The main environmental factor with potential to affect detection was the vineyard style: *French-style* plots contain small vines at high density, while *Australian-style* plots contain taller vines planted at lower density.

All inspectors had similar levels of experience participating in a range of biosecurity-related searches. We did not seek quantitative measures of their experience level.

UAV model (fixed wing or multi-rotor) and flight height were also expected to influence detection rate.

Site selection The experiment was conducted on a private vineyard with permission. Vineyard rows were selected across three different aspects, capturing the two different vineyard styles (French and Australian). Thirteen rows (totalling 2228 vines) were selected and marked for search.

Targets We sprayed Surround®, a compound approved for sun protection in vineyards, on individual vine leaves to simulate the appearance of a fungal disease similar to downy mildew. We wore protective clothing during the application process and used a plastic paint tray behind the selected leaves to control the number of leaves affected, by reducing splash onto other vine leaves.

Placing targets in plots Rows within the French/Australian strata were randomly assigned a prevalence of 5 or 10% vines affected. Within each 200-vine row, individual vines were selected at random for simulated disease at the assigned prevalence.

An equal number of individuals within each row were assigned a disease severity of 1, 2 or 4 leaves. These leaves were selected haphazardly across the outer and upper leaves of the vine, which would be apparent to the UAVs.

Plastic markers numbering the vines in multiples of 15 or 20 were used to help orient searchers and supervisors.

Searchers Four DEDJTR staff members with experience in biosecurity searches were recruited to individually inspect the vines.

Both fixed wing and multi-rotor UAVs with the same multispectral camera were deployed over the plots at 100 m and 50 m altitude, respectively. Unsupervised and supervised classification of the images was undertaken without prior knowledge of the location of all the 'infected' plants.

Implementing searches All participants searched all of the vine rows in the same order throughout the day. A sufficient delay was set between inspections to ensure that participants could not observe finds made by others in rows they had not yet searched.

Each participant was paired with a supervisor (sometimes another participant who had already searched that row; Figure 1). The participating searcher walked the length of each vine row at their own pace, inspecting vines for the simulated disease. When a diseased vine was detected, the searcher alerted the supervisor and reported the number of leaves that appeared diseased. The supervisor used a stopwatch to record periods of general inspection and resting/counting diseased leaves. They also recorded the vine ID number and number of diseased leaves of each vine where disease was detected.

2. DETECTION DOG SEEKING PLANTED HAWKWEEDS

Studies by Moore *et al.* (2011) and Hauser *et al.* (2012) show that human searchers have a variable capacity to detect invasive hawkweeds (*Hieracium* species) on the Bogong High Plains of Victoria. Hawkweeds can be rapidly detected while in flower, but detection can be severely hampered in the presence of other yellow-flowering species or dense vegetation, and when the hawkweed is not flowering.

The Victorian Department of Economic Development, Jobs, Transport and Resources (DEDJTR) have consequently investigated the capacity for a trained dog to detect the scent of hawkweeds. Initial training exercises were successful, and the agency sought to evaluate one trained dog's detection rate in typical

Figure 1. A DEDJTR staff member inspects vines for a simulated plant disease, while a supervisor times their search and records disease detections.

survey conditions. We designed a series of search trials that would allow us to build a statistical model linking time taken to detect a target hawkweed with influencing factors, such that the model could be compared with the human search models built by Moore *et al.* (2011) and Hauser *et al.* (2012).

Influencing variables We aimed to measure the effects of three target species (*Hieracium aurantiacum*, *H. praealtum* and *H. pilosella*, all of which occur on the Bogong High Plains) and abundance on detection rate.

We also anticipated that environmental factors might influence detection via the diffusion of scent from targets. We measured vegetation type around each target, wind speed and direction, temperature and humidity.

Only one handler-dog pair was recruited for the evaluation, so we were not able to test for differences or variation in detection among handlers or dogs. We anticipated that the time of day and week could influence their energy levels and detection success.

Site selection Our study used four 2 ha plots, in the same area that Hauser *et al.* (2012) tested the detectability of human searchers in 20 × 20 m plots. We decided on this markedly larger plot size to capture the full potential detection distance of a dog and allow for rapid, wide-ranging work where it was preferred by the handler.

The plot locations were selected to align with the hectare grid used by managers to plan their operations. We sought to capture variation in vegetation (classified and mapped as grass, open heath, and closed heath) within each plot, and some variation in difficulty of terrain among plots.

Targets The three hawkweed species were cultivated in pots by DEDJTR staff under permit. We aimed to test whether it was truly the hawkweed scent and not the scent of the soil or pot that might alert the dog to our targets. Therefore DEDJTR additionally cultivated other alpine plants to act as control targets in the experiment.

All targets were re-potted with the same fresh soil and clean, identical black plastic pots two days prior to trial commencement. This both controlled variation among targets, and ensured that hawkweed seeds and other plant material did not contaminate the soil. The non-hawkweed control plants were handled and potted before the hawkweeds were touched to prevent scent transfer. Numbered metal tags were pinned into the soil of each pot to uniquely identify each individual.

Placing targets in plots The study area was not known to contain hawkweeds, and so we expected to have full control over the density and distribution of hawkweed targets.

In placing the targets, we were mindful that the dog would primarily respond to the scent and not the sight of targets. Nevertheless, the dog's handler might be informed by visual cues during the search. We used two different methods of securing targets to the ground. First, we dug pot-sized holes and positioned the target plants at ground level; this produced a visually accurate result but disturbed the soil, potentially affecting the scent cue. Second, we secured outer pots above the ground using weed mat pins and nestled the potted plants inside; this approach produced more visually obvious targets above the ground but minimally disturbed the soil.

In order to obscure the visual cues provided to the handler, we placed 25 targets in each plot trial and altered their location, and the ratio of hawkweeds to control targets, in each trial. Each trial had 3, 12 or 21 hawkweeds (an equal number of *H. aurantiacum*, *H. praealtum* and *H. pilosella*) with the remainder of targets being randomly selected control target species. We stratified plots by their mapped vegetation, allocating an equal number of controls and hawkweeds (and, where possible, an equal number of each hawkweed species) to each vegetation type. Within a vegetation type, targets were randomly assigned GPS co-ordinate locations.

Maintaining biosecurity while handling State prohibited weeds and preventing scent contamination were high priorities throughout the trial. Potted plants were encased in Ziploc bags at all times (storage, transport, placement) to prevent soil and plant fragments escaping. All control targets were placed before the hawkweeds were retrieved from lidded boxes and directly handled to prevent hawkweed scent transfer.

Target-placing teams walked long and disparate paths around plots and between targets to avoid creating human scent trails to targets. A minimum of a half-day between target placement and search was scheduled, to allow the targets to diffuse their scent into the surrounding environment.

In the human search trial, Hauser *et al.* (2012) achieved replication by rotating a large number of searchers through a large number of plots. By contrast, this trial relied on a single dog-handler pair and only four large plots. We achieved statistical replication by reusing the same plots on multiple occasions throughout the experiment and arranging a different set of target species for each trial. Across a total of 15 trials, the dog-handler pair was exposed to 185 control targets and 190 hawkweed targets.

Searchers We evaluated one single search team, consisting of a trained dog and its handler. We secured an ethics permit in order to perform our trials and minimise the dog's exposure to harm. We relied on the dog's handler-trainer to evaluate the safety of the terrain for the dog, to allow the dog rest at any time, and to perform regular checks for bruises, cuts and ticks. As a biosecurity measure, the dog was brushed twice daily during trials to control the risk of spreading hawkweed material via dog fur.

Implementing searches Hawkweed densities were assigned to plots and plot trials were scheduled throughout the experiment so that the dog-handler pair was exposed to different hawkweed densities in different plots at different times of the day.

At the beginning of each trial, we assembled a weather station in one corner of the plot to collect temperature, humidity and wind data at 5 min intervals throughout the trial.

The speed and style of search by the dog-handler team in the alpine terrain was relatively unknown prior to the trials. We invited the dog-handler pair to search each plot using their preferred path and pace, taking rest breaks whenever they chose (Figure 2).

Two researchers followed the dog-handler pair, using a stopwatch to record time intervals of active search, non-search activities, and the tag numbers of found targets. GPS units were secured to the handler and to the dog in order to record their search paths.

3. DETECTION DOGS AND HUMAN SEARCHERS SEARCHING FOR *IN SITU* HAWKWEED INFESTATIONS

Following the detection dog trials conducted on the Bogong High Plains (see section 2), the NSW Office of Environment and Heritage (OEH) secured funding to train two different dogs to detect hawkweed (specifically *H. aurantiacum*, the most prevalent species in NSW; Cherry *et al.* 2016).

The evaluation team used the findings of the Victorian study to inform their experimental design. First, the two dogs did not have any prior training to detect other kinds of targets. By contrast, the dog trained for the Victorian evaluation had prior training to detect cats, foxes, hares and rabbits, and this was thought to

Figure 2. Hawkweed search trials: the dog and handler seek planted targets in Victoria.

have caused occasional distraction during the searches.

Second, analyses revealed some uncertainty around the dog's ability to distinguish between potted hawkweeds and potted control targets in the Victorian evaluation. To eliminate this possibility, the NSW evaluation used only naturally occurring hawkweed infestations as targets.

Third, a human search team was recruited to survey the same evaluation plots as the dog-handler teams. This maximized our capacity to directly compare the two search styles and detection rates.

Influencing variables Although many characteristics of the target could affect detection rate, we did not have control over these. We focused solely on target size in this experiment.

We anticipated that the same environmental factors could influence detection rate as in the Victorian evaluation: we noted vegetation type around each target, wind speed and direction, temperature and humidity.

With two dog-handler pairs and one human search team, we had some capacity to test for differences between teams. Again we anticipated that the time of day and week could influence the teams' energy levels and detection success.

Site selection OEH staff selected an area where *H. aurantiacum* had been found earlier in the season. These infestations were left untreated so that they could be used in this experiment.

OEH manage infestations using a 50 m × 50 m grid, and seven grids containing hawkweeds were selected as experimental plots. These plots were notably smaller than the 2 ha plots selected in Victoria. The Victorian evaluation indicated that the dog-handler team worked quickly and did not always cover the plot thoroughly, thus overlooking a number of hawkweed targets. In NSW, smaller plots provided consistency with the management scale and an opportunity for finer-scale search.

There was some variation in difficulty of terrain among the plots, due to slope as well as vegetation type. Neither was thoroughly mapped due to time constraints, although GIS vegetation and elevation layers are available for future analyses.

Targets Targets were infestations already occurring in the plots after discovery earlier in the season. Besides withholding herbicide treatment, we did not have control over target characteristics. On detection during the experiment, we estimated each infestation's length and width in cm, and counted the number of rosettes. Target size may influence detection rate, and it was also useful for matching which infestations were detected by more than one team.

Since targets were naturally occurring, we did not have control over their locations beyond strategic selection of plot boundaries. During the design phase, 12 hawkweed infestations were known to occur across the eight plots. Additional infestations were detected during the experiment, leading to a total of 27 different targets.

Searchers The dogs evaluated in NSW were trained exclusively to detect the one target species in this experiment, *H. aurantiacum*, as plant material without pots. One handler was the same handler-trainer who participated in the Victorian evaluation. The other handler had a background in weed management. We relied on the handlers to ensure the health and safety of the dogs throughout the trials.

The human search team comprised eight people; some were weed management staff and others were volunteers with prior experience in the program. They performed line searches typical of the NSW surveillance program (Figure 3).

Implementing searches We assigned teams to search plots so that different plots were visited first by different teams, in case the search path of one team interfered with the detection rate of subsequent teams. We assigned teams to plots to maintain distance between teams wherever possible, such that handlers and human searchers would have difficulty observing the finds made by other teams.

At the beginning of each day, we set up the same weather station used in the Victorian evaluation, in the vicinity of the dog's search plots. The station was within 100 m of the dog's search path at all times, and collected temperature, humidity and wind data at 5 min intervals.

As before, the handler was permitted to direct their dog through the plots using their preferred speed and path, taking rest breaks whenever they deemed appropriate (Figure 4).

One supervisor followed the human search team and two supervisors followed the dog-handler teams. They used a stopwatch to record time intervals of active search, non-search activities, and waypoints for any infestations found. GPS collars were attached to each dog to record their search path and provide live feedback via a handset; this assisted the handlers in covering each plot thoroughly. Similarly, a GPS was held by a human search team member to record the group's path.

Figure 3. Hawkweed search trials: a human team seek *in situ* infestations in New South Wales.

Figure 4. Hawkweed search trials: a dog and handler seek *in situ* infestations in New South Wales.

DISCUSSION

The three case studies here, in addition to the ones published by Moore *et al.* (2011) and Hauser *et al.* (2012), demonstrate a variety of detection experiment designs. Differences among the designs are driven by the survey structure, the influential factors being tested, and the logistics and constraints of the experiment. The vineyard disease study relied entirely on an imitation target, while various hawkweed studies have used different combinations of real rosettes in pots, naturally occurring rosettes in the ground and (in Hauser *et al.* 2012) imitation flowers. The target type has affected the relative control over hawkweed placement and the plant forms tested. They also affect the biosecurity protocols needed to ensure the experiment does not spread the target population further.

The variables expected to influence detection rate can vary markedly among search and searcher types. Vineyards provide a comfortable, linear environment to conduct human inspections. Alpine plots can be more difficult and variable to traverse due to slope and thick vegetation. In all cases, bad weather may increase the difficulty of search. A dog is also expected to be affected by terrain, and its detection rate is more likely to be affected by subtle changes in temperature, humidity and wind.

Searchers are likely to differ in their detection ability, and many searchers must be tested in order to quantify and understand this variation. Hauser *et al.* (2012) achieved this most effectively by recruiting 29 participants and having them search small plots individually. While it does not include any replication, the eight person team tested in this paper's NSW evaluation was a more accurate reflection of training and behaviour during hawkweed searches. Logistical and financial constraints have prevented more than two dogs or two UAVs from being tested at any one time.

Trade-offs must also be made in the layout of each experiment. Only Hauser *et al.* (2012) and the Victorian evaluation in this paper were able to stratify and randomise target placement across vegetation types to thoroughly test the effects of target type and background vegetation on detection rate. A common challenge across most of these studies is the abundance and placement of targets. To place many targets increases the replication in the study and therefore the strength of statistical inference. However, the populations considered typically occur at low densities and increasing their abundance in experimental conditions could falsely boost searchers' morale and detection rate. Locations without targets can seem a waste of testing time but are necessary for moderating a searcher's expectations.

Once detection data are collected, we have used them to build time-to-detection models (Moore *et al.* 2011, Hauser *et al.* 2012, Hauser *et al.* 2015). These have been an effective tool in the hawkweed program for providing feedback on and adjusting search intensity and comparing scenarios among target types, background vegetation types, and between human searchers and dogs. In the latter case we also seek additional information on the number of hours that human and dog-handler teams are fit to work each day, the costs of having them in the field and the costs of maintaining access to the searchers. We are investigating alternative methods for assessing the detection capabilities of dogs and UAVs such as distance sampling (Buckland *et al.* 2001) and cluster and classification techniques (Fielding 2007).

Novel search technologies such as trained dogs and UAVs are exciting avenues of inquiry. We must be mindful that their capacity is likely to change over time as technologies improve and methodologies are refined. Indeed, we observed marked differences in performance between the Victorian and NSW evaluations of hawkweed detection dogs.

We must also acknowledge that conducting detection experiments can be costly, and assess this against the potential benefits of this pursuit at any time. The relative costs and benefits will vary from study to study. Experiments by Hauser *et al.* (2012) and the Victorian detection dog evaluation in this paper drew on substantial financial resources and personnel, yielding large data sets and informative statistical models. The subsequent NSW evaluation was conducted on a smaller scale but made crucial adjustments that simultaneously improved the realism of the data and dramatically reduced its quantity (from 200–1000 observations in Victoria down to less than 50 in NSW). Further research is needed to understand the value of lower-cost small-scale detection experiments to environmental management.

Detection experiments are a sound, scientific tool that can assist us in planning environmental surveys and reliably interpreting survey data. The case studies discussed in this paper demonstrate that a variety of designs are possible for addressing the broad range of survey scenarios that occur in environmental management. Furthermore, different designs might be applied to different stages of a single management program, depending on the knowledge, priorities and resources available at the time.

ACKNOWLEDGMENTS

CEH was supported by the National Environmental Research Program Environmental Decisions Hub. Vineyard disease detection research is funded by the Plant Biosecurity CRC (project #2135), and we thank Michael Dhillon for permission to use his vineyard for the field trial. The detection dog research has been funded by the Victorian Department of Economic Development, Jobs, Transport and Resources and the NSW Office of Environment and Heritage. We thank Hillary Cherry and Angela Constantine for comments and discussion that improved this manuscript.

REFERENCES

Britton, J.R., Pegg, J. and Gozlan, R.E. (2011). Quantifying imperfect detection in an invasive pest fish and the implications for conservation management. *Biological Conservation* 144, 2177-81.

Buckland, S.T., Anderson, D.R., Burnham, K.P., Laake, J.L., Borchers, D.L. and Thomas, L. (2001). 'Introduction to distance sampling: estimating abundance of biological populations'. (Oxford University Press, New York).

Bulman, L.S., Kimberley, M.O. and Gadgil, P.D. (1999). Estimation of the efficiency of pest detection surveys. *New Zealand Journal of Forestry Science* 29, 102-15.

Bulman, L.S. (2008). Pest detection surveys on high risk sites in New Zealand. *Australian Forestry* 71, 242-4.

Cherry, H., Hauser, C., Constantine, A., Primrose, K. and Tasker, K. (2016). It takes a village: detection dogs, partnerships and volunteers aid hawkweed eradication in mainland Australia. Proceedings of the 20th Australasian Weeds Conference, eds R. Randall, S. Lloyd and C. Borger, pp. 164-170. (Weeds Society of Western Australia, Perth).

Christy, M.T., Yackel Adams, A.A., Rodda, G.H., Savidge, J.A. and Tyrrell, C.L. (2010). Modelling detection probabilities to evaluate management and control tools for an invasive species. *Journal of Applied Ecology* 47, 106-13.

Epanchin-Niell, R.S., Haight, R.G., Berec, L., Kean, J.M. and Liebhold, A.M. (2012). Optimal surveillance and eradication of invasive species in heterogeneous landscapes. *Ecology Letters* 15, 803-12.

Fielding, A.H. (2007). 'Cluster and classification techniques for the Biosciences'. (Cambridge University Press, Cambridge UK).

Hauser, C.E., Moore, J.L., Giljohann, K.M., Garrard, G.E. and McCarthy, M.A. (2012). Designing a detection experiment: tricks and trade-offs. Proceedings of the 18th Australasian Weeds Conference, ed. V. Eldershaw, pp. 267-72. (Weed Society of Victoria, Melbourne).

Hauser, C.E., Garrard, G.E. and Moore, J.L. (2015). Estimating detection rates and probabilities, *In* 'Biosecurity surveillance: quantitative approaches', eds F. Jarrad, S. Low Choy and K. Mengersen, pp. 151-66. (CAB International, Wallingford, UK).

Mehta, S.V., Haight, R.G., Homans, F.R., Polasky, S. and Venette, R.C. (2007). Optimal detection and control strategies for invasive species management. *Ecological Economics* 61, 237-45.

Moore, J.L., Hauser, C.E., Bear, J.L., Williams, N.S.G. and McCarthy, M.A. (2011). Estimating detection-effort curves for plants using search experiments. *Ecological Applications* 21, 601-7.

Regan, T.J., McCarthy, M.A., Baxter, P.W.J., Panetta, F.D. and Possingham, H.P. (2006). Optimal eradication: when to stop looking for an invasive plant. *Ecology Letters* 9, 759-66.

Stringer, L.D., Suckling, D.M, Baird, D., Vander Meer, R.K., Christian, S.J. and Lester, P.J. (2011). Sampling efficacy for the red imported fire ant *Solenopsis invicta* (Hymenoptera: Formicidae). *Environmental Entomology* 40, 1276-84.

Support tools for making better investment decisions about early invaders

Kate Blood[1] and Bec James[2]
[1] Department of Environment, Land, Water and Planning, PO Box 7, Beaufort, Victoria 3373, Australia
[2] Department of Environment, Land, Water and Planning, Private Bag 15, Ferntree Gully Delivery Centre, Victoria 3156, Australia
(kate.blood@delwp.vic.gov.au)

Summary The Victorian Government has designed a package of tools to help public land managers determine weed management priorities. A decision making framework, a series of six guides and the Victorian environmental weed risk database assist public land managers through a logical process to work out the highest priority early invaders to eradicate locally.

The Weeds at the Early Stage of Invasion (WESI) project focuses on high risk invasive plants that threaten biodiversity, and have not become locally abundant and widespread. Project staff work with Department of Environment, Land, Water and Planning (DELWP) and Parks Victoria staff looking after public land anywhere in Victoria.

Using the package will help public land managers make better investment decisions for early invaders and assist with successful localised eradication (extirpation) of high-risk weeds at the early stage of invasion. The tools are available at www.delwp.vic.gov.au/early-invaders.

Keywords Early invaders, environmental weeds, eradication, extirpation, risk.

WESI PROJECT

The WESI project commenced in late 2010 and is currently funded to June 2017. It has been funded through the Weeds and Pests on Public Land (WPPL) Program and had a consistent project team (1.5 full-time-equivalent staff). It has been able to evolve over the project period through adaptive management and with the support of its steering group.

The project was created to help shift some invasive species investment towards early invaders where a better return on outcomes for long-term biodiversity can be gained.

Decision making framework The range of activities that can be carried out during the management of early invaders were grouped logically into a six-step decision making framework.

Once the original framework was formed, it went through continual review and tweaking to improve it.

A very simplified version of the framework outlines the six basic steps (Figure 1), while an expanded

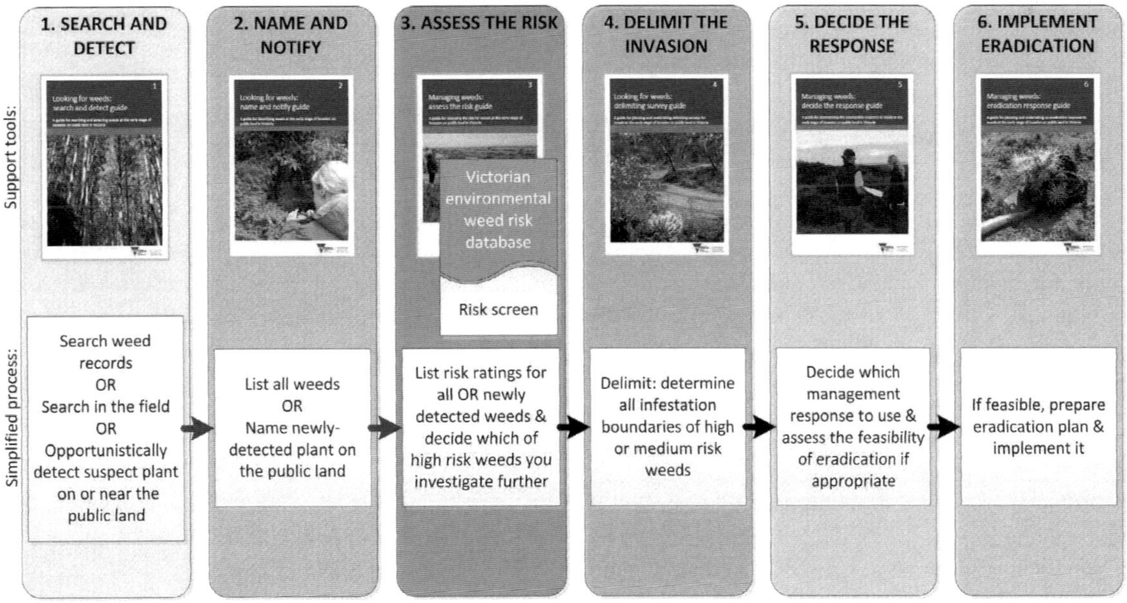

Figure 1. Decision making framework.

version illustrates three scenarios a public land manager might face:

(a) work out highest priority early invaders on the public land;
(b) work out what priority a new invader is on the public land; and
(c) work out highest priority weeds outside the public land to search for within.

EARLY INVADER TOOLS

The WESI project has developed a number of tools to assist in the management of early invaders.

Guide series An existing 'how to' guide appropriate for Victoria to describe how to undertake the various activities described in the decision making framework could not be located within existing literature. There are many papers describing various components or case studies.

To be able to take tools out to field staff that were understandable and practical, it was realised early in the project that new tools needed to be developed. Instead of a large single 'how to' manual, a separate guide that described each step in the framework was created.

Experienced users can use the guides as a refresher for the various activities described, and to extend their knowledge for new information. They can easily select the guide relevant to their needs. Inexperienced weed practitioners can read the guide series in their logical order.

The guides are thoroughly cross-referenced and have a consistent glossary, diagrams and templates across the series. Wellbeing, safety and hygiene for fieldwork are included. The guide series is available at www.delwp.vic.gov.au/early-invaders

Search and detect guide The aim of this guide (Sheehan, James and Blood 2016) is to assist users with the process of searching for and detecting weeds in the early stage of invasion on public land. It can assist in identifying target species and weed spread pathways. It provides advice on looking for weeds opportunistically during the completion of routine tasks, and information on the development of more structured approaches to searching.

This guide helps the user to choose the appropriate search type and method, and contains a ten step guide on how to plan and conduct a search. Search scenarios describe typical situations public land managers will face and the steps to follow with helpful hints. Weed sources and pathways are described to assist thinking about places to search. Templates can be copied or adapted for field recording and specimen collecting. There is an equipment checklist and many useful information sources and references.

Name and notify guide One of the most common requests staff make is for assistance with weed identification. Helping staff be more able to identify plants will help to reduce barriers to further action or inappropriate actions through misidentifications.

The aim of this guide (Blood and James 2016a) is to assist users with the process of identifying weeds. This guide describes how to record weed information, identify and name weeds, write weed names and who to notify. It contains field recording and specimen collecting templates, and useful information sources and references. A photographing and videoing guide contains many useful hints.

Assess the risk guide The main aim of this guide (Blood, James and Panetta 2016) is to assist users with the process of determining the risk a weed poses at a site.

During the process of bringing together the information for this guide, a large gap was recognised – a way of assessing the risk of a weed at a location in the absence of existing weed risk ratings i.e. a full Victorian Weed Risk Assessment or a rating in one of the five existing environmental weed advisory lists (e.g. Adair *et al.* 2008) for the State.

An **environmental weed risk screen** was developed by Dr F. Dane Panetta (Panetta 2016) to fill this gap (Blood, Panetta and James 2016).

The *Assess the risk guide* explains the risk ratings available, how to use the Victorian environmental weed risk database (see below), and how to use the environmental weed risk screen.

This guide contains the environmental weed risk screen score sheet template and a completed example, and many weed data sources.

Victorian environmental weed risk database To accompany the *Assess the risk guide*, an Excel database was created to bring together two main sources of risk ratings: the full Victorian Weed Risk Assessment scores; and the ratings from the five existing environmental weed advisory lists.

It is a decision support tool that can assist in generating lists of detection priorities for existing and potential species in and around a piece of public land at a particular date, helping to provide a structured approach to detection surveys.

The database is available at www.delwp.vic.gov.au/weed-risk-ratings.

Delimiting survey guide Delimiting (working out how far a weed has spread) is one of the most important activities to complete when considering eradication as a management response. This guide (James and Blood 2016) assists with planning and undertaking a delimiting survey, and analysing and reporting the results. It contains an eight step delimiting survey guide, a search plan template with a completed example, a field recording template and a survey equipment checklist.

A number of pilots were used to formulate and test the instructions including a survey for buffel grass (*Cenchrus ciliaris* L.) and white-spined Hudson pear (*Cylindropuntia pallida* (Rose) F.M.Knuth) in the Victorian Mallee.

Decide the response guide The aim of this guide (Blood and James 2016b) is to assist users with the process of deciding which management approach to take and determine if eradication is a feasible option. It brings together eradication concepts from a number of sources (e.g. Panetta and Timmins 2004, Dodd *et al.* 2014) into a logical process. The user is led through the weed management matrix that draws on risk and delimiting survey information collected already. It then tests the suitability for eradication, and finally uses the feasibility of eradication score sheet.

This guide puts weed eradication into the broader public land management picture and helps the user make decisions in the real world. The guide includes a case study, characteristics of 'more feasible to eradicate' weeds, and information resources and references.

Eradication response guide Once eradication is determined to be suitable and feasible for an early invader, this guide (Adair, James and Blood 2016) will help with the development and implementation of a 'park-scale' eradication response. This guide provides a summary of the important points that others have learnt in planning or undertaking eradication programs.

The guide describes many treatment techniques, contains an eradication response plan template with a completed example, and a flowchart for choosing appropriate chemicals.

DISCUSSION

The WESI project has and continues to run a number of pilot activities in different parts of Victoria to test and refine aspects of the decision making framework. These include activities in the: Mallee focusing on buffel grass (James, Blood and Raleigh 2016) and white-spined Hudson pear; Wimmera focusing on white-spined Hudson pear, Klein's cholla (*Cylindropuntia kleiniae* (DC.) F.M.Knuth), and bridal veil (*Asparagus declinatus* L.); Central Highlands Eden; Otways Eden; Glenelg Eden; and Tarago Reservoir (Melbourne Water). Future pilots include a Victorian high country pilot.

In September 2015, a pilot Mallee weed identification course was conducted at Mildura focusing on early invaders in the north west of Victoria. Using the positive feedback received, it has provided a model to continue capability building including training activities in other parts of the State. Future training will include additional information on treatment of the featured species.

Each pilot is being evaluated to inform and improve the project and the tools.

The pilots provided an opportunity to help staff understand the decision making framework and content of the guides during development. These and future pilots will continue into 2017. Other capability building activities are being carried out (Blood and James 2016c) including one-on-one and small group activities that focus on skills, knowledge and awareness to assist with the use of the WESI tools. This includes weed identification, reporting weeds, hygiene, surveying in the field, photographing weeds, and using social media. The effective use of technology is an important focus of these activities especially mobile technology.

CONCLUSION

Social media and email are the main communication pathways being used with staff to share information and nudge them to keep using the tools and information available to make better investment decisions. The final evaluation of the project will be made in 2017. It is expected that continued support for staff will be required to adopt and use the tools available.

ACKNOWLEDGMENTS

The WESI project's existence is due to the future-thinking (then) program leader, Anne Dennis, and Kate McArthur. Thank you.

We would like to acknowledge the consistent support and guidance of the WESI Steering Group: Simon Denby, Stefan Kaiser, Sally Lambourne, Kate McArthur (DELWP); Dr Nigel Ainsworth and Daniel Joubert (DEDJTR); Ben Fahey (Parks Victoria); Associate Professor Melodie McGeoch (Monash University); and former members Dr David Cheal and Penny Gillespie.

Technical input was greatly appreciated from Jackie Steel and Aaron Dodd (DEDJTR), Annette Muir (DELWP), and Simon Kennedy. Guide co-authors are thanked for their contributions: Dr F. Dane Panetta (Bioinvasion Decision Support); Dr Matt Sheehan (Wild Matters); and Dr Robin Adair (Australis Biological).

We appreciate the enthusiastic participation of the many DELWP and agency staff during the pilots and development phase.

REFERENCES

Adair, R., Cheal, D. and White, M. (2008). 'Advisory list of environmental weeds of coastal plains and heathy forests bioregions of Victoria'. (Department of Sustainability and Environment, Melbourne).

Adair, R., James, R. and Blood, K. (2016). 'Managing weeds: eradication response guide. A guide for planning and undertaking an eradication response to weeds at the early stage of invasion on public land in Victoria'. (Department of Environment, Land, Water and Planning, Victoria).

Blood, K. and James, B. (2016c). Training and nudging weed practitioners towards better on-ground results. Proceedings of the 20th Australasian Weeds Conference, eds R. Randall, S. Lloyd and C. Borger, p. 319. (Weeds Society of Western Australia, Perth).

Blood, K. and James, R. (2016a). 'Looking for weeds: name and notify guide. A guide for identifying weeds at the early stage of invasion on public land in Victoria'. (Department of Environment, Land, Water and Planning, Victoria).

Blood, K. and James, R. (2016b). 'Managing weeds: decide the response guide. A guide for determining the appropriate response to weeds at the early stage of invasion on public land in Victoria'. (Department of Environment, Land, Water and Planning, Victoria).

Blood, K., James, R. and Panetta, F. D. (2016). 'Managing weeds: assess the risk guide. A guide for assessing the risk for weeds at the early stage of invasion on public land in Victoria'. (Department of Environment, Land, Water and Planning, Victoria).

Blood, K., Panetta, F.D., James, B. (2016). Assessing weed risk at the local scale. Proceedings of the 20th Australasian Weeds Conference, eds R. Randall, S. Lloyd and C. Borger, p. 320. (Weeds Society of Western Australia, Perth).

James, B., Blood, K. and Raleigh, R. (2016). Buffel grass (*Cenchrus ciliaris* L.), an early invader in Victoria. Proceedings of the 20th Australasian Weeds Conference, eds R. Randall, S. Lloyd and C. Borger, pp. 337-8. (Weeds Society of Western Australia, Perth).

James, R. and Blood, K. (2016). 'Looking for weeds: delimiting survey guide. A guide for planning and undertaking delimiting surveys for weeds at the early stage of invasion on public land in Victoria'. (Department of Environment, Land, Water and Planning, Victoria).

Panetta, F.D. (2016). 'Environmental weed risk screen for Victoria: background and development'. (A report prepared for the Department of Environment, Land, Water and Planning, Vic).

Sheehan, M., James, R. and Blood, K. (2016). 'Looking for weeds: search and detect guide. A guide for searching and detecting weeds at the early stage of invasion on public land in Victoria'. (Department of Environment, Land, Water and Planning, Victoria).

Dealing with 'new' alien plants: risk assessment and risk management

F. Dane Panetta
Faculty of Veterinary and Agricultural Sciences, The University of Melbourne, Parkville, Victoria 3010, Australia
(dane.panetta@gmail.com)

Summary Decisions on how to deal with newly detected alien plant incursions need to be made against a background of uncertainty, initially with respect to the weed risk posed. Then, should a high risk be indicated there will be further uncertainty regarding how best to manage it. If a coordinated management strategy (e.g. eradication) is favoured, its feasibility and likelihood of success will depend *inter alia* upon the allocation of sufficient resources, but what constitutes 'sufficient' is difficult to estimate.

Weed risk assessment (WRA) is challenging, in part owing to an incomplete understanding of the mechanisms of invasiveness and, especially, impact. Desktop post-border WRA procedures are available, but there appears to have been little attempt to estimate the invasiveness or potential impact of new incursions based on real-time field observations. I discuss how such observations, in conjunction with weed history (where available), could be used to determine appropriate incursion responses. A recent categorisation of species in relation to biological impedance to eradication is recommended as a qualitative aid to estimating the resources required for eradication. Implications for the assessment of containment feasibility are also considered.

Keywords Containment, dispersal, eradication, feasibility, incursion, invasiveness, impact, weed history.

INTRODUCTION

Co-ordinated management strategies, such as eradication or containment, can prevent or substantially delay the impacts of new alien plants, particularly if detection occurs soon after plant introduction (Wilson *et al.* 2016). However, implementation of such strategies is generally expensive and labour-intensive. Constraints on the resources available for managing new incursions mean that difficult decisions need to be made. Decision makers first must take into account the risk posed by a new incursion: does the species represent a threat sufficient to warrant the investment of scarce resources? Weed risk assessment (WRA) poses a number of challenges, the most fundamental being that the plants that will become serious weeds comprise only a small proportion of all introductions (Hulme 2012). This is why a plant's history as a weed, in conjunction with a high probability of climatic suitability in the area at risk of invasion, remain the most powerful predictors of its weed risk. Obviously, this approach will not be an option for plants that have no history of weediness.

If a species is considered to pose a high risk, the second issue faced by decision makers is how to manage it. Here the key question is: how likely it is that a coordinated management strategy will be successful within the constraints of the resources that could be expected to be available? The factors that influence eradication feasibility are by now well known, characterised and, in many cases, quantified (Panetta 2015). A distinction can be made between factors that are extraneous to the targeted species (e.g. societal attitudes towards the plant and the infrastructure available for strategy implementation) and those relating to the target and where it is found (e.g. intrinsic biological features, detectability and the accessibility of its populations). The second group of factors comprises the 'technical feasibility' of eradication. Similar considerations pertain where the management goal is containment (Wilson *et al.* 2016).

The dynamics of alien plant invasions are usually much less dramatic than those observed for other invasive organisms, but timely action is still essential because positive management outcomes are more likely where smaller spatial extents of targeted species are involved (Pluess *et al.* 2012). Accordingly, incursion management decisions need to be made under conditions of considerable uncertainty. In many cases uncertainty arises because weed risk is not well known, but in almost all situations it is difficult to gauge with confidence whether a coordinated management strategy is likely to succeed, given the available (or anticipated) resources. The aim of this paper is to explore ways in which these underlying uncertainties can be made more explicit and can be better addressed when dealing with detections of new alien plants, especially where such plants are new to a location or region.

POST-BORDER WRA

If alien plant incursions are to be managed proactively, decisions need to be made before incursions develop into full-blown invasions. The first decision to make is whether a newly detected species poses sufficient risk to warrant further action. A number of desktop systems are available to assess post-border weed risk (e.g. Weiss *et al.* 2004, Setterfield *et al.* 2010) but in these procedures invasiveness is estimated largely on the basis of plant features (e.g. reproductive ability and dispersal mechanisms). Furthermore, impact estimates are based upon effects that could be observed at invasion stages that might better be considered as post-incursion. In this section I will describe a simple approach in which weed history is combined with observations made in the field in order to assess weed risk in a new location. The scope of the present exercise will be confined to weeds of natural ecosystems for two reasons:

1) these systems are generally more complex than agro-ecosystems, meaning that it is more difficult to assess risk, and
2) funding for incursion management in natural ecosystems is often more difficult to source than for management in agro-ecosystems, since the economic imperative associated with the latter is much clearer.

Weed history Online resources are invaluable sources of information on weed history. A particularly useful online compilation is 'A Global Compendium of Weeds' (Randall 2012). In this source records for environmental weeds are identified specifically, but how to evaluate the strength of evidence for weediness is an important issue that remains to be resolved. Owing to the matter of cascading references (Randall 2012) the total number of references listed here is not a particularly good metric.

Panetta (2016) has suggested that the number of world regions in which a species has been documented as an environmental weed (Figure 1) is a useful proxy for adaptation to a broad range of environmental conditions, both non-biological (e.g. climate and soils) and biological (e.g. different plant communities). However, the relationship between the number of world regions in which a species is an environmental weed and the strength of this factor as a predictor is unclear. Is a single region sufficient? (NB in Randall's compendium where a weed occurs in a single region this is almost always Australasia.) Presumably, as the number of regions increases a larger number of plant communities would have been invaded and impacted, and a broader range of abiotic conditions encountered.

Because most single-region recordings of species in Randall (2012) correspond to Australasian occurrences, a cut-off of two regions was used to distinguish between low and high levels of evidence for weed history (Panetta 2016), but this remains unsubstantiated. Detailed evidence of weed history, including description and quantification of impacts, may be gleaned from original sources listed in Randall (2012) as well as the scientific literature. Species that elsewhere have transformed the character, condition, form, or nature of ecosystems (Richardson *et al.* 2000) are often the ones that pose the highest risk (e.g. Fried and Panetta 2016).

Climate matching is an important adjunct to the use of weed history as a predictor of risk, but whether this is essential depends on scale: at the local scale, pathways of introduction may be sufficiently informative. For example, where a plant has been introduced for ornamental or amenity purposes, its appearance in a natural ecosystem indicates sufficient climatic pre-adaptation to enable it to escape cultivation and to persist, at least in the short term. If it has been introduced for purposes of production, its selection, importation and widespread use are all based on an assumption of climatic pre-adaptation (Panetta 2016). Where risk is being assessed at the level of a country or continent, however, climate matching will need to be undertaken as part of the procedure for estimating the total area at risk.

Field observations 'Likelihood' in the Risk = Likelihood × Consequence equation refers to the probability that an introduced plant will establish and spread,

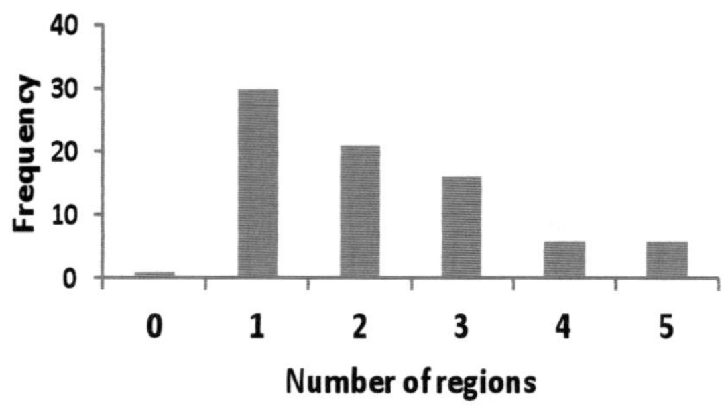

Figure 1. The number of world regions in which species (n = 80) sampled from the Victorian Weed Advisory Lists occur as environmental weeds, according to Randall (2012). World regions comprise Africa, Australasia (i.e. Australia and New Zealand), Caribbean, Europe, North America, Pacific, South America and Sub Antarctic. From Panetta (2016).

and 'consequence' encompasses all of the negative impacts that could result from its spread (Daehler and Virtue 2010). Post-border risk assessment has a significant advantage over its pre-border counterpart in that there is an opportunity to quantify spread *in situ* and to make at least qualitative observations regarding impact. Unfortunately, there appears to have been little attempt to exploit this opportunity.

Recent publications (Blood *et al.* 2016, Panetta 2016) detail a risk assessment screen that can be used by public land managers to determine when a newly detected species poses sufficient risk to warrant its delimitation, prior to making a decision on how best it might be managed. Operating through a score sheet template, the screen comprises *measures* of invasiveness (spread) and impact and *filters* that assist in the interpretation of field observations.

Invasiveness measures relate to whether reproduction occurs by vegetative means, by seed or by both means, plus the distance to which spread has occurred (Table 1). Potential indicators of impact, based upon both the potential to alter vegetation structure and the cover attained are also provided. Given the challenges associated with quantifying (let alone predicting) impact, this part of the screen remains a work in progress. A key consideration is that seedling density may be more effective in capturing impact potential for plants that have longer juvenile periods (e.g. some shrubs and most trees). Furthermore, it remains to be determined whether the use of qualitative measures for impact is superior to quantitative measures, the latter requiring the establishment of threshold (cut-off) values as part of a binary decision-making procedure.

There are two types of filters, relating to time and disturbance respectively (Figure 2). Time-related filters are designed to capture the opportunity that a species has had to express its potential for spread and for generating impacts. These are based upon residence time (how long a plant has been present, if known specifically or otherwise able to be estimated) and how long it would take for an individual to reproduce (i.e. the length of its juvenile period – this will be more important for plants that have long juvenile periods, such as some shrubs and many trees).

The disturbance filter is constructed to reflect the observation that many of the environmental weeds that pose the greatest risk to natural ecosystems are ones that have the ability to establish in relatively intact vegetation. It uses information about whether a species is restricted to edges or areas that have been recently disturbed, or is associated with other plants that are commonly considered to be indicators of disturbance.

In Figure 3 weed history evidence and field-derived measures of plant performance are assembled in a matrix that shows recommended management actions, together with levels of uncertainty associated with these recommendations. Here weed history evidence is treated qualitatively owing to the unresolved

Table 1. Scoring for invasiveness, modified from Panetta (2016).

Evidence of reproduction	
Has the plant reproduced in place by joined vegetative structures?	Yes = 1 No = 0
Has the plant reproduced by seeds or detachable vegetative structures?	Yes = 2 No = 0
Pre-reproductive period	
How long does it take for a new individual to produce seeds or other propagules?	< 1 yr = 3 1–2 yrs = 2 > 2 yrs = 1 Default[A] = 1
Evidence of spread	
For connected vegetative spread	≤ 1 m = 0 > 1 m = 1
For spread via seed or detachable vegetative propagules	≤ 10 m = 1 10–50 m = 2 > 50 m = 3

[A] For shrubs and trees.

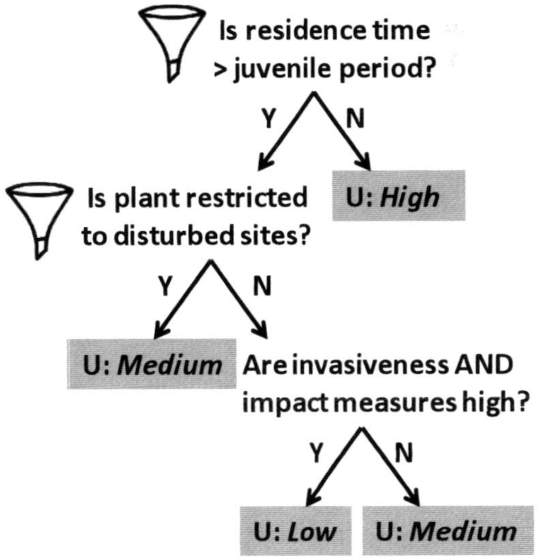

Figure 2. Flow diagram showing how time- and disturbance-related filters can be used to characterise the level of uncertainty (U) associated with weed risk assessments that are based on field-derived measures of invasiveness and impact.

issue raised in the previous section, i.e. the number of world regions versus strength of weed history as a predictor.

Site-based evidence	Weed history evidence: None/little	Weed history evidence: Abundant
Low	Monitor	Consider management feasibility?
High	Delimit and consider management feasibility	Delimit and consider management feasibility

Figure 3. Risk assessment matrix based on different combinations of levels of evidence from weed history and site-based measures. Levels of uncertainty associated with recommended risk management actions range from lowest (no shading) to highest (heavy shading). Highest uncertainty accompanies a situation where the management action adopted will likely depend on attitudes towards risk (see concluding remarks).

requiring more frequent site visits than slow-maturing perennials. Since eradication cannot be declared until seed banks are exhausted, seed persistence is a strong determinant of the minimum time required for an eradication program. Panetta (2015) used cut-off values of two years for juvenile period length and three years for seed persistence to categorise species in terms of their biological impedance. Different attribute combinations yield four species categories, which can be ranked from lowest to highest biological impedance (Figure 4).

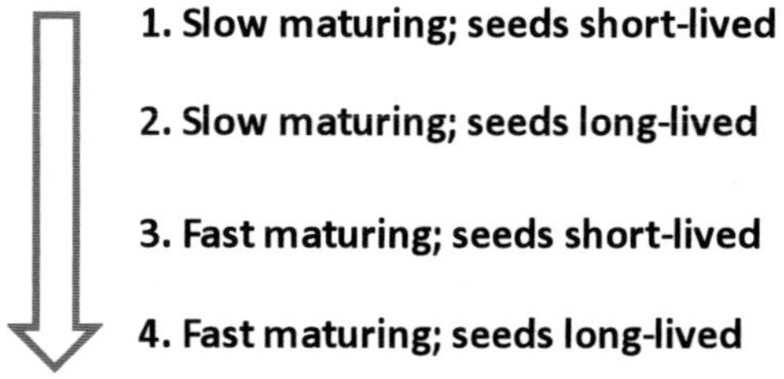

Figure 4. Ranking of plants according to biological eradication impedance posed at a similar stage of invasion. The arrow indicates order of increasing impedance.

ERADICATION FEASIBILITY

Factors that determine the technical feasibility of eradication fall under four broad categories: logistic considerations, species detectability, species biological characteristics and control effectiveness (Panetta and Timmins 2004). These factors will influence the amount of impedance posed by a targeted species – where higher levels of eradication impedance prevail, it will be more difficult (and hence more costly) to achieve eradication. In this section I will focus first on the biological components of eradication impedance, which together provide a biological rationale for categorising and ranking species in terms of eradication feasibility. Next I will consider how dispersal processes may be expected to interact with this categorisation.

Biological impedance Key features of this component of impedance are the duration of the juvenile period and for how long propagules (generally seeds, but sometimes other vegetative structures) persist in the field. Juvenile period length will determine how frequently infestations must be visited in order to prevent seed production, with rapidly cycling annuals

Examples of species in each category are given in Panetta (2015), but to illustrate the extremes, many plants that were introduced for ornamental and amenity purposes, e.g. camphor laurel (*Cinnamomum camphora* (L) Presl.), broad-leaved Brazilian pepper (*Schinus terebinthifolius* Raddi) and privets (*Ligustrum* spp.) demonstrate the least impedance (Category 1) and a number of weeds of crops and pastures, e.g. parthenium weed (*Parthenium hysterophorus* L.) and parasitic species in the genera *Phelipanche/Orobanche* and *Striga* demonstrate the most (Category 4). An extensive survey that would provide the basis for estimating the relative frequencies of each category in regional weed floras could prove very useful.

Dispersal While it is possible for a plant to cause significant impact without spreading (Wilson *et al.* 2016), the most problematic species will be ones that not only generate impact but also have (often multiple) dispersal mechanisms that assist them in moving beyond sites of introduction. The scale of management needs to approximate that of a plant's dispersal kernel if eradication is to be a realistic goal (Fletcher and Westcott 2013).

Most important are the mechanisms that promote long distance dispersal, since these facilitate the most rapid spread (Nathan 2006). Human-mediated dispersal (HMD) (dispersal by humans and their agents, e.g. machinery and livestock) is particularly effective in transporting weeds for long distances, often with sufficient propagule numbers to overcome barriers to establishment. From an eradication perspective HMD also provides multiple opportunities to manage spread, ranging from maintaining a very high standard of hygiene for machinery, to trace-forward activities that can identify previously undetected infestations. HMD, therefore, is a 'two-edged sword' that can be exploited but at the same time requires very careful management. Where long distance dispersal occurs largely via other mechanisms, e.g. dispersal by wild animals or physical vectors such as water, spread will be prevented most effectively by reducing levels of propagule production (Panetta and Cacho 2012). Note that prevention of reproduction is a key management goal in weed eradication efforts.

Impedance × dispersal interactions Unmanaged long distance dispersal will decrease the likelihood of achieving eradication. Maintaining vehicular and personal weed hygiene during search and control operations is therefore critical if operators are not to contribute unwittingly to further spread of the targeted species (Bocking *et al.* 2008). Preventing, or at least minimising, the reproduction of species whose major vectors are essentially unmanageable should reduce the risk of further spread from *known* sites. However, if additional foci of infestation (present either before or arising during the course of an eradication program) are not detected and managed, they will in turn contribute to further spread. The negative impact of further spread upon the probability of eradication will likely increase in order from Category 1 to Category 4 species (Figure 4), a function of combining shorter life cycles with greater persistence of the propagules that are produced. Reflecting both factors, a large component of the cost of eradication is attributable to the need to exhaust weed seed banks (Panetta and Cacho 2014).

IMPLICATIONS FOR CONTAINMENT

Containment is often considered as a fallback strategy when eradication appears to be unattainable, given budgetary and other constraints (Hulme 2006). Containment is not always the best option if an eradication program is not performing well, but it does become proportionally cheaper than eradication for plants that have smaller dispersal distances or more persistent seed banks (Fletcher *et al.* 2015). A containment program may require separate approaches to core, as opposed to outlier, populations. The establishment of barrier zones around core populations is not favoured when dispersal cannot be managed effectively, resulting in the possibility of frequent barrier zone breaches. Equally, eradication of outliers should not be attempted when a plant has a short juvenile phase, combined with long-lived seeds (i.e. Category 4 species) (Panetta and Cacho 2014). Major factors contributing to containment impedance include: time to reproduction; aspects of detectability (whether the species is conspicuous at any stage, plus the probability of passive detection, given where it occurs in relation to human habitation or activity); and the relative importance of human mediated dispersal (Wilson *et al.* 2016).

CONCLUDING REMARKS

Without timely intervention, opportunities for proactive management of new alien plant incursions will be lost. Direct observations on invasiveness and impact, supplemented by information available on the weed history of the species of interest may provide a reliable basis for a rapid assessment of the weed risk posed by a plant. Where such an assessment is made in support of decisions relating to intervention at a local scale, the need to establish climate suitability is moot; climate-matching would, however, be required to assess the risk posed over very much larger areas. Until now there appears to have been little attempt to develop even semi-quantitative methods that can be employed to integrate field observations into weed risk assessments for new incursions.

The attitudes of decision makers and managers towards risk influence the choices that are made regarding new incursions of alien plants. The predominating preference is for investment in the control of widespread species, because resulting productivity gains are more or less certain here. Perhaps counterintuitively, this is actually a manifestation of *risk aversion*. The future benefits of managing a specific incursion are far less clear: without intervention the species might ultimately have no serious impacts; the incursion response might fail; and even if eradication is achieved, future incursions might occur. However, Finnoff *et al.* (2007) argue that a *risk-neutral* approach should be adopted, since an acceptance of the uncertain benefits of proactive management will in most cases increase overall benefits.

Plants having rich histories as weeds elsewhere, yet low levels of site-based evidence of invasiveness or impact, are somewhat problematic. Might they be 'sleepers'? They should be relatively easy to eradicate, yet a risk-averse response (*sensu* Finnoff *et al.*) would be to 'wait and see'. However, very small incursions can be managed proactively with relatively little cost and a high likelihood of success. Harris and Timmins (2009) have argued in favour of targeting numerous incursions (provided that management costs for each are relatively low) in order to 'capture' the species that would ultimately cause the most damage.

Uncertainty around the amount of resources required to achieve a desired outcome from a co-ordinated management program may be reduced through consideration of the biological characteristics of the targeted species, as well as the dispersal mechanisms likely to be involved. For a given stage of invasion (including pre-invasion!), proactive management programs for species in which long pre-reproductive periods are combined with low seed persistence will be both less resource-demanding and have a higher likelihood of success per dollar invested than those in which the opposite characteristics pertain.

Successful containment, but failed eradication, of a relatively large incursion of branched broomrape (*Phelipanche ramosa* F.W.Schulz, formerly *Orobanche ramosa* L.) (7535 ha gross area) is an example of a targeted plant whose rapid reproduction was coupled with very long-lived seeds (Panetta 2015). Improved risk assessment procedures and better-informed decisions on risk management can be expected to yield a greater level of success overall in incursion response, but there remains a critical need for 'more runs on the board' in support of future investment in this area.

REFERENCES

Blood, K., James, R. and Panetta, F.D. (2016). Looking for weeds: assess the risk guide. A guide for assessing the risk for weeds at the early stage of invasion on public land in Victoria. Department of Environment, Land, Water and Planning, Victoria. URL: http://www.depi.vic.gov.au/environment-and-wildlife/weeds-and-pests/early-invaders

Bocking, J., Galway, K. and Brooks, S.J. (2008). Weed spread prevention: simple activities for field operations. Proceedings of the 17th Australian Weeds Conference, eds R.D. van Klinken, V.A. Osten, F.D. Panetta and J.C. Scanlan, pp. 461-3. Queensland Weeds Society, Brisbane.

Daehler, C.C. and Virtue, J.G. (2010). Likelihood and consequence: reframing the Australian Weed Risk Assessment to reflect a standard model of risk. *Plant Protection Quarterly,* 25, 51-5.

Finnoff, D., Shogren, J.F., Leung, B. and Lodge, D. (2007). Take a risk: preferring prevention over control of biological invaders. *Ecological Economics* 62, 216-22.

Fletcher, C.S. and Westcott, D.A. (2013). Dispersal and the design of effective management strategies for plant invasions: matching scales for success. *Ecological Applications* 23, 1881-92.

Fletcher, C.S., Westcott, D.A., Murphy, H.T., Grice, A.C. and Clarkson, J.R. (2015). Managing breaches of containment and eradication of invasive plant populations. *Journal of Applied Ecology* 52, 59-68.

Fried, G. and Panetta, F.D. (2016). Comparing an exotic shrub's impact with that of a native life form analogue: *Baccharis halimifolia* L. versus *Tamarix gallica* L. in Mediterranean saltmarsh communities. *Journal of Vegetation Science* 27, 812-23.

Harris, S. and Timmins, S.M. (2009). Estimating the benefit of early control of all newly naturalised plants. Science for Conservation 292, pp. 25. Department of Conservation, Wellington, New Zealand.

Hulme, P.E. (2006). Beyond control: wider implications for the management of biological invasions. *Journal of Applied Ecology* 43, 835-47.

Hulme, P.E. (2012). Weed risk assessment: a way forward or a waste of time? *Journal of Applied Ecology,* 49, 10-9.

Nathan, R. (2006). Long-distance dispersal in plants. *Science* 313, 786-8.

Panetta, F.D. (2015). Weed eradication feasibility: lessons of the 21st century. *Weed Research* 55, 226-38.

Panetta, F.D. (2016). Environmental weed risk screen for Victoria: background and development. A report prepared for the Department of Environment, Land, Water and Planning, Victoria. URL: http://www.depi.vic.gov.au/environment-and-wildlife/weeds-and-pests/early-invaders

Panetta, F.D. and Cacho, O.J. (2012). Beyond fecundity control: which weeds are most containable? *Journal of Applied Ecology* 49, 311-21.

Panetta, F.D. and Cacho, O.J. (2014). Designing weed containment strategies: an approach based on feasibilities of eradication and containment. *Diversity and Distributions* 20, 555-66.

Panetta, F.D. and Timmins, S.M. (2004). Evaluating the feasibility of eradication for terrestrial weed incursions. *Plant Protection Quarterly* 19, 5-11.

Pluess, T., Jarošik, V., Pyšek, P., Cannon, R., Pergl, J., Breukers, A. and Bacher, S. (2012). Which factors affect the success or failure of eradication campaigns against alien species? *PLoS ONE* 7, e48157.

Randall, R.P. (2012). 'A global compendium of weeds', Second edition, (Department of Agriculture and

Food, Western Australia). URL: http://www.researchgate.net/publication/235869421_A_Global_Compendium_of_Weeds_Second_Edition.

Richardson, D.M., Pyšek, P., Rejmánek, M., Barbour, M.G., Panetta, F.D. and West, C.J. (2000). Naturalization and invasion of alien plants: concepts and definitions. *Diversity and Distributions* 6, 93-107.

Setterfield, S.A., Ferdinands, K., Rossiter-Rachor, N. and Douglas, M.M. (2010). The Northern Territory's Weed Risk Management System: an application of the National Post-Border WRM Protocol. *Plant Protection Quarterly* 25, 71-4.

Weiss, J., Edgar, B., Hunt, T. and Morfe, T. (2004). Victoria's noxious weed review: roll out not fall out. Proceedings of the 14th Australian Weeds Conference, eds B.M. Sindel and S.B. Johnson, pp. 707-10. Weed Society of New South Wales, Sydney.

Wilson, J.R., Panetta, F.D. and Lindgren, C.J. (2016). 'Detecting and responding to alien plant incursions'. (Cambridge University Press, Cambridge).

UAVs and machine learning for classification of *Parkinsonia aculeata* L., and planning for integration into weed management activities

Sam Atkinson[1], Joshua Kelcey[1], Harriet Davie[2] and Susanna Beech[2]
[1] Astron Environmental Services, 129 Royal Street, East Perth, Western Australia 6004, Australia
[2] Roy Hill Iron Ore, 5 Whitham Road, Perth Airport, Western Australia 6105, Australia
(sam.atkinson@astron.com.au)

Abstract *Parkinsonia aculeata* L. is a Weed of National Significance and a declared pest in Western Australia. Roy Hill Iron Ore manage a population of parkinsonia in the Pilbara region of Western Australia through survey, mapping, control and monitoring activities.

In 2015 Roy Hill and Astron collaborated to use a machine learning algorithm to automatically classify and map individual parkinsonia plants amongst other vegetation using multispectral imagery captured by an unmanned aerial vehicle (UAV). The project objective was to more efficiently and effectively locate parkinsonia plants across large areas in comparison to conventional foot searches, and to provide a map to efficiently direct weed control technicians.

UAV imagery was captured over a 173 hectare area within a known parkinsonia population and used to segregate vegetation entities into uniquely identified objects. A subset of these objects were sampled in a ground survey using electronic annotation. The data from the ground survey were used to train a machine learning algorithm in an attempt to classify parkinsonia, and other species at the trial site.

An accuracy assessment of the classification recorded that 74% of the ground-truthed parkinsonia samples were correctly identified by the classifier.

In future the classifier could be applied to much larger areas and used to direct operational weed control activities, which could simultaneously ground-truth classification accuracy and provide a dataset suitable for refining the classifier if warranted. The trial outcomes and planning how to integrate UAV based weed identification techniques into full scale weed control programs will be presented.

Keywords UAV, weed detection, parkinsonia, Weed of National Significance, surveillance, efficiency.

Unmanned helicopter system for miconia weed surveys

Torsten Merz[1], Stefan Hrabar[1], Farid Kendoul[1] and Mick Jeffery[2]
[1] CSIRO Robotics Group, PO Box 883, Kenmore, Queensland 4069, Australia
[2] Biosecurity Queensland, Department of Agriculture and Fisheries, PO Box 20, South Johnstone, Queensland 4859, Australia
(torsten.merz@csiro.au)

Summary We present the design and evaluation of a system for aerial weed surveys based on a small unmanned helicopter. The system consists of an aircraft, a user interface, and software for flight planning and data visualisation. It is specifically designed for detecting *Miconia calvescens* DC., an invasive tree species threatening the Australian rainforest. The aim of our work is to provide a safer and more cost-effective alternative to traditional methods. The system has been successfully deployed to survey several hundreds of hectares of rainforest near Cairns.

Keywords UAV, aerial survey, *Miconia calvescens*, weed eradication, rainforest.

INTRODUCTION

Miconia calvescens DC (miconia) is a small rainforest tree native to tropical America which was originally brought into Australia for its ornamental values during the 1970s and 1980s. The pest potential of this species was later recognised after receiving reports of its invasiveness from overseas, leading to its classification as a Class 1 species in Queensland in 1996. In 2003, a National Eradication Program commenced with the intention of completely removing miconia from the wet tropical rainforests of North Queensland, and the subtropical rainforests of northern New South Wales.

One of the primary seed dispersal vectors of miconia is frugivorous birds. Dispersal models for miconia have been proposed by Murphy *et al.* 2008. Seed dispersal has been recorded up to 2 km from parent plants. By drawing a dispersal buffer around one plant, this can generate a search area of over 1250 hectares. This area is often dense rainforest within steep terrain which is extremely challenging to cover thoroughly by the use of ground teams only. The National Eradication Program started trialling helicopters manned with spotters in 2007, and has continued to utilise this technique until late 2015. This has been discontinued due to two serious helicopter accidents in the past five years. Long-term monitoring, however, is essential to evaluate the progress of an eradication program (Panetta and Brooks 2008).

The benefits of UAVs (Unmanned Aerial Vehicles) for aerial weed surveys have already been shown (see e.g. Bryson *et al.* 2014 and Dehaan *et al.* 2012). The focus of our work is the automation of unmanned aircraft to enable safe operations in mountainous environments, with minimal human interaction. Moreover, we aim for a seamless integration of all stages of a survey and a solution optimised for detecting miconia trees in the Australian rainforest. The system is generalisable for surveying other weeds.

MATERIALS AND METHODS

Aircraft selection We required an unmanned aircraft which can fly up to 5 km from the launch point, and at up to 1000 m altitude above main sea level under ISA (International Standard Atmosphere) conditions. Moreover, it should require at most two people for operation, and be deployable within one hour of arrival at a survey site. We have considered different options for unmanned aircraft and decided to use a conventional single-rotor helicopter with piston engine (Figure 1). The benefits of such a vehicle include: (1) easy launch and recovery; (2) suitability for flights close to the canopy; (3) a proven record of manned helicopter surveys and existing experience with unmanned helicopters; and (4) longer endurance compared to electric multirotors. A fixed-wing aircraft offers longer range but has launch and recovery limitations, and must fly at higher altitudes over hilly terrain to avoid having

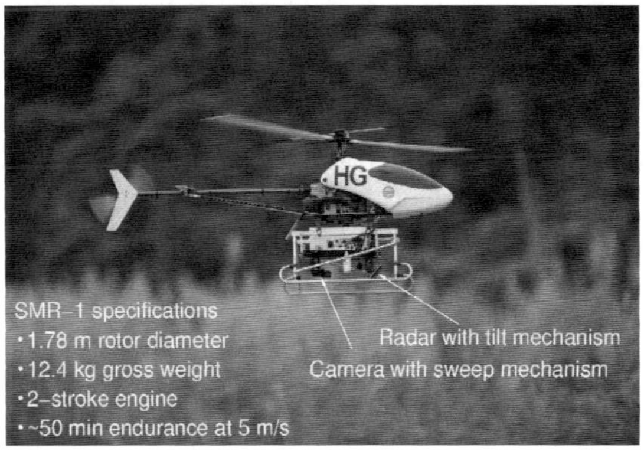

Figure 1. CSIRO SMR-1 helicopter configured for weed survey flights over mountainous rainforest.

to fly aggressive manoeuvres. Flying high increases sensor resolution and stabilisation requirements, limits surveys of narrow gullies, and increases the risk of encountering other air traffic. Additionally, a rotorcraft's ability to hover simplifies certain aspects of the automation.

Sensor selection Miconia is a small tree with large leaves with purple underside and three prominent veins. These features can be seen in high-resolution aerial images, however the view can be obstructed by larger trees, and the purple underside is not always visible. Multiple images from different view angles increases the chance of detecting miconia. To our knowledge, aerial images captured outside the visual spectrum do not improve the detection ability.

We developed an affordable high-resolution wide-angle camera system that is light enough to be mounted on a small unmanned helicopter. The camera system comprises a 5MP camera with global shutter which is triggered by the flight computer, a sweep mechanism, and an image recording computer. The camera is moved to five different sweep angles per second, and images are saved on a removable solid-state drive (SSD). The corresponding camera pose (required for geo-referencing) is stored on a USB stick which is attached to the flight computer. No images are transmitted to the ground.

From experiments with aerial images with various Ground Sample Distance (GSD) values, we determined a maximum GSD of 1.5 cm for detecting miconia trees. The camera is fitted with a 25 mm lens, which limits the distance of the helicopter to the leaves to 109 m. The field of view for one sweep is 80 × 16 degrees. Apart from increasing the field of view, the sweep mechanism allows the capture of images perpendicular to the ridge surface when flying along ridge contours, and to compensate for helicopter roll.

Helicopter automation Our helicopter is designed to be safely operated with minimal human interaction. During flight, the helicopter must perform the duties of a pilot without having to rely on the expertise of a remote pilot. This relaxes requirements on communication links and simplifies the user interface. The helicopter user interface is a small box with two switches: a push button switch with multiple functions depending on the state of the mission (e.g. starting a mission, aborting a mission, invoking an aircraft avoidance manoeuvre), and a toggle switch for terminating a flight in an emergency. Helicopter flight plans are generated with a flight planning tool before the flight. A flight plan is stored on a USB stick which also serves as flight recorder storage. The helicopter also includes a flight termination system which ensures that it stays within a pre-defined area, and under 122 m above ground level (AGL) to comply with Australian aviation safety regulations. The assumption is that in the event of a non-recoverable fault, the helicopter can crash anywhere in the area without causing severe damage to the environment.

Flight plans are generated using commercially available terrain data from aerial LIDAR (Light Detection and Ranging) surveys with an average sample density of two points per square meter. For areas where no LIDAR data is available we use 1 arcsec resolution terrain data from the Shuttle Radar Topography Mission (SRTM) which requires to fly at a higher altitude to accommodate for the lower resolution.

A plan is a sequence of linear and circular curve segments which allow the generation of paths required for low-altitude flight over hilly terrain (Figure 2). Our guidance and control system ensures that the path is followed with less than 5 m control error in winds up to 10 m s^{-1} (if sufficient power is available). The nominal cruise speed is 5 m s^{-1} (ground speed) which may be reduced to stay within the flight envelope and performance limits. Flight control is based on cascaded, decoupled Proportional-Integral-Derivative (PID) control for attitude, velocity, and position. The navigation solution is based on a single-frequency GPS receiver without differential correction, a barometric altimeter, a magnetometer and a MEMS-based (Micro Electromechanical Systems) inertial measurement unit.

We developed a radar-based obstacle avoidance system to accommodate for mapping errors. If the system detects an obstacle it will abort the mission. The radar is also used for landing, aircraft avoidance manoeuvres, barometric altimeter corrections, and in terrain-following flight mode. The latter is engaged when a mission is aborted, returning to base on a direct path whilst staying below 122 m AGL. The aircraft avoidance manoeuvre involves a rapid, safe

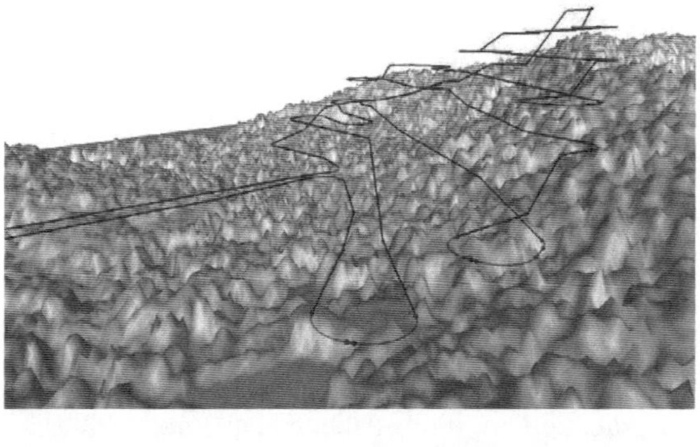

Figure 2. Helicopter flight plan generated by the flight planning tool for the El Arish site.

vertical descent until the radar senses an obstacle. The helicopter will then hover at low but safe altitude above the obstacle (e.g. canopy or ground). When the airspace is clear, the operator pushes the context-sensitive push button switch on the interface box, and the helicopter returns to base. The radar is mounted on a tilt mechanism to accommodate for the multiple radar orientations needed for the different modes of operation.

The helicopter automatically monitors its health, including engine, electric power, sensors, trajectory tracking errors, flight envelope, and mission execution. If it detects a problem, it aborts the mission.

Flight planning The goal of a flight is to capture images covering a given survey area with a given upper bound GSD. Flight plans are complex due to the topography of the terrain and the requirement to fly close to the canopy. An interactive flight planning tool based on the Rapidly Exploring Random Trees (RRT) algorithm was developed for generating plans which both satisfy the GSD constraint and a 30 m safety distance to obstacles. The tool runs on all common operating systems. It requires the survey area boundary, terrain data, a takeoff and landing location, and a 2D map for visualization purposes. The user can also define no-fly zones (e.g. houses) if needed.

In areas with steep slopes which are common for miconia infestations, flight plans typically follow contour lines (Figure 2). This is also a more energy efficient way to survey an area than crossing ridges. The planning process must also consider airspace constraints, however this capability is currently not integrated in the planning tool.

Flight operations Before commencing a flight it must be ensured that: (1) regulatory requirements are met; (2) the aircraft is serviceable; (3) the takeoff and landing area is suitable. The second requirement involves a daily inspection of the aircraft in addition to adhering to the maintenance schedule. For takeoff and landing our helicopter requires an area with approximately 5 m clearance in all directions with a firm approximately levelled surface and good GPS reception. For manual takeoff and landing these requirements can be relaxed.

During flight the operator must ensure separation from other aircraft and that the helicopter does not fly into bad weather. Separation from other aircraft can be achieved with airspace observers who monitor the mission boundaries and invoke the aircraft avoidance manoeuvre if necessary. The observers can also prevent the helicopter from flying into bad weather by aborting the mission.

We fitted our helicopter with a strobe light which significantly increases the visual line-of-sight distance. However, due to tall trees, hilly terrain and low-altitude flight it can be challenging to find locations from where the helicopter can be seen at all times.

Apart from that, the operations are quite simple: fuelling the helicopter, inserting the USB stick and the SSD, connecting the charged batteries of the helicopter and the interface box, waiting for the system self-checks to complete, starting the engine, and pushing the push button switch on the interface box when the launch area is clear. The helicopter will then take off and fly the mission. After landing the system automatically shuts down and the SSD and the USB stick can be removed.

Image analysis Our approach was to develop a method to analyse images which is (at least) comparable to spotting miconia from on-board a helicopter, but in a safer and more comfortable environment. We developed a visualisation tool that allows for easy access of images and geo-referencing of projected terrain points. Apart from the images, it requires flight data from the USB stick, terrain data, and a 2D map. It runs on all common operating systems. The geo-referencing error is less than 10 m which is accurate enough to locate a tagged tree for manual removal.

Images can also be displayed with the terrain data for better orientation in the survey area and to be able to consider topographic information in the analysis. The tool can display all images showing the same feature and enables the comparison of leaf sizes.

RESULTS AND DISCUSSION

To demonstrate the feasibility of the system, we conducted aerial surveys at two sites south of Cairns (Australia) where miconia was found in the past: El Arish and Harvey Creek. We surveyed the first site in May 2014 where we covered 68 ha and logged 1.2 flight hours. The Harvey Creek site was surveyed in October 2015. At this site we covered 293 ha and logged 6.3 flight hours. At the El Arish site, seven miconia plants (range of sizes from 2 to 4 m tall) were found based on the data our helicopter delivered. In comparison, two weeks earlier during a manned helicopter flight with three spotters, only two plants were detected in the same area. No trees have been found at the Harvey Creek site.

All surveys were conducted without incidents and we demonstrated that our system offers a feasible and safer alternative to manned helicopter surveys. Moreover, the images may also be useful for path planning and other surveys.

Figure 3. Flight paths of the unmanned helicopter at the Harvey Creek site.

Figure 4. Image captured by the unmanned helicopter showing a miconia tree.

A direct comparison with spotting from on-board a helicopter is difficult. Typically, more time is spent looking at recorded images. This reduces the risk of missing a miconia tree but on-board spotters have a wider field of view and can get more views of a suspicious tree if needed. Geo-referencing of trees is more accurate with our system compared to the position estimates that an on-board spotter can provide.

The system can be much improved by automating the image analysis. For instance, the Harvey Creek survey produced 66,000 images which had to be analysed manually. The development of an automated image analysis solution is challenging due to the small amount of aerial images showing miconia in an Australian environment.

Regulatory restrictions may limit the utilisation of the system. Under the current Australian aviation safety regulations, our helicopter must be operated by a certified pilot within visual line-of-sight. For flights beyond line-of-sight, a certified pilot with instrument rating and an area approval are required. This increases the costs of operations and takes more time.

We are currently planning to survey another 2500 ha of rainforest. As there are more areas for which no LIDAR data is available we will investigate more in depth solutions based on SRTM data. We also plan to work closely together with the aviation safety regulator to find ways to better utilise the potential of our system especially for operations beyond visual line-of-sight.

ACKNOWLEDGMENTS

The initial work was co-funded by the Queensland State Government Smart Futures Fund, Boeing Research & Technology – Australia, Insitu Pacific Ltd., Queensland University of Technology (QUT) and CSIRO as part of Project ResQu. The authors wish to thank Dennis Frousheger, Brett Wood and Matthew Dunbabin for their technical support in the development of the helicopter system.

REFERENCES

Bryson, M., Reid, A., Hung, C., Ramos, F.T. and Sukkarieh, S. (2014). Cost-effective mapping using unmanned aerial vehicles in ecology monitoring applications. Experimental Robotics: the 12th International Symposium on Experimental Robotics, pp. 509-23.

Dehaan, R.L., Weston, L.A. and Rumbachs, R. (2012). The design and the development of a hyperspectral unmanned aircraft mapping system for the detection of invasive plants. Proceedings of the 18th Australasian Weeds Conference, ed. V. Eldershaw, pp. 103-7. (Weed Society of Victoria Inc.).

Murphy, H.T., Hardesty, B.D., Fletcher, C.S., Metcalfe, D.J., Westcott, D.A. and Brooks, S.J. (2008). Predicting dispersal and recruitment of Miconia calvescens (Melastomataceae) in Australian tropical rainforests. *Biological Invasions* 10 (6), 925-36.

Panetta, F.D. and Brooks S.J. (2008). Evaluating progress in weed eradication programs. Proceedings of the 16th Australian Weeds Conference, eds R.D. van Klinken, V.A. Osten, F.D. Panetta and J.C. Scanlan, pp. 418-20. (Queensland Weeds Society).

Evaluating the Object-based Image Analysis (OBIA) approach for mapping African lovegrass (*Eragrostis curvula* (Schrad.) Nees) in the Bega Valley, NSW, Australia

Laurie A. Chisholm[1], Christopher J. Owers[2] and Nicholas Gill[3]
[1] Centre for Sustainable Ecosystem Solutions, School of Earth and Environmental Sciences, University of Wollongong, Wollongong, New South Wales 2522, Australia
[2] Geoquest Research Centre, School of Earth and Environmental Sciences, University of Wollongong, Wollongong, New South Wales 2522, Australia
[3] Australian Centre for Cultural Environmental Research, Department of Geography and Sustainable Communities, University of Wollongong, Wollongong, New South Wales 2522, Australia
(lauriec@uow.edu.au)

Summary The use of remote sensing for weed mapping has well-known tradeoffs between spectral and spatial resolution using current remote sensing platforms. Often restricted spectral capabilities limit the ability to discriminate particular plant types that are often spectrally similar, for example, similar shades of 'green', particularly in pastoral settings. Regardless of the spectral capabilities of any sensor, the discrimination of a single plant species remains challenging. Linking particular plant traits to a specific parameter of an image can assist towards discrimination. For example, targeting temporal patterns related to a plant's growth cycle, or, similarly, acquiring imagery at a time when the background can be separated from the target plant.

In addition, the use of an object-based image analysis (OBIA) approach to image classification, instead of the more typical pixel-based approach, has been shown to make significant traction with mapping individual species. OBIA seeks to identify spatially and spectrally homogeneous 'objects' which are created by grouping pixels of similar traits together, then allows a combination of ancillary data (morphology, contextual) to be incorporated to assist the classification results.

This paper will present preliminary results from the application of the OBIA approach to mapping African lovegrass (*Eragrostis curvula* (Schrad.) Nees) in the Bega Valley, NSW using World-View 2 satellite imagery acquired at a time to maximise contrast between the plant and the background pastoral setting. Focus will be given as to how the various elements of the approach were used to take advantage of plant traits, the different indices tested, and classification success.

Keywords African lovegrass, *Eragrostic curvula*, object-based image analysis, mapping.

Using EO-1 Hyperion satellite hyperspectral imagery to detect the pasture weed Paterson's curse (*Echium plantagineum* L.) in southern Western Australia

Daniel L. McIntyre and Robert J. Corner
Department of Spatial Sciences, Curtin University, GPO Box U1987, Perth, Western Australia 6845, Australia
(daniel.mcintyre@nt.gov.au)

Summary This paper investigated the utility of satellite hyperspectral (EO-1 Hyperion) imagery to detect and map the agricultural weed Paterson's curse (*Echium plantagineum* L.) in southern Western Australia. A matched filter classification was produced from image derived end-members and validated using a density independent (logistic regression and receiver operating characteristic analysis) approach. The probability (soft classification) map derived from logistic regression had an area under curve statistic of 0.87 and the threshold (hard classification) map had a user's and producer's accuracy of 81% and 83% respectively. The results demonstrated that Paterson's curse in a non-peak flowering condition can be accurately mapped from EO-1 Hyperion imagery. The most likely contributor to the accuracy was Paterson's curse remaining green into late spring when most pasture species were senescent. Future satellite hyperspectral sensors with a higher signal-to-noise ratio and the ability to image larger areas are likely to result in further improvements to mapping Paterson's curse with remote sensing.

Keywords Remote sensing.

INTRODUCTION

The most recent estimate of the financial impact of agricultural weeds in Australia is $3.4–4.4 billion per year (Sinden *et al.* 2004). The impacts of agricultural weeds include loss of production due to weeds outcompeting desired species, loss of value of products due to contamination (e.g. seeds in cotton), control costs, loss of value of products due to herbicide residue, and new production methods that are required in heavily affected weedy areas (Nordblom *et al.* 2001).

Paterson's curse (*Echium plantagineum* L.) is an annual herbaceous weed of particular concern in agricultural regions of Australia. Plants of the species contain pyrrolizidine alkaloids, which can have deleterious effects on animals and livestock, and infestations can reduce the productivity of pastures by replacing more desirable pasture species (Parsons and Cuthbertson 2001). Paterson's curse has a higher survival rate than other pasture species (e.g. *Trifolium subterraneum*) during periods of moisture stress and remains green in late spring and early summer when those species are senescent (Piggin 1976). In densely infested areas, the purple flowers of Paterson's curse makes it one of the most conspicuous weeds of roadsides, pastures, and disturbed lands during the spring months (September-November) in Western Australia (Dodd *et al.* 1993).

Reliable and cost effective methods for mapping agricultural weeds such as Paterson's curse can play an important role for successfully managing the species. Survey methods for Paterson's curse in Western Australia have been based on on-ground visual assessment of the approximate area of Paterson's curse for a given property, with no information on the spatial distribution. This approach involves a significant investment in time and resources to survey all affected properties in the south-west of Western Australia at a regional scale. Remote sensing offers a potentially good source of data for mapping and monitoring Paterson's curse, given the limitations and cost of other survey methods; in addition, the phenological properties (prominent flowers and delayed senescence) of Paterson's curse make it a potentially amenable species for detection with remote sensing.

The objective of this study was to evaluate the accuracy of satellite hyperspectral (EO-1 Hyperion) imagery for detecting Paterson's curse in southern Western Australia. This study is the first application of satellite hyperspectral imagery for Paterson's curse, and provides an opportunity to foreshadow the performance of planned satellite hyperspectral sensors for mapping Paterson's curse.

The study area is located in the wheatbelt region of Western Australia, approximately 85 km north of Perth. The study focussed on three properties located east of the town of Bindoon in the Chittering Shire. The primary land use around Bindoon is dominated by dry-land agriculture (mainly cropping), with some conservation reserves and natural environments dominated by remnant *Eucalyptus* woodland. The climate is hot-summer Mediterranean, characterised by long warm dry summers and winter dominant rainfall.

MATERIALS AND METHODS

The Hyperion sensor on the Earth Observation 1 (EO-1) satellite has the same 30 m spatial resolution as the Landsat 7 and 8 multispectral satellite sensors, but a hyperspectral resolution of 196 unique bands (Jupp and Datt 2004). The large number of bands allows materials to be mapped with conventional hyperspectral classification approaches such as matched filtering and linear spectral un-mixing. A Hyperion image was acquired on November 2, 2006. Conditions during the time of overpass (02:30 GMT/10:30 Western Australian Standard Time) were clear and cloud free. The dimensions of the image were 45 km in the along-track direction (length) and 7.65 km in the across-track direction (width).

The image was atmospherically corrected and spectrally subset to include 49 bands from the visible and near infrared, from 428 nm to 917 nm. This range reduced the dimensions of the dataset, and focused on the parts of the spectrum that are most relevant to Paterson's curse flowering and growth. The image was rectified to a level 1G (terrain corrected) Landsat 5 image using a second-order polynomial, resulting in an overall root mean square (RMS) error of 9.15 m.

An end-member signature representing dense flowering Paterson's curse was derived directly from the image. Three separate end-member signature were extracted from plots comprised of 3 × 3 pixels (90 × 90 m, or 0.81 ha) and combined as one endmember. This endmember was input into a matched filter classification (Boardman 1998).

Field data for accuracy assessment were collected on October 18, 2006, between the daylight hours of 08:00 and 17:00. A total of 85 randomly generated plots containing Paterson's curse were surveyed. Each plot measured 90 × 90 m (3 × 3 pixels). At the centre of each plot, four oblique digital photos were captured, each orientated towards one of the four cardinal compass directions. The approximate percentage cover of Paterson's curse for each plot was visually estimated post-survey from the oblique photos. A total of 100 plots that didn't contain Paterson's curse were used for accuracy assessment, but were not surveyed in the field.

The accuracy of the matched filter classification was assessed a threshold independent approach, which consisted of logistic regression (Pohar *et al.* 2004) and receiver operating characteristic (ROC) analysis (Zweig and Campbell 1993). The approach provides a quantitative measure of the predictive accuracy of a model by comparing the probability values derived from logistic regression with a corresponding validation dataset, comprised of presence and absence data. The predicted probabilities were also used to produce a hard (binary) image output based on the maximum efficiency method (Lippitt *et al.* 2008): a threshold value was determined at the location on the ROC plot where there was the greatest difference between the false positive rate and true positive rate across all possible threshold values of the validation dataset.

RESULTS

High matched filter scores were generally present in pasture and cropped areas of the image and known areas infested by Paterson's curse. Matched filter scores were generally low for areas of Eucalypt woodland, water bodies and fallow/cleared areas. Other areas of Eucalypt woodland had a 'speckle' effect of matched filter values caused by the spatial and spectral heterogeneity.

The hard binary output from the ROC threshold of the matched filter classification is shown in Figure 1 for the three properties. Small areas of *Eucalyptus* (sclerophyll) woodland are incorrectly classified as Paterson's curse in the northern part of property A (Figure 1a). Some areas of crop were misclassified in the eastern part of the property (A). In the south and south-west area of the property, some areas of pasture (B) were classified as Paterson's curse, where these areas are known not to contain Paterson's curse. The main area of Paterson's curse in the eastern section of property B (Figure 1b) was correctly classified (C). The areas on property C (Figure 1c) classified as Paterson's curse were mostly open areas of pasture, although not all of these areas were surveyed on the ground, and could have contained some Paterson's curse. There were some instances of *Eucalyptus* woodland (D) that were misclassified as Paterson's curse.

The results from the accuracy assessment are shown in Table 1. The logistic regression model showed good agreement with the validation data. The model λ of 86.95 rejects the null hypothesis that the matched filter score is not linearly related to the log-odds of presence/absence of Paterson's curse. The area under curve (AUC) statistic of 0.87 from the ROC analysis also indicated good agreement between the matched filter output and the validation data. The resulting matched filter cut-off value represented the top 43% of the dataset, in which 81% of Paterson's curse records were correctly identified (user's accuracy) and 17% of non-Paterson's curse records were misidentified (83% producer's accuracy).

Table 1. Chi-square and AUC statistics and user's and producer's accuracy.

N	λ	AUC	UA	PA
184	86.95	0.87	81%	83%

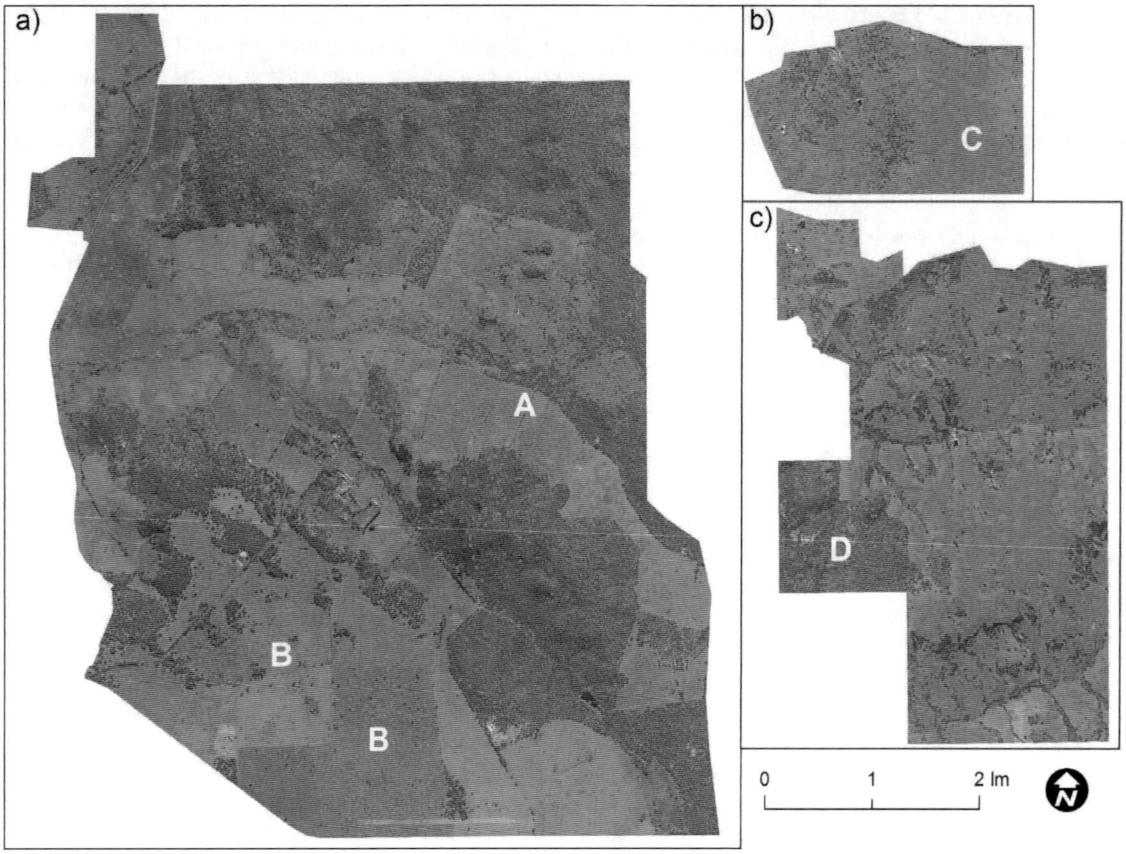

Figure 1. Paterson's curse (purple) mapped on the three properties.

DISCUSSION

The result from the matched filter classification indicates that Paterson's curse in a non-peak flowering condition can be accurately mapped with EO-1 Hyperion data. The most likely contributor to the accuracy was Paterson's curse remaining green into late spring when most pasture species were senescent. The user's and producer's accuracy (81% and 83%) was similar to other non-woody weeds that have been mapped from airborne hyperspectral imagery (e.g. Mundt *et al.* 2005, Mirik *et al.* 2013).

The most likely sources of classification error were from cropped areas and pasture that retrained greenness longer into the season as well as lower density (patchy) areas of Paterson's curse that could not be reliably discriminated from senescent pasture. The Hyperion sensor is limited in its ability to detect weeds at low density due to the coarser spatial resolution and lower signal-to-noise ratio compared to airborne sensors. Within the context of this research, the level of accuracy is acceptable, given that the rationale for using EO-1 Hyperion data was to map the distribution of Paterson's curse over a larger area in a more cost effective manner than airborne sensors, and to highlight areas that can be further investigated in detail (e.g. airborne hyperspectral or field visit).

Despite the limitations of the Hyperion sensor (primarily the low signal-to-noise ratio), the previous limitations on detecting Paterson's curse with medium spatial resolution data (Bulman 2004) were overcome to an acceptable extent by the higher spectral resolution of the Hyperion sensor. This application of Hyperion to Paterson's curse represents an improvement over the previous method of visual estimation on the ground, given that it can map the spatial destruction of Paterson's curse over a large area, with less field time. Next-generation satellite hyperspectral platforms such as EnMAP, with a large swath width (30 km) and a higher signal-to-noise ratio than Hyperion (Segal *et al.* 2010) have the potential to provide greater consistency in image quality and processing, and be more amenable with operational requirements for mapping Paterson's curse over regional scales.

ACKNOWLEDGMENTS

The authors are grateful to the owners of the three properties for permission to conduct fieldwork on their land. This research was partially funded by an Australian Research Council Linkage grant (LP0454890): Development of New Generation Tools for the Regional-Scale Mapping of Noxious Weeds.

REFERENCES

Boardman, J.W. (1998). Leveraging the high dimensionality of AVIRIS data for improved subpixel target unmixing and rejection of false positives. *AVIRIS 1998 Proceedings*, JPL, California.

Bulman, D. (2004). The application of remote sensing imagery to weed mapping: a case study for Paterson's curse (*Echium plantagineum* L.). Unpublished Ph.D thesis, Monash University, Clayton.

Dodd, J., Martin, R.J. and Howes, M.K. (1993). Management of agricultural weeds in Western Australia. Perth, Western Australia, Department of Agriculture Western Australia.

Ishii, J. and Washitani, I. (2013). Early detection of the invasive alien plant *Solidago altissima* in moist tall grassland using hyperspectral imagery. *International Journal of Remote Sensing* 34, 5926-36.

Jupp, D.L.B. and Datt, B. (2004). Evaluation of the EO-1 Hyperion hyperspectral instrument and its applications at Australian validation sites 2001–2003. CSIRO Earth Observation Centre Report 2004/06, CSIRO Atmospheric Research, Canberra.

Lippitt, C.D., Rogan, J., Toledano, J., Sangermano, F., Eastman, R.J., Mastro, V. and Sawyer, A. (2008). Incorporating anthropogenic variables into a species distribution model to map gypsy moth risk. *Ecological Modelling* 210, 339-50.

Mirik, M., Ansley, J.R., Steddom, K., Jones, D.C., Rush, C.M., Michels Jr., G.J. and Elliott, N.C. (2013). Remote distinction of a noxious weed (musk thistle: *Carduus nutans*) using airborne hyperspectral imagery and the Support Vector Machine classifier. *Remote Sensing*, 5, 612-30.

Mundt, J.T., Glenn, N.F., Weber, K.T., Prather, T.S., Lass, L.W. and Pettingill, J. (2005). Discrimination of hoary cress and determination of its detection limits via hyperspectral image processing and accuracy assessment techniques. *Remote Sensing of Environment* 96, 509-17.

Nordblom, T., Smyth, M., Swirepik, A., Sheppard, A. and Briese, D. (2001). Benefit-cost analysis for biological control of *Echium* weeds species (Paterson's curse/Salvation Jane). 45th Annual Conference of the Australian Agricultural and Resource Economics Society (AARES), 23–25 January, Adelaide.

Parsons, W.T. and Cuthbertson, E.G. (2001). 'Noxious weeds of Australia', 2nd edition. (CSIRO Publishing, Collingwood).

Piggin, C.M. (1976). Factors affecting seedling establishment and survival of *Echium plantagineum* L., *Trifolium subterraneum* L. and *Lolium rigidum* Gaud. *Weed Research* 16, 267-72.

Pohar, M., Blas, M. and Turk, S. (2004). Comparison of logistic regression and linear discriminant analysis: a simulation study. *Metodološki Zvezki* 1, 143-61.

Segal, K., Guanter, L. and Kaufmann, H. (2010). Simulation of spatial sensor characteristics in the context of the EnMAP hyperspectral mission. *IEEE Transactions of Geoscience and Remote Sensing* 48, 3046-54.

Sinden, J., Jones, R., Hester, S., Odom, D., Kalisch, C., James, R. and Cacho, O. (2004). The economic impact of weeds in Australia. CRC for Australian Weed Management Technical Series No. 8. (CRC for Australian Weed Management, Adelaide).

Zweig, M.H. and Campbell, G. (1993). Receiver-Operating Characteristic (ROC) plots: a fundamental evaluation tool in clinical medicine. *Clinical Chemistry* 39, 561-77.

A decade of weed surveys across south-eastern Australia – changes in herbicide resistance levels and identification of new cases of resistance

Peter Boutsalis, Christopher Preston and Gurjeet Gill
School of Agriculture, Food and Wine, University of Adelaide, PMB 1, Glen Osmond, South Australia 5064, Australia
(peter.boutsalis@adelaide.edu.au)

Summary Random weed surveys have been conducted in south-eastern Australia since 2005 funded by the GRDC. Three regions in South Australia and three in Victoria were surveyed with at least 150 paddocks in each region chosen at random before crop harvest. Seeds were collected from several species including *Lolium rigidum* Gaudin., *Bromus* spp., *Avena* spp., *Hordeum* spp., *Sisymbrium orientale* L., *Brassica tournefortii* Gouan, *Sonchus oleraceus* L. and *Raphanus raphanistrum* L. Each region was surveyed at five year intervals. This research has highlighted that in some regions, resistance to ALS- and ACCase-inhibiting herbicides, trifluralin, triallate and glyphosate in *L. rigidum* have increased.

Resistance to 2,4-D and sulfonylurea herbicides was identified in *S. oleraceus* and *S. orientale*. New cases of resistance to atrazine and diflufenican were also identified in *S. orientale* from the Eyre Peninsula. In *Bromus* spp. and *Hordeum* spp., a higher incidence of resistance to ALS-inhibiting herbicides than to ACCase-inhibiting herbicides was identified.

These surveys have raised awareness of herbicide resistance as a serious issue for the Australian grains industry. The detection of new cases of resistance from these surveys serves as a warning of future increases and contributes to the adoption of management practices to delay resistance.

Breaking Bad – 10 years into a projected 30 year weed eradication program on World Heritage Listed Lord Howe Island

Susan M. Bower
Lord Howe Island Board, GPO Box 5, Lord Howe Island, New South Wales 2898, Australia
(sue.bower@lhib.nsw.gov.au)

Summary The Lord Howe Island Weed Eradication Program is possibly one of the most ambitious island weed eradication programs on an inhabited island, in the South Pacific – if not globally – given the density, distribution and diversity of weed populations that were present on the island at the program's commencement.

The program aims to search all weed management blocks, at two year intervals to continue to deplete target weeds. Novel techniques to survey and eradicate weeds on inaccessible terrain have been employed.

Since program inception in 2004, in the first 10 year period, the number of weeds, of up to 25 species has reduced by 80% (all life stages) and matures by 90% comparing first and last, measured treatments across 1164 hectares. The top two weed species removed include cherry guava (*Psidium cattleyanum* Sabine var. *cattleyanum*) (Figures 1–3) and ground asparagus (*Asparagus aethiopicus* L.). Six weeds have been declared eradicated, however they were initially limited in occurrence. Over two million individual weeds have been removed over the ten year period at a cost of AUD $6.4 million.

For the broader dispersed species the program is in the 'active' eradication phase (Panetta 2007) with mature weeds still being removed, albeit at a significantly reduced rate than at commencement. The strong downward trend in weeds removed per weed block suggests that the program will reach the 'containment and monitoring' phase for widespread species within the next decade; given sustained investment and technical solutions for weeds on cliff lines. However, the duration and lead in time makes the program vulnerable to eradication fatigue.

Eradication is considered the optimum investment if the goal is to provide long term protection to Lord Howe Island World Heritage listed ecosystems from weed threat.

Keywords Eradication, grid search, helicopter lance spray, cherry guava, *Psidium cattleyanum*, weeds on islands.

INTRODUCTION

Lord Howe Island (LHI) is an isolated oceanic island in the Tasman Sea, located 780 km north-east of Sydney (31°33'29"S, 159°05'12"E), New South Wales (NSW), Australia. The main island and its associated offshore islands are known as the Lord Howe Island Group (LHIG) and were inscribed on the World Heritage List in 1982 in recognition of the Group's superlative natural landscapes and scenery, its rich terrestrial and marine biodiversity (UNESCO; see http://whc.unesco.org/en/list/186).

The main island is 11 km long and 0.6–2.8 km wide covering 1455 ha and rises to 875 m (Mt Gower). The island has 87% native vegetation cover with 75% of the LHIG protected in the Permanent Park Preserve (PPP), similar to a national park but managed by the Lord Howe Island Board (LHIB) under the *Lord Howe Island Act 1953*.

Prior to its discovery in 1788 and subsequent settlement in 1833, the LHIG remained isolated from human influences. The key threats to the biota of LHI are climate change, plant diseases, introduced pest animals and invasive weeds (Hutton and Auld 2004, DECC 2007).

The impact of cherry guava on island ecosystems is well reported (ISSG 2013) and the risk to LHI along with other invasive weeds was quantified through baseline weed density and distribution mapping across 460ha by Le Cussan (2002a,b, 2003a,b), which advised an island wide time driven (30 year) eradication program. The program commenced in 2004 based on methodology from Raoul Island, New Zealand.

Over 670 introduced plants have been recorded from LHI, mostly in the settlement. Of these, 270 have invasive characteristics and/or have naturalised (DECC 2007). The LHI Weed Management Strategy 2006 initially targeted 25 species for eradication based on risk and feasibility. Currently up to 68 are prioritised initially assessed by the NSW Department of Primary Industries (DPI) Weed Risk Management (WRM) criteria and declared noxious for LHI Authority under the NSW *Noxious Weeds Act 1993*, to allow an 'all tenure' management approach.

The LHI weed eradication program forms part of a multi species recovery plan (LHI Biodiversity Management Plan, DECC 2007), complementing a range of other eradication/restoration initiatives being implemented coincidently. Here, we report on the results, including setbacks and learnings in the first 10 years of a strategy to eradicate highly invasive weeds. Data

Figure 1. Cherry guava – a tasty food plant, native to South America was introduced to LHI in the late 1890s. As with many pacific islands, it has demonstrated to be serious ecosystem invader.

Figure 2. Cherry guava thicket from Stevens Reserve in the 1990s. Cherry guava can colonise intact to disturbed plant communities and persist as a seedling bank.

Figure 3. Outbreaks of cherry guava in the southern mountains have been detected and removed. These patches may be indicative of prior dispersal by feral pigs.

on cherry guava is provided as an example of what has been achieved. The output data for other major weeds follows similar trends to that of cherry guava (LHIB 2016). Future planned pest animal eradications and the importance of continued investment to trend target weeds to zero density/eradication for improved biodiversity outcomes on LHI is also discussed.

METHODS

To manage and eradicate weeds, the island is mapped into nine main landscape units – and further divided into 395 weed blocks corresponding to a Geographic Information System (GIS) and MS Access database (Figure 4). Weed blocks are physically marked on ground with blue marker tape or are aligned to lease boundaries. In the PPP blocks range from 0.38 to 62 hectares.

Weed searching is undertaken in a systematic grid pattern (including teams on ground, on rope or aerial search and control methods). On ground weed searching is demarked by string lines, used to retrace the grid edge. Team members must be evenly spaced so all land can be visually surveyed (average width of 5 m, decreasing with dense vegetation). All search effort is recorded on geographic positioning systems (GPS).

Target weeds in the PPP are subject to simultaneous search as most occupy similar habitats (to avoid

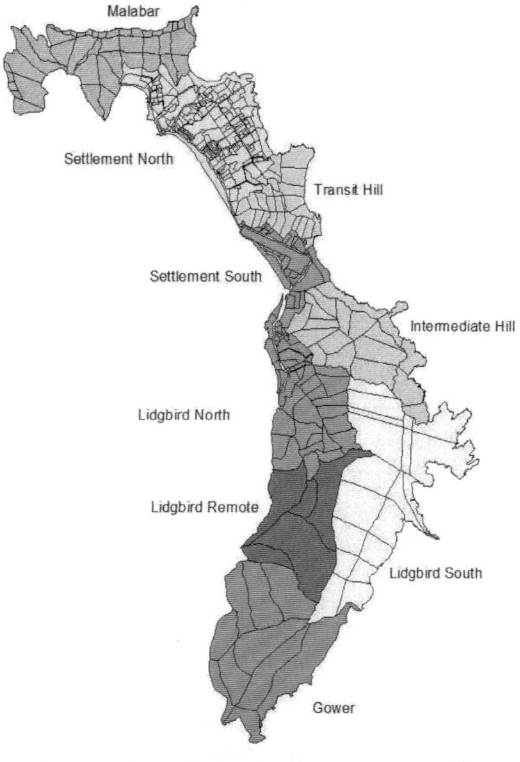

Figure 4. Landscape units and weed blocks – LHI.

replacing one weed with another), have comparable invasive capacity and environmental persistence (e.g. bird dispersed, persist in low light, form seedling banks, as opposed to weed seed banks). Several species with different reproductive traits are targeted during their active growth phase prior to fruiting.

About 80% of the island (1164 ha), where the target weeds are known to occur, is searched on ground. The search intervals of weed blocks should be no more than 24 months to treat at least 500 ha per year (approximately half of known weed habitats on island), based on reproductive biology in that target weeds will not mature within 24 months (from seed bank or juvenile). Three species, with limited distribution and/or shorter stage to reproductive maturity require annual search effort (bitou bush *Chrysanthemoides monilifera* (L.) Norlindh subsp. *rotundata* (DC) Norl., glory lily *Gloriosa supberba* L. and climbing asparagus *Asparagus plumosus* Baker, in migratory seabird nesting sites).

Isolated mature weed plants are marked in the field (Figure 5) and mapped on GPS and entered into the database as an 'INF' (for *infestation)*; to increase awareness of the need for repeated search efforts for new recruits. Marking isolated mature plants for focused searching is increasingly important as numbers reduce.

Figure 5. Flagging an 'INF' *Infestation* marker for an isolated 'outlier' mature cherry guava.

Most weed species are controlled by hand 'crowning out' or herbicide application – a mix of water with glyphosate and metsulfuron methyl for cut stump, scrape paint or foliar spray (Figure 6) (with additives of Pulse® or emulsified oil). Picloram gel is being trialled on blue passionflower – (*Passiflora caerula* L.). A helicopter with front mounted lance spray apparatus was developed to improve access to weeds on cliffs in 2015 (Figure 7). To accelerate search efforts in remote terrain in the southern mountains, teams are deployed by heli-winch over multiple days.

Figure 6. Dense widespread infestations of ground asparagus on Transit Hill were foliar sprayed with back packs. These dense infestations are now depleted. Grid search continues in accessible terrain, at least every 24 months, to remove any remaining matures and extirpate seed banks.

Figure 7. The helicopter mounted – forward lance spray apparatus provides simultaneous detection and treatment of outlier weeds (ground asparagus and bitou bush) on sheer cliffs. The spray pods have a 200 litre capacity. With the lance extending beyond the rotor blades, downwash is limited and foliar spray treatment is targeted.

Weed search effort is entered into Microsoft Access – 'the LHI Weeds Database' (hours of labour, numbers of weeds removed per weed plant life stage – seedling, juvenile or mature) per block. The database is used to monitor change in weed populations, hours of search effort per management unit, forecast treatment priorities and model trends.

To measure progress towards eradication after the first ten years, two methods were employed:

1. Scrutiny of the LHI weeds database on hectares treated, abundance of weeds removed and search effort.
2. Repeat landscape scale, weed density and distribution mapping as per J. Le Cussan (2002/03).

Mapping weed density and distribution across four landscape units (covering 460 ha) in 2002/3 measured the extent of weed threat across LHI. Repeat survey effort in 2013/14 replicated the method and extent of survey (Figure 8).

Parallel transects were set on a compass bearing (north–south or east-west) across each landscape unit. Transects were spaced 100 m apart and on each, at 20 m intervals, a circular sample plot four m in diameter (12.566 m^2) was surveyed for weeds. Target weeds were categorised according to height – juvenile (ground to knee height), shrub (knee to head height) and mature (above head).

A GPS waypoint was taken at each plot and corresponding field data downloaded into the Environmental Systems Research Institute (ESRI) Geographic Information System (GIS) package ArcView 3.3. Distribution and density maps were generated using Independent Distance Weighted (IDW) surface interpolation method in the ArcMap Spatial Analyst (ESRI) (Figure 3).

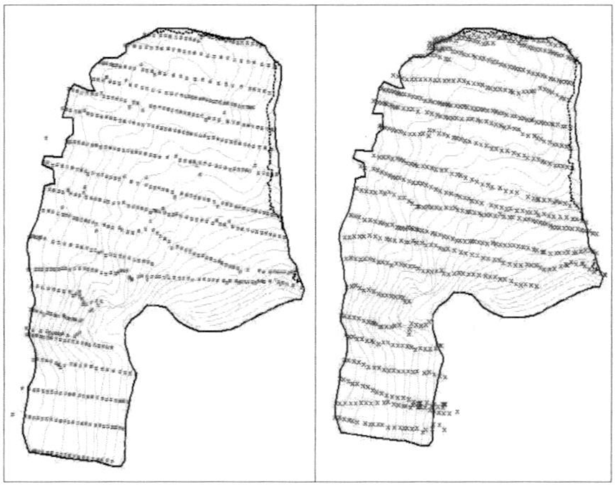

Figure 8. Lidgbird North weed mapping survey effort 2002–03 (left) and 2013–14 (right) with over 700 sample plots.

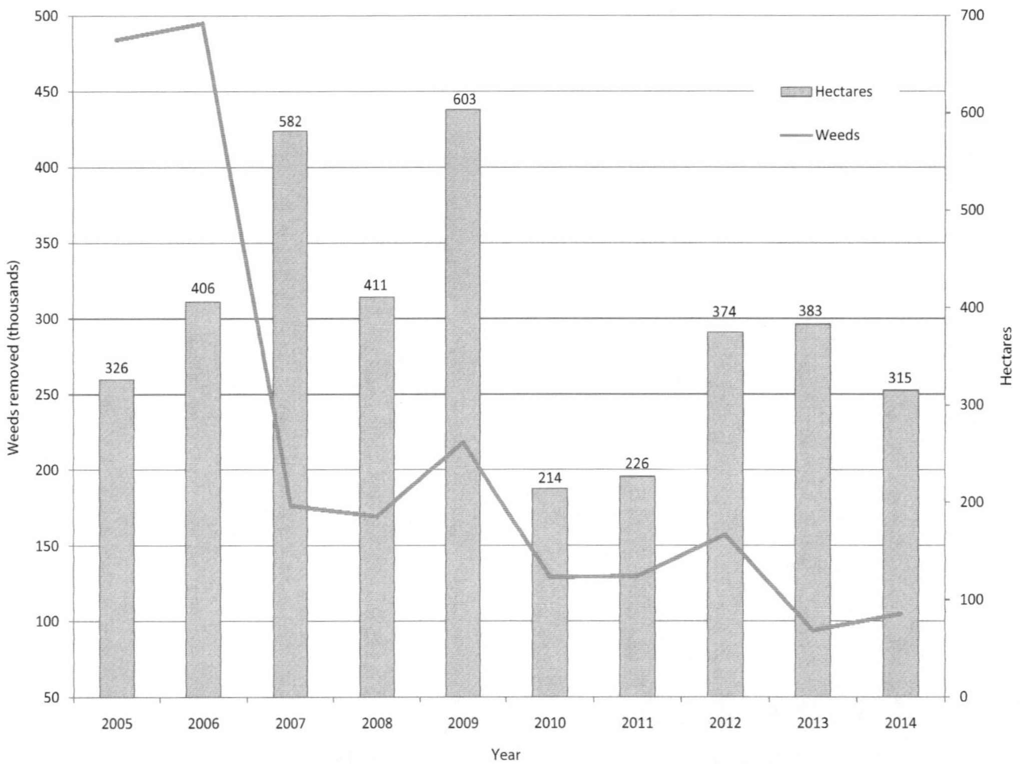

Figure 9. Numbers of weeds removed and area treated (ha) LHI 2005–2014.

RESULTS

The 2006 LHIWMS advised a minimum 24 month search interval which equates to near 500 ha yearly search effort. Figure 9 shows yearly search effort and the cumulative number of 25 weed species removed (all life stages).

Over the ten year period a total of 3941 ha have been searched, with the yearly effort from 214 to 603 ha. The extent of hectares searched per year reflects funding regimes. By year ten, weed blocks should have received five complete visits but have averaged 3.9 visits. However a strong downward trend is evident (all weeds). To deliver the preferred 24-month search frequency a yearly search target of 500 ha is required.

Outputs from the LHI Weed Database show the top ten weeds removed over ten years. Weed extent is based on detection in weed blocks (regardless of abundance) and sum of block size (Table 1).

The most abundant weed, cherry guava (first recorded on LHI in 1898), has been removed from 165 weed blocks across 1064 ha. The second most abundant weed, ground asparagus (first recorded in 1930), has been removed from 192 weed blocks across 1018 ha. Both species have island scale invasion potential and cumulative biodiversity impacts (occupying differing strata – ground and tree layer). Ground asparagus has 32 years less occupancy on LHI yet has established an invasion range on par with cherry guava.

Weed density mapping for landscape unit, Lidgbird North, indicates reductions in cherry guava at all life stages (Figure 10) with the map on the left comprising mostly mature plants and the map on the right mostly juvenile plants hidden amongst dense crofton weed (*Ageratina adenophora* Spreng.). These density changes are evidenced for most target weeds across island where repeat visitation has been measured.

Database outputs also show a strong downward trend in all life stages (Figure 11). Sum of weed mapping plot data of cherry guava from Lidgbird North, indicates reductions in all life stages up to 90% (Table 2). These figures compare to database outputs for the same landscape unit using number of plants removed from first and last treatment, which shows decline in mature plants of 96% (Table 3).

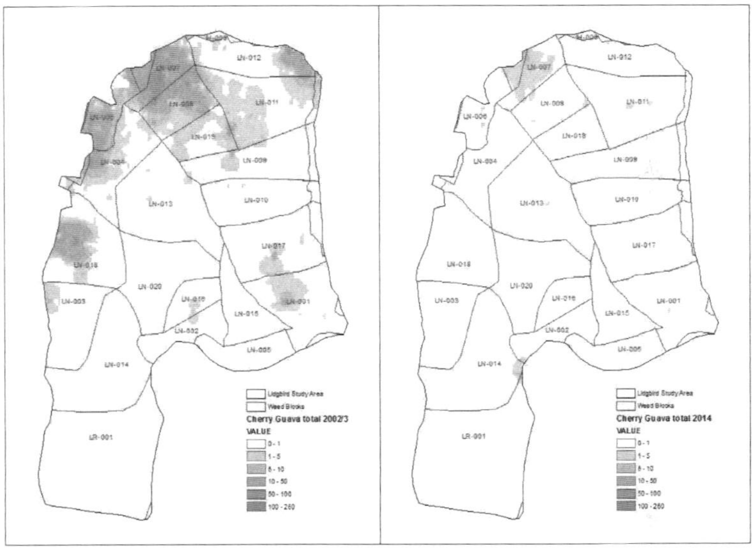

Figure 10. Lidgbird North – cherry guava density mapping 2002–03 (left) and 2013–14 (right); all life-stages.

Table 1. The top ten weeds removed over ten years.

Top 10 weeds removed (all life stages)	No.	Ha
Cherry guava (*Psidium cattleyanum* var. *cattleyanum*)	704,266	1064
Ground asparagus (*Asparagus aethiopicus*)	665,831	1018
Ochna (*Ochna serrulata* Hochst.)	458,168	648
Bridal creeper (*Asparagus asparagoides* (L.) Druce)	110,794	328
Sweet pittosporum (*Pittosporum undulatum* Vent.)	84,729	394
Climbing asparagus (*Asparagus plumosus* Baker)	53,840	260
Cotoneaster (*Cotoneaster glaucophyllus* Franch.)	26,211	266
Glory lily (*Gloriosa superba* L.)	13,655	55
Night jasmine (*Cestrum nocturnum* L.)	13,380	81
Bitou bush (*Chrysanthemoides monilifera* subsp. *rotundata*)	3,459	215

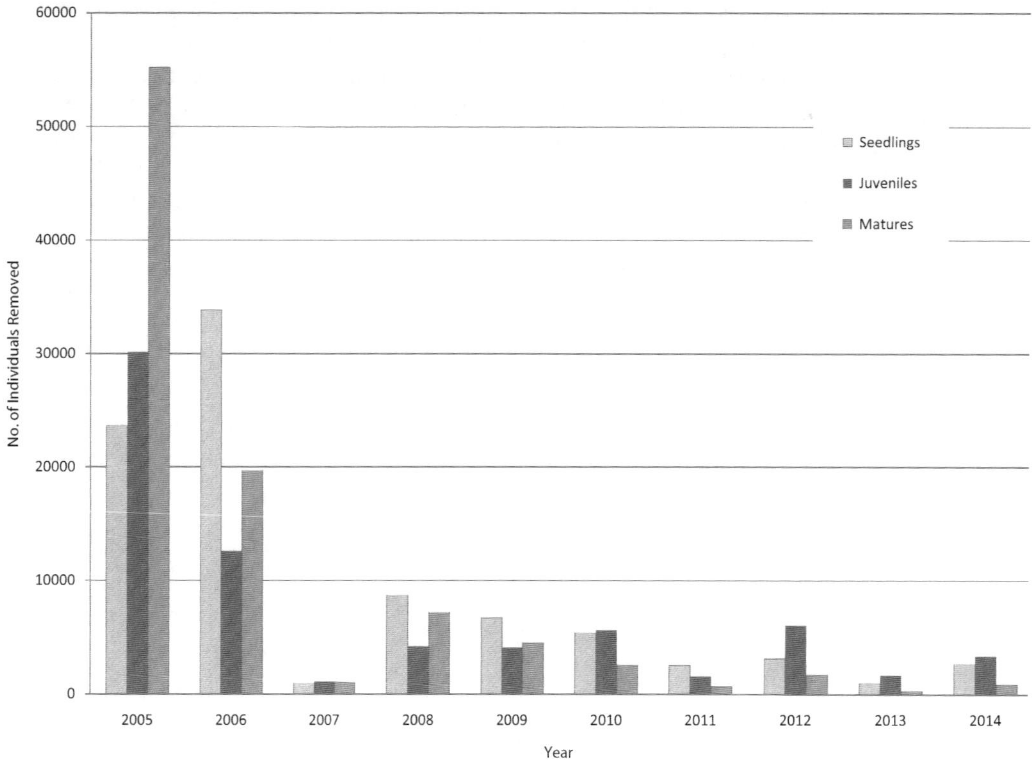

Figure 11. Cherry guava removed from Lidgbird North, 2005–14.

In 2014 mature cherry guava were represented by shrub sized plants (<2 m) with few inflorescence or buds compared to large fruit bearing trees (2–5 m) requiring chainsaw treatment which were common up to 2006 (Table 3).

Table 2. Cherry guava density mapping in 2002–03 and 2013–14 (Lidgbird North) sum of plots 0.92 ha.

Size Class	2002–03	2013–14	% reduced
Mature (over head)	1,008	7	99.3
Shrub (knee to head)	548	39	92.9
Juvenile (to knee)	385	149	62

Table 3. Cherry guavas removed comparing first and last treatment 2004–2015 LHI weeds database (Lidgbird North) 129 ha.

Treatment	First	Last	% reduced
Matures	78,277	3,005	96.2
Juvenile (immature)	45,340	11,074	76
Seedling (ankle high)	51,086	6,819	86.6

DISCUSSION

Species identified for eradication on LHI have been considered against criteria identified by Panetta and Timmins (2004) and LHI BMP (2007). For eradication to be successful it is critical to remove all mature weed plants, search and control at a rate to prevent reproduction, and continue removal of new recruits to extirpate/exhaust weed seed banks (Panetta 2007). Weed eradication programs often require ten years or more to achieve their objective and may require a period of gradual elimination to precede eradication (Holloran 2006).

Subject to adequate resourcing, the LHI weed eradication program fulfils these criteria by virtue of being surrounded by hundreds of kilometres of ocean with restrictions on plant importation, delimitation of area (the island), a grid search method that removes all life stages of target weeds each sweep and legislative powers that enable an all tenure approach. Furthermore the LHI weed data management systems provide a sound measurable basis for monitoring change in weed populations with effort, predicting future costs and priorities.

With a ten year lead in period out of a predicted 30 year program period, weeds have been depleted to such an extent that eradication capacity has significantly improved.

On first impressions when arriving at LHI it is difficult to conceive the previous extent and impact that ecosystem-modifying weeds imposed (Figure 12). Since program commencement, the number of weeds of up to 25 species has reduced by 80% (all life stages) and matures by 90% comparing first and last, measured treatments across 1024 hectares. Over 2 million weeds have been removed, with six species of limited distribution – declared eradicated. Another 36 are nearing eradication. In 2014 up to 45 exotic species of limited extent and mostly confined to the settlement were listed noxious to enable early intervention.

The program has had setbacks which are conceivable considering the prior sheer volume and distribution of the main weeds (Table 1) and the complexities of program roll out in an isolated environment with rugged cliff lines and resource limitations. Setbacks include:

- Failure of some control methods early in the program, requiring large areas to be re-treated;
- Fluctuation in funding impacting search regimes (3.9 return visits compared to a preferred 5 visits);
- Misreporting – now remedied with GPS;
- Lead in and growth of eradication ethic amongst all stakeholders; and
- Lead in phase for development of technology for weeds on steep and remote terrain.

The setbacks have provided valuable lessons to drive improvements and innovation such as the development of a front mounted heli lance sprayer enabling treatment of weeds from cliffs and heli winching teams to remote locations.

The LHIWMS 2006 predicted search effort would reduce post removal of mature plants. Analysis of effort per weed landscape unit comparing first and last treatment did not support this, except for the Transit Hill area (42% reduction in search time). This site exhibited widespread asparagus species that were initially treated by spraying and now only requires hand control.

Despite not receiving the targeted 24-month repeat treatment of weed blocks in the PPP, the program is still delivering a strong downward trend in weed density and distribution. The 24-month search interval is based on weed species reproductive traits and relies on effective on-ground grid search effort to remove all weeds from a block. Repeat treatments do detect previously missed mature plants, albeit at significantly reduced numbers. Ongoing island wide search effort is required to continue to remove remaining mature target weeds and extirpate weed seed banks to reach a containment and monitoring phase. Now is the time to 'Invest and Not Rest' (Jae De Clouett pers. comm. following a return visit from Little Barrier Island NZDOC weed eradication program).

The past ten year dataset is being analysed to model future abundance of weeds, management effort and timeframe for eradication (Baker *et al.* in prep).

As weeds have reduced on LHI a ground swell of 'eradication' awareness has grown through recognition of the human effort and visual change in the landscape. Community acceptance of delivering island eradication programs is growing, particularly where it can be shown to be beneficial to the island way of life (including tourism).

Figure 12. A view from the summit of Mt Gower. Pest plant and animal eradications on LHI are working to protect and improve the resilience of the islands World Heritage values. Photo: J. Shick.

The biodiversity benefits in continuing the program to eradicate highly invasive weeds on LHI are of global significance. Island comprise only 5% of the earth's surface, collectively they support 20% of all known bird, reptile and plant species and near half of all endangered species. Islands represent the greatest concentration of both biodiversity and species extinctions (http://www.islandconservation.org/why-islands/) – yet their isolation and limited size provides unique opportunities for conservation.

A draft LHI Weed Management Strategy 2016 has been prepared to drive the next decade of weed eradication focusing on:

- Prevention, early intervention (Weeds on Alert), eradication – immediate or suppression towards zero density and control;
- Removal of outlier incipient (nascent foci) weeds;
- Repeat weed search, minimum 500 ha y^{-1}
- New innovations – weed detector dogs, drones, Herbicide Ballistic Technology (Leary *et al.* 2013);
- Build and sustain local capacity;
- Seek committed project partners and secure long term – multi year funding;
- Apply island conservation management priorities including biosecurity measures.

It has been fortuitous this project was initiated well prior to the planned rodent eradication program. Rodents are significant seed and seedling predators (Auld and Hutton 2004). Weeds currently in low numbers and likely to spread post rodent eradication have been pre-emptively listed noxious and targeted for removal prior to rodent eradication.

The continuation of the program to eradicate invasive weeds will have complementary ecological benefits, building on other previous eradication programs on Lord Howe Island; pig *Sus scrofa* Linnaeus, cat *Felis cattus* Linnaeus and goat *Capra hircus* Linnaeus), current (African big-headed ant *Pheidole megacephala*) and future (rodent). The multi-species recovery actions being implemented by the LHIB are aimed at protecting and enhancing the islands unique biodiversity and World Heritage Values for the long term.

ACKNOWLEDGMENTS

Jenni Le Cussan for her vision and technical skills in initiating a program to eradicate weeds from LHI. Field staff, volunteers (including the Friends of LHI) and local community for their effort and support. Commonwealth and State Government agencies and the Lord Howe Island Board in their ongoing commitment towards World Class conservation goals on Lord Howe Island. The pertinent comments from the review team for this paper are also acknowledged.

REFERENCES

Auld, T.D. and Hutton, I. (2004). Conservation issues for the vascular flora of the Lord Howe Island Group. *Cunninghamia* 8, 490-500.

Baker, C. *et al.* (in prep). Modelling spread of weeds to inform decision-making on islands. Paper in preparation for Ecological Society of Australia conference 2016

DECC (2007). Lord Howe Island Biodiversity Management Plan. NSW Department of Environment and Climate Change, Sydney.

Island Conservation (URL: http://www.islandconservation.org/why-islands/)

ISSG (2013). 100 of the world's worst invasive alien species. Invasive Species Specialist Group (ISSG), Species Survival Commission, World Conservation Union (IUCN), Auckland, NZ (see http://www.issg.org/worst100_species.html)

Holloran, P. (2006). Measuring performance of invasive plant eradication efforts in New Zealand. *New Zealand Plant Protection* 59, 1-7.

Leary, J. (2013). (see http://www.cal-ipc.org/symposia/archive/pdf/2012/3_Leary.pdf)

Le Cussan, J.M. (2002a). Density distribution of bridal creeper, *Myrsiphyllum asparagoides* and two other weeds on Malabar Range, Lord Howe Island. Unpublished internal report to the Lord Howe Island Board.

Le Cussan, J.M. (2002b). Density distribution of 5 major weeds on Transit Hill, Lord Howe Island. Unpublished internal report to the Lord Howe Island Board.

Le Cussan, J.M. (2003a). Density distribution of 6 major weeds on Intermediate Hill, Lord Howe Island. Unpublished internal report to the Lord Howe Island Board.

Le Cussan, J.M. (2003b). Density distribution of 6 major weeds on the north-western slopes of Mt Lidgbird, Lord Howe Island. Unpublished internal report to the Lord Howe Island Board.

LHIB (2006). Weed Management Strategy for Lord Howe Island. Lord Howe Island Board, Lord Howe Island.

LHIB (2016). Lord Howe Island Weed Eradication Program – results 2004–2014. Lord Howe Island Board, Lord Howe Island, NSW.

Panetta, F.D. and Timmins, S.M. (2004). Evaluating the feasibility of eradication for terrestrial weed incursions. *Plant Protection Quarterly* 19 (1), 5-11.

Panetta, F.D. (2007). Evaluation of weed eradication programs: containment and extirpation. *Diversity and Distributions* 13, 33-41.

UNESCO (2016). Lord Howe Island Group. UNESCO/WHC URL: http://whc.unesco.org/en/list/186

A risk-based approach for prioritising the threat of exotic weeds to Australia

Sandra Parsons, Alexander Brazenor and Bertie Hennecke
Australian Bureau of Agricultural and Resource Economics and Sciences (ABARES), PO Box 858, Canberra, Australian Capital Territory 2601, Australia
(Bertie.Hennecke@agriculture.gov.au)

Summary Globalisation and increasing trade increase the risk of new weed species entering Australia and impacting negatively on agricultural production and the environment. At the same time ongoing limited resources and budgets require decision makers to carefully determine where to invest for maximum return.

The development of the Australian Weeds Threat Prioritisation Process aims to assist decision making by informing governments, industries and other stakeholders of exotic weeds (not known to be present in Australia) that have the greatest potential to threaten Australian agriculture, environment and social amenity.

The process targets the 'worst of the worst' species that are exotic to Australia. A set of questions based on previous, well-established risk-assessment criteria and the scientific literature are applied to highlight species with the highest threat level from a pool of exotic weeds species. A number of 'stop/go' questions limit the number of exotic weed species to which a more detailed evidence-based assessment of invasiveness, potential distribution and impacts is required.

The prioritisation process is not designed to identify unknown species that are potentially serious weeds. The pragmatic process focuses on prioritising known invasive weed species that exist outside of Australia and might meet the national significance test and therefore pose a threat if they were to enter the country.

The identification of priority exotic species that show a high potential to meet Australia's national significance criteria enables governments and industries, individually or collectively, to develop a strategic approach to minimise the threat the species poses to Australia.

Keywords Weeds, exotic, prioritisation, national significance.

Introducing non–indigenous plants to the Western Australian rangelands – what can weed risk assessment contribute?

Christine Munday[1], Margaret Byrne[1], Geoff Moore[2] and Clinton Revell[2]
[1] Department of Parks and Wildlife Western Australia, Dick Perry Ave., Kensington, Western Australia 6151 Australia
[2] Department of Agriculture, Locked Bag 4, Bentley DC, Western Australia 6983, Australia
(christine.munday@dpaw.wa.gov.au)

Summary What place can weed risk assessment (WRA) have in the decision to introduce species with agricultural potential to novel environments?

Northern Western Australia (WA) is a vast area containing a wide range of environments that have evolved unique communities of plants and animals. This has been recognised with the listing of National Diversity Hotspots in both the Kimberley and the Pilbara. Many areas have also undergone development for mining, agriculture and tourism.

There is current interest in increasing and intensifying the pastoral industry in northern WA. Beef cattle production has developed by grazing extensive areas of native and supplemented vegetation. In this highly seasonal environment where natural vegetation may provide limited nutrition for domestic animal production it is hoped that the introduction of fodder plants into dryland pastures and irrigated fodder production will broaden the feed base and result in increased output. However, many species introduced into agriculture in northern Australia have become weeds (Lonsdale 1994) with resulting costs in time, money and to natural environments. More recently Driscoll *et al.* (2014) outlined the potential weed risk from new varieties of pasture species which may show increased weedy characteristics.

Future Farm Industries Cooperative Research Centre developed an Environmental Risk Framework and WRA to examine species with agricultural potential in southern Australia (Stone *et al.* 2012). This informed the selection of species for research and promotion. This framework is now being adapted for use in the rangelands of northern Western Australia to provide weed risk information for decision makers including government bodies, land owners pastoralists and the wider community.

Keywords Weed risk assessment, non-indigenous species, rangelands.

REFERENCES

Driscoll, D.A., Catford, J.A., Barney, J.N., Hulme, P.E., Inderjitg, Martina, T.G., Pauchard, A., Pyšek, P., Richardson, D.M., Rileyn, S. and Visserm, V. (2014). New pasture plants intensify invasive species risk. PNAS, Volume 111, No. 46, DOI10.1073/pnas.1409347111

Lonsdale, W.M. (1994). Inviting trouble: introduced pasture species in Northern Australia. *Australian Journal of Ecology* 19, 345-54.

Stone, L., Munday, C. and Bettink, K. (2012). Environmental Weed Risk Assessment Protocol. Future Farm Industries CRC, 20 pp. http://www.futurefarmonline.com.au/about/weedrisk.htm

Weed risk assessment in countries with porous borders: a case study in Bhutan

Dorjee[1], Stephen B. Johnson[2], Anthony Buckmaster[3] and Paul O. Downey[1]
[1] Institute for Applied Ecology, University of Canberra, Canberra 2617, Australia
[2] New South Wales Department of Primary Industries (NSW DPI), Locked Bag 6006, Orange, New South Wales 2800, Australia
[3] Invasive Animals Cooperative Research Centre (CRC), University of Canberra, Canberra 2617, Australia
(dorjee.dorjee@canberra.edu.au)

Summary Weed risk assessment systems have been successfully tested and implemented in Australia, NZ and the USA. They have not been implemented in developing countries, particularly those with porous borders. The assumption is that effective management of the movement of goods and people across borders cannot easily occur in these latter situations. To test this we used the Australian Weed Risk Assessment (AWRA) system and applied it to Bhutan, a developing country with an open and porous border with India, as a case study.

Firstly, we examined the risk and determined that there were two risks, being the risk that a species could be (i) introduced, and (ii) establish if introduced. The AWRA system only really addressed the second risk. Assessment of the introduction pathways relative to the porous nature of the border showed that the first risk could not be effectively managed. As an alternate approach, we propose a modified risk assessment model tailored for Bhutan, in which specific sectors that intentionally import the bulk of plant species are targeted (i.e. the nursery, forestry and agriculture sectors) to undertake an ARWA approach. Secondly, a modified weed risk management (WRM) approach is used to assess (i) plant species which have the ability to be dispersed into Bhutan from India as well as to assess new incursions and retro assess existing ones to determine their risk and priority for management. We consider this hybrid risk assessment approach to be a step in helping improve the biosecurity concerns in developing countries with open and porous borders.

Keywords Developing countries, open and porous border, modified risk assessment model, retro assess, hybrid weed risk assessment approach, biosecurity.

The weed potential of *Erica* species for Tasmania and other areas of southern Australia

Michael R. Noble
Department of Primary Industries, Parks, Water and Environment, PO Box 303, Devonport, Tasmania 7310, Australia
(Michael.Noble@dpipwe.tas.gov.au)

Summary The plant genus *Erica* contains species found mostly in South Africa, but also Europe, Macaronesia and other parts of the world. There are approximately 820 *Erica* species, none are native to Australia. These shrubs and small trees often have attractive flowers and have been introduced to Australia via the ornamental plant trade. Two *Erica* species, Spanish heath (*Erica lusitanica* Rudolphi) and tree heath (*Erica arborea* L.), are recognised globally as the two most widespread invasive *Erica* species to date. Both these are unfortunately naturalised in southern Australia, including Tasmania. There are up to 18 *Erica* species naturalised in southern Australia. At least six of these are naturalised in Tasmania. Besom heath (*Erica scoparia* L.) was first recorded as naturalised in Tasmania (and Australia generally) in 1983, and it is proving to be a particular concern. In Tasmania it shows every sign of being at least as bad an invasive as Spanish heath.

In preparation for potential weed declarations, the Tasmanian Government risk-assessed besom heath and 15 other *Erica* species considered to pose a potential risk. Results indicated each of those species assessed are of weed concern to Tasmania. Of particular concern are those *Erica* species of European origin. Investigation of the biology of these plants indicates just how well suited to naturalisation and spread as invasive species many *Erica* species are. Climatic and other physical similarities between their places of origin and southern Australia adds to their suitability here.

Keywords *Erica*, heath, naturalised, weed, invasive, species.

Innovative field nurseries to assess weed risk in the rangelands of Western Australia

Geoff A. Moore and Clinton Revell
Department of Agriculture and Food, Western Australia, Locked Bag 4, Bentley DC, Western Australia 6983, Australia
(geoff.moore@agric.wa.gov.au)

Summary There is growing interest in intensifying beef cattle production in the rangelands of northern Western Australia through 'mosaic agriculture' to complement the extensive grazing of native vegetation. Irrigated fodder production and the introduction of improved forage plants (dryland) can broaden the feed base of rangeland grazing systems. With agricultural development there can be unwanted consequences and one of these is that some pasture plants have become agricultural and environmental weeds in certain situations.

Taking an innovative approach a series of four field nurseries have been established in key environments (soil × climate) in the West Kimberley and Pilbara to obtain data on the persistence and/or spread of a wide range of commercial pasture and fodder grasses and legumes. The trials were established under irrigation to simulate a worst case scenario as weed spread in a rangelands context is likely to be episodic, for example, following a cyclone. The results should help to inform debate and provide decision-makers with the information to find the balance between agricultural development and the protection of high value environmental assets. This research is a work in progress and the preliminary results will be presented at the conference.

Keywords Weed risk assessment, irrigated agriculture, assessment of weed potential, pastures, environmental weeds.

INTRODUCTION

There is growing interest in intensifying beef cattle production in the rangelands of northern Western Australia through 'mosaic agriculture' to complement the extensive grazing of native vegetation. Mosaic agriculture refers to, 'irrigation schemes in which small patches of irrigation occur within a region rather than irrigation of one large contiguous area' (Cook *et al.* 2008). Irrigated fodder production and the introduction of improved forage plants (dryland) can broaden the feed base of rangeland grazing systems.

With agricultural development there can be unwanted consequences and one of these is that some pasture plants have become agricultural and environmental weeds in certain situations (Lonsdale 1994, Panneta *et al.* 2001). The challenge is to find the right balance between agricultural development and minimising the risk of weed invasion of high value environmental assets.

Diversification permits are required to grow non-indigenous plants on pastoral leases in WA. In this context, non-indigenous refers to plants which are not native to Western Australia. The current assessment procedure includes a desk-top assessment of weed risk by the state government departments involved. For example, the Department of Agriculture and Food WA (DAFWA) currently uses the environmental weed risk assessment system developed by the Future Farm Industries CRC (Stone *et al.* 2012), with a modified potential distribution system based on climate. However, there is currently a paucity of field data from previous trials or agricultural developments to calibrate the desktop modelling.

This project is collecting data on the persistence and/or spread of a wide range of commercial pasture and fodder grasses and legumes by re-visiting 'old trial sites' and from a series of 'weed risk' field nurseries in key environments (soil × climate) in the West Kimberley and Pilbara. This paper focuses on the series of field nurseries. All of the species being evaluated are commercial pasture and fodder options which are widely used in similar environments in the Northern Territory and Queensland. The only non-commercial species are two native legumes from the Fortescue region in the Pilbara. In addition, as a safeguard, protocols were carefully developed to monitor for and minimise the risk of movement of seed off-site.

MATERIALS AND METHODS

Field nursery trials have been established at four sites which represent key soil-climate environments in the West Kimberley and Pilbara: Birdwood Downs near Derby, Wallal Downs south of Broome, Woodie Woodie in the Pilbara and Gogo Station in the Fitzroy Valley (Table 1). The average annual rainfall at these sites varies from 300 to 600 mm per annum and the soils vary from red-brown sands with 'Pindan' vegetation (*Acacia ancistrocarpa* Maiden & Blakely and

A. eriopoda Maiden & Blakely dominated associations on sandplain) to black cracking clays associated with the floodplains of the Fitzroy River. Two sites had native vegetation; while the other two sites were previously cleared and had been used for dryland agriculture (Table 1). The trials were established over a 13 month period from April 2015 to April 2016.

The sites were cleared of vegetation, levelled and cultivated to ensure a fine soil seed-bed suitable for sowing and the installation of a trickle irrigation system. Sites were weed-free at sowing following cultivation and spraying with a knockdown herbicide (glyphosate 480 a.i. at 2 L ha^{-1}). The treatments were sown by hand and the seeding depth adjusted from 10 mm to 50 mm to match the seed size. Seeding rates varied depending on species, seed size and were also adjusted for seed-lots with low germination. For the 'plus Rhizobia' treatments the seed was inoculated and lime pelleted the previous day. Where possible the treated seed was kept cool prior to seeding.

Plots were sown as 3 m rows and irrigated with a trickle irrigation system to assist seedling establishment and provide the best possible conditions for growth and seed production. During the seedling stage the trials were irrigated daily to maintain soil moisture near the soil surface. The frequency of irrigation was reduced as the plants grew and developed. The high to extreme temperatures in the late spring and early summer resulted in a high evaporative demand and regular irrigation was required to reduce heat and moisture stress on the plants.

At each site there are two replicated trials. The 'grass' field nursery trials have 23 entries and include a range of warm season (C4) annual and perennial grasses (Table 2). Each entry has plus and minus (+/–) complete fertilizer sub-treatments and each combination is replicated three times. The 'legume' field nursery trials have 23 entries and include a range of tropical legumes, plus the temperate perennial legume – *Medicago sativa* (lucerne) (Table 3). Each entry has +/– fertilizer and +/– rhizobia sub-treatments and each combination is replicated three times (note: Birdwood and Gogo legume trials do not include the 'plus fertiliser and minus Rhizobia' treatment).

The soils are low in many of the essential nutrients for plant growth. All of the sites are low in organic carbon, with a range from 0.32 to 0.6% organic C in the topsoil (0–10 cm). The sites at Birdwood Downs, Wallal Downs and Woodie Woodie are low in the essential nutrients phosphorus, potassium and sulphur, while zinc is also deficient at Wallal Downs. At these sites the Colwell P levels ranged from 2 to 8 mg kg^{-1}. The fine-textured soil at Gogo has an alkaline soil reaction and is sufficient to high in the key nutrients, except for sulphur which is marginal. A complete fertiliser was applied for the plus fertiliser treatments which included nitrogen for the grasses, while nitrogen was not applied to the legume treatments.

A protocol has been developed to minimise the risk of spread from the field nursery sites. The following guidelines are being implemented to minimise the risk of movement of seed off-site:

- Trial area fenced to exclude stock and wildlife
- Regular monitoring (4× times per year) of a 50 m buffer around the trial area to check for and control any plants that have established outside the trial area
- Any plants remaining in the trials after a given period (~two years) will be killed with a knockdown herbicide
- There will be on-going monitoring (2× per year) of the site for a further two years to check for and control any plants within the trial area and surrounding buffer. (This monitoring will be extended for 18 months after the last recorded germination of any species).
- Other factors which reduce weed risk are the low fertility of the soils (without added fertiliser).

At the field nursery at Birdwood Downs the site was fenced with wallaby proof ring-lock fencing, but this proved inadequate to keep out small wallabies, so chicken wire was added which proved effective in keeping the wallabies out.

Table 1. Summary of site details for the established field nurseries.

Pastoral station	Vegetation	Soil type	Annual rainfall (mm)	Date established
Birdwood Downs (Derby)	Cleared pasture	Red-brown sand (Pindan)	600	April 2015
Wallal Downs	Native vegetation (Pindan)	Red-brown sand	350	April 2015
Woodie Woodie (Warrawagine)	Native vegetation (spinifex)	Red-brown loam	300	Sept 2015
Gogo	Cleared – dryland cropping	Black cracking clay (vertisol)	570	April 2016

Table 2. A list of species and cultivars in the grass field nursery trials.

Botanical species	Common name	Cultivar/variety
C4 annual grasses – fodder		
Cenchrus americanus (L.) Morrone	pearl millet	Pearler Maxa Lawrence
Sorghum bicolor (L.) Moench	sorghum spp. hybrids	Sweet Jumbo LPA Superdan 2 Stargrazer BMR Rocket Sugargraze
Zea mays L.	forage maize	HM 102
C4 perennial grasses		
Chloris gayana Kunth	Rhodes grass	Callide Katambora Finecut Mariner
Megathyrsus maximus (Jacq.) B.K.Simon & S.W.L.Jacobs	panic grass	Gatton G2
Digitaria eriantha Steud.	digit/pangola grass	Premier
Panicum coloratum L.	Bambatsi panic	Bambatsi
Digitaria milanjiana (Rendle) Stapf	Jarra grass	Jarra
Bothriochloa pertusa (L.) A.Camus	bluegrass	Floren
Urochloa mosambicensis (Hack.) Dandy	sabi grass	Sabi grass
Cenchrus ciliaris L.	buffel grass	Gayndah Biloela
Cenchrus setiger Vahl	Birdwood grass	–

Once the plants are well established and have flowered and set-seed the irrigation frequency and duration will gradually be reduced over a period of a few weeks to simulate a naturally drying soil profile at the end of the 'wet' season. The sites will then be monitored for plant persistence and recruitment from seed over the next two years. Large perennial plants were also cut back to reduce the biomass and leaf area and consequently the moisture stress on the plants as the irrigation was reduced. This helps to replicate the rangeland context where there is essentially uncontrolled grazing, whether by stock, wildlife or feral animals. Reducing the biomass also improves the potential for perennial plants to persist over the dry season. The sites were maintained largely free from background weeds while the plants were establishing and the irrigation system was in place. This proved more difficult at Birdwood Downs where there was a large seed-bank of annual and perennial grasses and broadleaf weeds.

RESULTS

In terms of weed risk, the importance of data collection increases once the irrigation has been turned-off. The irrigation was turned-off at Derby (Birdwood Downs) and at Wallal Downs in mid-April 2016 and will be turned-off at Woodie Woodie in late June 2016, while the field nursery trials at Gogo are still establishing.

There was good to excellent establishment of most lines across all four sites, and the few treatments with poor establishment were generally re-sown. Even though it is comparatively early stages of the trials, there have been some interesting observations and results.

Lablab and cowpea were the strongest performing legumes at both Derby and Wallal Downs sites over

Table 3. A list of species and cultivars in the legume field nursery trials.

Botanical species	Common name	Cultivar/variety
Temperate legumes		
Medicago sativa L.	lucerne	SARDI 10
Tropical legumes		
Centrosema pascuorum Mart. ex Benth.	centro	Cavalcade
Lablab purpureus (L.) Sweet	lablab	Highworth Rongai
Vigna unguiculata (L.) Walp.	cowpea	Ebony
Macroptilium bracteatum (Nees & Mart.) Marechal & Baudet	burgundy pea/bean	B1
Chamaecrista rotundifolia (Pers.) Greene	roundleaf cassia	Wynn
Clitoria ternatea L.	butterfly pea	Milgarra
Macroptilium atropurpureum (DC.) Urb.	siratro	Aztec atro
Leucaena leucocephala (Lam.) De Wit.	leucaena	Wondergraze Tarramba
Stylosanthes hamata (L.) Taub.	Caribbean stylo	Amiga Verano
Stylosanthes scabra Vogel	shrubby stylo	Seca Siran
Stylosanthes guianensis (Aubl.) Sw. var. *guianensis*	common stylo	Hughes Beefmaker
Stylosanthes guianensis (Aubl.) Sw. var. *intermedia* (Vogel) Hassl.	fine stem stylo	V8
Stylosanthes seabrana B.L. Maass & 't Mannetje	Caatinga stylo	Caatinga
Desmodium intortum (Mill.) Urb.	greenleaf desmodium	–
Desmanthus virgatus (L.) Willd.	desmanthus	Marc
Cullen graveolens (Domin.) J.W.Grimes	cullen	native
Cullen cinereum (Lindl.) J.W.Grimes	cullen	native

the dry season and grew through to large plants at maturity, produced plentiful seed and by early summer had senesced. However, in spite of seed lying on the soil surface, there were no seedling recruits at either site, but rather the plots were invaded by 'weeds' (i.e. non-sown species). Early growth of the other legumes in the Derby field nursery was slow and compromised by being preferentially grazed by the wallabies until the fence was upgraded. The best performed legumes over the summer 'wet' season at Derby were the two varieties of Caribbean stylo and Wynn cassia.

At Wallal Downs the early growth of most legumes apart from lablab and cowpea was also slow until late spring even though they were regularly irrigated across the dry season and night temperatures over winter are mild due to the proximity to the coast. This was the only site where leucaena has grown well and after 12 months the largest plants were 180–200 cm in height. However, at this site the leucaena plants were attacked by termites once the stems started to get woody (i.e. about the diameter of a pencil or larger). Large termites (*Mastotermes* spp.) hollowed out the taproot and stem of some leucaena seedlings at Wallal Downs resulting in the death of plants. Conventional mud tunnels and termites are also common on leucaena plants. The continuity of the mud tunnels was deliberately broken on most of the remaining plants so that persistence can be assessed in the medium-term.

Butterfly pea grew well following the out-of-season rainfall at Wallal in early May, but it had grown slowly over summer even with trickle irrigation. Growth of siratro was strong at both the Woodie Woodie and Wallal Downs trials with the trailing stems more than 2–3 m across, but growth was patchy at Derby.

The extreme temperatures at Woodie Woodie in early January, (i.e. six consecutive days with maximum temperatures >44°C) resulted in the death of 96% of the Highworth lablab, 100% of the Rongai lablab and 98.5% of the cowpea plants (20–30 cm in height). All (100%) of the maize plants were also killed (bleached) by the extreme temperatures.

Most of the warm season C4 annual grasses and the C4 perennial grasses had good plant vigour and competed well with any 'weeds', so there have been comparatively few non-sown species in the grass trials across all of the sites. However after the annual grasses had senesced at Derby these rows were invaded by broadleaf weeds.

DISCUSSION

In the WA rangelands the spread of weeds is often episodic so the field nursery trials were established under irrigation to simulate the worst case scenario, i.e. if plants establish and set seed following a tropical cyclone or an extremely high rainfall year. The irrigation was gradually turned off over the 'wet' season once the plants were well established, and most species had flowered and set-seed. Persistence of annual species relies on regeneration from the seed-bank, while with perennial species individual plants can persist for two to many years or regenerate from the seed-bank.

The native vegetation is adapted to the environment and the method of adaptation depends on the plant growth habit. The herbaceous perennial grasses and forbs tend to respond rapidly to rainfall in certain periods (e.g. early wet season) and then persist by going into a semi-dormant to dormant state for an extended period (Petheram and Kok 2003). The woody perennial plants have conservative growth patterns in line with the long-term rainfall. However, the native ephemeral annual herbs and grasses are adapted to the climate by having extremely rapid seedling growth and plant development following good rainfall. For instance, the annual herb *Cleome viscosa* L. (tickweed, mustard bush) a common volunteer plant at both the Wallal Downs and Woodie Woodie trial sites displays rapid growth following good rainfall. Tickweed plants germinated, grew to a height of 60–70 cm, flowered and set-seed in a period of four to six weeks, while the sown legumes alongside which were already established grew slowly and were still comparatively small plants. On the whole, the annual and perennial grasses tend to be more competitive than the legumes, so there is a much lower density of volunteer plants within the grass field nurseries. The warm season annual grasses (hybrid sorghum, pearl millet, maize) had the highest seedling and early vigour of all the sown species, while *Cenchrus ciliaris* recruits from the naturalised populations at Wallal Downs and Woodie Woodie displayed both good vigour and an ability to adapt and set-seed under a range of environmental conditions.

The growing of non-indigenous plants in a rangeland context can be a contentious issue, with many interest groups with a wide range of viewpoints. The results of these studies will help to inform the debate so that a balance can be achieved between agricultural development on one hand and protecting high value environmental assets on the other.

ACKNOWLEDGMENTS

We acknowledge the financial support from the WA State Government's Royalties for Regions program directed through the Department of Agriculture and Food WA in the Northern Beef Futures project and the Meat and Livestock Australia Donor Company. The authors also thank Mr Tony Shimmin, Mr Tim Wiley, Ms Yao Jun Tang and Mr Tyler Loud for capable support with the fieldwork.

REFERENCES

Cook, F.J., Xevi, E., Knight, J.H., Paydar, Z. and Bristow, K.L. (2008). Analysis of biophysical processes with regard to advantages and disadvantages of irrigation mosaics. CRC for Irrigation Futures Technical Report (07/08).

Lonsdale, M.W. (1994). Inviting trouble – introduced pasture species in northern Australia. *Australian Journal of Ecology* 19, 345-54.

Pannetta, F.D., Mackey, A.P., Virtue, J.G. and Groves, R.H. (2001). Weed risk assessment: core issues and future directions, *In* 'Weed risk assessment', eds R.H. Groves, F.D. Panetta and J.G. Virtue, Chapter 19, pp. 231-40. (CSIRO Publishing, Melbourne, Australia).

Petheram, R.J. and Kok, B. (2003). 'Plants of the Kimberley Region of Western Australia.' Revised edition, 562 p. (University of Western Australia Press).

Stone, L., Munday, C. and Bettink, K. (2012). Environmental weed risk assessment protocol. Future Farm Industries CRC.

Weed surveying and mapping in urban and urban-fringe bushland: the Hobart experience

Adam Muyt
PO Box 90, South Hobart, Tasmania 7004, Australia
(adammuyt@yahoo.com.au)

Summary The City of Hobart's Bushland and Reserves Unit undertook a detailed weed survey and mapping project within bushland reserves in the city from 2012 to 2015. The primary aim of the project was to identify the most serious, and potentially serious, invasive plants within several urban and urban-fringe reserves, with the main objective then being to develop a comprehensive weed management program in these areas in subsequent years.

The project involved surveys along pre-determined transects, at varying intervals depending on the site and terrain. This approach was necessary due to the dense vegetation and rugged nature of much of the terrain, with limited access in certain areas. The information was digitally captured and then integrated within Council's digital mapping system, for use in planning on-ground weed management.

Keywords Mapping, rugged terrain, urban and urban-fringe, weed survey.

INTRODUCTION

The City of Hobart is a municipality covering the central business district (CBD) and surrounding inner suburbs of Hobart, the capital of Tasmania. The City occupies 77.9 square kilometres (7790 hectares), and has a population of approximately 49,000 (Australian Bureau of Statistics, 2011). Many residents live in suburbs backing onto, or threaded with, substantial tracts of native bushland.

Physically the municipality is characterised by a narrow coastal verge adjoining the foothills, slopes and ridges of kunanyi-Mt Wellington, the 1271 metre high peak dominating the city's landscape. Native vegetation in the City is dominated by a mixture of woodland and forest types, together with smaller areas of lowland grassland and montane, sub-alpine and alpine vegetation.

The municipality retains considerable stands of bushland within just a few kilometres of the CBD. This is largely a consequence of two factors: the steep and rugged terrain limiting wholesale clearing and development; and, efforts in the 19th and 20th centuries to secure the capital's water supplies through protecting large tracts of water catchment from development (Hobart Water 2009, Futurepast Heritage Consultants 2012).

The City has been responsible for managing various stands of native vegetation going back to the 1860s, initially with an emphasis on water catchment management and recreation activities. Fire management was added in the 1940s but it was not until the late 1980s – early 1990s that works focused on ecological management began, which eventually lead to the creation of a bushland unit within Council in 1993–94 (Hickie and Toonen 2014) This unit has since developed a sophisticated bushland management program encompassing strategic planning, asset management, recreation infrastructure, weed, fire and wildlife management, ecological restoration and revegetation works and community 'bushcare' and interpretation activities (HCC 2008).

The City's Bushland and Reserves Unit manages 4589 ha of native vegetation, of which 2966 ha occurs within the municipal boundaries (38% of the total area of the City) and 1623 ha in adjacent municipalities. These latter areas are largely a consequence of historical land purchases and grants aimed at protecting the City's water supplies (HCC 2008).

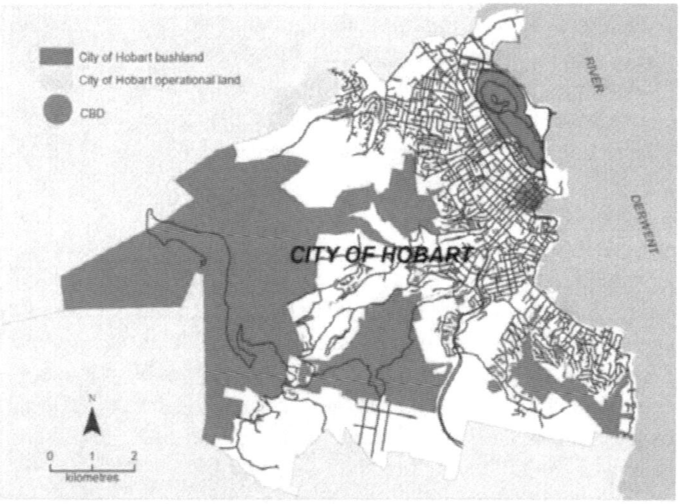

Figure 1. City of Hobart, showing extent of bushland reserves.

WEEDS AND WEED MANAGEMENT IN HOBART RESERVES

A total of 320 non-indigenous species, sub-species, varieties and cultivars have been recorded in Hobart's public open space parks and reserves, which is 35% of the total of 917 taxa recorded for such areas (City of Hobart 2016). This is broadly similar to the percentage for Tasmania as a whole (Matthew Baker, Tasmanian Museum and Art Gallery, pers. comm.). Around 120 taxa are considered serious, or potentially serious, environmental weeds in Hobart's bushland.

Programmed weed controls have been undertaken in the City's bushland reserves over the last 20+ years, primarily by Council works crews and community 'bushcare' groups, supplemented at times by bush regeneration contractors and occasional correctional teams or labour-market program work teams (e.g. Green Corp/Green Army).

Weed populations have been significantly reduced in this period, however, like any urban/peri-urban bushland, weeds continue to be a major management focus due to factors such as abundant weed sources in adjacent gardens and private bushland, garden dumpings, physical disturbances (historical and contemporary) and weed seedbanks in reserves (a particular issue in Hobart with persistent Fabaceae species such as gorse (*Ulex europaeus* L.), Scotch broom (*Cytisus scoparius* L.) and Montpellier broom (*Genista monspessulana* L.).

Council's weed management program has been shaped by several factors over the years such as meeting legislative obligations to control/eradicate certain species, restoring and rehabilitating bushland, community requests to control weeds, reducing impacts on recreational activities and reserve amenities, and aesthetic considerations. Another major determinant influencing weed management in the City's reserves for the last twenty years has been, and continues to be, fire.

FIRE AND WEED MANAGEMENT

Much of Hobart's native vegetation requires fire to maintain ecosystem functionality but with housing abutting so much bushland, wildfires pose a serious threat to life and property. The City of Hobart has therefore developed a detailed prescribed burning regime across its reserve system, with the dual aims of maintaining broad ecosystem functionality while reducing the threat from, and impacts of, any fires that occur. As fire can have a major influence on weed populations, occurrence and abundance, this burn program has incorporated both pre and post-burn weed controls as a basic and fundamental management requirement.

All of the significant reserves in the City have a Fire Management Plan (FMP), which cover a range of issues such as fire history and fire regimes, weed management, threatened species, fauna management, cultural and heritage values, access and recreation. Each FMP identifies the prescribed burn regime for each Vegetation Management Unit (VMU) in a reserve, a VMU being a distinct management area within a reserve. All FMPs are reviewed and updated every eight to ten years.

FMPs have been developed by consultant botanists and ecologists, with City of Hobart staff contributions largely limited to providing background and supplementary information, including the occurrences of certain weeds. However, virtually no information on weeds had been integrated within Council's spatial data and mapping programs – instead, the information had been compiled and presented in hardcopy form, supplemented by written descriptions for each VMU within each FMP.

WEED SURVEY AND MAPPING PROJECT

With three FMPs due for review or development within the next few years, a decision was made in early 2012 by the Bushland and Reserves Unit to develop a better understanding of the weed issues across the reserve system, particularly in the three areas where the FMPs were to be developed, and in areas where no detailed weed surveys had ever been undertaken.

Recognising that Council's bushland management field staff had the most in-depth knowledge of weeds and the reserves in the municipality, it was decided that they would conduct the weed surveys, capturing digital data using a hand-held Trimble®, Juno 3D and 5B GPS devices with onboard TerraSync™ software. This approach was seen as an ideal means for the Unit to develop technical capabilities and the skills of staff, and ensure data on weed occurrences was incorporated within Council's GIS spatial data base (using ArcGIS® software by ESRI™).

As virtually all weed control records were in hardcopy form only, the project's initial focus was on creating a digital record (summary) of previous weed control works over the last two decades, arranged by VMUs for each reserve. Once this was completed the focus shifted to the field component.

Initial surveying was undertaken in Bicentennial Reserve, a 250 ha reserve located approximately four kilometres south of the CBD in the suburbs of Mt Nelson and Sandy Bay. This reserve had never been comprehensively surveyed for weeds, and because of its urban and peri-urban elements, challenging terrain and dense vegetation, it presented the ideal location to trial and develop survey and mapping methods, train staff and develop their skills, and monitor the progress of the survey work.

The surveying was undertaken by following designated transect lines across each VMU. Before going into the field these, along with map imagery, were loaded into the Trimble®. Field staff would then follow the designated transect line through the VMU, looking for, and recording weeds, weed numbers and the size of each infestation into the Trimble®. Each day's field data was then uploaded in the office, for eventual incorporation into Council's ARCGIS mapping program.

Spatial data was collected in vector format – points, lines and polygons. Accuracy varied according to the topography and density of the vegetation but was generally within 3–5 metres. This was not considered a problem as the level of accuracy was noted and the plant species documented (mostly shrubs, which from experience, can usually be relocated with this level of accuracy). Each day's field data would then be uploaded back in the office, for eventual incorporation into Council's spatial database. The data was post-processed with Trimble GPS Pathfinder® Office.

In more accessible sections, transects were evenly spaced at 10–20 m intervals. However, in steeper terrain and areas of very dense vegetation, such transect spacing was deemed impractical so a more flexible approach was developed. In certain VMUs this meant 50–100 m distances between transect lines, in other cases, only a portion of a VMU could be surveyed.

Originally it was intended to have one person do the surveying but it soon became apparent that this was not practical as it was difficult and time consuming for one person to be looking for weeds, and entering detailed data. It was also seen as potentially unsafe to have staff working alone in difficult terrain. It was therefore decided to utilise two staff for surveying, except where VMUs were open and easily traversed.

Another early development was the decision to have staff remove weeds as they surveyed, provided plants were easy to hand-pull or cut down. A record was still made of such plants, as a means of monitoring weed occurrences and documenting controls.

The surveying at Bicentennial Reserve took four to six hrs per day, one or two days per week, over the course of a year. Related office based work usually occupied a further six to eight hrs per week.

Once the survey at Bicentennial Reserve was completed, the data was tested by sending works crews out to locate selected weed occurrences. Once it was verified that the methodology was sound, the data was used to shape the weed management works program in the Reserve for the 2013–2016 period. A review was also undertaken to assess how the surveying had proceeded, and what elements required modifying before moving the project to other reserves.

The major lesson from the review was the recognition that comprehensive surveying across the whole of a large, physically challenging reserve, was very time consuming; to conduct surveys on a similar scale in other large reserves, was therefore deemed to be impractical. Instead, selective surveying would be used in other large reserves.

The project then moved into surveying Ridgeway Park and Waterworks Reserve, two contiguous reserves occupying the foothills and lower slopes of kunayi-Mt Wellington south-west of the CBD. In these reserves the areas surveyed were selected on several factors such as proximity to weed populations on adjacent private properties, past management history, staff knowledge, gaps in existing knowledge, and accessibility. These surveys took around nine months to complete. The project then surveyed several smaller bushland reserves in Fern Tree and South Hobart before concluding at McRobies Gully and parts of Knocklofty Reserve. These surveys took a further nine months to complete. In each instance a similar survey methodology and resource allocation operated to that used in Bicentennial Reserve.

The entire project ran for three and a half years, with the surveying and mapping component taking around two years and nine months to complete. Total staff resources used came to approximately 3600 h, equivalent to around 19 h per week for the life of the project. The total cost was approximately $144,000, an average of around $41,000 per annum, almost all of which consisted of staff time and resources (labour and on-costs). At the conclusion of the project a total of approximately 1000 ha of municipal reserve land had been surveyed and mapped.

CONCLUSIONS

Clearly the project was a snapshot in time, one that will require repeating at some level in future if long-term weed management is to be effective.

That said, the project is considered to have been particularly useful for Council, with several positive outcomes:

- Provided detailed information on weed occurrences in several Council reserves where the level of existing knowledge was poor, inconsistent or non-existent.
- Weed data has been captured digitally for the first time by Council, allowing the information to be readily incorporated into FMPs, and making it easily accessible to staff when planning/implementing prescribed burning and weed management programs.

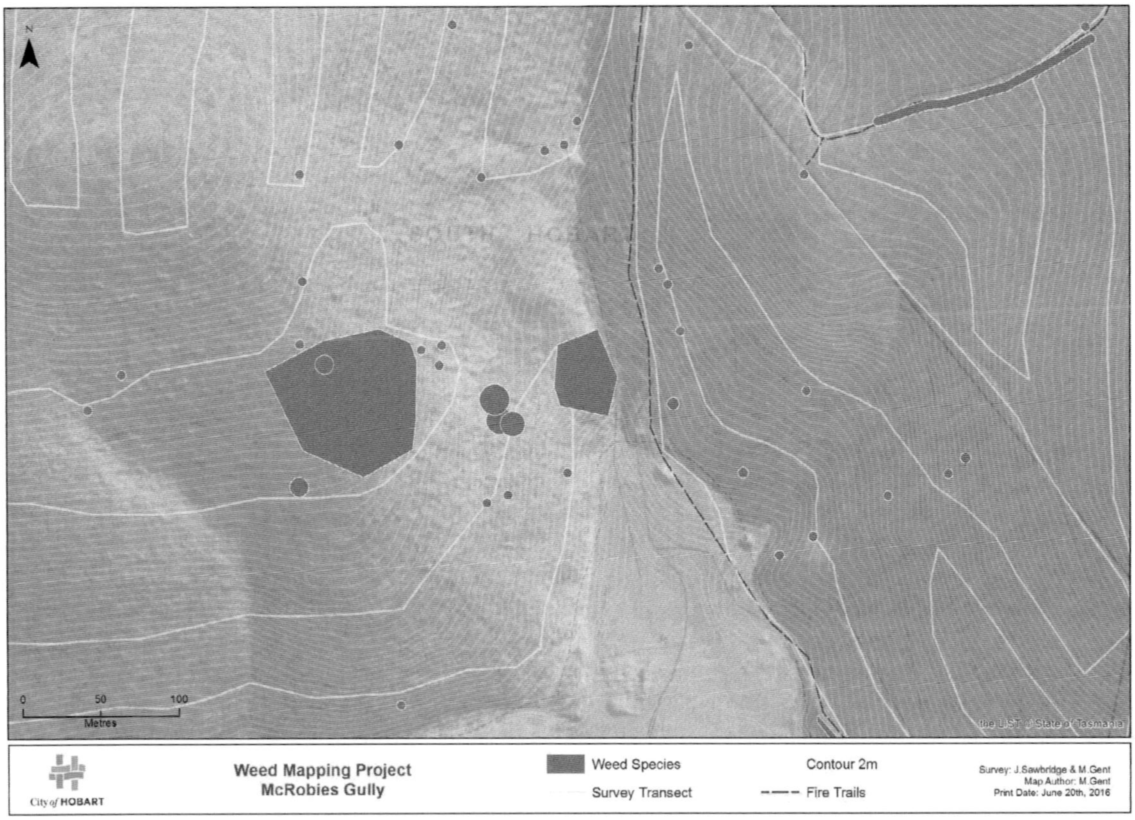

Figure 2. A map (June 2016) showing the location of weeds and the survey transects (in yellow) used at McRobies Gully, City of Hobart.

- Developed a broader and more detailed knowledge of reserves and management issues within Council's bushland management team.
- Many hundreds of weeds were removed during surveying, saving time and resources in the longer term.
- Staff interest and skills have been significantly developed and enhanced, particularly with technical aspects of surveying and mapping.
- The methodology used provides a broad template and basis for further weed surveys in the municipality in years to come.

However, the project did have some limitations, most notably that not all areas could be surveyed on foot due to the difficult terrain and dense vegetation. This means there are still gaps in determining the possible extent of weeds in some parts of certain reserves. Some of these areas may only be able to be surveyed on foot after a prescribed burn or wildfire passes through, thereby opening the vegetation for a period of time.

One option for future weed surveys in Hobart (and similar rugged areas) is to selectively use an Unmanned Aerial Vehicle (UAV, or drone), accepting that this method also has potential limitations such as range, clarity of image, seasonal timing, identifying target species, cost and the requirement for a certain level of expertise to both use the machines and interpret imagery.

ACKNOWLEDGMENTS

Several members of the City of Hobart Bushland and Reserves Unit have contributed to this project, in particular Martina Smith, Melanie Gent and John Sawbridge, without whose dedication, the project could not have succeeded. Martina's technical nous provided the structure for the field work undertaken by Mel and John. Special thanks to these three for expanding our knowledge and understanding of weed issues and weed management in the City of Hobart.

REFERENCES

Australian Bureau of Statistics (2011). http://www.hobartcity.com.au/Hobart/Demographics

Baker, M. (2016). Pers. comm. Weed Taxonomist, Tasmanian Herbarium, Hobart.

City of Hobart (2016). Bushland Reserves Flora Lists.

Futurepast Heritage Consultants (2012). Hobart Mountain Water Supply Conservation Management Plan.

HCC (2008). Hobart City Council, Bushland Management Strategy 2007–2017.

Hickie, J. (2014). Pers. comm. Open Space Planner, City of Hobart.

Hobart Water (2009). Draft Wellington Park Drinking Water Catchment Management Strategy.

Toonen, P. (2014). Pers. comm. Bushland Operations Manager, City of Hobart.

Chilean needle grass (*Nassella neesiana*) control – the ACT experience

Steve Taylor[1], Jenny Conolly[2] and Ani Gruber[1]
[1] ACT Parks and Conservation Service, 500 Cotter Road, Weston, Australian Capital Territory 2611, Australia
[2] City Services, Wattle St, Lyneham, Australian Capital Territory 2602, Australia
(Steve.Taylor@act.gov.au)

Summary Mapping Chilean needle grass (*Nassella neesiana*) infestations using the *Collector* application (app) with *ArcGIS On-line* has made it easier to recognise factors associated with successful control: winter control, use of flupropoante for initial control of large infestations, thorough follow-up control, and field staff with good plant identification and weed control skills. Rapid spread of Chilean needle grass is associated with: proximity to large infestations, mowing/slashing, incomplete control, grazing and high levels of people/vehicle movements.

Keywords Control, *Nassella*, weed, WoNS.

INTRODUCTION

Chilean needle grass (*Nassella neesiana* (Trin. & Rupr.) Barkworth) is an invasive grass from South America. It is a Weed of National Significance (WoNS) and a legislated Pest Plant in the Australian Capital Territory (ACT) (Must be Contained and Prohibited). It favours heavier soils, wetter areas, mown areas and readily fills gaps in disturbed and un-disturbed native grassland. It has coarse leaves with many fine hairs. It takes on a distinctive straw colour after frosts in winter. Upright awned seed heads with a purple corona give it a characteristic look. Cleistogene seed or 'stem/basal seed' are also present. Chilean needle grass is readily spread by mowers/vehicles/animals, bushwalkers and water.

Faithfull (2012) notes very high rates of spread for Chilean needle grass from mowing. This has been observed across the ACT in both urban and rural areas.

Chilean needle grass is widespread across urban areas in the ACT. Infestations are also starting to be detected and controlled in the upland grasslands of Namadgi National Park. Chilean needle grass is a significant threat to native plant diversity in a wide range of grassland and woodland areas, and as such has a 'Very High' risk rating in the ACT. It is able to invade intact native grassland and smothers the surrounding grasses and wildflowers (Figures 1 and 2).

Figure 1. Chilean needle grass (yellow-green grass) at centre, invading frosted-off kangaroo grass (*Themeda triandra*) at St Marks Grassland Reserve, 2015.

MATERIALS AND METHODS

Photopoints and mapping have revealed some common factors that affect successful control of this invasive grass. Although these observations have not been tested statistically, we are confident that these are the key factors that dictate success or otherwise with control work.

Mapping of infestations and photo points are the main ways weed control efforts are monitored in the ACT.

The *Collector* app (Esri 2016a) which syncs with *ArcGIS On-line* (Esri 2016b) has been used for mapping since 2015. When the *Collector* app is used in streaming mode, very accurate mapping is achieved (similar to tracking with a global positioning system (GPS)). Patterns have emerged from control data dating back to 2009. The data is stored on *ArcGIS On-line* and allows adaptive management. The figures below include screenshots from the *Collector* app. All are orientated north. The hatched polygons are either areas of weed infestations or where weed control has occurred. The faded green background to the screenshots denotes nature reserve or other reserve.

RESULTS

Case study 1 – Crace Grasslands Nature Reserve
Crace Grasslands (105 ha) is natural temperate grassland with large patches of phalaris and Chilean needle grass. There are also rare and endangered wildflower species such as button wrinklewort (*Rutidosis leptorhynchoides* F.Müell.). There are cattle grazing in the southern part of the reserve and eastern grey kangaroo (*Macropus giganteus* Shaw) grazing over the entire reserve. Tracks are slashed for vehicle access.

A 2001 survey of Chilean needle grass showed infestations (orange shaded polygons in Figure 3) developing across the north-western part of the reserve, cantered on spread from the grounds of a now demolished Defence Department building.

Mapping treated areas using the *Collector* app in 2015 showed that the Chilean needle grass had spread greatly but at varying densities. The density ratings before thorough control in 2015 ranged from 1–10% in least disturbed areas to >50% in most disturbed and mown areas. The only areas with isolated or no Chilean needle grass detected were the waterlogged areas dominated by *Phalaris aquatic* L., some of the kangaroo grass (*Themeda triandra* Forssk.) areas and some of the driest hill tops or rises.

There had been control work spot spraying Chilean needle grass with glyphosate 360 g L^{-1} at 1 L 100 L^{-1} water over the period 2001 to 2015, so the infestation densities could have been worse without this work. However, mapping shows that follow-up control was missed in some years. The 2015 control work used flupropionate 745 g L^{-1} (residual grass herbicide) in the higher density areas to contain the spread (Australian Pesticides and Veterinary Medicines Authority – APVMA permit 9792).

The rate of spread observed at Crace Grassland supports observations elsewhere of rates of spread exceeding 20% per year (Faithfull 2012). In some

Figure 2. Chilean needle grass monoculture in woodland. at Stirling Ridge Reserve, 2016.

Figure 3. Chilean needle grass at Crace Grasslands Nature Reserve – orange shaded polygons are from the 2001 survey, the larger green hatched polygons are treated areas in 2015. Pink spots are protected plants.

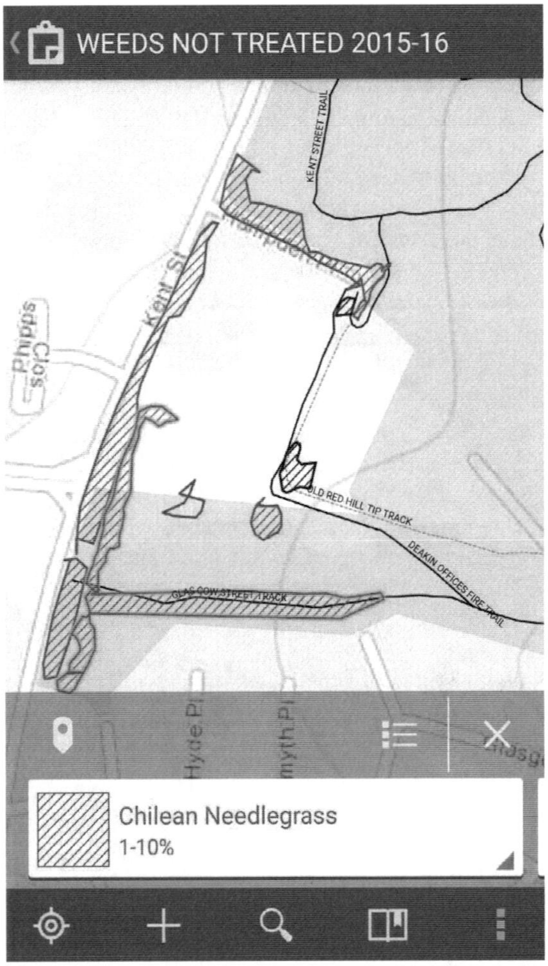

Figure 4. Chilean needle grass at Red Hill Nature Reserve.

parts of Grace Grasslands the rate of spread well exceeded this level.

Case Study 2 – Red Hill Nature Reserve Red Hill Nature Reserve (210 ha) is an *Environment Protection and Biodiversity Conservation Act 1999* (EPBC) listed yellow box (*Eucalyptus melliodora* A.Cunn. ex Schauer) – red gum (*Eucalyptus blakelyi* Maiden) grassy woodland. Unlike many other ridges in the urban area, clay soils support this community at summit areas. The reserve adjoins suburbs that have extensive areas of Chilean needle grass on nature strips, backyards, and open space.

Most of the infestations on Red Hill are found in disturbed and slashed areas at the reserve perimeter. However, patches have readily formed well within the reserve in rough spear grass (*Austrostipa scabra* (Lindl.) S.W.L.Jacobs & J.Everett), snow grass (*Poa sieberiana* Spreng.) and kangaroo grass (*Themeda triandra* Forssk.) areas. These are quickly controlled due to early reporting and mapping. A Parkcare group, the Red Hill Regenerators, and a Weeds Officer who lives nearby has helped in this regard.

Control in buffer areas and disturbed areas is by spot spraying with flupropanate 745 g L^{-1} at 150 mL and glyphosate 360 g L^{-1} at 335 mL 100 L^{-1} water (APVMA permit 9792). Spot spraying glyphosate 360 g L^{-1} at 1 L 100 L^{-1} water is used for scattered plants in high conservation value sites, and for follow-up control of other areas.

Areas are mapped with the *Collector* app in streaming mode for greatest accuracy of mapped boundaries (Figure 4). The mapped areas were synced with *ArcGIS On-line*.

Kangaroo grazing intensity varies greatly. There are many walkers, bike riders, and frequent water utility vehicle access. As a result infestations often begin adjacent to tracks and trails and in over-grazed areas.

The spot spraying of Chilean needle grass before the first spring slashing or mowing has reduced the density in the mown areas, but the extent has increased. These observations are based on photo points and mapping. Figure 4 shows how the infestations favour slashed areas at the perimeter of the reserve. Rigorous follow-up control has reduced the density of the infestations.

Case Study 3 – Jaramlee Reserve Jaramlee Reserve (112 ha) is a remnant native grassland and woodland that has had a long grazing history. It was transferred to government land management in 2015. The former land manager did not control Chilean needle grass when it first invaded in the 1990s. As a consequence it spread across the reserve and Figure 5 shows some of the primary control areas, which varied in density from 1 to 10% to >50%.

Use of flupropionate was restricted due to concerns about its impact on the endangered golden sun moth (*Synemon plana* Walker) larvae that feed on grass roots including those of Chilean needle grass. There was also a general restriction on any herbicide use during the golden sun moth flight season (October to December).

Seedlings from cleistogene or basal seed and seed banks were the main reasons for a high level of follow-up control required at Jaramlee. The basal seed may have been the reason for the apparent re-sprouting of a number of plants that appeared killed by glyphosate.

To avoid the basal seed development and subsequent reinfestation (which can develop in under six months), more frequent follow-up control is required (when relying on glyphosate). In the worst case this can be two to three times per year. However, it is extremely difficult to identify Chilean needle grass from look-a-like species outside of the winter months, making twice yearly spraying difficult. The golden sun moth flight season restriction also makes twice yearly control difficult to achieve.

Despite these complexities there has been effective initial control at a number of sites. Figure 5 shows Chilean needle grass control polygons (green). Other control polygons include serrated tussock (brown), and Paterson's curse (purple).

Case Study 4 – Oakey Hill Nature Reserve Oakey Hill (61 ha) is surrounded by urban areas with large Chilean needle grass infestations. It is also a popular recreation area like Red Hill. But unlike Red Hill it has less yellow box-red gum grassy woodland.

Figure 5. Chilean needle grass control at Jaramlee Reserve.

Infestations tend to be concentrated in the mown areas, but it spreads from these areas along walking trails and animal tracks. A mown area at the reserve summit is heavily infested and seed washes down into the surrounding woodland where new infestations establish (Figure 6).

Flupropanate 745 g L^{-1} has shown to be more effective halting spread in mown areas than glyphosate 360 g L^{-1}, due to its residual properties.

Case Study 5 – Dudley St Reserve Dudley St Grassland is a very small (2 ha) urban native grassland that has become highly degraded due to constant mowing. It is surrounded by very dense (>50%) infestations of Chilean needle grass, so there is a ready supply of seed to re-infest treated areas.

The lower lying kangaroo grass is degraded with one section becoming dominated by Chilean needle grass.

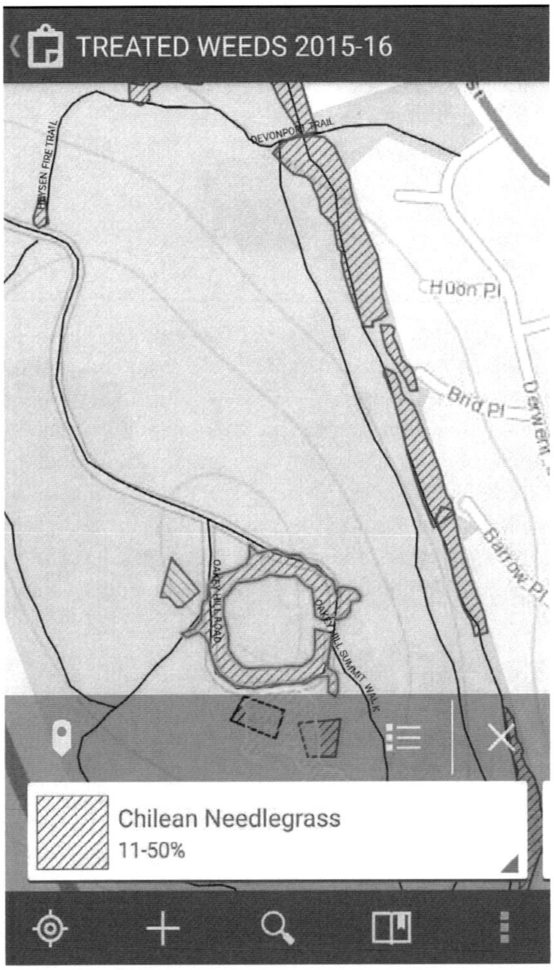

Figure 6. Chilean needle grass control at Oakey Hill Nature Reserve.

The grassland is mowed by machines used in areas infested with both Chilean needle grass and African lovegrass (*Eragrostis curvula* (Schrad.) Nees). Lack of hygiene leads to seed being spread between sites. Spread of invasive grasses like Chilean needle grass by slashers is well documented (Snell *et al.* 2007).

DISCUSSION

Overgrazing and associated animal movement spreads Chilean needle grass. This is a difficult subject as kangaroo culling is contentious.

The main factors affecting Chilean needle grass management have been identified from the case studies discussed above.

The pressure to mow or slash grassland reserves in urban areas is great, due to perceptions of amenity, fire danger and concern about snakes. There is a strict wash-down policy on slashers operating in nature reserves, but this policy is not applied in urban open space. This has lead to a rapid spread of invasive grasses across urban Canberra.

Employing field staff and contractors with very good grass and weed identification skills is required for successful control.

The following factors have been associated with spread of Chilean needle grass: Mowing-slashing, lack of machinery and equipment hygiene, poor grass identification skills, in-complete control, and competing winter duties of land managers so optimal control time is missed. Factors associated with effective control are: winter control, use of flupropionate 745 g L^{-1} for residual control, good follow-up control, and accurate mapping to allow adaptive management.

REFERENCES

Esri (2016a). Collector for ArcGIS, url http://www.esri.com/products/collector-for-arcgis

Esri (2016b). ArcGIS On-line, url http://www.esri.com/software/arcgis/arcgisonline

Faithfull, I.G. (2012). Biodiversity impacts of Chilean needle grass *Nassella neesiana* on Australia's indigenous grasslands. PhD thesis, Victoria University.

Snell, K., Grech, C. and Davies, J. (2007). National best practice manual Chilean needle grass. Victorian Department of Primary Industries and National Chilean needle grass Taskforce.

Mexican feather grass (*Nassella tenuissima*) control – the ACT experience

Jenny Conolly[1] and Steve Taylor[2]
[1] City Services, Wattle St, Lyneham, Australian Capital Territory 2602, Australia
[2]ACT Parks and Conservation Service, 500 Cotter Rd, Weston, Australian Capital Territory 2611, Australia
(jenny.conolly@act.gov.au)

Summary There have been two main biosecurity incidents involving Mexican feather grass (*Nassella tenuissima*) in the Australian Capital Territory (ACT), the first in 2008–09 and the most recent in December 2015. Successful control relies on: strict hygiene during removal and disposal of plants, removal of all contaminated plant material and mulch, use of flupropionate 745 g L^{-1} to suppress seedlings, covering the sites in deep mulch, thorough and regular follow-up control to locate seedlings and good public engagement.

Keywords Biosecurity, containment, Mexican feather grass, *Nassella tenuissima*.

INTRODUCTION

Mexican feather grass (*Nassella tenuissima* (Trin.) Barkworth) is a prohibited and notifiable pest plant in the ACT. Using the NSW Weed Risk Management System (Johnson 2009), with ACT data, gives a risk of 'Very High' for this species.

The increased fire danger observed with serrated tussock (*Nassella trichotoma* (Nees) Hack. ex Arechav.) (Osmond *et al.* 2008) could also apply to Mexican feather grass, as it also has very fine leaves and produces masses of fine-awned seed. It also has low palatability. Potential distribution in the ACT could be as high as that observed for serrated tussock.

During 2008, a Victorian nursery imported stipoid grass seed. The seed was either mislabelled or contaminated with Mexican feather grass seed, which was then propagated and distributed. The plants were sold under various stipoid type names. But this was not the first time it has been on sale in Australia. It has been recorded for sale in NSW and Victoria as far back as 1996 (Queensland Government 2016).

Online overseas seed companies have also marketed the plant under various names including *Stipa tenuissima*, *Stipa tenaccissima*, elegant spear grass, white tussock, Texas tussock grass and ponytail grass (NSW DPI 2016).

There was a national response to the 2008–09 illegal importation. The Victorian Government supplied valuable information to the ACT Government regarding numbers of plants sold to the ACT nursery industry.

Plants were tracked down to as many households as possible. There was also one naturalised infestation discovered in Namadgi National Park.

Not all plants were recovered, and in December 2015, seven years after the initial response, two more infestations originating from gardens were discovered. These resulted from plants purchased in 2008–09.

MATERIALS AND METHODS

Enough time has passed since the 2008–09 biosecurity response for us to see what has helped with eradication and containment efforts. Four case studies are discussed below.

Identification Mexican feather grass is a tufted perennial grass native to Texas, Mexico, Argentina and Chile. It can grow to 70 cm tall (Figure 1) and has very narrow leaves (<1 mm wide) with a 1.5 mm to 4 mm long ligule. It has open panicle flower/seed head (University of Queensland 2016).

The numerous flower spikelets have a very long awn (Figure 2) of 50 to 90 mm long (University of Queensland 2016). When mature and not in seed it

Figure 1. Mexican feather grass spread from a garden onto a nature strip at Fraser in 2016.

looks very similar to serrated tussock. At flowering it can look like a cross between serrated tussock and the native rough spear grass (*Austrostipa scabra* (Lindl.) S.W.L.Jacobs & J.Everett).

Figure 2. Mexican feather grass seed at Fraser in 2016.

The 2008–09 Response A well-publicised media alert across the ACT lead to many reports of Mexican feather grass. Most turned out to be other tufted or tussock grass species – spear grasses (*Austrostipa* spp.), poa tussock (*Poa labillardierei* Steud.), serrated tussock, and snow grass (*Poa sieberiana* Spreng.). Eight sites were confirmed and control work was undertaken. Four case studies are looked at below as they provide valuable information on why control was successful. One case study is from the recently discovered 2015 infestation.

All the infestations in the case studies and recent control work have been mapped in the field using the *Collector* (Esri 2016a) application (app) with *ArcGIS On-line* (Esri 2016b). The figures below include screenshots from the *Collector* app. All are orientated north. The hatched polygons are either areas of weed infestations or where weed control has occurred. The faded green background to the screenshots denotes reserved land.

Case Study 1. Gordon This site was part of the original 2008–09 biosecurity response. It was at a home garden in the southern ACT suburb of Gordon. The householder responded to a media alert after returning from a long summer holiday. She reported not being able to enter the front door of her house, as windblown seed from six or seven Mexican feather grass plants had accumulated covering most of the doorway. The plants had been purchased from a local nursery, possibly in 2007.

Initial control required removal of mature plants, hundreds of seedlings, contaminated mulch and soil, and vacuuming loose seed (Figure 3). This proved to be time-consuming due to the large volume of seed and the ability of seed to drill into cracks in pavers.

The contaminated waste was deeply buried (>2 m deep) at the waste and recycling facility at Mugga Lane. Disposable overalls and gloves were also buried.

A search of surrounding gardens and an open space area was also conducted. Only a few seedlings were found at those locations.

Follow-up control has been on-going since then. The first few years required twice a year control to remove seedlings. The home owner helped greatly by spot spraying with glyphosate 360 g L^{-1} and manual removal. No seedlings were allowed to mature. This meant the weeds officers did not have to use flupropionate 745 g L^{-1} to deal with germinating seed.

It took six years using this approach before no more seedlings were detected. At least 2000 seedlings had been removed over the six years. The site and surrounding area will continue to be monitored annually.

Case Study 2. Holder infestation Holder is a suburb in the Weston Creek part of Canberra. The infestation at this household spread from four plants purchased from a local nursery, possibly in 2007.

A thick cover of fine eucalypt mulch at this site limited initial spread. The plants were discovered by an off-duty ACT Parks and Conservation Weeds Officer walking her dog.

The owner of the house was happy to have the plants removed because he noticed they had started to spread to other parts of his garden and he understood the importance of controlling invasive weeds.

Figure 3. Mexican feather grass seed removal at Gordon in 2008.

A search of a nearby park and neighbour's gardens did not uncover any more plants.

As with the Gordon site, follow-up control was initially twice a year (early spring and autumn). Seedlings kept being discovered for seven years.

Flupropanate 745 g L^{-1} had to be used in the garden because the home owner was not as diligent as the Gordon householder with on-going control.

Case Study 3. Bendora Dam, Namadgi National Park. This was a very interesting infestation on a number of levels. It was in a disturbed understorey in montane forest in Namadgi National Park.

The plants originated from the edge of the Ranger's house garden. It was thought that they were planted when builders from Victoria renovated the house. None of the builders could remember planting the grass and none had properties with look-a-like serrated tussock or stipoid grass like infestations.

This infestation was successfully contained to 0.2 ha (Figure 4), because it could be monitored continuously so seedlings were removed as they appeared. It took seven years of rigorous follow-up control to reach the stage of only one or two seedlings in 2015–2016.

As with the other sites there will be on-going monitoring for seedlings. The monitoring will include searching surrounding forest, with the focus on disturbed and more open understorey.

Case Study 4. Fraser This is the first of the post 2008–09 national response infestations to be discovered (Figure 5). A retired weeds officer reported a suspicious grass at the local preschool. He was the manager of the ACT Parks and Conservation Service Weeds Unit when the 2008–09 response occurred.

At the preschool, seedlings had spread into adjacent garden beds. Prevailing wind helped prevent spread off-site. A local Fraser resident donated the plants used in the garden. The resident was pleased to hear from the Parks Service because the Mexican feather grass plants in his garden were spreading along the nature strip and into drains (Figures 6 and 7) despite his best efforts removing the plants on a regular basis. Unfortunately the resident had taken a couple of trailers uncovered loads of plants and seed to the green waste recycler. No plants have been detected there yet.

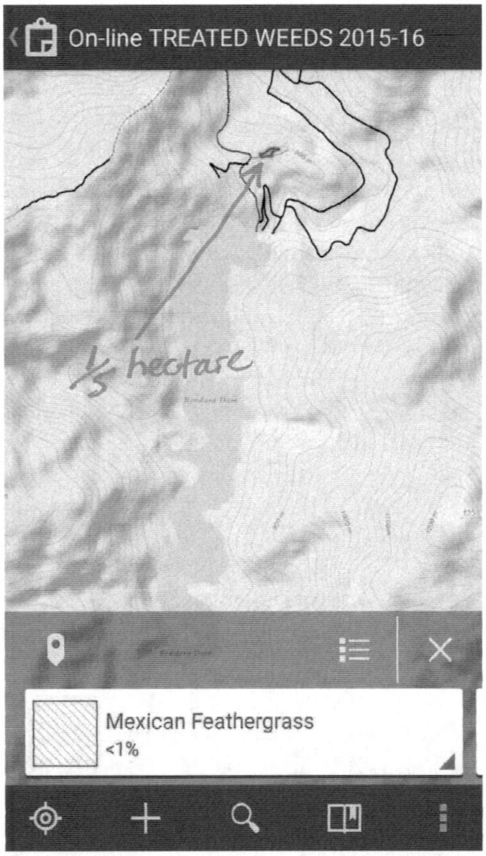

Figure 4. Namadgi National Park Mexican feather grass Control Area.

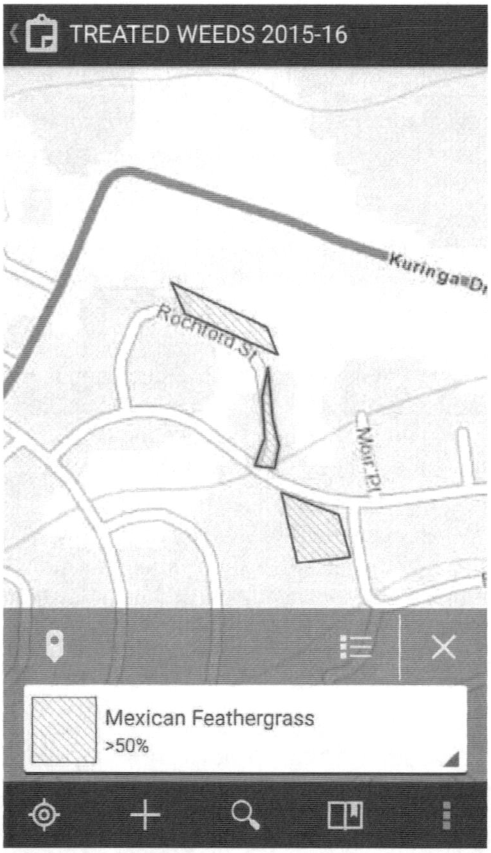

Figure 5. Fraser Mexican feather grass control areas at Fraser.

A search of surrounding areas found several plants in parkland and an area of grassy open space adjacent to a yellow box (*Eucalyptus melliodora* A.Cunn. ex Schauer) – red gum (*Eucalyptus blakelyi* Maiden) grassy woodland. Drains and creeks were also searched and fortunately no plants were detected.

Plants had spread 600 m from the garden (Figure 5). There was also a scattered infestation of serrated tussock overlapping with the outlier Mexican feather grass. These were all spot sprayed with a mix of flupropanate 745 g L^{-1} at 300 mL and glyphosate 360 g L^{-1} at 335 mL (APVMA permit 9792) per 100 L water.

The plants in the garden were physically removed along with seed heads and contaminated mulch. These were bagged for deep burial. The area was then sprayed with the mix of flupropanate 745 g L^{-1} and glyphosate 360 g L^{-1} and covered in a deep layer of mulch.

The Fraser infestations originated from four to six plants purchased sometime between 2007–09.

As of March 2016, the flupropanate 745 g L^{-1} used to prevent germination seems to be working as follow-up inspections only found a few seedlings. At other sites where flupropanate 745 g L^{-1} had not been used, there were dozens of seedlings only a couple of months after initial control.

Figure 6. Seedlings removed from roadside at Fraser in 2016.

Figure 7. Seedling spread to drain at Fraser in 2016.

DISCUSSION

Mexican feather grass produces masses of viable seed on par with serrated tussock. The longer awn appears to make it more likely to be accidentally spread plus creates larger masses of seed.

It readily invades disturbed areas but also invades sparse understory and inter-tussock spaces.

Intensive control for at least seven years is needed to significantly reduce the seed bank. Successful control requires: preventing seedlings from maturing, frequent monitoring and use of the residual grass herbicide flupropanate 745 g L^{-1}.

It is too soon to declare eradication but all sites are contained.

Factors that we have observed that affect success or otherwise in Mexican feather grass control are listed below:

Helps control – reporting suspicious plants early, use of flupropanate 745 g L^{-1} to reduce seed bank, accurate mapping with *Collector* app to ensure follow-up control, deep mulch cover in treated areas to prevent movement of soil off site and suppress germination and good grass identification skills.

Increases spread – on-line sales of Mexican feather grass material, careless disposal of contaminated material, inadequate resourcing of follow-up control, and slashing/mowing before control work.

REFERENCES

Esri (2016a). Collector for ArcGIS, url http://www.esri.com/products/collector-for-arcgis

Esri (2016b). ArcGIS On-line, url http://www.esri.com/software/arcgis/arcgisonline

Johnson, S. (2009). NSW Weed Risk Management System, url http://www.dpi.nsw.gov.au/agriculture/pests-weeds/weeds/wrm-system/weed-risk

NSW Department of Primary Industries (2016). Mexican feather grass, url http://weeds.dpi.nsw.gov.au/Weeds/Details/162

Osmond, R., Veebeck, M., McLaren, D.A., Michelmore, M., Wicks, B., Grech, C.J. and Fullerton, P. (2008). Serrated tussock – National best practice manual. Victorian Department of Primary Industries.

Queensland Government (2016). Mexican feather grass, url https://www.daf.qld.gov.au/__data/assets/pdf_file/0017/64160/IPA-Mexican-Feather-Grass-Risk-Assessment.pdf

University of Queensland (2016). LucidMobile Environmental Weeds of Australia App, url https://play.google.com/store/apps/details?id=org.lucidcentral.mobile.ewa&hl=en

Determining hybrids between invasive and native *Carpobrotus* N.E.Br. species in South Australia: field, morphological and molecular insights

Michelle Waycott[1,2], Hellmut Toelken[2], Ed Biffin[2], Korjent Van Dijk[1], Chris Brodie[2], Jessica Burdon[1,2], Ainsley Calladine[2], Kat Ticli[2], Chelsea Tothill[2], Helen Vonow[2] and Jürgen Kellermann[1,2]

[1] The University of Adelaide, School of Biological Sciences, Adelaide, South Australia 5001, Australia

[2] State Herbarium of South Australia, Department of Environment, Water and Natural Resources, GPO Box 1047, Adelaide, South Australia 5001, Australia

(michelle.waycott@adelaide.edu.au)

Summary *Carpobrotus edulis* (L.) L.Bolus ssp. *edulis* has been listed in the Metropolitan Adelaide and Northern Coastal Action Plan as a priority environmental weed in the region. This has led to coastal on-ground works targeting *C. edulis* for weed control since 2009. In many of the same areas there are native species of *Carpobrotus*, and putative hybrids have been observed. Taxonomic work, in South Australia, Australia and around the world, has uncovered that there are real difficulties in differentiating native species from hybrids with introduced species.

Plants being used in restoration work, purported to be identified as the native species, *C. rossii* (Haw.) Schwantes, may potentially be hybrid material between the introduced and native species (*C. edulis* × *C. rossii*). Ultimately there is a significant concern that hybrids have become incorporated into restoration plantings. If so, this will be an issue for seed collectors and growers of plant materials used for restoration programs.

Our project has expanded the detailed understanding of *Carpobrotus* taxonomy, formulated by Hellmut Toelken for all of the Australian taxa, by utilising a molecular genetic approach to identify hybrids among populations and species in representative material from South Australia. We detected an extremely high-level of hybridisation among native and introduced *Carpobrotus* from the Adelaide area using a ddRAD Next Generation Sequencing approach. Back-crossing between hybrids and parental types was also detected. A revised taxonomy is being developed to support the ongoing identification of hybrids in the Adelaide region, however hybridisation is likely to be occurring on a larger spatial scale.

Keywords *Carpobrotus*, molecular detection, Next Generation Sequencing, hybridisation, introgression, South Australia, taxonomy.

Prospects for the biological control of *Jatropha gossypiifolia*: *Stomphastis* sp. as a potential agent from South America

Dianne B.J. Taylor and Kunjithapatham Dhileepan
Biosecurity Queensland, Department of Agriculture and Fisheries, Ecosciences Precinct, GPO Box 267, Brisbane, Queensland 4001, Australia
(di.taylor@daf.qld.gov.au)

Summary *Jatropha gossypiifolia* L., commonly known as bellyache bush, is a serious weed of rangelands and riparian zones of northern Australia and has been identified as a Weed of National Significance. Bellyache bush has been a target of biological control in Australia since 1997, with one agent released to date. A renewed biological control effort, involving exploration in South America identified a number of potential biological control agents, the most promising of which, a small leaf-mining moth *Stomphastis* sp. (Lepidoptera: Gracillariidae), was imported from Peru into quarantine for further research in 2014. Newly emerged larvae mine directly into a leaf and remain there until pupation. Preliminary results suggest that *Stomphastis* sp. can complete development on bellyache bush and congener *J. curcas* L. (also a target for biological control) only. The leaf-miner has a short generation time and high fecundity, which bodes well for its future as a biological control agent

Keywords Bellyache bush, *Jatropha*, *Stomphastis*, leaf-miner, biological control.

INTRODUCTION

Jatropha gossypiifolia L. (Euphorbiaceae), commonly known as bellyache bush, is a serious weed of rangelands and riparian zones of northern Australia and has been identified as a Weed of National Significance (Australian Weeds Committee 2012, Bebawi *et al.* 2007). Well established infestations occur in the Burdekin, Fitzroy, Walsh, Palmer, Flinders and Gregory River catchments in Queensland, the Daly McArthur, Roper and Victoria river catchments in the Northern Territory, and the east Kimberley in Western Australia (Randall *et al.* 2009). Bellyache bush forms dense thickets, reducing the usefulness of land for pastoral and grazing purposes. All parts of the plants are toxic to stock. Dense infestations along the Burdekin River have reduced the carrying capacity to zero (Randall *et al.* 2009). Bellyache bush also reduces biodiversity, affects fire regimes and increases erosion along creek and river banks (Csurhes 1999, Bebawi *et al.* 2007). Based on morphological and phenological differences, a number of biotypes have been identified; Queensland Bronze, Queensland Green and Queensland Purple in Queensland, Darwin Purple and Katherine Green in the Northern Territory and Kununurra Green in Western Australia (Bebawi *et al.* 2007)

Biological control is an important component of the long-term management strategy for bellyache bush in Australia. Bellyache bush has been a target for biological control since 1997. Native range surveys in Mexico, northern South America, and the Caribbean resulted in the release of one agent, the seed feeding jewel bug *Agonosoma trilineatum* (F.) (Heteroptera: Scutelleridae), which failed to establish (Heard *et al.* 2012).

As part of a renewed biological control effort, recent exploration was undertaken in central South America, specifically Peru, Bolivia and Paraguay. Field surveys in 2012 and 2013 identified 11 insect species, one mite species and a leaf rust (Dhileepan *et al.* 2014). Of these species, a yet to be described leaf-mining moth *Stomphastis* sp. (Lepidoptera: Gracillariidae), was prioritised for further studies. In this paper we discuss current research on this potential biological control agent.

MATERIALS AND METHODS

Native range studies Widespread and severe leaf damage was observed by a leaf-mining microlepidopteran on *J. gossypiifolia* in the San Martin province in northern Peru and occasionally on *J. curcas* L. (Dhileepan *et al.* 2014). In the Manuel María Caballero province of central Bolivia the leaf miner was common on *J. gossypiifolia* and *J. excisa* Griseb., and uncommon on *J. curcas* and *J. clavuligera* Müll.Arg. Leaf-mining damage was not observed on *J. curcas* or castor oil (*Ricinus communis* L.) co-occurring with *J. gossypiifolia* (Dhileepan *et al.* 2014).

Similar to *S. thraustica* (Meyrick), a well-known pest of *J. gossypiifolia* and *J. curcas* in Asia and Africa, *Stomphastis* sp. is currently being described as a new species (Dhileepan *et al.* 2014). This is the first time a *Stomphastis* species has been recorded in South America.

Insect source In November 2014 more than 500 *Stomphastis* sp. larvae and pupae collected from

12 sites in Peru, were imported into our quarantine facility in Brisbane. As adults emerged from the imported leaf material they were collected and transferred to gauze covered cages (90 × 100 × 50 cm) containing potted *J. gossypiifolia* plants (mixture of biotypes). Female *Stomphastis* sp. readily oviposited on *J. gossypiifolia* plants under caged conditions and a colony was able to be quickly established in quarantine. The culture was established and maintained in a quarantine glasshouse kept at 27°C, 60% relative humidity and natural photoperiod. Once a colony had been established, lifecycle studies and host specificity testing were initiated. Newly emerged adults were used in all life cycle studies and host specificity tests.

Host specificity testing The host test list was compiled using the centrifugal phylogenetic method (Wapshere 1975), which proposes testing plants of increasingly distant relationship to the target until the host range is circumscribed, and is based on the host test list approved for *Agonosoma trilineatum* (Heard *et al.* 2009). The list contains 37 species from Euphorbiaceae and closely related families (Table 1). Since the testing of *A. trilineatum*, Euphorbiaceae has been split into four families (see Stevens 2001–). Phyllanthaceae Martynov contains most of the old Euphorbiaceae–Phyllanthoideae (with the exception of *Drypetes* and relatives, now part of Putranjivaceae Endlicher), while Picrodendraceae Small contains most of the taxa from Euphorbiaceae–Oldfieldioideae (Köhler & Webster). Species included in the test list approved for *Agonosoma trilineatum* but omitted from this proposed list were either difficult to source or were replaced by species deemed more suitable. Species from unrelated families were also removed. *Microstachys chamaelea* (L.) Hook.f., a recorded host of *S. thraustica*, was also added to the test list.

Host specificity testing is currently in progress. Initially, all test plants are being subjected to no-choice oviposition/larval development trials, with at least five replicates conducted for each species. Larval development trials using newly emerged larvae are not possible since the larvae enter the leaf directly from the egg. Thus larval development could only be monitored on plants on which eggs were laid. Twenty unsexed newly emerged adults are released into a gauze-covered cage (90 × 45 × 45 cm) containing a single test plant. For each round of testing, ten unsexed adults are released into a cage containing one *J. gossypiifolia* plant. Test plants and *J. gossypiifolia* plants are checked weekly for evidence of eggs and larval mines. Test plants are maintained until a new generation of adults is recovered from bellyache bush, at which time they are destroyed. Test species on which larvae develop will be subjected to further testing such as choice trials with bellyache bush.

RESULTS

Life cycle Adult *Stomphastis* sp. are small moths (less than 1 cm long) and live for an average of ten days under quarantine conditions. Females lay eggs singly on leaves; usually on the underside, near a leaf vein. One female moth can lay in excess of 100 eggs. Newly emerged larvae mine directly into the leaf and remain in the leaf as they develop. Larvae exit the leaf to pupate, which they do mainly on the leaf, though some pupate in other locations such as a branch crevice. A generation from adult to adult takes around 22 days under quarantine conditions, but has been completed in as little as 18 days.

Host specificity testing No-choice host specificity testing of *Stomphastis* sp. has been completed for 31 non-target species so far (Table 1) with another six or seven species remaining. Eggs have been laid on many of the non-target species (17), however larval development has only occurred on bellyache bush and its congener *J. curcas*. Damage to non-target species other than *J. curcas* has been limited to first instar mines.

DISCUSSION

The results from the host specificity testing of *Stomphastis* sp. conducted so far are extremely promising with the insect only completing development on bellyache bush and its congener *J. curcas*. It is likely that the oviposition that occurred on many of the non-target test species are false positive results prompted by the confined conditions under which the tests are necessarily conducted (Marohasy 1998, Heard 2000). The non-target species on which eggs were laid have been shown to be unsuitable hosts by the absence of any larval development (with the exception of *J. curcas*). The only damage recorded on non-target species other than *J. curcas* has been limited to first instar feeding.

Complete development of *Stomphastis* sp. on *J. curcas* is not unexpected and nor is it a hindrance to the moth being released for the biological control of bellyache bush. *Jatropha curcas* has invaded parts of northern Australia and is also an approved target for biological control. Eggs were laid, but no development occurred, on the congeners *J. podagrica* Hook. and *J. multifida* L., two species occasionally grown in Australia as ornamentals.

Stomphastis has completed development on all of the biotypes identified in Australia. Further work, comparing the suitability of the various biotypes, is in progress.

Table 1. Host specificity test list for *J. gossypiifolia* including status for no-choice oviposition and larval feeding and development trials with *Stomphastis* sp.

Test plant	Status[A]	Tested
Euphorbiaceae: Crotonoideae		
Jatropha gossypiifolia L.	Target	Y
Jatropha curcas L.	I	Y
Jatropha podagrica Hook.	O	Y
Jatropha multifida L.	O	Y
Aleurites moluccanus (L.) Willd	N	Y
Baloghia inophylla (G.Forst.) P.S.Green	N	Y
Beyeria lechenaultii (DC.) Baill.	N	Y
Beyeria viscose (Labill.) Miq.	N	Y
Codiaeum variegatum (L.) A.Juss.	E	Y
Croton acronychioides F.Muell.	N	Y
Croton insularis Baill.	N	Y
Croton verreauxii Baill.	N	Y
Dimorphocalyx australiensis C.T.White	N	N
Endospermum myrmecophilum L.S.Sm.	N	N
Manihot esculentum Crantz	C	Y
Manihot grahamii Hook.	I	Y
Euphorbiaceae: Acalyphoideae		
Acalypha eremorum Mull.Arg.	N	N
Alchornea ilicifolia (J.Sm.) Muell.Arg.	N	Y
Macaranga tanarius (L.) Müll.Arg.	N	Y
Mallotus philippensis (Lam.) Muell.Arg.	N	Y
Omphalea celata P.I.Forst.	N	N
Ricinus communis L.	I	Y
Euphorbiaceae: Euphorbioideae		
Euphorbia pulcherrima Willd. ex Klotzsch	O	Y
Homalanthus nutans (G.Forst.) Guill.	N	Y
Microstachys chamaelea (L.) Hook.f.	N	N
Phyllanthaceae		
Antidesma bunius (L.) Spreng.	N	Y
Actephila lindleyi (Steud.) Airy Shaw	N	Y
Breynia oblongifolia (Mull.Arg.) Mull.Arg	N	Y
Bridelia exaltata F.Muell.	N	Y
Cleistanthus hylandii Airy Shaw	N	Y
Flueggea virosa (Willd.) Voigt	N	Y
Glochidion ferdinandi (Muell.Arg.) F.M.Bailey	N	Y
Phyllanthus cuscutiflorus S.Moore	N	Y
Picrodendraceae		
Choriceras tricorne (Benth.) Airy Shaw	N	N
Dissiliaria baloghioides F.Muell. ex Baill.	N	Y
Petalostigma pubescens Domin	N	Y
Putranjivaceae		
Drypetes deplanchei (Brongn. & Gris) Merr.	N	Y

[A]N = native; O = ornamental; C = crop; I = invasive

Stomphastis sp. has a short generation time and high fecundity, which bodes well for its future as a biological control agent. We anticipate completing the host specificity testing by the end of 2016. If the insect is found to be host specific an application to release will be made.

ACKNOWLEDGMENTS

This project was funded by Meat and Livestock Australia and Queensland Government's War on Western Weeds funding initiative. We thank Stefan Neser (Plant Protection Research Institute, Pretoria, South Africa), Damian Rumiz (Museo de Historia Natural Noel Kempff Mercado, Universidad Autónoma Gabriel René Moreno, Santa Cruz de la Sierra, Bolivia), and Victor Hugo Sanchez (Instituto National de Investigacion Agraria, Tarapoto, Peru), for their help with the overseas field studies. Tony Pople provided valuable comments on earlier versions of the manuscript.

REFERENCES

Australian Weeds Committee (2012). Weeds of National Significance (2012). Department of Agriculture, Fisheries and Forestry, Canberra, ACT.

Bebawi, F.F., Vitelli, J.S., Campbell, S.D., Vogler, W.D., Lockett, C.J., Grace, B.S., Lukitsch, B. and Heard, T.A. (2007). The biology of Australian weeds 47. *Jatropha gossypiifolia* L. *Plant Protection Quarterly* 22, 42-58.

Csurhes, S.M. (1999). 'Bellyache bush (*Jatropha gossypiifolia*) in Queensland. Pest Status Review Series – Land Protection'. Queensland Department of Natural Resources, Brisbane, Qld.

Dhileepan, K., Neser, S. and De Prins, J. (2014). Biological control of bellyache bush (*Jatropha gossypiifolia*) in Australia: South America as a possible source of natural enemies. Proceedings of the XIV International Symposium on Biological Control of Weeds, eds F.A.C. Impson, C.A. Kleinjan, and J.H. Hoffmann, pp. 5-10 (Kruger National Park, South Africa).

Heard, T.A. (2000). Concepts in insect host-plant selection behaviour and their application to host specificity testing. Proceedings of session: Host-specificity testing of exotic arthropod biological control agents—the biological basis for improvement in safety, eds R.G. Van Driesche, T.A. Heard, A. McClay, R. Reardon, pp. 1–10. USDA Forest Service, Forest Health Technology Enterprise Team, Morgantown.

Heard, T.A., Chan, R.R., Wilmot Senaratne, K.A.D., Palmer, W.A., Lockett, C. and Lukitsch, B. (2009). *Agonosoma trilineatum (Heteroptera: Scutelleridae) a biological control agent of the weed bellyache bush, Jatropha gossypiifolia (Euphorbiaceae). Biological Control* 48 (2), 196-203.

Heard, T.A., Dhileepan, K., Bebawi, F., Bell, K. and Segura, R. (2012). *Jatropha gossypiifolia* L. – bellyache bush. *In* 'Biological control of weeds in Australia: 1960 to 2010', eds M. Julien, R.E. McFadyen and J. Cullen, pp. 324-33. (CSIRO Publishing, Melbourne).

Marohasy, J. (1998). The design an interpretation of host specificity tests for weed biological with particular reference to insect behaviour. *Biocontrol News and Information* 19, 13N-20N.

Randall, A., Campbell, S., Vogler, W., Bebawi, F. and Madigan, B. (2009). Bellyache Bush (*Jatropha gossypiifolia*) Management Manual – current control options and management case studies from across Australia. Queensland Primary Industries and Fisheries, Department of Employment, Economic Development and Innovation, Brisbane.

Stevens, P.F. (2001–). Angiosperm Phylogeny Website. Version 12, July 2012. Available at: http://www.mobot.org/ MOBOT/research/APweb/. [Accessed 6th June 2016]

Wapshere, A.J. (1975). A protocol for programmes for biological control of weeds. *Pest Articles and News Summaries* 21, 295-303

The Jatropha webber (*Sciota divisella*): a potential biological control agent for *Jatropha gossypiifolia* (bellyache bush) from India

Elizabeth L. Snow, Kunjithapatham Dhileepan and Dianne B.J. Taylor
Biosecurity Queensland, Department of Agriculture and Fisheries, Ecosciences Precinct, Boggo Road, Dutton Park, Queensland 4102, Australia
(liz.snow@daf.qld.gov.au)

Summary *Jatropha gossypiifolia* L. (bellyache bush) is a deciduous shrub from tropical America that was introduced into Australia as an ornamental and has since become a serious weed of rangelands and riparian zones in northern Australia. It forms dense thickets that degrade habitat, including suppressing pasture growth and rendering flood plains unproductive. All parts of the plant are toxic and cattle deaths have been attributed to bellyache bush. Opportunistic surveys in India identified the moth *Sciota divisella* (Duponchel) (Lepidoptera: Pyralidae) as a potential biological control agent for bellyache bush and in December 2014 the insect was imported into the quarantine facility at the Ecosciences Precinct at Dutton Park in Brisbane. The larvae of this moth feed on the leaves and fruits and, in the absence of these, the larvae bore into the stem from the shoot-tip. The moth causes defoliation, fruit loss and shoot die-back. Host testing of the insect has begun with no-choice testing of larvae on 40 species to ascertain whether the agent is specific to bellyache bush. Larval feeding and complete development occurred only on bellyache bush, on the closely related *J. curcas* (another approved target for biological control) and on *Euphorbia grantii* a highly toxic, uncommon exotic ornamental). Aside from these two seemingly acceptable non-target hosts, if the agent proves to be host specific then approval to release will be sought through the Australian government.

Keywords *Jatropha gossypiifolia*, bellyache bush, *Jatropha curcas*, *Sciota divisella*, Jatropha webber.

INTRODUCTION

Jatropha gossypiifolia L. (bellyache bush), a Weed of National Significance (WoNS), is a deciduous shrub from Central and South America, possibly native to the drier islands of the Caribbean and coastal Venezuela (*Heard et al.* 2009). The plant was introduced into Australia as an ornamental and has since become a serious weed of wet and dry tropical rangelands and riparian zones in northern Australia. It forms dense thickets that degrade habitat, including suppressing pasture growth and rendering flood plains unproductive (Dhileepan *et al.* 2014). All parts of the plant are toxic and livestock deaths have been attributed to bellyache bush, particularly during times of drought (Csurhes 1999, Oliveira *et al.* 2008). The foliage of bellyache bush varies in colour from deep red-purple to bright green and depends on leaf age and plant biotype (Queensland green, bronze or purple, Darwin purple or Katherine green and Kununurra green) (Taylor and Dhileepan 2014). Bellyache bush requires ongoing chemical treatments making chemical control expensive in extensive grazing areas (Bebawi *et al.* 2007a). The plant is capable of a high rate of reproduction with seed production being estimated as high as 12,000 seeds per plant per year. In addition, the plant has the capacity to flower and produce seed year round, with seeds that are viable for up to four years (Bebawi *et al.* 2007b). To keep the weed in check, control must be regular, ongoing and widespread. For these reasons, biological control is highly desirable for bellyache bush in Australia.

Bellyache bush has been a target for biological control since 1997. Native range surveys in Mexico, central and northern South America, and the Caribbean resulted in the release of the seed-feeding jewel bug (*Agonosoma trilineatum* F.) in 2003, which failed to establish (Heard *et al.* 2012). The leaf rust, *Phakopsora arthuriana* Buriticá & J.F. Hennen was also identified and host specificity testing is in progress at CABI in the UK.

Recent opportunistic surveys in India identified *Sciota divisella* (Duponchel) (Lepidoptera: Pyralidae) (Figure 1) feeding on bellyache bush, *J. curcas* L. and *Euphorbia grantii* Oliv., all members of the family Euphorbiaceae. The moth was subsequently imported into the QC3 quarantine facility at the Ecosciences Precinct in Brisbane to determine its host range and study its biology.

This paper outlines the progress of host testing, preliminary biological observations, taxonomy and possibilities for future research.

MATERIALS AND METHODS

Insect source The first importation of *S. divisella* larvae from Tamil Nadu, India, was in December 2014. This colony failed to establish in quarantine. A second importation of 192 larvae and 41 pupa occurred in July 2015, with a further importation of 150 larvae in October 2015 to augment the colony.

Insect culture Moths were reared in quarantine glasshouse facilities at the Ecosciences Precinct at Boggo Road, Dutton Park, Queensland. Six to eight bellyache bush plants (mixed biotypes) were placed in cages (900 × 888 × 560 mm) covered with polyester voile mesh (45 gsm) and adult moths were introduced to the cages. The number of adults used was dependent on availability of moths from culture. A container of Gatorade® with a cotton wick was also supplied to provide a source of sugar and salt for female feeding. The moths were cultured in controlled climatic conditions of 22°C (night) and 26°C (day) temperature and 60% relative humidity. During late autumn 2016, day lengths were extended over winter with the use of ultra-violet lighting between 5–7am and 5–7pm to encourage egg laying and larval development.

Test plants Forty test plant species from the Euphorbiaceae and closely related families were included in the host test list. *Jatropha* is the only representative of the tribe Jatropheae in Australia with five introduced species – *J. gossypiifolia, J. curcas* L., *J. mulifida* L., *J. integerrima* Jacq. and *J. podagrica* Hooker. The latter three are ornamental plants and only *J. gossypiifolia* and *J. curcas* are naturalised.

Host specificity tests Eggs (less than a day old) were harvested from bellyache bush by cutting the egg batch from plant surface and placing in a Petri dish with moistened filter paper to await eclosion. Ten newly emerged larvae were then placed onto each test plant using a single bristle paint brush. There are a minimum of five replicates for every species. No-choice larval survival and development tests and no-choice oviposition tests are being done first and only those test plant species on which the moth lays eggs and or those on which the larvae feed and complete development will be subjected to choice oviposition and larval feeding and development tests.

RESULTS

Insect culture The mean number of adults (± SEM) from a colony cage was 83 ± 10 (n = 23) (mean number of adults to start colony cage was 103 ± 17). There was no diapause phase during the culturing of the insect under controlled conditions and one generation took approximately six weeks to complete. The adults laid eggs on leaves and stems of bellyache bush plants, but eggs were most commonly found on the underside of leaves in clumps or in rows along vein margins and on the main trunk, toward the apex. Plants were extensively damaged during the insect culturing process with leaf-webbing and stem damage caused by feeding (Figure 2).

Host specificity Host specificity tests are partially completed for 25 species (Table 1) and tests are in progress for the remaining 15 test plant species. Larval feeding and complete development has occurred only on the target plant (bellyache bush), on the closely related *J. curcas* (another approved target for biological control) and on *Euphorbia grantii* (highly toxic, uncommon exotic ornamental). At this stage there has been no development on any of the other 22 species tested.

Insect identification The species was originally identified as *Morosaphycita morosalis* (Saalmüller). Specimens were sent to Marianne Horak, (CSIRO, NRCA, Canberra) for verification of species identity due to the complex and cryptic nature of this sub

Figure 1. Adult *Sciota divisella*.

Figure 2. Larval damage to bellyache bush leaves and stems.

Table 1. Host range tests completed. (Rep = number of replicates currently completed, Dev = agent development).

Test Species	Rep	Dev
CROTONOIDEAE		
Jatropheae		
Jatropha gossypiifolia L.	13	Y
Jatropha curcas L.	3	Y
Jatropha multifida L.	3	N
Jatropha podagrica Hook.	2	N
Crotoneae		
Croton insularis Baill.	3	N
Croton acronychioides F.Muell.	1	N
Manihotaeae		
Manihot esculenta Crantz.	1	N
Adenoclineae		
Endospermum myrmecophilum L.S.Smith	2	N
Codiaeae		
Codiaeum variegatum Blume	1	N
Ricinocarpeae		
Beyeria lechenaultii (DC.) Baill.	1	N
Beyeria viscosa (Labill.) Miq	2	N
EUPHORBIOIDEAE		
Euphorbieae		
Euphorbia pulcherrima Willd.	2	N
Euphorbia grantii Oliv.	1	Y
Hippomaneae		
Homolanthus nutans (G.Forst) Guill.	3	N
ACALYPHOIDEAE		
Acalypheae		
Ricinus communis L.	1	N
Macaranga tanarius (L.) Muell.Arg	1	N
Mallotus phillipensis (Lam) Muell.Arg	2	N
PHYLLANTHACEAE – PHYLLANTHOIDEAE		
Wielandieae		
Actephila lindleyi (Steud.) Airy Shaw	4	N
Bridelieae		
Cleistanthus hylandii Airy Shaw	2	N
Phyllantheae		
Flueggia virosa (Willd.) Voight	2	N
Breynia oblongifolia Muell.Arg.	2	N
Glochidion ferdinandi (Muell.Arg.) F.M.Bailey	1	N
Glochidion sumatranum Miq	1	N
Picrodendraceae		
Petalostigma pubescens Muell.Arg.	2	N
Putranjivaceae		
Drypetes deplanchei (Brongn. & Gris.) Merr.	1	N

family and genus. She advised that after consulting with Dr David Lees at the Natural History Museum in London, that the species is *S. divisella* and not *M. morosalis*. Genetic barcoding will be used to confirm this identification.

DISCUSSION

The Jatropha webber shows potential as an effective biological control agent for bellyache bush based on significant damage to leaves, stems and reproductive parts of the plant (when present). Development of the moth on *J. curcas* was expected given observations in the home range of the insect and the close relationship of *J. curcas* to bellyache bush. Further research will be conducted to assess the impact and implications of development on any other alternative host plants. At this stage the development has only been on plants that are closely related and are either weeds or exotic and uncommon ornamentals; and none on any Australian native Euphorbiaceae plants. The short generation time and lack of a diapause phase may be advantageous to eventual field establishment by allowing populations to build up more quickly and provide more prolonged periods of control of the weed under field conditions. While host range studies are being conducted, further research in quarantine will assess how well the moth reproduces on plants with no foliage as plants in the field experience periods of leaf loss. During these periods, the importance of possible alternative hosts such as *J. curcas* may also need to be considered.

Studying the viability of seeds that have been damaged by larval feeding on fruits would also be valuable, given high seed production is characteristic of this weed. The extreme climate where bellyache bush has established in Australia may present challenges for agent establishment in terms of excessive temperatures, variable humidity and unreliable rainfall and their effect on the various life stages of the insect.

This moth species was initially thought to be *M. morosalis* as this has been documented as a pest of *J. curcas* in India. The revised species *Sciota divisella* is a known pest of *J. curcas* in India (Durairaj *et al.* 1999) and in Benin, West Africa (Datinon *et al.* 2013). There is also reference to *S. divisella* on a Euphorbia species in arid zones (Singh and Harsh 1996). Sharma and Srivastava (2011) reported a similar lepidopteran pest as *M. morosalis*, which is likely to also be a misidentification. Sharma and Srivastava (2011) do not give details of how their moth was identified. It is likely that a species complex exists and further genetic work may help elucidate the taxonomy of the moth. Should the agent prove to be host specific (beyond the three exotics recorded so far) then approval to release will be sought through the Australian government.

ACKNOWLEDGMENTS

This project was funded by Queensland Government's War on Western Weeds initiative. We thank Dr A. Balu and S. Murugesan (Institute of Forest Genetics and Tree Breeding, Coimbatore, India) for their help with the overseas field studies. We are extremely grateful to Dr Marianne Horak from the CSIRO National Research Collection in Canberra and Dr David Lees from the Natural History Museum in London for their efforts in determining the taxonomy of the insect. Thanks also to Dr Tony Pople for his comments on manuscript.

REFERENCES

Bebawi, F.F., Vitelli, J.S., Campbell, S.D., Vogler, W.D., Lockett, C.J., Grace, B.S., Lukitsch, B. and Heard, T.A. (2007a). The biology of Australian weeds 47. *Jatropha gossypiifolia* L. *Plant Protection Quarterly* 22, 42-58

Bebawi, F.F., Lockett, C.J., Davies, K.M., Lukitsch, B.V. (2007b). Damage potential of the introduced biological control agent (*Agonosoma trilineatum* F.) on bellyache bush (*Jatropha gossypiifolia* L.). *Biological Control* 41, 415-22.

Csurhes, S.M. (1999). Bellyache bush (*Jatropha gossypiifolia*) in Queensland. Pest Review Series – Land Protection. Queensland Government Department of Natural Resources.

Datinon, B.D., Glitho, A.I., Tamo, M., Amevoin, K., Goergen, G. and Douro Kpindou, O.K. (2013). Inventory of major insects of *Jatropha curcas* L. (Euphorbiaceae) and their natural enemies in southern Benin. ARPN *Journal of Agriculture and Biological Science* 9, (10) 711-18.

Dhileepan, K., Neser, S. and De Prins, J. (2014). Biological control of bellyache bush (*Jatropha gossypiifolia*) in Australia: South America as a possible source of natural enemies. Proceedings of the XIV International Symposium on Biological Control of Weeds, pp. 5-10, eds. F.A.C. Impson, C.A. Kleinjan and J.H. Hoffman, Kruger National Park, South Africa.

Durairaj, S., Srinivasan, G., Balu, A. and Raja Rishi, R. (1999). Ecofriendly methods of control the *Jatropha* shoot and fruit borer *Sciota divisella* (Duponchel) (Lepidoptera: Pyralidae). Paper presented in the National Symposium on biological control of insects in Agriculture, Forestry, Medicine and Veterinary Science, 21 January 1999, Bharathiyar University, Coimbatore, India (Abstract).

Heard, T.A., Chan, R.R., Senaratne, K.A.D., Palmer, W.A., Lockett, C. and Lukitsch, B. (2009). *Agonosoma trilineatum* (Heterophera: Scutelleridae) a biological control agent of the weed bellyache bush, *Jatropha gossypiifolia* (Euphorbiaceae). *Biological Control* 48, 196-203.

Heard, T.A., Dhileepan, K., Bebawi, F., Bell, K.L. and Segura, R. (2012). *Jatropha gossypiifolia* L. – bellyache bush. *In* 'Biological Control of Weeds in Australia', pp. 324-33.

Oliveira, L.I., Jabour, F.F., Nogueira, V.A. and Yamasaki, E.M. (2008). Experimental poisoning by the leaves of *Jatropha gossypiifolia* (Euphorbiaceae) in sheep. *Pesquisa Veterinária Brasileira* 28 (6), 275-78.

Sharma, R.P. and Srivastava, C.P. (2011). Assessment of biological and morphometric parameters of Jatropha Leaf Webber Cum Fruit Webber, *Pempelia morosalis* (Saalm Uller) along with associated natural enemies in eastern Uttar Pradesh of India. *Journal of Entomology* 8 (1), 88-94.

Singh, M.P. and Harsh, L.N. (1996) *Denticera divisella* Duponchel) (Pyralidae: Lep) infesting *Euphorbia antisyphilitica* in the arid zones. *Entomon* 21, 263-4.

Taylor, D.B.J. and Dhileepan, K. (2014). Application to import *Morosaphycita morosalis* into quarantine for evaluation as a weed biological control agent for bellyache bush *Jatropha gossypiifolia*. Department of Agriculture, Fisheries and Forestry, Queensland Government.

Host suitability testing of a gall forming fly *Polymorphomyia basilica* Snow (Diptera: Tephritidae), a potential biological control agent of Siam weed *Chromolaena odorata* (L.) R.M.King & H.Rob. (Asteraceae) in South Africa

Nontembeko Dube[1] and Costas Zachariades[1,2]
[1] Agricultural Research Council Plant-Protection Research Institute, Private Bag X 6006, Hilton 3245, South Africa
[2] School of Life Sciences, University of KwaZulu-Natal, Private Bag X01, Scottsville 3209, South Africa
(duben@arc.agric.za)

Summary *Chromolaena odorata* (L.) R.M.King & H.Rob., commonly known as Siam weed or devilweed, is a widely distributed tropical shrub native to the Neotropics, from the southern USA to northern Argentina, and is among the world's worst weeds. A stem-galling tephritid fly, *Polymorphomyia basilica* Snow, from the northern Caribbean islands, was considered as a potential biological control agent for the invasive alien weed *Chromolaena odorata* in South Africa. The tephritid *Cecidochares connexa* Macquart causes similar damage and has been used to good effect on the Asian/West African biotype of *C. odorata*, but does not develop on the different, southern African *C. odorata* biotype. Positive biological characteristics of *P. basilica* include a high rate of increase, long-lived and mobile adults, the ability of females to produce viable offspring without repeated mating, the ability of adults to eclose from galls on dry stems and the production of several generations per year. Host range was investigated in single-choice adult tests and in no-choice tests using single pairs of adults, under laboratory conditions. Oviposition and larval development through to adulthood occurred on three other South American and on one South African species in the same tribe as *C. odorata*, but at a lower and slower rate. The ability of *P. basilica* to develop on indigenous species triggers concern; nevertheless, false positive results are common under quarantine conditions and further trials will be conducted. The poor offspring survival on non-target plants tested so far forecasts the suitability of *P. basilica* for release in South Africa. Its ability to develop on the Asian/West African biotype will also be determined.

Keywords *Chromolaena odorata*, *Polymorphomyia basilica*, biology, host range.

Biological control of parthenium weed: field collection and redistribution of established biological control agents

Jason T. Callander and Kunjithapatham Dhileepan
Biosecurity Queensland, Department of Agriculture and Fisheries, Ecosciences Precinct, GPO Box 267, Brisbane, Queensland 4001, Australia
(jason.callander@daf.qld.gov.au)

Summary Parthenium (*Parthenium hysterophorus* L.), a noxious weed of grazing areas in Queensland (Qld), is a Weed of National Significance in Australia. Eleven biological control agents (nine insect species and two rust pathogens) have been released against parthenium in Australia. The majority of these agents have become established and have proven effective against the weed in central Qld.

Parthenium is spreading further into southern Qld, where many of the widespread and effective biological control agents in central Qld are not present. Hence, a program to redistribute these agents from central Qld to the south and south east of the state has been initiated. The seed-feeding weevil (*Smicronyx lutulentus* Dietz), the stem-boring weevil (*Listronotus setosipennis* (Hustache)), and the root-boring moth (*Carmenta* nr. *ithacae* Beutenmüller), the winter rust (*Puccinia abrupta* var. *partheniicola* Dietel & Holway), and the summer rust (*Puccinia xanthii* var. *parthenii-hysterophorae* Seier, Evans & Romero) have been identified as priority agents for redistribution.

So far, field collected winter rust, the *Listronotus* weevil, the *Smicronyx* weevil, the *Carmenta* moth and the leaf-feeding beetle (*Zygogramma bicolorata* Pallister) have been field released at ten sites in south and south east Qld. Field releases of the winter rust from glasshouse cultures were also made at several sites. So far there is evidence of field establishment of the winter rust and the *Smicronyx* weevil in some of the release sites.

Cultures of the summer rust and *Listronotus* weevil have recently been established at the Ecosciences Precinct for future mass rearing and field releases. Redistribution of field collected and glasshouse cultured biological control agents and monitoring their establishment status in the field will continue over the next three years.

Keywords *Parthenium hysterophorus*, biological control, redistribution, Queensland, Australia.

INTRODUCTION

Parthenium hysterophorus L. (Asteraceae), commonly known as parthenium, is a Weed of National Significance in Australia (Dhileepan and McFadyen 2012). The noxious weed predominately occurs in Queensland (Qld) affecting 17 million hectares of grazing land and reducing pasture production by as much as $16.5 million dollars (AUD) annually (Chippendale and Panetta 1994).

Parthenium is an annual herb, which has the ability to germinate and grow year-round, provided it receives adequate rainfall. Like many weeds, parthenium is a prolific seed producer, with a single mature plant capable of generating in excess of 150,000 seeds (Dhileepan 2012).

Biological control of parthenium in Australia was first initiated in 1977. Since then, 11 agents have been released in parthenium infested areas of central and northern Qld including nine insect species, the seed-feeding weevil (*Smicronyx lutulentus* Dietz), the stem-boring weevil (*Listronotus setosipennis* Hustache), the root-boring moth (*Carmenta* nr. *ithacae* Beutenmüller), the leaf-feeding beetle (*Zygogramma bicolorata* Pallister), the sap-feeding plant-hopper (*Stobaera concinna* (Stål)), the leaf-mining moth (*Bucculatrix parthenica* Bradley), two stem-boring moths (*Epiblema strenuana* Walker) and (*Platphalonidia mystica* Razowski & Becker) and the stem-boring weevil (*Conotrachelus albocinereus* Fiedler) and two rust pathogens, the winter rust (*Puccinia abrupta* Diet. & Holw. var. *partheniicola* Jackson Parmelee), and the summer rust (*Puccinia xanthii* var. *parthenii-hysterophorae* Seier, Evans & Romero). Most of these agents have established in central Qld and most have proven effective against the weed there (Dhileepan and McFadyen 2012).

Parthenium is spreading into southern Qld and northern New South Wales, where many of the widespread and effective biological control agents, established in central Qld, have not yet extended. On this basis, the seed-feeding weevil, the stem-boring weevil, and the root-boring moth, the winter rust, and the summer rust have been identified as agents for redistribution into southern Qld.

MATERIALS AND METHODS

Identification of central Queensland collection sites Based on ten years of survey data, sixteen sites, historically shown to harbour parthenium biological control agents, were identified as potential collection sites in central Qld. Sites were visually surveyed for the presence of biological control agents and collections were made in November and December 2015 and in February/March and March/April 2016.

Stem and root boring agents were collected as larvae-infested stem and root material. Affected parthenium plants were uprooted and excess soil around the roots, leaves and flowers were removed and the plants were cut into 20 cm pieces before being placed upon a 2 cm layer of sand inside ventilated plastic containers. Cut plant pieces were covered with moistened paper towel to help keep plant material fresh to sustain the growth of larvae inside. Containers were transferred to the laboratory for emergence of adults.

Adult seed-feeding weevils were collected from maturing parthenium plants. Plants were visually inspected for weevils and occupied flower stems were removed and placed into ventilated cylindrical containers lined with moistened paper towel.

Glasshouse culturing A culture of the winter rust was established at the Ecosciences Precinct (ESP) shade house in autumn/winter of 2015 from rust infected leaf material collected at Helidon Spa. Six medium/large parthenium rosettes were placed into an inoculation chamber (50 × 50 × 50 cm, aluminium frame covered with plastic). Rust infected leaves were suspended, on a wire grid, above the healthy parthenium rosettes. The inside of the inoculation chamber was sprayed with a light mist of water and then sealed for 48 hours. Plants were removed from the chamber and grown at 15–23°C for about ten days, until rust pustules were observed. Rusty leaves were harvested three weeks post inoculation, when rust pustules were ripe, and used to inoculate new parthenium rosettes.

Glasshouse cultures of the stem-boring weevil and the summer rust were recently established at ESP following collections made in March/April 2016.

The stem-boring *Listronotus* adults were collected from emergence containers and transferred to oviposition cages (50 × 90 × 100 cm aluminium frame, covered with gauze) containing six parthenium plants, at pre-flowering stage, for one week. Adult weevils burrow into the soil during daylight hours and thus are hard to recover when exchanging old plants for new ones. To circumvent losing adults in this way, a pot tray was fitted around each potted parthenium plant, lined with gauze and filled with sand. Plants were inspected for oviposition, adults recovered from trays around pots, and exchanged with new plants weekly. Infested plants were kept inside the glasshouse at 27°C for 4–6 weeks before being uprooted, trimmed and placed in emergence containers (as described above). Fresh parthenium leaves were placed on top of the moistened paper towelling as a food source for emerging adults, making it easier to collect them from the emergence containers.

Summer rust was collected from the Burdekin River in northern Qld in May 2016 and sent to ESP for establishment of a glasshouse culture. The summer rust was cultured using the same method as described above for the winter rust, except inoculation and growing of the infested plants was done in a temperature controlled glasshouse at 27°C.

Field release sites in southern Queensland Selection criteria were formulated for the identification of potential biological control release sites in south and southeast Qld. Sites were required to have at least a moderate infestation of parthenium, which was temporally persistent. Sites were not to be subjected to insecticide and herbicide sprays and accessible for the duration of the project, and therefore were in localities where eradication was not possible or where chemical control was not feasible.

RESULTS

Biological control agents in southern Queensland Preliminary surveys in parthenium infested sites in south Qld (e.g. Ceder Vale, Kamorooka, Womillia Creek, Bowood and Amby-Springfield Road) and in south east Qld (e.g. Kilcoy, Junction View, Helidon Spa, Summerset and Biggenden) indicated the presence of only the *Zygogramma* beetle, the *Epiblema* moth and the leaf-mining *Bucculatrix* moth. There was no evidence of other agents including the seed-feeding *Smicronyx* weevil, the stem-boring *Listronotus* weevil, the root-boring *Carmenta* moth and the summer rust. The winter rust, however, was found to be already established at Helidon Spa.

Collection sites in central Queensland Of the sixteen sites visited between November 2015 and April 2016 (Figure 1), only eight sites consistently had abundant biological control agents for collection and redistribution in southern Qld (Table 1). At Lotus Creek, Clermont and Old Orion Road the *Smicronyx* weevil was particularly plentiful; the *Zygogramma* beetle was collected in high numbers from Clermont in November 2015 but was found sparsely in subsequent collections; the *Carmenta* moth was recovered in high numbers from Sandhurst Bridge, Lotus Creek and Mooliyember Creek; the *Listronotus* weevil was the

Table 1. Parthenium biological control agents (SW = *Smicronyx* weevil; EM = *Epiblema* moth; LW = *Listronotus* weevil; CM = *Carmenta* moth; ZB = *Zygogramma* beetle; PM = *Platphalonidia* moth) present at collection sites in central Queensland surveyed in 2015/16.

Collection sites	Times surveyed	Agents present
Gracemere	3	SW, EM
Mount Hay	3	SW, EM
Wycarbah	3	SW, EM
Apis Creek	2	LW, SW
Lotus Creek	4	CM, LW, SW, EM
Carfax yard	1	SW
Clermont	4	CM, SW, EM, ZB
Morebridge	1	SW
Gaylong	3	EM
Gordon Road	3	
Sandhurst Bridge	4	CM, LW, SW, EM
Wyntoon	4	CM, LW, SW, EM
Old Orion Road	4	CM, SW, EM
Rolleston	4	CM, LW, SW, EM, PM
Mooliyember Creek	4	CM, LW, SW, EM, PM
Hutton Creek	3	CM, LW, SW, EM, PM

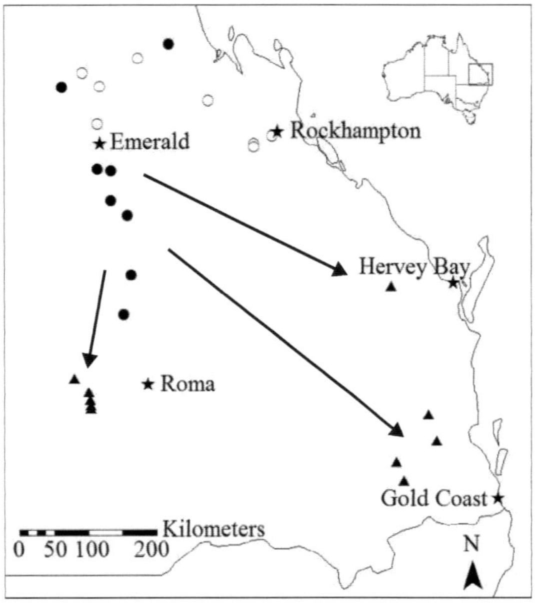

Figure 1. Map of central Queensland collection sites (circles indicate all collection sites, solid circles indicate key collecting sites) and southern Queensland release sites (solid triangles). Stars indicate nearby major cities or townships.

dominant agent collected from Mooliyember Creek; the *Epiblema* moth was present at most sites in high numbers; the *Platphalonidia* moth was recovered from sites south of Emerald and only at the end of the season. No biological control agents were recovered from parthenium at Carfax yard, Morebridge, Gaylong and Gordon Road sites in 2015–16.

Field releases of agents to date In consultation and collaboration with community and regional council groups, ten release sites were identified which met the above criteria (Figure 1).

Approximately 3000 field collected *Smicronyx* weevils were released across five sites and about 400 *Listronotus* and *Carmenta* infested plants (collect from central Qld) were supplied and distributed at three sites in southern Qld between November 2015 and March 2016 (Table 2).

Winter rust was released at eight sites in southern Qld. Forty-five potted, winter rust infected parthenium rosettes were planted in the Mitchel region at Ceder Vale (n = 5), Kamorooka, Womillia Creek, Bowood and along the Amby-Springfield road (n = 10 at each site). In addition, 100 winter rust infected parthenium leaves, collected from Helidon Spa, were scattered directly onto suitable parthenium rosettes at each of the abovementioned sites, as well as sites at Kilcoy, Summerset and Junction View.

Table 2. Parthenium biological control agents released at each site (WR = Winter rust; SW = Smicronyx weevil; LW = Listronotus weevil; CM = Carmenta moth; ZB = Zygogramma beetle) in south and south east Queensland in 2015/16.

Region	Sites	Agents supplied/ released
South Qld	Ceder Vale	WR
	Kamorooka	WR, LW, CM, SW
	Womillia Creek	WR, LW, CM, SW, ZB
	Bowood	WR
	Amby-Springfield Road	WR
South East Qld	Kilcoy	WR, SW, ZB
	Junction View	WR
	Helidon Spa	SW
	Summerset	WR
	Biggenden	LW, CM, SW

To date, the winter rust appears to have established at both Summerset and Ceder Vale sites and the *Smicronyx* weevil has established at both Helidon Spa and Kilcoy sites. Monitoring of the other release sites will continue.

Despite parthenium being able to grow all year round, provided adequate rainfall, the peak growing season for the weed is between spring and autumn. For this reason, neither the *Listronotus* weevil nor the summer rust, cultured at ESP, have been field released during winter. While both these agents have mechanisms that enable them to persist over winter, the colder winter temperatures and sporadic frosts that occur in southern Qld are not conducive to the development and thus establishment of either agent. Cultures will continue through winter and then be ramped up to mass-rearing with releases planned for November 2016.

DISCUSSION

With the growing presence of parthenium weed in southern Qld and the absence of many effective agents, there was a need to expedite the spread of the highly effective agents from central Qld. Field collection and subsequent release of collected agents allows for a rapid and efficient means to get some of these more effective agents from central Qld down into parthenium infested areas of southern Qld. However, field collection alone will not be adequate to meet this challenge due to variability in the availability of parthenium and its biological control agents between central and southern Qld.

For example, due to a very dry spring, collections in November of 2015 were made of a very small range of agents which were present in low numbers. By December this had improved, and by February/March a greater variety of agents were collected and in greater numbers, although mainly between Emerald and Injune. Sites between Rockhampton and Emerald did not receive as much spring rainfall as areas south of Emerald, delaying the growth of the weed and ultimately delaying the buildup of agents in those sites (Table 1). Both the Lotus Creek and the Clermont collection sites situated were on the fringes of damns, and thus had their own source of water, minimizing their reliance on local rainfall.

By February/March many of the parthenium plants in southern Qld had matured (largely due to limited rainfall and dry conditions). Ideally the best time to release the insect agents and the summer rust into southern Qld is early in the growing season (from spring onwards), when parthenium grows vigorously, and these agents have the best conditions to establish. This also allows time for the agents to complete several generations and prepare for winter (i.e. winter diapause). Unfortunately the optimal time to collect insect agents is later in the season once they have had time to build in numbers. The winter rust is, however, adapted to thrive during winter conditions and thus releases of this agent are best suited to the cooler months.

This necessitates the need for the establishment and mass rearing of one or more agents, to supplement the field release program. Although the summer rust was not present at any of the collection sites in central Qld, it is believed that the climatic conditions in southern Qld may be a better match for this agent. Future efforts will focus on the mass rearing and releasing of the summer rust and the *Listronotus* weevil in southern Qld between spring and autumn. During winter, the collection and redistribution of the winter rust will be the focus of work, as this is the only agent active then.

ACKNOWLEDGMENTS

The study was funded by the Rural Research and Development for Profit Program of the Australian Government (Department of Agriculture and Water Resources), the Meat & Livestock Australia and the Queensland Government (Department of Agriculture and Fisheries). We thank Queensland Murray Darling Committee (QMDC), South East Queensland Catchments, Junction View Pest Management Group and North Burnette Regional Council for their continued assistance in identifying potential release sites in southern Queensland. In particular we thank Tom Garrett (QMDC) and Glenn Proctor (North Burnett regional council) for their aid in the releasing biological control agents in their areas. We also thank Rachel McFadyen for her expertise and assistance with field collections and releases and Di Taylor and Boyang Shi for their technical support.

REFERENCES

Chippendale, J.F. and Panetta, F.D. (1994). The cost of parthenium weed to the Qld cattle industry. *Plant Protection Quarterly* 9, 73-6.

Dhileepan, K. (2012). Reproductive variation in naturally occurring populations of the weed *Parthenium hysterophorus* (Asteraceae) in Australia. *Weed Science* 60, 571-6.

Dhileepan, K. and McFadyen, R.C. (2012). *Parthenium hysterophorus* L. – Parthenium. *In* 'Biological control of weeds in Australia: 1960–2010', eds M. Julien, R.E. McFadyen and J. Cullen, pp. 648. (CSIRO Publishing, Melbourne).

Silt, snags and snapping handbags – managing *Cabomba caroliniana* in the tropics of the Northern Territory

Thomas Price and Christopher Collins
Department of Land Resource Management, PO Box 496 Palmerston 0830, Northern Territory, Australia
(thomas.price@nt.gov.au)

Summary Cabomba (*Cabomba caroliniana* Gray), a submerged aquatic plant from the southern United States, is one of the world's most serious aquatic weed species and recognised in Australia as a Weed of National Significance (WoNS). In 2004 cabomba was located in the Darwin River, posing a significant threat to the quality of the Darwin region's drinking water, tourism, recreation and the aquatic ecology of the iconic rivers and billabongs of the Northern Territory. In response, an eradication program was initiated in 2004. This program has reduced the size of the infestation from over 11 km of river down to an area less than 1 km using methods such as drawdown, shading and chemical control with 2,4-D n-butyl ester. The program remains ongoing and continues to challenge with the unique environmental conditions of the Northern Territory. A change of herbicide has been necessitated and interim results of carfentrazone-ethyl treatment indicate that after stagnation of the program eradication is back on track.

Keywords Cabomba, aquatic weeds, WoNS carfentrazone.

INTRODUCTION

Eradication of the WoNS species *Cabomba caroliniana* continues to pose a challenge to the Northern Territory (NT). An infestation in the Darwin River has been subject to an eradication program since 2004. Significant investment since this time has resulted in the reduction of infestation size from 11 km of river to less than 1 km. However this small infestation continues to pose a significant social and economic threat to water quality of the Darwin Region potentially impacting tourism and the ecology of the NT's iconic wetlands. This paper outlines the history of management practices to date, the subsequent challenges faced and the future direction of the program including the use of carfentrazone-ethyl.

HISTORY OF CABOMBA IN DARWIN

Cabomba was first found in Marlow Lagoon, a man made recreational lake near Darwin in 1996. The Marlow Lagoon infestation was successfully eradicated by the Northern Territory Government though combinations of shading, drawdown and eventually in 2002 though the successful application of 2,4-D n-butyl ester (2,4-D) (NRETA 2006). The success at the Marlow Lagoon site supported a multi-stakeholder decision to commence a cabomba eradication program in the Darwin River following the discovery in 2004 (Van Oosterhout 2009).

A large publicity campaign initiated wide community response exposing multiple instances of cabomba as an ornamental plant including an infestation in Pine Creek, 200 km south of Darwin. The Pine Creek infestation resisted control and was found to be producing viable seed (NRETA 2006). Experiments determined fragments of cabomba from Darwin River were also able to produce viable seed. Subsequent trials with sediment from cabomba infested areas revealed two instances of plants germinating from seed (Anon. 2006). Longevity of viability was never established as supplies of seed were depleted before this was concluded (Wingrave 2011).

The upper reaches of Darwin River form the Darwin River Dam, the main water source to Darwin City. The dam regulates flow throughout the year but flooding events and runoff from tributaries ensures variable but sustained flow during the wet season (summer) discharging into Darwin Harbour. During the dry season (winter) the river forms a series of billabongs with minimal flow. The system is intercepted sporadically with subterranean springs which are able to inject substantial amounts of water into the larger billabongs. This prohibits the use of drawdown and effective containment as discussed by Dugdale *et al.* (2013). The Darwin River site remains the only current infestation in the NT and a significant priority for eradication.

2,4-D application Subsurface injection of 2,4-D has been the most successful tool used in the elimination of cabomba in Darwin River (APVMA permit PER11145).

In late 2004 following the discovery of cabomba, 2,4-D in a diatomaceous earth suspension was applied directly into plants though hand-wands and calibrated, dinghy mounted booms.

Initial treatments produced an estimated knockdown of 99.99% of the living cabomba with the stringent environmental protocols at that time revealing no significant effect on the flora or fauna in the Darwin River system. However, by mid wet season 2005 the cabomba was estimated to have returned to 60% of pre-treatment levels (NRETA 2006).

Small contained billabong infestations of cabomba were successfully controlled with combinations of drawdown, shading and application of 2,4-D. Shading was utilized in the larger billabongs but constrained by the uneven depth, snags and the impending but unpredictable arrival of the wet season and associated flow.

Between 2004 and 2007 cabomba was eliminated from the top 8 km of the known area, leaving only the largest, at 2.4 km long, and most downstream billabong, Lok Landji, containing the weed.

Cyclone Carlos, in February 2011, produced a once in 100 year flooding event. The huge initial volume of water and the corresponding sustained flow contributed to a natural flushing of sediment from Lok Landji into Darwin Harbour effectively killing cabomba which has minimal tolerance to salinity. Since 2011, there has been no emergence in once recurrent infestation sites toward the upstream regions of Lok Landji leaving only the downstream half of the billabong with cabomba remaining.

In mid-2011 the Darwin River Dam wall was raised 1.3 metres ensuring future flood mitigation and allowing the dam an additional 20% capacity. There has only been one minor flooding event since the wall was raised.

Between 2011 and 2013 cabomba appeared sporadically throughout the downstream end of Lok Landji billabong with weekly visual 'seek and destroy' patrols aiming to eliminate small isolated plants with spot treatments. The program stagnated with no real mechanism seemingly available to nail the decisive eradication blow.

CHALLENGES FACED

A number of confounding factors contributed to the apparent regression in the cabomba program including herbicide efficacy, physical and environmental factors.

Herbicide constraints In October 2006 the APVMA undertook a review of 2,4-D high-volatile esters and the active constituent was consequently deregistered. All products containing 2,4-D esters were removed from sale across Australia on the grounds they posed an 'unacceptable risk for off target damage' (APVMA 2006).

The NT permit PER11145 remained valid as it was understood risk to crops and off-target plants would be mitigated by subsurface injection in diatomaceous earth suspension (Van Oosterhout 2009). In response all remaining Australian product was purchased for use under the permit (Wingrave 2011).

By late 2013 cabomba was appearing in high concentrations throughout the downstream end of Lok Landji. Cabomba was noted to resist 2,4-D treatment, consequently repeat treatments became commonplace. Successive low rainfall wet seasons and no flooding meant that seasonal flushing did not occur. Likewise water clarity remained higher for longer allowing light to penetrate deeper and treatment periods were extended throughout the year.

During 2014 new areas of infestation began to appear midstream in water >3 m. The reserves of herbicide had diminished and questions regarding the efficacy of the product were raised. Samples of 2,4-D were sent for testing however results indicated minimal deterioration of the active ingredient. Irrespective there remained doubt as to the uniformity of deterioration between drums. Correspondence with manufacturers estimated the product to have a shelf life of five to six years. In October 2015 the program used the last remaining 2,4-D ester in supply.

Subsurface flowering Weekly surveillance and control was deemed necessary to prevent the plants from producing flowers, negating the essential step to them forming seed. Whilst surface flowers were preventable occasionally, cabomba was observed producing subsurface flowers. Irregularly, dislodged floating strands were also witnessed to produce flowers. This posed questions regarding the possibility that flowers could be pollinating under water. Again, samples of silt from regular infestation sites were sent, in 2015, to an aquatic plant specialist to grow out in a carefully controlled quarantine environment. To date no cabomba has been found to germinate.

Crocodiles Surveying the depths of the billabong for cabomba also presents a challenge. *Crocodylus porosus* pose a real threat. Four estuarine crocodiles in the vicinity of 3.8 to 4.4 metres long have been trapped within freshwater Darwin River between 2010 and 2014 (Nichols 2015). Although measures are taken to mitigate encounters the hazard restricts entry into the water and hampers the ability of personnel to use aquatic weed control methods such as diving and manual removal.

A device cast into the water consisting of opposing rake heads on an anchor rope revealed substantial infestations prospering beyond the field of vision at between four to six metres. This suggested the spotting and spraying method from the dinghy was ineffective for complete control. The downside of benthic sampling 'rake' was that it caused substantial cabomba fragmentation with the potential to make the distribution of cabomba propagules significantly worse.

FUTURE DIRECTION

The stagnation of the eradication program necessitated a review of methodology. Suitable herbicides were researched as well as alternative methods such as dredging, raising the salinity or pH (Bickel 2012).

Dr Tony Dugdale was engaged to review the program in October 2015. Recommendations were that Shark™ (240 g L^{-1} carfentrazone-ethyl) be used as either spot spray application or as a half billabong treatment. Shark was registered in Australia for treatment of cabomba in 2011 and had shown great promise reducing cabomba in ponded waterbodies such as those to protect the Ramsar listed Myall Lakes (Inkson *et al.* 2014) and Glenbrook Lagoon in the Blue Mountains (Day *et al.* 2014). Being a contact herbicide rather than systemic the most reliable results occur with sustained exposure concentrations of 2 ppm (Day *et al.* 2014). The spring fed Lok Landji maintains a small amount of flow posing an issue with exposure time and off-target movement. Additionally it meant the label specifications of Shark, restricted to contained, non-flowing waterbodies, would not apply. The APVMA granted an off-label permit for spot application of cabomba in Lok Landji in November 2015- PER81710 (DLRM 2016).

In late 2015 flumioxazin also gained registration for use on cabomba in Australia. Similarly a permit to use flumioxazin was granted in December 2015, PER81721.

LOK LANDJI CARFENTRAZONE TREATMENT

Spot treatment of Shark began in December 2015 and has shown promise reducing and degrading cabomba. Shark has shown a long lag between application and degradation, taking over three weeks before significant weakening of stems, loss of colour and vigour became apparent.

Fragmentation of weakening stems has been prolific. Attempts are being made to collect the pieces however the ability of these fragments to effectively form new plants remains unknown.

Environmental monitoring A new regime of environmental monitoring protocols were required with the use of new herbicides in flowing waterbody. Collaboration between NT Government departments developed a regime to assess exposure against concentration over time, and to measure water quality pre- and post-treatment to ascertain any changes in chemistry resulting from herbicide application. At this early stage, there has been no trace of carfentrazone when sampled after 48 hours post- treatment (DLRM 2016).

Pre- and post-treatment sampling of pH, conductivity and dissolved oxygen revealed no change after Shark spot treatment. Differing from the 2004 2,4-D treatments where a significant drop in the dissolved oxygen had to be rectified with the use of aerators (Wingrave 2010).

The future A permit for half billabong treatment with Shark is currently being considered in conjunction with development of associated monitoring protocols (DLRM 2016). The aim is to raise the concentration across the bottom half of the billabong to 2 ppm as done in the Glenbrook Lakes (Day *et al.* 2014). By treating half the waterbody the need to determine the exact location of the cabomba will also become redundant. To maximise exposure time, treatments are to commence when the flow of water into and from the billabong is minimal. It is anticipated, as with ponded waterbodies, there will be a required three month interval between carfentrazone treatments. Flumioxazin will be used as a knockdown spot treatment should cabomba emerge between applications of Shark.

While program successes have been substantial, the prospect of eradication has at times seemed distant. Positive results with carfentrazone reinvigorate the expectation that eradication of cabomba from the NT may again become a reality in the near future.

REFERENCES

Anon. (2006). Cabomba seed production and research in Darwin (November 2004–2006). Northern Territory Government Internal Document, Darwin.

APVMA (2006). Preliminary review findings (environment). Part 1: 2,4-D esters. Volume 1: Review summary. (Australian Pesticides and Veterinary Medicines Authority, Canberra).

Bickel, T.O. (2012). Ecology of the submersed aquatic weed *Cabomba caroliniana* in Australia. Proceedings of the 18th Australasian Weeds Conference, ed. V. Eldershaw, pp. 21-4. (Weed Society of Victoria, Melbourne).

Day, C., Petroeschevsky, A., Pellow, B., Bevan, J., O'Dwyer, T., St Lawrence, A. and Smith, G. (2014). Managing a priority outlier infestation of *Cabomba caroliniana* in a natural wetland in the Blue Mountains: could this be eradication? Proceedings of the 19th Australasian Weeds Conference, ed. M. Baker, pp. 147-52. (Tasmanian Weed Society, Hobart).

DLRM (2016). Darwin River Cabomba eradication program interim report 2016–2016. (Department of Land Resource Management, Northern Territory Government, Darwin).

Dugdale, T.M., Butler, Clements, K.L. and Hunt, T.D. (2013). Survival of cabomba (*Cabomba caroliniana*) during lake drawdown within mounds of stranded vegetation. *Lake and Reservoir Management*, 61-7.

Inkson, T., Petroeschevsky, A., Officer, D. and Schmitzer, T. (2014). Cabomba control for the protection of the Ramsar listed Myall Lakes. Proceedings of the 19th Australasian Weeds Conference, ed. M. Baker, pp. 328-32. (Tasmanian Weed Society, Hobart).

Nichols, T. (2015). Personal communication. (T. Price, interviewer).

NRETA (2006). Cabomba eradication program report for 2004/05. (Northern Territory Government Department of Natural Resources Environment and the Arts, Darwin).

Van Oosterhout, E. (2009). Cabomba control manual. (NSW Department of Primary Industries, Orange, NSW).

Wingrave, S. (2010). Final Cabomba Report. (Northern Territory Government Department of Natural Resources Environment The Arts and Sport, Darwin).

Wingrave, S. (2011). Personal communication. (T. Price, Interviewer).

The need for innovation in aquatic weed management

Marcus R. Girvan
Boffa Miskell Ltd on behalf of Land Information New Zealand, 4 Hazeldean Road,
PO Box 110, Christchurch 8140, New Zealand
(marcusg@boffamiskell.co.nz)

Summary Land Information New Zealand (LINZ) administers approximately 8% of New Zealand, including the beds of many of its iconic lakes. Aquatic weeds have plagued a number of these lakes and waterways for over 60 years, with many control programmes having never achieved any level of enduring control. Successful aquatic weed management requires new, effective tools and innovative ways of utilising existing tools.

With a limited choice of effective herbicides, and high costs associated with mechanical control methods, LINZ continues to explore new ways of managing aquatic weeds in a variety of challenging environments, from high public use lakes to those experiencing high sediment inflows. Hornwort (*Ceratophyllum demersum* L.) continues to be one of New Zealand's worst aquatic pests, reaching the peak of its biomass in the height of summer, choking waterways which impact on recreational users, aesthetic values, hydro-generation and biodiversity.

With hornwort not present in the South Island, *Lagarosiphon major* thrives in the cooler, low-nutrient waters, with similar impacts to that of hornwort. With ever decreasing public tolerance to the use of herbicides to control and limit the spread of aquatic weeds, alternative control options are sought to manage these weeds more effectively, more efficiently and for longer.

Keywords Aquatic weeds, lagarosiphon, hornwort.

Cyperus rotundus L.: an ancient food staple but now designated the world's worst weed

John Dwyer
47 Miller Street, Fitzroy North, Victoria 3068, Australia
(thedwyers@ozemail.com.au)

Summary The cosmopolitan perennial sedge, *Cyperus rotundus* L. (nutgrass), has been associated with humans as a food staple for millennia, as is demonstrated by recent research based on the extraction of dental calculus from pre-historic skeletons in Central Sudan. The dental calculus paper asserted that *C. rotundus* tubers formed part of the diet of Aboriginal people in Australia. But accounts have varied as to the tubers used by Aboriginal people, and there is no conclusive evidence that *C. rotundus* was so used. Opinions have also differed as to whether *C. rotundus* is an introduced species in Australia. *C. rotundus* was designated the World's Worst Weed, but it is not the worst weed in Australia. Although a weed of cultivation and gardens throughout the mainland, associated with disturbed sites, *C. rotundus* does not seem to be an invasive species. The plant's long association with humans is likely to have played a part in extending its range.

Keywords World's worst weed, prehistoric food staple, archaeological findings, Aboriginal use, weed status in Australia.

INTRODUCTION

Some plants long associated with humans and now often thought of as mere weeds have had extensive periods in ages past when they were used for food and medicine. *C. rotundus* is an outstanding example. This persistent perennial plant, having been in association with humans for thousands of years, has in modern times been designated the worst weed in the world, as a weed of 52 crops in 92 countries (Holm *et al.* 1977). But there is much more to *C. rotundus* than that.

Cyperus rotundus is not a grass but a sedge. Both the common and the botanical names come from the nut-like underground tubers: *Cyperus* from *kuperos* or *Cypeiros*, a Greek name for some sedges, and *rotundus*, Latin for the round tubers. There are about 600 species of *Cyperus* world wide (Walsh and Entwisle 1994), found mainly in the tropics and sub-tropics, with very few species in temperate regions. That said, *C. rotundus* is found in New Zealand from Auckland to the far north, usually as a weed of gardens (Calder 1952).

Sedges are identified by their triangular stems, the dark green, sheathed leaves, and the characteristic flowers in very narrow, sharply pointed spikelets at the end of the rays. *C. rotundus* and the related *Cyperus esculentus* L. (yellow nutgrass), may grow together in mixed populations, and can be difficult to distinguish before they flower (Auld and Medd 1997). The inflorescence of *C. rotundus* is a loose umbel (flowers in a head with stalks arising at the same point) of brown to purple flattened persistent spikelets, while that of *C. esculentus* is a yellowish green. *C. rotundus* has an extensive underground system of roots, rhizomes, tubers and basal bulbs, making it very hard to eradicate (Maiden 1920, Whittet 1962).

This paper considers the role of nutgrass in prehistoric times, and whether it was used as food by Aboriginal people in Australia. Opinions have differed as to whether it is indigenous. Its present weed status is reviewed.

DISCUSSION

The medicinal uses of *C. rotundus* *Cyperus rotundus* was used by humans as medicine in ancient times. Both Dioscorides (c40–90 AD) and Pliny (23–70 AD) wrote of its medicinal properties and its use to treat a range of ailments (Gunther 1934, Holland 1601).

A recent Indian study of the benefits of 'Nagarmotha' (*C. rotundus*) lists numerous current pharmacological uses, whether diuretic, carminative (anti-flatulence), emmenagogue (promoting menstruation), anthelmintic (killing parasitic worms), stimulant, analgesic, hypotensive, anti-inflammatory, anti-dysenteric, or anti-rheumatic. On top of all this, the tuber's essential oil is used in perfumery, soap making, and insect repellents. As an insecticide, *C. rotundus* was shown to be more effective than organophosphate and carbamate, and the essential oil is effective against a range of bacteria and fungi (Imam *et al.* 2014). Long used in traditional medicines in India, modern alternative medicine recommends using the plant to treat fever and inflammation, reduce pain, and relax muscles, and other uses (van Wyk and Wink 2005).

***Cyperus rotundus* as food** *Cyperus rotundus* has been in association with humans from remote pre-history to the present: as a food for thousands of years in prehistoric times, and as a troublesome weed in modern times. Abundant remains of *C. rotundus* tuber, thought to have been collected as food, were found at the 18,000 year old site of Wadi Kubbaniya, near Aswan, Egypt (Hardy *et al.* 2016). At later sites at Al Khiday, 25 kilometres south of Omdurman (the ancient city opposite Khartoum) on the White Nile in Central Sudan, use of the tubers as food has been demonstrated by recent research based on the extraction of chemical compounds and microfossils from dental calculus derived from prehistoric skeletons, that has shed light on plants used as food in ancient Nubia. The complex of burial sites has yielded dental calculus samples from pre-Mesolithic, Neolithic, Late Meroitic and Mesolithic ages, covering a time span of more than 7000 years (Buckley *et al.* 2014). The periods stretched from the pre-agricultural fisher-hunter-gatherer based economy, through the early Neolithic with its incipient agriculture, through the sedentary culture known as the Kartoum Mesolithic that existed approximately 8000 years ago (Haynes 1996), and on to the fully developed agricultural context of the Meroitic, the Nubian civilization centred on the ancient town of Meroe that flourished for 1000 years from about 750 BC to about AD 350 (Shinnie 1996). *C. rotundus* material was found on the teeth of people from all of these periods, and was probably a carbohydrate staple for millennia, as well as being used for perfume and medicine (Buckley *et al.* 2014).

Kloot (1979) suggests that nutgrass was cultivated in South Australia in the late 19th century. He describes the cultivation of yellow nutgrass for its sweet edible tubers called, 'groundnuts', or 'chuffas' that were roasted and eaten as snack food much as peanuts of potato chips are used today. Demand was so great that there was a shortage of bulbs with action taken by the Central Bureau of Agriculture to distribute seed. Kloot suggests that 'due to honest misidentification and a little bit of fraud most of the so-called "*C. esculentus*" that was being circulated was probably *C. rotundus*', leading, he thinks to the sudden end of the fad. He concluded, 'regardless of what was being planted as *C. esculentus*, only *C rotundus* has survived.' (Kloot 1979)

Was *C. rotundus* the 'Yalka' of Aboriginal people? The scientists who examined the dental calculus of the ancient Sudanese skeletons also argued that recent hunter-gatherers in tropical regions ate *C. rotundus* tubers as a carbohydrate staple, and that in some agrarian societies the tubers served as a famine food. They pointed out that the nutritional value is enhanced by the presence of lysine, an a-amino acid essential for humans as a building block for protein (Buckley *et al* 2014). They also said that despite its bitter taste *C. rotundus* was one of three tuber staples among Aboriginal populations in Central Australia, citing Latz (1995).

The plant discussed by Latz (1995) was, however, not *C. rotundus* but another species, *Cyperus bulbosus* Vahl. (yalka or onion grass); although Latz (1995) did contain a passing reference to *C. rotundus* as a synonym for *C. bulbosus*. Kloot (1979) identified *C. bulbosus* as native to South Australia. Other sources such as Clarke (2011) confirm that *yalka* was *C. bulbosus*. Clarke does say in a footnote that *yalka* was formerly often identified as *C. rotundus*. For example, Basedow (1925) described *C. rotundus* as one of the commonest vegetables, eaten in very large quantities in the central and west-central regions, called *yelka* (sic) by the Arunndta (sic) people. His observation that the bulbs have a sweetish, nutty flavour confirms the error. Maiden (1920) wrote of *C. rotundus* that 'aborigines roasted and ate the tubers in many parts of the continent, and continue to do so in South Australia.' From Low (1997) we have another Aboriginal name *Nalgoo* for *C. bulbosus,* and confirmation that it was previously misidentified as *C. rotundus*. To add to the uncertainty, Auld and Medd (1997) wrote that tubers of *C. esculentus* were eaten raw or roasted by Aboriginal people. They made no mention of *C. bulbosus*. Yet another *Cyperus* species, *C. victoriensis* C.B.Clark, found in all mainland States of Australia, was a food plant eaten by Central Australian Aborigines according to Gott (1993), but she did not give the name used by them. The suggestion that nutgrass was used as food by Aboriginal people in Australia has not been established.

Is *C. rotundus* indigenous to Australia? Botanists have differed in their opinions on whether *C. rotundus* is native or introduced in various parts of Australia. Some authorities claim it to be indigenous only to India, while others suggest the origin to be more widespread, including northern and eastern Australia.

Cyperus rotundus might be one of a number of tropical or subtropical plants indigenous in northern Australia but introduced in the south; or an introduction from other parts of the world. One reason why *C. rotundus* has been seen as probably indigenous to Australia is the fact that it has been a troublesome weed of cultivation since the days of the first European settlers (Lamp and Colett 1984). Michael (1972) suggested that *C. rotundus* had been introduced at least into southern Australia, and referred to the 1858 Report of the Sydney Botanic Gardens as the earliest record known to him of *C. rotundus* as a weed problem.

Michael seems to have regarded *C. rotundus* as one of a number of subtropical or tropical plants probably indigenous in northern Australia but introduced in the south; and left open introduction of *C. rotundus* from other parts of the world.

A paper by the clerical botanist Woolls (1867) 'Plants introduced accidentally' discusses *C. rotundus,* under the name *C. hydra* of the West Indies, as an introduced troublesome weed. Maiden (1920) accepted that *C. hydra* was a synonym for *C. rotundus.* The name 'hydra' invoked the nine-headed monster encountered by Hercules: as soon as he struck off one of its heads, two shot up in its place. One American study reported that one tuber produced 600 plants and spread 10 feet in one year (Ridley 1930).

In 1923 the plant was proclaimed a noxious weed in Victoria, having previously been declared a Thistle in 1890 under the then artificial system for proscribing noxious weeds by designating them as Thistles under the *Thistle Act* (Parsons 1981). Its noxious weed status was continued in a proclamation of 1974 but following the introduction in 1994 of a new system for dealing with noxious weeds under the *Catchment and Land Protection Act* it was not included in any category (Anon. 1998).

Ewart (1930) was convinced the plant was native to Victoria: 'Speaking from an experience of 25 years, the proclamation [as a noxious weed] of this native plant has served no useful purpose.' Currie (1945) wrote that the plant was 'indigenous to Australia in common with other parts of the world, [and] a pest in all States, but causes most harm in Queensland.' Willis (1973), was more cautious, describing the plant as 'probably indigenous in northern Victoria but now widespread with settlement… all States except Tasmania, cosmopolitan in most warm countries.'

The 25 species of *Cyperus* recorded in South Australia were carefully considered by Kloot (1979). Kloot sought to distinguish between native and introduced species, but suggested that the status of *C. rotundus* seemed to be a matter of opinion. It might be said that the differences of opinion as to whether *C. rotundus* was a native demonstrated the fragility of the category, which has come under sustained criticism in recent years (Chew and Hamilton 2001).

Kloot (1979) noted that the first recorded introduction of *C. rotundus* to South Australia was as a garden plant, of unspecified 'Australian' origin, into the Botanic Gardens in 1884. He concluded that *C. rotundus* was one of several species probably native to tropical eastern Australia, but introduced to South Australia. This theory would not be accepted by some botanists today: the consensus seems to be that while some species of *Cyperus* may be indigenous to Australia, *C. rotundus* is an introduced species. Walsh and Entwisle (1994) regard *C. rotundus* as an introduced plant naturalized in all mainland States, as do Randall (2002) and Richardson *et al.* (2011). The plant may be indigenous in some parts of Australia, but has been introduced to many places where it is now a weed of cultivation, gardens and disturbed sites.

How serious a weed is *C. rotundus* in Australia?
Cyperus rotundus seems to be a weed of cultivation rather than of bushland, 'an unmitigated nuisance in gardens and arable land' as Maiden (1920) put it. Lazarides *et al.* (1997) describe it as a troublesome weed of cultivation and gardens throughout mainland Australia. Parsons (1981), however, was of the opinion that *C. rotundus* was not of great importance in Victoria, occurring in roadsides, parklands, channel and drain banks, and occasionally in orchards, vineyards and market gardens. Although it has been a proclaimed noxious weed in Victoria as noted earlier, and in South Australia, it is a serious problem only in Queensland and northern New South Wales.

Cyperus rotundus seems not to be an invasive environmental weed. Weber (2003) listed the related *Cyperus eragrostis* Lam. as an invasive species, but not *C. rotundus*. It was not included in Australian publications on environmental weeds such as Blood (2001), Muyt (2001) or Carr, *et al.* (1992). Despite its importance as a weed in India, Asia, the Pacific Islands, Africa, South America, North America and the Middle East, *C. rotundus* is not the worst weed in Australia.

The plant may well have been assisted in extending its range by its association with humans as a source of food and medicine for millennia in ancient times. When humans were using *C. rotundus* for their own purposes it is most unlikely that they called it a weed.

REFERENCES

Anon. (1998). Report on weeds in Victoria. Government Printer, Melbourne.

Auld, B. and Medd, R. (1997). 'Weeds: an illustrated guide to the weeds of Australia'. (Inkata Press, Melbourne).

Basedow, H. (2012). 'The Australian Aboriginal'. 2nd, ed. David Welch, Virginia, Northern Territory.

Blood, K. (2001). 'Environmental weeds: a guide for SE Australia'. (Jerram and Associates, Mt Waverley).

Buckley, S., Usai, D., Jakob, T., Radini, A. and Hardy, K. (2014). Dental calculus reveals unique insights into food items, cooking, and plant processing in prehistoric Central Sudan. *PLOS ONE* 7, 1-10.

Calder, J. (ed.) (1952). Weeds of New Zealand. (Whitcombe and Tombs, Christchurch).

Carr, G., Yugovic, J. and Robinson, R (1992). 'Environmental weed invasions in Victoria'. Department of Conservation and Environment and Ecological Horticulture, Melbourne.

Chew, M. and Hamilton, A. (2011). The rise and fall of biotic nativeness. *In* 'Fifty years of invasion ecology: the legacy of Charles Elton'. (Wiley-Blackwell, Chichester).

Clarke, P. (2011). 'Aboriginal people and their plants'. 2nd ed. (Rosenberg, Sydney).

Currie, G. (1935). 'Survey of weed problems in Australia'. (CSIRO, Melbourne).

Ewart, A. (1930). 'Flora of Victoria'. (Government Printer, Melbourne).

Gott, B. (1993) Use of Victorian plants by Koories, *In* Flora of Victoria, Vol. 1, eds D. Foreman and N. Walsh. (Inkata Press, Melbourne).

Gunther, R. (ed.) (1934). 'The Greek herbal of Dioscorides'. (Oxford University Press, Oxford).

Hardy, K. and Kubiak-Martens, L. (eds) (2016). 'Wild harvest: plants in the hominin and pre-agrarian worlds'. (Oxbow Books, Oxford).

Haynes, J. (1996). Nubia. *In* 'The Oxford companion to archaeology', ed. B. Fagan. (Oxford University Press, Oxford).

Holland, P. (trans.) (1601). The historie of the world, commonly called the natural historie of C. Plinius Secundus. (Flip, London).

Holm, L., Plucknett, D., Pancho, J and Herberger, J. (1977). 'The worlds worst weeds'. (University Press of Hawaii, Honolulu).

Imam, H., Zarnigar, S., Seikh, G., and Lone, A. (2014). The incredible benefits of Nagarmotha (*Cyperus rotundus*). *International Journal of Nutrition Pharmacology, Neurological Diseases* 4, 23-7.

Kloot, P. (1979). The native and naturalized *Cyperus* species in South Australia. *Journal of the Adelaide Botanical Gardens* 1, 333-41.

Lamp, C and Collet, F. (1984). 'A field guide to weeds in Australia'. (Inkata Press, Melbourne).

Latz, P. (1995). 'Bushfires and bush tucker'. (IAD Press, Alice Springs).

Lazarides, M., Cowley, K. and Hohnen, P. (1997). 'CSIRO handbook of Australian weeds'. (CSIRO Publishing, Collingwood).

Low, T. (1997). 'Wild food plants of Australia'. 2nd ed. (Angus and Robertson, Sydney).

Maiden, J. (1920). 'The weeds of New South Wales Part 1'. (Government Printer, Sydney).

Michael, P. (1972). The weeds themselves – early history and identification. *Proceedings of the Weed Society of New South Wales* 5, 3-18.

Muyt, A. (2001). 'Bush invaders of South-East Australia'. (R.G. and F.J. Richardson, Meredith).

Parsons, W. (1981). 'Noxious weeds of Victoria'. (Inkata Press, Melbourne).

Richardson, F.J., Richardson, R.G. and Shepherd, R.C.H. (2011). 'Weeds of the South East', 2nd ed. (R.G. and F.J. Richardson, Meredith).

Ridley, H. (1930). 'The dispersal of plants throughout the World'. Reeve, Ashford.

Shinnie, P (1996). Meroe. *In* The Oxford companion to archaeology, ed. B. Fagan. (Oxford University Press, Oxford).

Van Wyk, B. and Wink, M. (2005). 'Medicinal plants of the World'. (Briza, Arcadia).

Walsh, N. and Entwisle, T. (1994). 'Flora of Victoria', Vol. 2. (Inkata Press, Melbourne).

Weber, E. (2003). 'Invasive plant species of the World'. (CABI Publishing, Cambridge).

Whittet, J. (1962). 'Weeds'. (Government Printer, Sydney).

Willis, J. (1973). 'A handbook of plants in Victoria Vol.1'. 2nd ed. (MUP, Carlton).

Woolls, W. (1867). Plants introduced accidentally. *In* 'A contribution to the Flora of Australia'. (White, Sydney).

Using 'WeedSearch' to assess the feasibility of eradicating 34 high-risk invasive plant species in Queensland

Steve Csurhes
Biosecurity Queensland, Queensland Department of Agriculture and Fisheries, GPO Box 267, Brisbane, Queensland 4001, Australia
(steve.csurhes@daf.qld.gov.au)

Summary The computer model 'WeedSearch' was used to estimate the cost of eradicating 34 high-risk plant species in Queensland. Eradication of six species is predicted to cost less than $100,000 per species. For 14 species, cost is predicted to be between $100,000 and $1M per species. For the remaining 14 species, eradication is predicted to cost more than $1M per species. The analysis allows species to be prioritized so that finite resources can be directed to species most vulnerable to eradication. Moreover, it identifies species for which eradication may no longer be a realistic management objective.

Keywords Invasive plant, eradication.

INTRODUCTION

The complete eradication of potentially high-impact invasive plant species is an appealing but often elusive management objective. Given finite resources and an ever-expanding list of candidate species, it is necessary to prioritise eradication targets and to ensure policy objectives remain practical and realistic. This paper presents a useful method for ranking candidates for eradication – a method that also provides a quick guide to the total quantum of funding and search-effort likely to be required for recently detected species.

MATERIALS AND METHODS

The study focused on 34 plant species considered to have significant long-term pest potential. All species are declared 'Class 1' pests in Queensland (as defined by the Queensland *Land Protection (Pest and Stock Route Management) Act 2002* and associated regulation). An additional eight species targeted for eradication under national cost-sharing arrangements were outside the scope of the study, as were six species believed to have been eradicated from Queensland and two species that have not yet been effectively delimited.

A computer model called 'WeedSearch' (Cacho and Pheloung 2007) was used to estimate cost and duration of eradication for 34 species. The model required 24 input parameters to be completed for each species. Parameters include an estimate of total search area, average plant density (per hectare), seed longevity, search speed, search frequency, population growth rate and administration costs (among others). Estimates of cost and search-effort are based on a premise that a defined area must be methodically and regularly searched to detect every specimen and to exhaust the soil seed-bank. The model estimates the number of years required to achieve eradication and the number of search-hours that must be invested to satisfy the mathematical calculations applied.

Operational biosecurity staff across Queensland were surveyed to collect input data. Distribution data was also gleaned from 'Pest Central', a database of information on the distribution of pests across the state. Data on various biological parameters required by the model were obtained from the literature. In cases where data did not exist, a reasonable estimate was made, often based on congeners or closely related taxa.

Search area and average plant density per hectare were particularly difficult parameters to estimate. For each site where a target had been detected, an arbitrary search area with a 200 metre radius was calculated (yielding a total search area of about 12 hectares for single specimens or small clumps of specimens). For species where populations were scattered across a larger area, GIS-based mapping was utilized to calculate total search areas (again using a 200 metre 'search buffer' around specimens). The 'search buffer' assumes propagules could be reasonably expected to disperse over this distance. Where data exists to indicate a larger 'search buffer', these data were used in the model.

RESULTS

Of the 34 species, eradication of six species is predicted to cost less than $100,000 per species. For 14 species, with larger populations, cost is predicted to be between $100,000 and $1M per species. For the remaining 14 species, eradication is predicted to cost more than $1M per species (Table 1).

DISCUSSION

Experience in Queensland and elsewhere clearly shows that the cost of eradication increases dramatically as the spatial size of a target population increases. Analysis of successful weed eradication projects from around the world by Waldendorp and Bomford (2004) estimated an average cost of $4270 for eradicating an infestation with a net area less than 0.1 ha, $19,700 for an area of 1 ha and $1,052,500 for an area of 400 ha.

Table 1. Cost of eradicating 34 weed species, as estimated by WeedSearch.

Species	Estimated search area (ha)	Estimated total cost ($)
Opuntia elatior (red-flower prickly pear)	12	57,200
Opuntia puberula (puberula cactus)	12	57,200
Salix cinerea (grey willow)	12	59,900
Salix alba (white willow)	12	59,900
Vachellia gerrardi (grey-haired acacia)	12	79,500
Senegalia nigrescens (knob-thorn)	12	79,500
Vachellia xanthophloea (yellow fever tree)	25	159,000
Chrysanthemoides monilifera subsp. *rotundata* (bitou bush)	460	163,500
Salix nigra (black willow)	12	179,700
Ulex europaeus (gorse)	25	190,400
Prosopis laevigata (smooth mesquite)	50	258,800
Opuntia leucotricha (Aaron's beard cactus)	62	272,500
Vachellia karroo (Karroo thorn)	50	318,000
Opuntia elata (Riverina pear)	600	471,000
Neptunia oleracea/plena (water mimosa)	22	695,200
Opuntia sulphurea (sulphur cactus)	2,012	766,700
Cylindropuntia rosea/tunicata (Hudson pear)	842	811,700
Solanum viarum (tropical soda apple)	867	867,000
Gmelina elliptica (badhara bush)	871	709,800
Senegalia insuavis (pennata wattle)	125	879,000
Senegalia rugata (soap-pod wattle)	137	1,072,500
Hygrophila costata (hygrophila)	317	1,132,800
Gymnocoronis spilanthoides (Senegal tea plant)	172	1,461,900
Acaciella glauca (redwood)	192	1,496,900
Opuntia microdasys (bunny ears cactus)	400	1,552,000
Cecropia peltata/palmate (Mexican bean tree)	800	1,612,800
Mimosa pigra (giant sensitive tree)	1,000	1,725,800
Alternanthera philoxeroides (alligator weed)	91	1,993,500
Pithecellobium dulce (madras thorn)	425	2,033,200
Acaciella angustissima (white-ball acacia)	375	2,637,000
Asparagus asparagoides (bridal creeper)	1,500	5,586,400
Nassella neesiana (Chilean needle grass)	14,000	14,780,000
Gleditsia triacanthos (honey locust tree)	62,600	22,729,500
Chromolaena odorata (Siam weed)	14,776	47,812,500

Similarly, based on an analysis of eradication programs undertaken by the California Department of Food and Agriculture, comprising 16 species and 50 infestations, Rejmánek and Pitcairn (2002) concluded that for infestations less than 1 ha in size, eradication was almost always possible; for infestations between 1 and 100 ha, approximately 30% were successfully eradicated; and for infestations between 100 and 1000 ha, only about 25% of eradication attempts were successful. Woldendorp and Bomford (2004) found that seven out of eight successful eradication programs involved species that had net infestation sizes less than 4 ha.

Estimated cost of eradication for 34 weed species in Queensland increases dramatically as search area increases, with 14 species estimated to require more than $1M per species. The high costs highlight the importance of detecting potentially serious weed species at a very early stage of population development, preferably when the population affects no more than a few hectares.

WeedSearch not only offers a method for ranking candidates for eradication, it also provides a guide to the total quantum of funding and search-effort likely to be required. However, the quality of the predictions is dependent on the quality of the input parameters, especially distribution data for each species. While the Department has invested heavily in hand-held data capture devices over many years of field operations, certain species are difficult and costly to delimit, especially once they have spread over more than a few hectares. While there is scope to collect more precise data in the future, as surveillance techniques evolve, delimitation will probably remain a challenge. The level of confidence associated with delimitation undoubtedly declines as the area infested increases.

It is worth noting that the predicted cost of eradication is heavily influenced by a species' detectability. For example, WeedSearch predicts that eradication of alligator weed will cost $1,993,500, despite a relatively small search area of 91 hectares. This is due to the fact that alligator weed grows within very dense aquatic vegetation at a number of sites in Queensland and the model takes into account the 'effective search width' – the distance (in metres) that an observer can visually detect the target species. In the case of alligator weed, the 'effective search width' is small. The more readily a target species can be detected growing among other vegetation, the lower the number of search hours required and the lower the cost. For example, the cost of eradication for visually conspicuous species such as *Opuntia sulphurea* is relatively modest ($766,700) despite a sizeable search area (2012 hectares). Seed longevity and other factors also interact in complex ways to influence eradication costs.

Despite a level of uncertainty associated with the model's predictions, the study offers evidence-based recommendations for future management objectives for 34 eradication targets. For example, eradication can continue to be pursued if cost is predicted to be affordable, whereas more realistic policy objectives, such as containment or biological control, can be adopted for high-cost or particularly cryptic or otherwise resilient species. In addition, results from this study can be compared to a similar study completed in 2009 (Panetta and Csurhes 2011), so that managers can see whether progress is being made towards eradication and at what rate.

With on-going refinement, there seems little doubt that the model will help ensure finite resources are directed at species most vulnerable to complete eradication. While not as sophisticated as the methodology applied to nationally cost-shared eradication projects, this study offers a relatively low cost, rapid, transparent and repeatable method to assess eradication feasibility for fairly large numbers of species – a method that is transferable to other state and territory jurisdictions.

ACKNOWLEDGMENTS

Dr Sheldon Navie from IVM Group was contracted to help collect data and analyse results. Operational staff from Biosecurity Queensland provided considerable information and assistance. Taxonomic advice and location data was provided by the Queensland Herbarium. Detection and eradication of high-risk invasive species is undertaken in close partnership with local governments across Queensland.

REFERENCES

Cacho, O. and Pheloung, P. (2007). WeedSearch: Weed Eradication Feasibility Analysis, Software Manual, CRC for Australian Weed Management Project 1.2.8, August 2007, University of New England, New South Wales.

Panetta, F.D. Csurhes, S.M., Markula, A. and Hannan-Jones, M. (2009). Predicting the cost of eradication for 41 Class 1 declared weeds in Queensland, *Plant Protection Quarterly* 26, 42-6.

Rejmánek, M. and Pitcairn, M.J. (2002). When is eradication of exotic pest plants a realistic goal? *In* 'Turning the tide: the eradication of island invasives', eds C.R. Vietch and M.N. Clout, Occasional Paper of the IUCN Species Survival Commission No. 27, pp. 249-53. (IUCN – The World Conservation Union, Auckland, New Zealand).

Waldendorp, G. and Bomford, M. (2004). Weed Eradication: strategies, timeframes and costs (Bureau of Rural Sciences, Australian Government).

Recreating the invasion of exotic conifers in New Zealand

Clayson J. Howell
Department of Conservation, PO Box 10-421 Wellington 6123, New Zealand
(chowell@doc.govt.nz)

Summary In little over 100 years, the extent of exotic conifers in New Zealand has progressed from isolated seedlings to occupy about 15% of the land area. Currently about half the area is commercial forestry dominated by two species with the balance comprising commercially worthless stands or scattered individuals of ten species that have largely self-established. Wilding conifers are now regarded as significant environmental weeds. Despite warnings of weed threats for 50 years, and sustained management for 40 years, the area occupied by wilding stands continues to increase.

I used three techniques to recreate the invasion and estimate the rate of increase each decade. The strengths and weaknesses of each technique are discussed along with departures from generalised invasion curves. Though much uncertainty remains, recreating a historic invasion has many uses. Spread rates can be used to predict future spread, understand impacts of investment and control strategies and lobby for operational funding.

Keywords Wilding conifers, invasion, spread.

INTRODUCTION

The conifer element of the indigenous vegetation of New Zealand comprises no species of Pinaceae but many have been introduced for forestry, amenity plantings, windbreaks and erosion control. Self-established seedlings of exotic conifers were first noted in the late 1880s (Smith 1903) and by 2015, 26 species have been recorded as naturalised.

The environmental weed potential of exotic conifers has long been recognised (Beauchamp 1962) and today these species are known as wilding conifers and affect around 1.7 million ha (Greenaway *et al.* 2015). Illustrating this dramatic invasion has been difficult as data have not been consistently collected. In this investigation, I compared three methods for graphically recreating the invasion of wilding conifers.

MATERIALS AND METHODS

Published national estimates of wilding conifer infestation area were compiled and the quoted area occupied was plotted against time.

Regional estimates of infestation area were collated from published and unpublished reports. Estimates of area were standardised by defining presence in 1 km^2 cells. Estimates were grouped by regions, and summarised by decades to 1995 then five yearly to 2015. I included areas where establishment was assisted (usually by aerial seeding) if the area is now considered an unmanaged stand. Where information was missing for any time period, a linear relationship was assumed. Categories of density were derived based on stems per ha. Where density information was not available, the ratio of sparse, moderate and dense stands was assumed using 2014 ratios.

Electronic records of specimens of the 12 most common wilding conifer species and 12 native beech and podocarp tree species that are common in montane forests were obtained from four national herbaria. Native and exotic species specimen locations were overlaid with a 10 × 10 km grid based on the NZTM projection. I plotted the cumulative number of cells occupied against time to create an invasion curve (*sensu* Delisle *et al.* 2003).

Finally, I calculated the total land area that could potentially be invaded by wilding conifers by summing the vulnerable land classifications in the Land Cover Database (LCDB) version 4.1. I used a criteria modified from an earlier report (North and Ledgard 2005).

RESULTS

Published estimates area for the area of wilding conifers were obtained from seven publications and a start point of 1 ha in 1895 was assumed. The resultant invasion curve depicts extremely rapid rise in affected area from the late 1990s (Figure 1).

In addition to the national figures, over 70 reports contained valuable, but highly variable observations and information on wilding conifers throughout New Zealand. The summarized estimates of invasion extent are illustrated in Figure 2.

A total of 290 useable herbarium records were obtained for wilding conifers, but 3418 records were obtained for the 12 native montane forest species. For wilding conifers, a period of increased recordings appeared in the late 1980s (Figure 3). A similar period was observed for native species in the 1970s. When the slopes of the logged cumulative area occupied were compared, the slope for exotic conifers was significantly steeper than for native montane species.

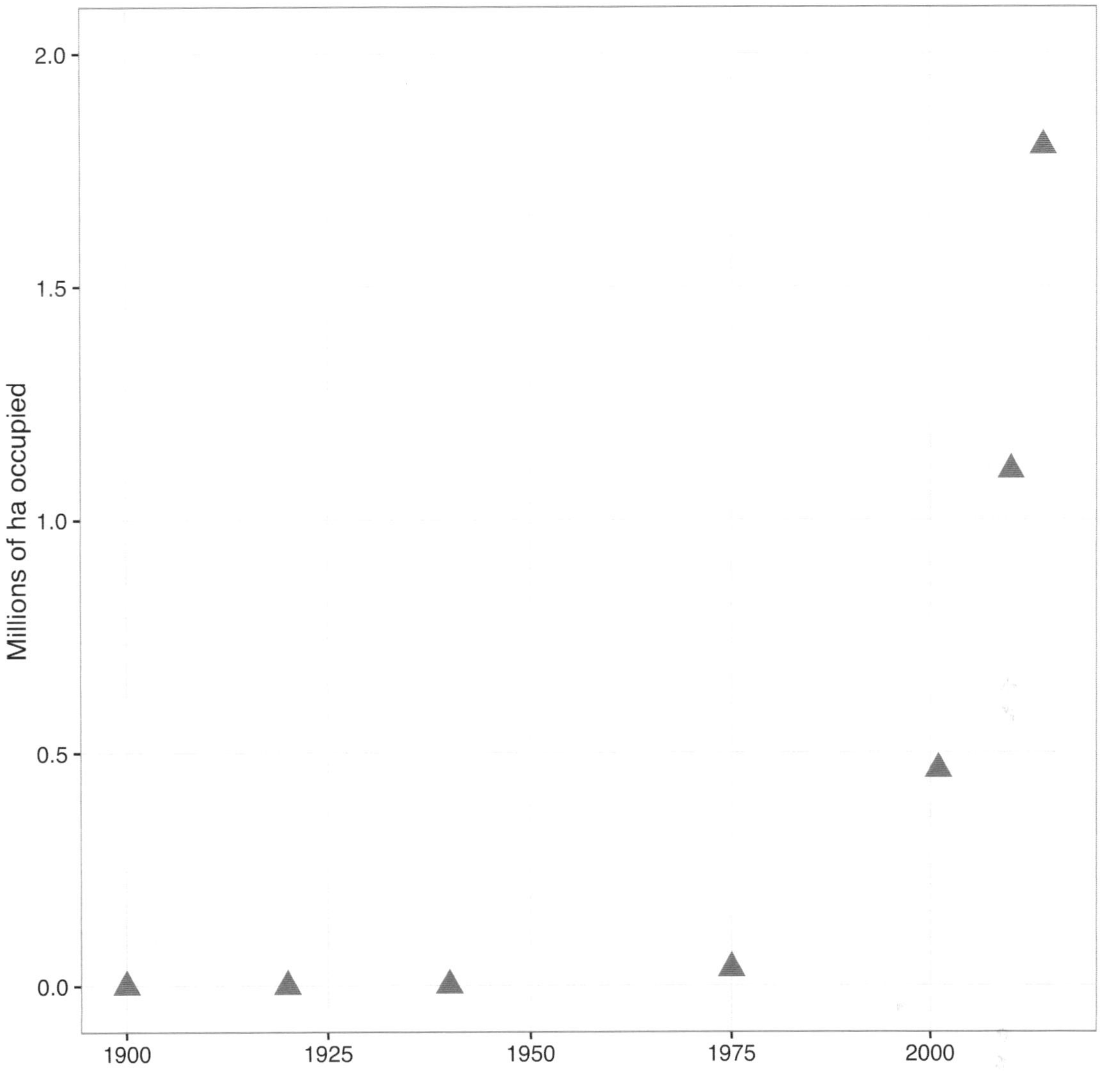

Figure 1. Approximate area occupied by wilding conifers using only published national estimates.

Using summed area of vulnerable lands, a figure of 7.3 million ha was derived as a possible upper limit for wilding conifer invasion.

DISCUSSION

Simply relying on published numbers of area infested has several problems. Firstly, the threat of wildings was debated from 1930 to 1970 and estimates did not include ongoing deliberate establishment for erosion control. Secondly, early estimates typically excluded very sparse wildings. The lower level of infestation for inclusion was typically one tree per ha. Stephens (2003) described the lightest level of infestation 0.001% cover. This equates to 0.1, 1 m^2 individuals per ha. In search and destroy operations from helicopters, operators frequently report densities below one plant per ha. It is now clear that although least some areas were prone to very sparse long distance spread by the 1960s (Buckley *et al*. 2005) and surveillance over very large areas is required because dispersal over distances of 10 km or greater is regularly observed.

Collating information from a large number of reports was very time-consuming and a large number of assumptions were required. However, I believe this is the most accurate reconstruction possible of the methods tested here. The total area fits a 6% growth model (Greenaway *et al*. 2015), but in some areas

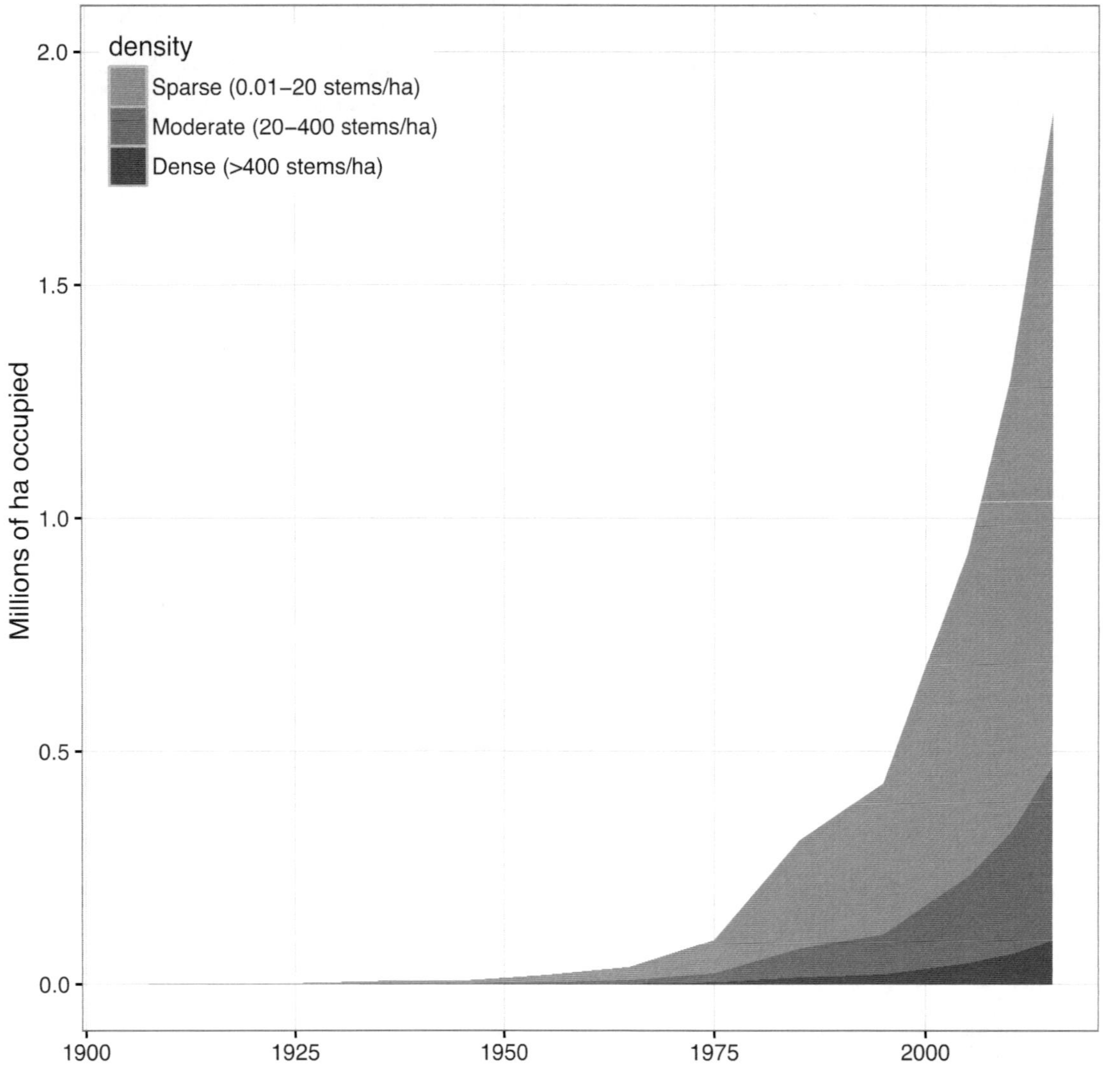

Figure 2. Approximate area occupied by wilding conifers in New Zealand using standardised regional reports.

and decades there is likely to have been significant departure from this.

Inclusion of some deliberately established areas of exotic is contentious (Dodet and Collet 2012), but I believe it is justified because such areas were never intended as commercial forestry, and are major contributors to wilding spread today. Exclusion of such stands initially, results in almost instantaneous expansion which is misleading. As well as promoting establishment, human intervention has also limited conifer growth through sustained control in some areas.

Using 2014 ratios of sparse, moderate and dense infestations in all eras was done out of necessity. As applied, such ratios fits with likely rates of progression of stands from sparse to moderate and ultimately dense infestations. It is possible that environmental changes such as changes in herbivore densities, mycorrhizal facilitation, or control intensity may have altered the rate of progression with time. However this level of complexity was not examined in any detail.

The estimates for area occupied here are significantly larger than the best estimate of area occupied using a remote sensing approach (Paul 2013). The main difference is likely to be related to the lower threshold of wilding conifer density I used for an area to be affected. The quality of images currently available for remote sensing does not allow reliable detection of small trees. Manual searching typically

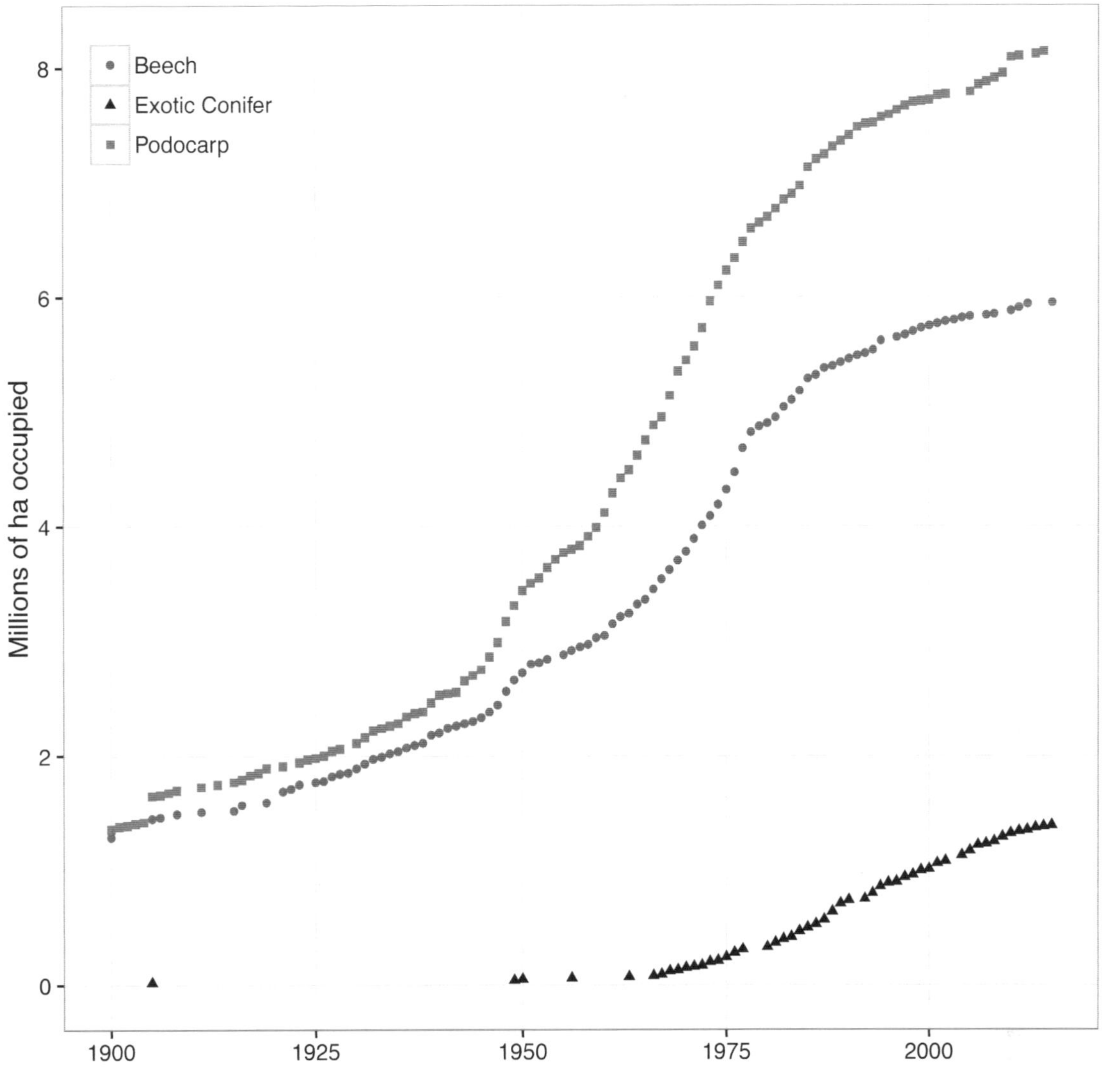

Figure 3. Approximate area occupied by wilding conifers and native tree species in New Zealand, using herbarium specimen occupancy of 10 × 10 km cells.

results in detection of smaller individuals, many more detections per ha, and a much greater overall area.

The increased collection of herbarium specimens of exotic conifers in the late 1980s coincides with the production of a major update to the published exotic flora (Webb *et al.* 1988) and is observed in other groups of exotic plants. Herbarium specimens are only a small fraction of presence records for New Zealand weeds (Howell and Terry 2016) so care needs to be taken when inferring distributions from specimens alone. However, comparison of collections of exotic conifers and montane native trees is informative. The methodology of Delisle *et al.* (2003) indicates that wilding conifers are spreading rapidly but comparing slopes on logged data penalises more widespread species. It seems likely that reductions in the collection rate of native species can be partly attributed to lack of interest in hundreds of specimens of the same species.

The potential area that could be invaded by wilding conifers of 7.3 million ha equates to approximately 30% of the land area of New Zealand. Using a spread model of 6% from 2015, this figure could be reached in as little as 30 years. However, it is likely that a logistic model will provide a better fit, as vulnerable land will become increasingly geographically isolated. Similarly, interventions to clear and maintain wilding-free areas will delay colonisation in some places. There are significant ecological, social and financial costs associated with continued spread of wilding conifers – it remains to be seen to what extent wilding conifers can be contained.

ACKNOWLEDGMENTS

Larry Burrows (Landcare Research) compiled the first figure of wilding conifer extent from published papers in 2013 that prompted this investigation. The Ministry for Primary Industries (MPI) funded national collection of data in 2014 on the extent of wilding conifer infestations. Nick Ledgard (Scion) authored many detailed reports on the status and management options for wilding conifer infestations for over 30 years. Kelvin Lloyd and Des Smith (Wildlands Consultants) have refined the classification of the categories of LCDB v. 4.1 that are vulnerable to wilding conifer invasion.

REFERENCES

Beauchamp, R. (1962). Birth of a forest? Regeneration of Mackenzie trees. The Press, Christchurch, July 1962. Reprinted in *Tree Grower* 4, 67-8.

Buckley, Y.M., Brockerhoff, E., Langer, L., Ledgard, N., North, H. and Rees, M. (2005). Slowing down a pine invasion despite uncertainty in demography and dispersal. *Journal of Applied Ecology* 42, 1020-30.

Delisle, F., Lavoie, C., Jean, M. and Lachance, D. (2003). Reconstructing the spread of invasive plants: taking into account biases associated with herbarium specimens. *Journal of Biogeography* 30, 1033-42.

Dodet, M. and Collet, C. (2012). When should exotic forest plantation tree species be considered as an invasive threat and how should we treat them? *Biological Invasions* 14, 1765-78.

Greenaway, A., Samarasinghe, O., Rees, T., Bayne, K., Velarde, S.J., Heaphy, M., Paul, T. and Kravchenko, A. (2015). Evaluating the (non-market) impacts of wilding conifers on cultural values. Report Prepared for the Department of Conservation, p. 83.

Howell, C.J. and Terry, J.A. (2016). The creation of a new Zealand weed atlas. Science for Conservation. Department of Conservation, Wellington, New Zealand, p. 21.

North, H. and Ledgard, N. (2005). Modelling wilding conifer spread and control. Report for Environment Canterbury. Scion, Christchurch, New Zealand, p. 27.

Paul, T. (2013). Extent and severity of wilding conifer presence in the New Zealand high country. New Zealand Forest Research Institute Limited (Trading as SCION), p. 20.

Smith, W.W. (1903). Plants naturalised in the County of Ashburton. *Transactions and Proceedings of the New Zealand Institute* 36, 203-25.

Stephens, R.T.T. (2003). Wilding conifer control: how important is it relative to other conservation actions? *In* Managing wilding conifers in New Zealand: present and future. Proceedings of a workshop held in conjunction with the annual conference of the New Zealand Plant Protection Society, eds R.L. Hill, S.M. Zydenbos and C M. Bezar. New Zealand Plant Protection Society, Christchurch, N.Z., pp. 27-39.

Webb, C.J., Sykes, W.R. and Garnock-Jones, P.J. (1988). 'Flora of New Zealand Volume IV. Naturalised pteridophytes, gymnosperms, dicotyledons', pp. 1365. (Botany Division DSIR, Christchurch).

Green alien space invaders: adding the spatial dimension to weed population models

Michael Renton
School of Plant Biology, The University of Western Australia, Crawley, Western Australia 6009, Australia
(michael.renton@uwa.edu.au)

Summary Weed management is an essential part of every agricultural system. However, weed management is difficult because the dynamics of weed populations are influenced by a mix of interrelated factors, many of which may be hard or impossible to observe directly and/or play out over long periods of time. Computational simulation modelling provides an important tool for integrating existing knowledge in order to help understand, predict and manage weed population dynamics. However many weed population models have treated weed populations as spatially homogeneous, ignoring the possibility of spatial heterogeneity in environments, management, densities and genetics. We hypothesise that in many situations accounting for spatial heterogeneity is essential.

This presentation will describe some examples where spatially explicit modelling of weed populations has been used to address real world weed management issues. The examples include predicting spread of invasive weeds in urban bushland and agricultural landscapes; evaluating whether occasional strategic inversion (mouldboard ploughing) can slow the evolution of herbicide resistance; exploring whether eradication could be a viable strategy for dealing with herbicide resistance; and investigating whether accounting for spatial heterogeneity changes predictions of how quickly resistance will spread through a weed population and impact crop yields.

In each of the examples, results show that the spatial dimension is essential for properly understanding and predicting the issue being examined. While space may not be the final frontier in the war against weeds, it is an important step in overcoming the green invaders. Spatial heterogeneity should be considered when using computational simulation modelling to help understand, predict and manage weed population dynamics.

Keywords Weed, simulation, model, spatial, invasion, resistance, evolution, spread.

Use of social media for sharing invasive species information

Kate Blood
Department of Environment, Land, Water and Planning, PO Box 7, Beaufort, Victoria 3373, Australia
(kate.blood@delwp.vic.gov.au)
Social media: @weedyk8

Summary Social media, like other forms of communication, can be used effectively to share information about invasive species. The popular contemporary social media include Facebook, Twitter, Instagram, YouTube and others. Invasive species practitioners working for government agencies have typically been prevented from using social media to talk about their work. For some agencies, this is changing and staff are being encouraged to share their work following guidelines for appropriate social media use.

Social media can be an effective form of communication for invasive species practitioners allowing people to reach a global audience of varied experiences and knowledge. Apart from shared information about invasive species, there are social and mental health benefits.

Keywords Facebook, hashtags, Instagram, Twitter, social media, Yammer, YouTube.

INTRODUCTION

The benefits of using social media can be many. This paper is not attempting to be a general social media user guide – there are plenty of these available online. It is highlighting some of the social media uses and benefits for sharing information specifically about invasive species.

SOCIAL MEDIA HINTS

Why are YOU using social media? It is worth thinking about why you are using social media and what your expectations are.

Depending on what you are aiming for will influence how you use social media. Is it to have influence, to be popular, to learn, to share, to have conversations, to spy on your kids, or to fill the time? Maybe a combination of many things.

These are four of my late father's (Doug Blood) principles that I try to consider often and are part of the reason I use social media and mobile technology:

- What are the possibilities?
- Collaboration.
- Don't be put off by challenges.
- Use technology and be an early adopter.

I use social media professionally to promote my work, increase awareness about invasive species, share my knowledge, learn about what others are doing in this field, expand my understanding, broaden my knowledge of the wider biodiversity and environmental field in which I work and live, and connect with others so I don't feel isolated.

Separating personal from professional After using social media for some time personally, I increasingly found that I was sharing information about invasive species and was finding a wider audience of people with similar interests. I was always cautious about communicating about my work and frustratingly avoided it for the most part.

I expressed my desire to be able to communicate about my work through my agency, Department of Environment, Land, Water and Planning (DELWP) and it coincided with a pilot for the use of social media by staff called #openmic. Through the #openmic pilot and with the permission of the steering group of the project I work on (Weeds at the Early Stage of Invasion), I was able to communicate about my work on social media.

I had already commenced separating my personal profiles on the different media and created a professional presence on the same media to keep them apart. I created a 'handle' that was available on all the media to keep things simple @weedyk8.

It is more time consuming monitoring and using the different accounts, however, it suits me to keep them separate.

Finding your voice and community It can take a while to 'find your voice' on social media. Your style, content, and how you let your personality come through in your writing and images takes time to develop.

Finding a community of people you connect and feel comfortable with can help you find your voice. When you can relax and be yourself within the framework of your agency's requirements is when the most fruitful engagement can take place.

When you find meaningful connections with others it can be beneficial to look at the people they follow to expand your network further. Searching on topics of interest using hashtags (see below) can open up further rich seams of connections. Following the feeds of people attending conferences relevant to your field and the conference hashtag e.g. #20awc can be interesting.

Building an audience Creating a social media profile and then not using it generally does not generate any connections. To build an audience takes time to actively find and connect with the people of mutual interest. What you put in to making connections and creating content will influence what you get out.

'Following' and 'liking' the people and agencies you are interested in will often result in a reciprocal 'follow' or 'like'. It is not compulsory to follow people who follow you first. Search on social media etiquette. It can be beneficial to unfollow and unlike people or agencies that are no longer of interest.

These are some of the people/agencies I follow on Twitter that post regular information about invasive species: @GoodenBen @TimDoherty_ @LoraxCate @ManuSaunders @ISCAustralia @tasbiophiliac @WoNS_weeds. There are many links in the appendix.

Creating your own content that you post through your own channels, and sharing relevant posts from others e.g. 'retweeting' on Twitter or 'sharing' on Facebook is a good way of building a following (if that is the aim).

My principles are to be respectful, factual, be responsible for what I post, limit my personal opinions and attribute sources of information I share when it is appropriate to share them.

Challenges The internet has lots of content about many topics that may be offensive or illegal. The same goes for social media. Through the effective use of 'blocking' tools, it is possible to exclude many people and bots (automated computer software applications) from your feed. You do have to be vigilant and monitor your feed to pick them up quickly.

Use of the word 'weed' does attract people and bots that are interested in drug use, hence the use of alternative terms such as 'invasive species'. I use both and block those followers I am not interested in.

Inevitably, you will be followed and linked to by people/sites that promote drug use. I try to be very vigilant and block all these users immediately. I try to remember every 4–6 weeks to review my follower list and block the ones I missed earlier.

Social media can be a time stealer. There is advice available on how to monitor and control the time spent using social media if it becomes an issue. There is advice available on avoiding interruption to sleep patterns. Try the screen-dimming function on mobile devices that you can set for a certain time each evening to reduce brain stimulation from screen glare.

Equally, social media can be a good distraction while you are having to wait in queues, when the television offering is a bit thin in the evenings, or while travelling.

For a new user of social media, it can be daunting and intimidating while you learn the lingo, avoid pitfalls and get with the flow.

For many people privacy concerns are the main reason for not using social media. I make the assumption that anyone anywhere is reading my material and looking at my images. I write and photograph accordingly. I don't post material that I want to keep private. I am conscience of the privacy of others including who I photograph. Regularly read and review the privacy settings on the social media you use.

Keep your wits about you to avoid scams and identity theft. Don't believe everything you read and check people out carefully before engaging with them if you feel concerned.

Which social media to use? It really depends on what you are after. I find a combination of Facebook, Twitter and Instagram suit me well. Our Department has an internal Yammer network that I use to promote our work and learn about what is going on around the State.

Try a few different media and see what combination suits you.

I first started using Facebook personally, then as my confidence grew I started using Twitter. As my professional use grew I created separate professional accounts. I wanted to try Instagram so created a personal account first, learnt how to use it and then created my professional Instagram account.

Search terms including hashtags Hashtags are a form of search term. The # symbol in front of a continuous string of letters or words makes it a live search term that can be clicked on to bring up posts in which others have used the same hashtag. It helps to find people that you can connect with or posts about the same topic you are interested in.

On Twitter, these are the hashtags I typically use:

- #invasivespecies. I sometimes abbreviate this to #invspp and I note others use #invsp. I like the longer term #invasivespecies as it removes any confusion about the intended topic.
- #weeds [expect to have to sift through lots of drug posts!].
- #weedid [for weed identification].
- #scicomm for science communication.

Tweeting from conferences There are heaps of hints on Twitter (links to papers and blogs) about Tweeting from conferences. It is good fun, shares with others who cannot attend, is a form of note taking that I then share back in meetings with people I work with, and it promotes sharing information and raising awareness about invasive species.

Pia Lentini (2013) has some good advice about Tweeting and conferences: https://pelentiniresearch.wordpress.com/2013/08/27/a-plea-to-make-conferences-more-twitter-friendly/

Hints on choosing hashtags for events When planning a conference, field trip or other event, think carefully about a hashtag that can be used to promote and use during the event.

Choose carefully and consider a term with not too many characters, it's easy to remember, and ALWAYS test first to see if it is already being used. Then promote it to make sure everyone uses the same hashtag.

Social media research This paper is by no means a comprehensive research paper on the use of social media for invasive species. It is based on my experiences. I have come upon this paper (Daume 2016) that looks at the use of Twitter to monitor invasive species. For those wishing to look further into this topic, the Daume (2016) paper has a list of references for those wanting a toe into the topic.

CONCLUSION

If you haven't used social media before, I hope this information encourages you to give it a try. If you only use social media personally, consider using it professionally if you have the permission to do so.

For me, I learn a lot through social media, broaden my networks and I find it is great for my mental wellbeing – I don't feel isolated working on a topic where there are few co-workers. Most of all, it is good to share my knowledge and observations of what I see in the field.

ACKNOWLEDGMENTS

I thank the Horticultural Media Association Australia member who wrote an article about using Twitter that helped me make the next leap into a larger social media world, and my nieces who teach me new things. I appreciate the opportunity DELWP has given me to share my work with others around the world.

The social media audiences I engage with are generally supportive and enter into communication in a positive and sharing spirit.

REFERENCES

Daume, S. (2016). Mining Twitter to monitor invasive alien species – an analytical framework and sample information topologies. *Ecological Informatics* 31, 70-82.

Lentini, P. (2013). https://pelentiniresearch.wordpress.com/2013/08/27/a-plea-to-make-conferences-more-twitter-friendly/ (viewed online 8/8/2016).

APPENDIX

These are some social media links that may be of interest. It is not a comprehensive list. Follow the threads and networks to explore your specific interests.

Kate Blood social media:
https://twitter.com/weedyk8
https://www.facebook.com/weedyk8
https://www.instagram.com/weedyk8/

Australasian Weeds Conference:
Twitter: https://twitter.com/austweedsconf
Facebook: Australasian Weeds Conference
Search terms: #20awc

Weed Societies in Australia:
Facebook: Tasmanian Weeds
Facebook: The Weed Society of Victoria Inc
Facebook: Weed Management Society of South Australia 5th Conference

These are some social media (and web links) recommendations made by members of the Aliens-L email listserver in June 2016 (Note: Listing here doesn't mean endorsement. There are also many links that can be found by seeing who each of these people/agencies follow and who follow them):

Pests and Weeds Victoria Australia Facebook group:
https://www.facebook.com/groups/1427695277483336/

Facebook: Bush Revegetation and Regeneration group
Facebook: NSW Native Plant Identification group
Facebook: Weeds in West Gippsland group

Facebook: Wet Tropics Management Authority Yellow Crazy Ant Program: https://www.facebook.com/groups/ycawettropics/

Facebook: Invasive Species South Africa https://www.facebook.com/invasivespeciessouthafrica/

Facebook: Cape Town Invasives
https://www.facebook.com/ctinvasives/

Twitter:
@InvasivesZA https://twitter.com/InvasivesZA

The Center for Invasive Species Prevention invasive species blog: www.cisp.us

CONICET (the National Scientific Council), and the Ministry of Science in Argentina Facebook pages: Marine-Coastal Exotic Species Guide for Argentina
Facebook: @GEAC.CONICET

https://www.facebook.com/MudpuppyNight/ have recent posts about alien crayfish.

https://twitter.com/wildlifeinwater

Inland Fisheries Ireland – Twitter feed specifically on aquatic invasive species in Ireland and internationally and promotes biosecurity issues with stakeholders: twitter.com/aisireland

National Invasive Species Committee of Palau: www.palaunisc.org

Pacific Invasives Initiative (PII) Facebook page: https://www.facebook.com/Pacific-Invasives-Initiative-PII-109320925778960/

http://www.facebook.com/pages/Kaitiaki-o-Ngahere/162171857214996

CABI has an invasive species blog and social media:
https://cabiinvasives.wordpress.com/
https://twitter.com/cabi_invasives
https://www.facebook.com/CABI.development

The IUCN Commission on Ecosystem Management (CEM) Thematic Group Ecosystems and Invasive Species Facebook: https://www.facebook.com/IUCN-CEM-Ecosystems-and-Invasive-Species-154657961556106/
Theme Leader on Twitter: @judithfish

Website: http://geaccenpat.wix.com/geac

The Smithsonian Institution
Biodiversity Heritage Library (BHL)
IUCN
California Academy of Science
The Nature Conservancy
News from Science (Magazine)
ISSG (Invasive Species Specialists Group)

Paradigm™ Arylex™ Active, a new post-emergent, broadleaf herbicide for use in winter cereals

Peter Nott
Dow AgroSciences Ltd, 20 Rodborough Road, French's Forest, New South Wales 2086, Australia
(nott@dow.com)

Summary Paradigm™ Arylex™ Active was registered in 2015 for post emergent broadleaf herbicide for use in wheat, barley, triticale and oats for hay. It contains the new auxinic (Group I) herbicide, Arylex (common name halauxifen-methyl), and florasulam, a Group B herbicide. Most weeds on the label require tank mixing with MCPA EHE, and Paradigm is compatible with a range of broadleaf herbicides. Paradigm has a broad label, with flexibility to tank mix required, and can be applied from GS 13 to 39. This paper summarizes field studies addressing selectivity, efficacy and rotational intervals to sensitive crops.

Keywords Paradigm™, Arylex™, Active, broadleaf weed control, winter cereals.

Rinskor™ active – a new herbicide for rice weed control in Australia

Gregory S. Wells[1] and Malcolm Taylor[2]
[1] Dow AgroSciences Australia Ltd., PO Box 838, Sunbury, Victoria 3429, Australia
[2] AgroPraisals Pty. Ltd., 283 Torgannah Road, Cobram, Victoria 3644, Australia
(wells1@dow.com)

Summary In 2016 Dow AgroSciences submitted registration application for Rinskor™ active (300 g L^{-1} florpyrauxifen-benzyl ester), a new herbicide for grass, sedge and broadleaf weed control in rice. This product provides good control of key weeds including barnyard grass (*Echinochloa crus-galli* (L.) P.Beauv.), water plantain (*Alisma plantago-aquatica* (L.), starfruit (*Damasonium minus* (R.Br.) Buchenau), arrowhead (*Sagittaria montevidensis* Cham. & Schltdl. subsp. *calycina* (Engelm.) Bogin), jerry jerry (*Ammania multiflora* Roxb.) and dirty Dora (*Cyperus difformis* L.) suppression, whilst being selective to rice. It will be registered for weed control in rice and represents a new option for rice weed management in Australia. New products are essential for the rice industry due to weed resistance to existing treatments and potential loss of products through industry regulation.

Keywords Rinskor, efficacy, selectivity, plant-back, weeds, control, rice.

INTRODUCTION

There have been no new herbicides released for weed control in rice for more than 10 years in Australia. Rinskor is a new rice selective herbicide that controls key weeds and exhibits short persistence in soil.

Rinskor was first introduced in 2015 (Weimer *et al.* 2015), where the general utility of Rinskor for rice and other crops was outlined. Soil persistence was reported to be as short as 56 days to replant soybeans safely in the southern United States (Miller *et al.* 2016).

This paper summarizes new research trials to confirm rice selectivity, plant-back safety and weed efficacy for registration of Rinskor in rice in Australia.

MATERIALS AND METHODS

Trials were conducted in southern New South Wales and northern Victoria on rice research farms. Replicated small plot trials using 3 or 4 replications were conducted from 2012 to 2015 to determine the selectivity, weed control efficacy and crop rotation safety. Applications were made with gas powered small plot booms delivering 100 L ha^{-1} of spray solution. Data were summarized using Agricultural Research Manager Summary Across Trials software (Gylling 2015) and analysed using Analysis of Variance (ANOVA) with Duncan's Multiple Range means comparison test and 5% significance level.

Rice selectivity Seven weed-free selectivity trials were conducted in summers of 2013 to 2015, covering 10 commonly grown rice cultivars (both medium and long grain types). Rinskor was applied at 30 to 90 g a.i. ha^{-1} at the three to five leaf stage of rice development. Crop oil concentrate was added at 2% v v^{-1}. Crop injury was measured by visual assessment where 0 = no injury and 100 = complete crop loss. Grain yields were taken at normal harvest time and expressed as t ha^{-1} and percent of the untreated control.

Weed efficacy Seventeen weedy trials were conducted to determine the efficacy of Rinskor on barnyard grass (14 trials), dirty Dora (9), water plantain (3), starfruit (7), arrowhead (7) and jerry jerry (1). Weed control was measured using visual assessment where 100 = complete control.

Weeds were treated at the small seedling stage up to 5 cm across and 5 leaf stage. Rinskor was applied at rates of 20, 30 and 40 g a.i. ha^{-1} to determine what rate was effective. Crop oil concentrate at 2% v v^{-1} was applied in all treatments. Weed efficacy was considered acceptable if control was more than 95% at final assessment.

Plant-back safety Five weed-free trials were conducted to determine the plant-back safety of Rinskor, either applied in fallow to bare soil, or in rice crops. Rinskor was applied at rates of 5 to 90 g a.i. ha^{-1} prior to planting lentil (*Lens culinaris* Medik.), chickpea (*Cicer arietinum* L.), field pea (*Pisum sativum* L.), clover (*Trifolium balansae* Boiss.) or subclover (*Trifolium subterraneum* L.). Planting times varied from 39 to 141 days after treatment. Rainfall (mm) was monitored at the nearest weather station. Crop injury was measured by visual assessment where 100 = complete crop loss. A positive standard treatment, with long term soil persistence was included to determine relative injury compared with Rinskor.

RESULTS AND DISCUSSION

Rice selectivity Rinskor applied at 30 to 90 g a.i. ha^{-1} resulted in average injury of less than 16% for all 10 varieties (Table 1). This was commercially acceptable for rice in Australia.

All varieties gave statistically similar or higher yield than weed free untreated control (NS $P = 0.05$, Table 2).

Weed efficacy Rinskor applied post-emergence at up to 5 leaf weed stage gave control of barnyard grass, water plantain, starfruit, arrowhead or jerry jerry and suppression of dirty Dora (Table 3). Rinskor gave similar high control of all weeds irrespective of rate, apart from dirty Dora, where highest rate gave best suppression.

Plant-back safety Minimum plant-back periods of six months were indicated for pasture and grain legumes after Rinskor application to rice (Tables 4 and 5).

CONCLUSIONS

Rinskor active is a new product that has acceptable rice selectivity, broad weed control spectrum and short soil persistence. Therefore it is a significant new product for weed management in rice in Australia.

ACKNOWLEDGMENTS

The authors would like to acknowledge Rice Research Australia Pty. Ltd. and NSW Department of Primary Industry for cooperation with field sites and research trials.

REFERENCES

Weimer, M.R., Yerkes, C., Schmitzer, P. and Mann, R.K. (2015). Introduction to a new arylpicolinate herbicide from Dow AgroSciences with utility in rice and other crops. Weed Science Soc. Amer., Lexington, Kentucky.

Miller, M.R., Norsworthy, J.K., Weimer, M.R., Huang, R., Lancaster, Z. and Martin, S. (2016). Environmental fate of Rinskor™ active: field dissipation and replant interval for soybean. Weed Science Soc. Amer., San Juan, Puerto Rico.

Gylling, S. (2015). Agricultural Research Manager, summary across trials. http://gdmdata.com/Products/Summary Across Trials (ST). Gylling Data Management Inc, Brookings, South Dakota.

Table 1. Mean rice injury (%) by Rinskor at 30 to 90 g a.i. ha^{-1}. (Brackets = replicates of data assessed).

	Rinskor rate (g a.i. ha^{-1})			
Variety	30	45	60	90
Doongarra (7)	1	0	2	2
Koshihikari (7)	8*	7	4	7
Kyeema (3)	0	0	3	3
Langi (3)	0	0	2	3
Opus (7)	1	0	0	2
Reziq (7-25)	8	0	9*	0
Sherpa (7-25)	7	0	7	0
Topaz (3-21)	11*	7	16*	5
YRK 5 (4)	0	0	0	0

* Significantly ($P = 0.05$) higher than untreated.

Table 2. Mean relative rice yield as percentage of untreated. (Values in brackets show the number trials harvested with each variety).

	Rinskor rate (g a.i. ha^{-1})			
Variety	30	45	60	90
Doongarra (2)	104	103	100	103
Koshihikari (1)	111	99	104	109
Kyeema (1)	102	97	99	96
Opus (2)	103	105	105	116
Reziq (5)	99	107	100	98
Sherpa (5)	102	104	104	104
Topaz (4)	104	99	100	101
YRM 70 (1)	102	102	100	101
YRL 127 (1)	104	103	102	96
YRK 5 (1)	101	99	101	94
All varieties	102	103	102	103

Table 3. Mean weed control (%) by Rinskor at 20 to 40 g a.i. ha^{-1}. (Values in brackets show the number of trials assessed).

	Rinskor rate (g a.i. ha^{-1})		
Weed	20	30	40
Arrowhead (3–7)	99	100	100
Barnyard grass (5–14)	98	96	99
Dirty Dora (4–8)	60[b]	80[a]	83[a]
Jerry jerry (1)		100	
Starfruit (3–7)	100	100	100
Water plantain (1–3)	100	100	100

Dirty Dora LSD ($P = 0.05$) = 12.7.

Table 4. Mean legume injury (%) by Rinskor at 5 to 15 g a.i. ha^{-1} and LSD values, after treatment of bare fallow soil.

Trial	155001ML	155001ML	152008HG	152008HG	142004HG
Site	Halbury, SA	Halbury, SA	Breeza, NSW	Breeza, NSW	Breeza, NSW
Replant (days)	36	57	63	85	39
Rain (mm)	93	131	83	115	60
Crop	Lentil	Lentil	Chickpea	Chickpea	Chickpea
Untreated control	0	0	0	0	0
Rinskor 5 g a.i. ha^{-1}	20	53	14	4	
Rinskor 10 g a.i. ha^{-1}	38	70	18	6	0
Rinskor 15 g a.i. ha^{-1}	54	80	33	20	
Arylex + Florasulam 5 + 5 g a.i. ha^{-1}	94	91	27	17	75^
LSD (P = 0.05)	11.3	15.3	9.9	6.3	

^ 90 g a.i. ha^{-1} clopyralid.

Table 5. Mean legume injury (%) by Rinskor at 30 to 90 g a.i. ha^{-1} and LSD values, after treatment of rice.

Trial	H43-14	H43-14	H43-14	H51-14	H51-14
Site	Cobram, Vic	Cobram, Vic	Cobram, Vic	Cobram, Vic	Cobram, Vic
Replant (days)	89	141	89	50	50
Rain (mm)	97	177	97	24	24
Crop	Subterranean clover	Subterranean clover	Field pea	Subterranean clover	Shaftal clover
Untreated control	8	20	0	7	6
Rinskor 30 g a.i. ha^{-1}	13	16	6	9	8
Rinskor 60 g a.i. ha^{-1}	11	10	0	38	24
Rinskor 90 g a.i. ha^{-1}	23	23	4	78	45
Quinclorac 600 g a.i. ha^{-1}	91	90	53	98	96
LSD (P = 0.05)	19.5	23.4	10.7	17.9	19.3

Rexade™ Arylex™ Active, a new post-emergence herbicide for control of grass and broadleaf weeds in wheat

Peter Nott
Dow AgroSciences Ltd, 20 Rodborough Road, French's Forest, New South Wales 2086, Australia
(nott@dow.com)

Summary Rexade™ Arylex™ Active is to be submitted for registration in 2016. It contains the new auxinic (Group I) herbicide, Arylex (common name halauxifen-methyl), and pyroxsulam, a Group B herbicide with grass and broadleaf activity. Arylex controls a range of broadleaf weeds, including some not controlled by pyroxsulam e.g. fumitory (*Fumaria* spp.). Rexade can be tank mixed with a range of broadleaf herbicides to provide broad spectrum control or suppression of important grass and broadleaf weeds in wheat and triticale. This paper summarises field studies addressing selectivity, efficacy and rotational intervals to sensitive crops.

Keywords Rexade™, Arylex™, Active, broadleaf weed control, winter cereals.

Increasing the efficiency of herbicides for management of dense infestations of wilding conifers

Carol A. Rolando[1], Stefan F. Gous[1], Robyn Gaskin[2] and David Horgan[2]
[1] Scion, Private Bag 3020, Rotorua, 3046, New Zealand
[2] Plant Protection Chemistry New Zealand, PO Box 6282, Rotorua, New Zealand
(carol.rolando@scionresearch.com)

Summary Introduced conifers are the backbone of commercial forestry in New Zealand worth c. $5 billion per year (NZ) in export earnings. However, invasion by wilding conifers is arguably New Zealand's most serious and intractable weed problem. Wilding conifers are thought to occupy ca. 1.8 million ha (4%), and may invade 20% of New Zealand's land area within 35 years. Although effective, the current control methods are costly and heavily dependent on herbicides, sometimes at rates that exceed label recommendations, particularly for management of dense infestations. Transformation of control requires significant rate reductions and new application systems to both reduce cost and risk of negative environmental impacts, while maintaining efficacy. The pathway to reducing herbicide rates, while maintaining or increasing biological efficacy, is by identifying adjuvants that enhance plant uptake and translocation of active ingredients and improving spray distribution throughout the canopy using targeted application systems. We will present the results of preliminary controlled studies conducted to understand factors affecting the uptake of triclopyr into *Pinus contorta* Douglas ex Loudon needles. The aim of these studies is to identify opportunities to reduce rates currently used in operational programmes. We will also outline the aims, objectives and challenges of a broader programme aimed at increasing the efficiency of chemical control across a range of wilding infestation levels.

Keywords Herbicide uptake, triclopyr, picloram, adjuvants, *Pinus contorta*.

Synergism between synthetic herbicides and *Nigrospora oryzae* (Berk. & Broome) Petch for the inundative biocontrol of pampas in New Zealand

Stanley E. Bellgard[1], Chantal M. Probst[1] and Von W. Johnson[2]
[1] Landcare Research Private Bag 92170 Auckland 1142, New Zealand
[2] Enform Tech NZ Ltd, 19 Laureston Ave, Auckland 2025, New Zealand
(bellgards@landcareresearch.co.nz)

Summary South American pampas (*Cortaderia* spp.) are serious weeds in New Zealand, particularly in pine forests. Literature documenting the synergy between specific herbicides and mycoherbicides, identify benefits, including; greater efficiency and lowered product rates. Understanding herbicide-microbial interactions is an important step in the selection of synergistic partners. Organic herbicides are residue-free alternatives to synthetic herbicides, however, they are non-selective in activity. In Experiment 1, we studied the effect of three herbicides on the viability and growth of *Nigrospora oryzae* (Berk. & Broome) Petch;

1. 0.25 strength Gallant™ Ultra,
2. Greenscapes® and
3. Weedfree®.

The biomass accumulation was measured over 24 weeks, and the colony forming units (CFUs) and spore formation sampled on five occasions. No stimulation of fungal growth was shown by any of the herbicide treatments. Gallant herbicide at 0.25 strength neither interfered with fungal growth or sporulation as did both the organic herbicides. Weedfree was more fungitoxic than Greenscapes. For Experiment 2, we undertook a glasshouse spray-trial of the co-application of 0.25 strength Gallant and the organic herbicides with and without *N. oryzae*. Gallant at 0.25 strength killed the test plants (with or without *N. oryzae*). The organic herbicides initially killed the laminar of the test plants, but the plants re-sprouted from the pseudostem. The co-application of *N. oryzae* with each of the organic herbicides resulted in the death of the test plants. These results suggest that the co-formulation of *N. oryzae* with synthetic herbicides has the potential to kill pampas more effectively than the effect of the fungal pathogen alone.

Keywords Pampas, *Cortaderia* spp., *Nigrospora*, haloxyfop-P-methyl.

INTRODUCTION

The South American pampas grasses, *Cortaderia selloana* (Schult. & Schult.f.) Asch. & Graebn. and *C. jubata* (Lemoine) Stapf (Poaceae) are serious ecological weeds in New Zealand, particularly of low-growing and relatively open indigenous plant communities in habitats such as dunelands, cliffs and wetland margins (Gosling *et al.* 1999). Pampas grass is also one of the worst weeds of pine plantations in New Zealand (Richardson *et al.* 1996). Both species are very invasive and can form dense impenetrable stands, and as pampas grass spreads, other desirable vegetation is smothered. Pampas grass poses a fire risk in summer, is home to rats and mice, and affects visibility on roadsides. The range of areas under threat includes forests, conservation areas, scrublands, reserves and roadsides. A reduction in herbicide residues in forest products has necessitated the study for lower- and residue-free alternatives for weed control in pine plantations. Previous studies have demonstrated synergy between synthetic and microbial herbicides for weed control (Peng and Wolf 2011). As Hoagland (1996) suggested, synergy may reduce the application rate of biological and synthetic herbicides required for effective weed control. Lowering mycoherbicide rates helps reduce the costs of biological control, and decreasing the rate of herbicides reduces the load of pesticide in the environment. Selected fatty acids such as pelargonic and caprylic acids are effective non-selective herbicides for a wide spectrum of annual weed species (Coleman and Penner 2008). Organic herbicides are available on the hardware-store shelf, and are advertised as soft alternatives to chemical herbicides because they do not have any residues.

As part of a natural enemy survey for pampas pathogens, Bellgard *et al.* (2011) recovered a rice pathogen from the laminar of pampas, *Nigrospora oryzae*. This study aimed to investigate the synergy between *Nigrospora oryzae* inoculum and three synthetic herbicides; 1) 0.25 strength Gallant™ Ultra and Greenscapes® (700 g L^{-1} Pelargonic acid) and Weedfree® (220 g L^{-1} pine oil and 560 g L^{-1} caprylic acid). In experiment 1, the *in vitro* activity of *N. oryzae* was studied to assess the effect of herbicides on; biomass accumulation, colony forming units, and spore production. The second experiment describes a glasshouse spray-trial to compare the performance of the three herbicides alone and together with *N. oryzae* in a fully replicated trial, using both *Cortaderia jubata* (ex Nelson) and *C. selloana* (ex Te Atatu) test plants.

METHODS

Experiment 1. *In vitro* interactions between *N. oryzae* and synthetic herbicides The design was fully randomised, with five replicates of each of the following combinations of 200 mL of each of three herbicides plus 300 mL *N. oryzae* (7.6×10^7 CFU mL^{-1}) and a negative control using 0.03% Tween on the 14/10/14. *Nigrospora oryzae* was grown in sterilised, clarified V8 juice liquid medium. Three measures were used to assess the effect of herbicides on the viability of *N. oryzae*:

1. Dry weight after 24 weeks,
2. Assessment of the number of colony forming units at five sampling times (4, 12, 16, 20 and 24 weeks), and
3. Spore formation at five sampling occasions (4, 12, 16, 20 and 24 weeks).

Mean dry matter accumulation was measured after samples were extracted under suction onto pre-weighed, sterilized, Whatman™ No. 1 filter papers. Filters and samples were weighed and then dried overnight at 60°C. The filter mass was then subtracted from the total dry mass to give the dry matter value. Mean numbers of colony forming units were determined by taking five, 50 µL replicate sub-samples from each pair of harvested flasks at each sampling occasion. The sub-samples were spread onto PDA plates amended with streptomycin sulphate (130 mg L^{-1}). A single dilution was established using 10 µL of sample to 90 µL of sterilized Reverse Osmosis (RO) water, and this was plated to a fresh plate of PDA with 100 ppm streptomycin sulphate (Leyronas and Raynal 2008). After four days, colonies were counted. Two replicates were harvested at each of the following occasions after inoculation; 4, 12, 16, 20 and 24 weeks. Spore numbers were estimated using a C-chip™ disposable haemocytometer. Five, 50 µL aliquots were removed from each pair of harvested flasks at each sampling occasion and placed in the haemocytometer. The spores in each of five replicate large squares were counted and the initial concentration calculated. Two replicates were harvested at each of the following occasions after inoculation; 4, 12, 16, 20 and 24 weeks.

Experiment 2. Glasshouse spray-trial *N. oryzae* was grown on wild rice which had been autoclaved with 10% V8 juice in 250 mL volumetric flasks. Inoculum was grown for two months at 18°C. The inoculum was removed from the flask, and combined with 500 mL of sterilized RO water. The suspension was blended at high speed in a Waring-type blender for five minutes. The suspension was made up to 1.5 L. The suspension was added to 1 L of 0.3% Tween (to achieve a final volume of 2.5 L). Five, 500 mL aliquots were added to each of the herbicides (in a 1:1 ratio), and applied to five replicates of each *C. jubata* (ex Nelson) and *C. selloana* (ex Te Atatu) plants using a hand held, Yates Plassay® Maxi 6 Pressurised spray unit. Five replicates of each species also received each of the herbicides, and one replicate set received only water (negative control). A positive control was also established, which only received the *N. oryzae* spore suspension in a 1:1 ratio with water.

RESULTS AND DISCUSSION

Experiment 1. *In vitro* interactions between *N. oryzae* and synthetic herbicides Biomass accumulation decreased over time as the inoculum resided in the herbicides. The organic herbicides based on fatty acid and fatty acid/pine oils were the most biocidal, with 0.25 strength Gallant apparently being the least biocidal (Figure 1). Both pelargonic and caprylic acid are known to have antibiotic properties (Gilbert and McBain 2003), and the anti-fungal activity of pine oil is well documented (Gilbert and McBain 2003).

After 24 weeks, there was a significant effect of herbicides for dry matter accumulation ($F_{1,3}$ 33.4, P <0.0001) of *N. oryzae* (Figure 1).

At the end of the 24 weeks, the number of CFUs in the control had increased by 15% (Table 1). Colony forming unit numbers were three to four times greater than spore numbers, as CFUs also contained a hyphal fragment fraction. The colony forming units showed a downward trend, with the organic herbicides displaying a strong biocidal effect (Table 1). After four weeks, CFU numbers had dropped by approximately 25% in the Weedfree® treatment and 15% in the Greenscapes® treatment (Table 1). Gallant™ Ultra had little or no effect after four weeks, with a drop in CFU numbers after 24 weeks of around 2% (Table 1). By the end of the 24 week period, the organic herbicides had reduced the CFU numbers by up to 33% (Table 1).

Spore numbers were also negatively impacted by the organic herbicides, with 0.25 strength Gallant having very little biocidal effect, and reducing spore numbers by a factor of six (Table 2). Spores were less sensitive than the CFUs to the biocidal effect, with the overall reduction in number after 24 weeks being between 13 and 15% (Table 2). After 24 weeks, the spore numbers had remained more or less static in the control (Table 2).

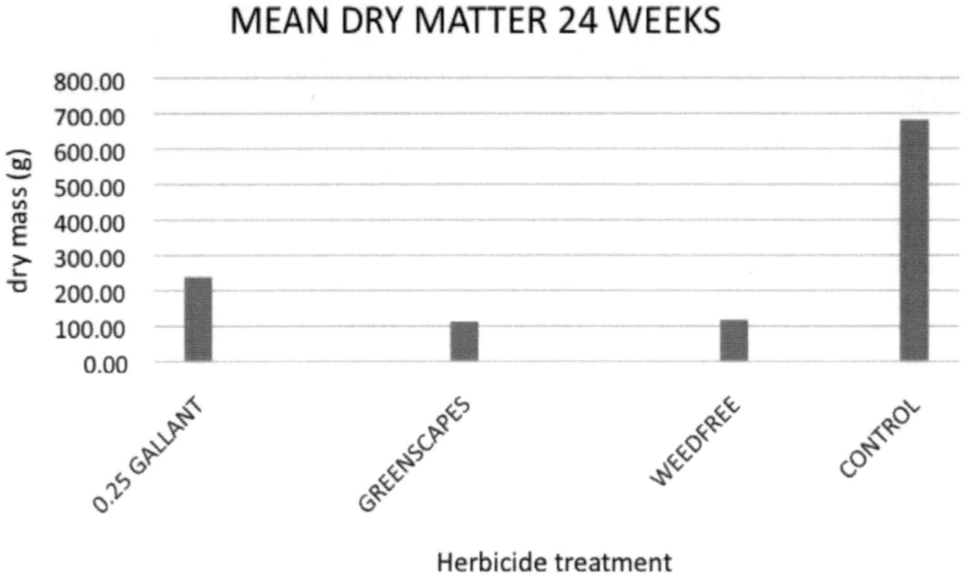

Figure 1. Fungal biomass after 24 weeks after addition of *N. oryzae* suspension to herbicide.

Table 1. Colony forming units CFUs per L after addition of *N. oryzae* suspension to herbicide (mean ± s.e.m. n = 5).

Herbicide	Initial concentration CFU per L	Concentration after 4 weeks CFU per L	Concentration after 24 weeks CFU per L
0.25 strength Gallant™ Ultra	7.6×10^7	$7.54 \times 10^7 \pm 2.2 \times 10^5$	$7.47 \times 10^7 \pm 8.21 \times 10^5$
Greenscapes®	7.6×10^7	$6.54 \times 10^7 \pm 4.22 \times 10^5$	$5.17 \times 10^7 \pm 5.18 \times 10^5$
Weedfree®	7.6×10^7	$5.72 \times 10^7 \pm 3.35 \times 10^5$	$4.98 \times 10^7 \pm 3.4 \times 10^5$
Control + Tween	7.6×10^7	$7.69 \times 10^7 \pm 8.96 \times 10^4$	$9.06 \times 10^7 \pm 8.69 \times 10^5$

Table 2. Spore numbers after addition of *N. oryzae* suspension to herbicide (mean ± s.e.m. n = 5).

Herbicide	Initial spore concentration CFU per L	Spore concentration after 24 weeks CFU per L
0.25 strength Gallant™ Ultra	$2.1 \times 10^7 \pm 1.6 \times 10^7$	$3.5 \times 10^7 \pm 1.6 \times 10^7$
Greenscapes®	$2.1 \times 10^7 \pm 1.6 \times 10^7$	$1.83 \times 10^7 \pm 2.6 \times 10^5$
Weedfree®	$2.1 \times 10^7 \pm 1.6 \times 10^7$	$1.78 \times 10^7 \pm 2.6 \times 10^5$
Control + Tween	$2.1 \times 10^7 \pm 1.6 \times 10^7$	$1.8 \times 10^7 \pm 2.2 \times 10^7$

Experiment 2. Glasshouse spray-trial The final spore suspension, as applied, was 2.0×10^7 CFU per L ($\pm 1.2 \times 10^5$ CFU per L) [s.e.m.] (Commenced 17th June 2015). The herbicides alone performed quite variably against *C. selloana* and *C. jubata*. After two weeks, the 0.25 strength Gallant™ caused a slight yellowing along the margins of the younger leaves. The initial impact of Weedfree and Greenscapes were very dramatic after two weeks on both *Cortaderia* species. However, the plants treated with the organic herbicides re-sprouted from the pseudo-stem. In comparison, by the end of the experiment, the 0.25 strength Gallant eventually killed the plants.

After two weeks, there was very little difference in the plants treated with *N. oryzae* alone compared to the uninoculated control for either *C. selloana* and/or *C. jubata* (Figures 2a,b). *N. oryzae* is a leaf spot pathogen and so, we were not expecting for the inundative application of *N. oryzae* alone to be able to kill pampas.

Figure 2. Response of a) *C. selloana* and b) *C. jubata* to *N. oryzae* (2.7×10^7 CFU per L) after two weeks (control on left).

In comparison, after two weeks, there was an observed synergy between the 0.25 strength Gallant and both the organic herbicides when co-applied with *N. oryzae* to *C. selloana*, compared to the control (Figures 3a,b,c). In a similar way, after two weeks, there was a very discernable difference in the *C. jubata* plants with the co-applied herbicides and *N. oryzae* inoculum compared to the control (Figures 4a,b,c).

The addition of *N. oryzae* to the herbicides increased the efficacy of all three herbicides. This synergism was associated with a reduced rate of Gallant. Also, the addition of the *N. oryzae* to the organic herbicides enabled both organic herbicides to kill pampas, which is beyond their biocidal capacity on their own. Previous studies by Peng and Byer (2005) reported an increase in pathogenicity of a leaf-spot pathogen *Pyricularia setariae* Y. Nisik. when added to the synthetic herbicide sethoxydim for the biocontrol of green foxtail.

Smith and Hallett (2006) reported that a range of chemical herbicides and adjuvants were strongly inhibitory to germination of *Microsphaeropsis amaranthi* Heiny & Mintz (Ell. and Barth.) conidia. They observed that *M. amaranthi* was compatible with a selection of post-emergent herbicides including; carfentrazone, chloransulfam and imazethapyr. However, most glyphosate products suppressed or abolished germination of *M. amaranthi* conidia.

Little biocidal interaction was observed between the 0.25 strength Gallant and *N. oryzae*, while the organic herbicides impacted upon the viability of *N. oryzae*. Successful synergy between herbicides and *N. oryzae* requires a spore with a resistant wall that protects it from the biocidal active ingredient, but does not impinge upon its infectivity. Enform Tech has applied its proprietary 'Surface Fermentation Technology' to produce a pure-spore formulation of *N. oryzae* with a hardened fungal spore which increases resilience in co-formulation with herbicides, improves product efficacy and extends shelf life.

ACKNOWLEDGMENTS

We acknowledge the research advocacy of L. Hayes and funding from Dr S. Fowler (BWII) Landcare Research. K. Richards PF Olsen Ltd. led a Sustainable Farming Fund application.

REFERENCES

Bellgard, S.E., Winks, C.J., Than, D.J., Smith, L. and Toft, R.J. (2011). Invertebrates and pathogenic microorganisms associated with pampas grass, *Cortaderia selloana* and *C. jubata* (Poaceae), in New Zealand. Landcare Research Contract Report Number: LC 140.

Coleman, R. and Penner, D. (2008). Organic acid enhancement of pelargonic acid. *Weed Technology* 22 (1), 38-41.

Gilbert, P. and McBain, A. (2003). Potential impact of increased use of biocides in consumer products on prevalence of antibiotic resistance. *Clinical Microbiology Reviews* 16, 189-208.

Figure 3. *C. selloana* responses to co-applied *N. oryzae* + herbicide (L–R): a) 0.25 strength Gallant, b) Greenscapes, c) Weedfree (control on left).

Figure 4. *C. jubata* responses to co-applied *N. oryzae* + herbicide (L–R): a) 0.25 strength Gallant, b) Greenscapes, c) Weedfree (control on left).

Gosling, D.S., Shaw, W.B. and Beadel, S.M. (1999). Review of control methods for pampas grasses in NZ. Dept. of Conservation Consultancy Report No. 290 (unpubl.) by Wildlands Consultants. 28 p.

Hoagland, R.E. (1996). Chemical interactions with bioherbicides to improve efficacy. *Weed Technology* 10, 651-74.

Leyronas, C. and Raynal, G. (2008). Role of fungal ascospores in the infection of orchardgrass (*Dactylis glomerata*) by *Epichloë typhina* agent of choke disease. *Journal of Plant Pathology* 90 (1), 15-21.

Peng, G. and Wolf, T.M. (2011). Synergy between synthetic herbicides and microbial herbicides for weed control. *Pest Technology* 5 (Special Issue 1), 18-27.

Peng, G. and Byer, K.N. (2005). Interactions of *Pyricularia setariae* with herbicides for control of green foxtail (*Setaria viridis*). *Weed Technology* 19, 589-98.

Richardson, B., Vanner, A., Ray, J., Davenhill, N. and Coker, G. (1996). Mechanisms of *Pinus radiata* growth suppression by some common forest weed species. *New Zealand Journal of Forestry Science* 26, 421-37.

Smith, D.A. and Hallett, S.G. (2006). Interactions between chemical herbicides and the candidate bioherbicide *Microsphaeropsis amaranthi*. *Weed Science* 54, 532-7.

Should we be concerned about rust-resistant biotypes of *Hypericum androsaemum* L. (tutsan) in south-eastern Australia?

Robin J. Adair[1], Huw Evans[2] and Michaela Jones[3]
[1] Australis Biological, PO Box 151, Bittern, Victoria 3918, Australia
[2] Central Tablelands Local Land Services, PO Box 510, Cowra, New South Wales 2794, Australia
[3] National Parks and Wildlife Service, PO Box 330, Oberon, New South Wales 2787, Australia
(robin.adair@australisbiological.com.au)

Summary *Hypericum androsaemum* L. (tutsan) invades pastures and natural ecosystems in high rainfall areas of south-eastern Australia. The rust fungus *Melampsora hypericorum* was detected in Australia in 1991 and drastically reduced the extent of *H. androsaemum* populations in some areas. The impact of the rust is variable and host resistance is present within populations of *H. androsaemum*. Dense populations of *H. androsaemum* reduce native species richness, particularly herbs, grasses and sub-shrubs. The selection of virulent strains of the tutsan rust or other biological control agents is required, particularly for the Blue Mountains region of NSW, where *H. androsaemum* threatens high value biodiversity assets, and is resistant to rust strains that have been highly effective in Victoria.

Keywords Biocontrol, rust *Melampsora*, resistance.

INTRODUCTION

Hypericum androsaemum L. (Clusiaceae) is a semi-deciduous, soft-wooded sub-shrub native to western and southern Europe and northern Africa. *H. androsaemum* is shade-tolerant and most commonly found on south-facing slopes of forested areas with high rainfall, warm temperatures and moderate to high humidity (Parsons and Cuthbertson 2001). Dense populations of *H. androsaemum* are reported to reduce the grazing value of pasture, and reduce or eliminate native plant species (Parsons 1973). However, there are no quantitative data available for impacts on biodiversity.

Herbicide and mechanical control options are available for *H. androsaemum*, but are mostly unsuitable for use in native vegetation due to non-target damage and the extent of invasion. Biological control is now the only long-term sustainable means of reducing the spread and impact of *H. androsaemum* (Gourlay *et al*. 2014).

In 1991, the rust fungus *Melamspora hypericorum* (DC.) J.Schröt. (tutsan rust), an unintentional introduction, was first detected in Australia at Apollo Bay in Victoria (Bruzzese and Pascoe, 1992). Although variable in its impact, the rust caused a spectacular decline of *H. androsaemum* in the Otway Ranges of Victoria (McLaren *et al.* 1997). Tutsan rust is a host specific, microcyclic fungus with only one known spore stage, uredinia (Casonato *et al.* 1999). Recorded hosts are a small group of closely related *Hypericum* species: *H. androsaemum*, *H. calycinum* and *H.* × *inodorum*, although verification is required for species other than *H. androsaemum* (Baker 1955, Casonato 1998). The impact of tutsan rust on *H. androsaemum* is highly variable. No significant relationship could be found between the presence of tutsan rust in the field and phylogenetic grouping based on the ITS sequences of *H. androsaemum* populations (Nel 2012) or random markers (Casonato 1998). By analogy with other rusts, the lack of infection in some populations of *H. androsaemum* is attributed to genetically determined variation in rust virulence and to host susceptibility, possibly due to host polyploidy (Gourlay *et al*. 2014) and multiple introductions into Australia (Casonato 1998, Casonato *et al*. 1999). Similarly, genetic variation in Australian populations of *M. hypericorum* has been detected and may represent more than one introduction of the rust (Casonato 1998, Nel 2012).

Identifying genetic markers for either host resistance (*r* gene) or pathogen virulence (*avr* gene) could explain variation in virulence of different strains of rust on different provenances of *H. androsaemum*, and lead to the selection of effective rust strains for biological control. However, multiple attempts to identify these markers have so far been unsuccessful (Casonato 1998, Nel 2012, Landcare Research 2011, H. Gourlay pers. comm. 2016) and the molecular methods used so far do not target specifically the genes for virulence in the fungus or resistance in the host. An international collaborative program led by Landcare Research NZ is attempting to find tutsan rust strains or other potential agents that have a greater impact on *H. androsaemum* in New Zealand (Landcare Research 2011, H. Gourlay, Landcare Research NZ, pers. comm.). Investment in further biological control in Australia is contingent on understanding the impacts of *H. androsaemum* in natural ecosystems (R. Carter, Central Tablelands Local Land Services, pers. comm. 2014). The impact of weeds on biodiversity values are not well documented in Australia and reports for *H. androsaemum* are anecdotal and unsubstantiated.

In this paper we investigate the impact of *H. androsaemum* on plant species richness in the Jenolan Caves area within the Jenolan Karst Conservation Reserve, New South Wales, an area heavily invaded by *H. androsaemum*.

MATERIALS AND METHODS

The study site was located on steep south or south-east facing slopes adjacent to the Jenolan River in Jenolan Karst Conservation Reserve, NSW. The vegetation community in the study site was sheltered grey gum forest or sheltered narrow-leaf peppermint forest. In November 2015, circular quadrats (4 m diameter) were haphazardly located in areas invaded by *H. androsaemum*. Sampling points were located in areas with uniform density of *H. androsaemum* and within patches exceeding 50 square metres. Sites with varying densities of *H. androsaemum* were intentionally selected. The positions of sample sites were recorded using GPS (UTM GDA94) and the aspect measured. At each sample site, all vascular plants were recorded and a Domin scale cover-abundance ranking was allocated to each species. The above-ground biomass of *H. androsaemum* was harvested from a 1 m^2 quadrat located in the centre of each sample area. Harvested material was air-dried at 40–60°C for several weeks in a closed polyhouse, before being weighed.

RESULTS

Hypericum androsaemum reduced native species richness and exotic species richness as cover-abundance of *H. androsaemum* increased. Polynominal functions best described the trend line between species richness and *H. androsaemum* cover, with declines in native species richness becoming evident when cover values of *H. androsaemum* exceeded 25% (cover class value 5 or higher) (Figure 1). Where *H. androsaemum* cover is greater than 90%, native species richness was reduced by around 66%, where mostly shade tolerant herbs e.g. (*Stellaria flaccida* Hook, *Urtica incisa* Poir., *Geranium* sp., *Hydrocotyle laxiflora* DC.), grasses (*Microlaena stipoides* (Labill.) R.Br.) and ferns (*Adiantum aethiopicum* L., *Pellaea falcata* (R.Br.) Fée, *Asplenium flabellifolium* Cav.) persisted at low cover-abundance levels. Increasing cover of *H. androsaemum* also lead to decreased cover of exotic plant species (Figure 2).

A strong polynomial relationship was found between the dry weight of *H. androsaemum* and native species richness ($r^2 = 0.831$). Native species richness declined with increasing biomass of *H. androsaemum*, but reached a basal asymptote of around 10 species^{-2} (Figure 3). Exotic species richness was weakly correlated with the dry mass of *H. androsaemum* ($y = 2E - 05x^2 - 0.0153x + 5.29$, $R^2 = 0.5088$), and also declined with increasing biomass of the latter.

Increasing cover of *H. androsaemum* had variable impact on life-form groups. Grasses, herbs and sub-shrubs showed weak trends (low r^2 values), with cover declining with increasing cover of *H. androsaemum*.

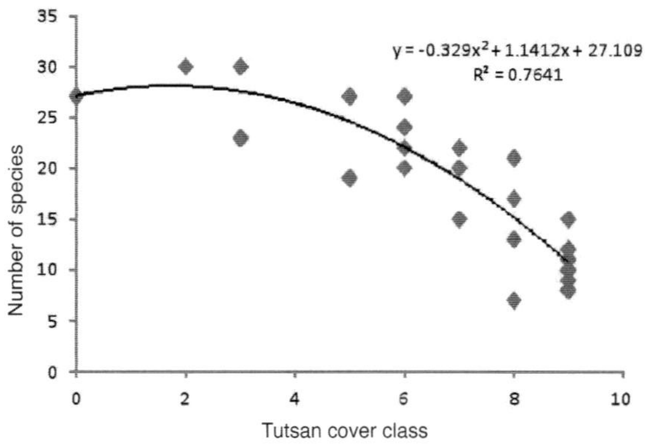

Figure 1. Native plant richness and cover class of *H. androsaemum*.

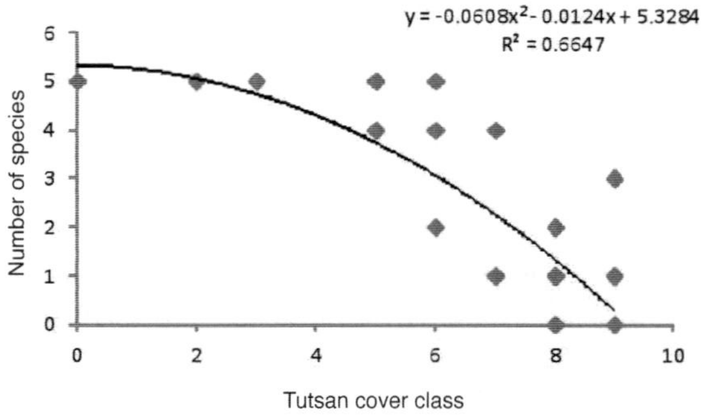

Figure 2. Number of exotic species and Domin cover class of *H. androsaemum*.

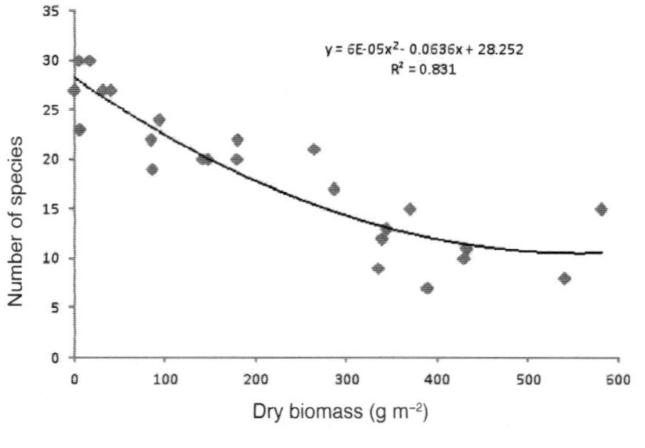

Figure 3. Plant species richness and dry biomass of *H. androsaemum*.

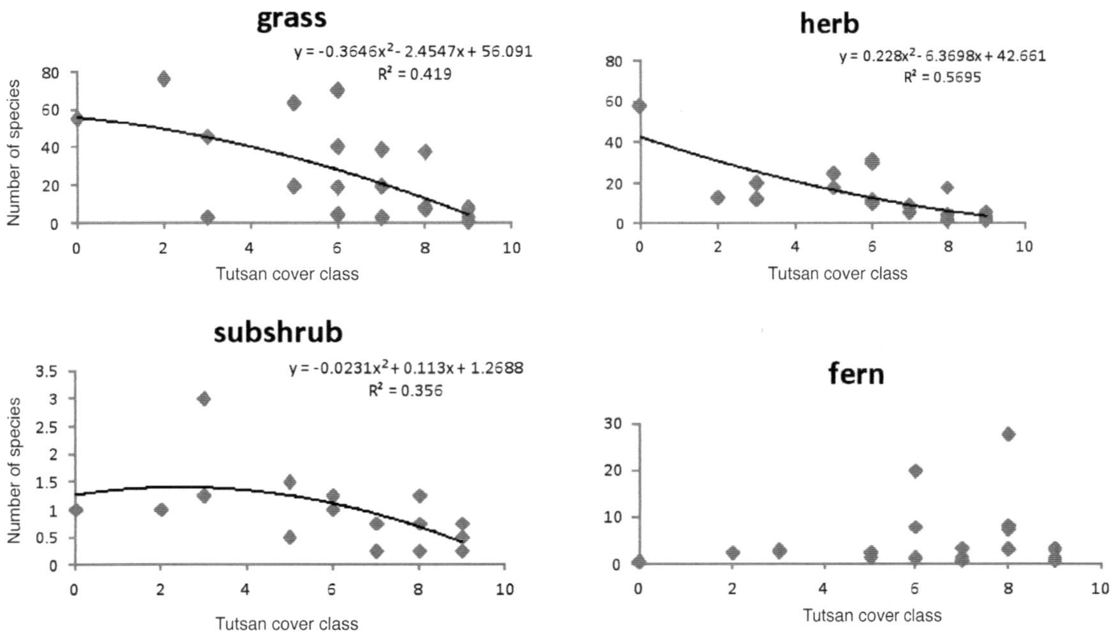

Figure 4. Richness of life forms and cover of *H. androsaemum*.

However, no clear trends were evident for shrubs, trees, ferns or lianes (Figure 4).

DISCUSSION

The discovery of tutsan rust in Victoria in 1991 and its subsequent impact on *H. androsaemum* is claimed as the most spectacularly successful example of weed biological control ever witnessed in Victoria (McLaren *et al.* 1997). In a short period of time infestations of *H. androsaemum* at Apollo Bay were drastically reduced in extent and density. Rusts generally have high levels of host specificity, long-distance dispersal capabilities and can severely reduce host fitness when virulent. Therefore, they have considerable potential as classical biological control agents. Weed species with narrow genetic variation make good targets for biological control with rusts (and other agents) as there is a lower probability for the development of host resistance.

The current strains of tutsan rust in Australia provide inadequate control to all populations of *H. androseamum*. In New South Wales, *H. androsaemum* demonstrates resistance to tutsan rust collected from areas in Victoria where control has been effective (Adair 2014). Regrettably, resistant populations of *H. androsaemum* are also emerging in Victoria (Adair 2014).

Our data indicate that native species richness in wet forests of the Blue Mountains is adversely affected by invasion of *H. androsaemum*, and that even low levels of biomass contribute to species decline. While *H. androsaemum* canopy cover could follow a similar trend it was not evident in our data and appeared to require a threshold cover level of around 50% before native species decline was evident. However, this could reflect lower sensitivity of categorical data (cover class) in detecting trends in the response variable (species richness). The mechanism of species depletion in areas invaded by *H. androsaemum* are not fully understood, but are most likely intense shading and lack of sufficient light to support ground stratum communities of grasses, herbs and sub-shrubs. Reduced availability of water and competition for nutrients may also be involved. While native species may persist at sub-threshold levels, *H. androsaemum* may form a complete canopy cover in susceptible habitats with time, therefore broadening its ecological impact. It is likely that the tree regeneration may be restricted to gaps in the canopy of *H. androsaemum* or areas where *H. androsaemum* is absent or has low cover, but this observation remains speculative until further data is available.

H. androsaemum is widely established along the Jenolan and lower Coxs River catchments in the Blue Mountains and forms dense and extensive cover on south-facing slopes in steep and rugged terrain. Infestations occur within the Greater Blue Mountains World Heritage Area, a biodiversity hotspot (Williams *et al.* 2011), specifically the Kanangra-Boyd Wilderness Area, the Warragamba Special Area, the Blue Mountains and Kanangra-Boyd National Parks, and the Jenolan Karst Conservation Reserve. *H. androsaemum* has spread from Jenolan Caves to the foreshores of Lake Burragorang (Sydney's main water supply), via the Jenolan and Coxs Rivers.

Recently, a population was discovered in the Hollander's River catchment, which drains to the Kowmung River, a Wild and Scenic river. Currently, *H. androsaemum* threatens restricted plant communities such the Jenolan Limestone Scrub Woodland, and the swamp communities (Carey 2007). In addition, the habitat of threatened or possibly threatened species such as *Trachymene scapigera* (Domin) B.L.Burtt, *Acacia clunies-rossiae* Maiden, *Euphrasia scabra* R.Br. and *Lastreopsis hispida* (Sw.) Tindale are at risk. The distribution of *H. androsaemum* is anticipated to expand within the Blue Mountains, where all south-facing slopes in high-rainfall areas are susceptible to invasion, further jeopardising the conservation status of native communities and species.

Conventional control methods cannot contain or suppress rust-resistant invasions of *H. androsaemum* in New South Wales due to the extent and severity of its distribution, and the difficult terrain it which it grows. Biological control is now the only option available and selection of new pathogens and other agents continues in New Zealand (e.g. New Zealand Landcare Research 2016). Attempts to locate an *avr* gene in tutsan rust have so far been unsuccessful, but *avr* genes have been found in *M. lini*, which infects cultivated flax (Dodds *et al.* 2007) and could guide the search in *M. hypericorum*.

In answer to the question we pose in the title, Australia does need to be concerned about the unabated spread of *H. androsaemum*, and further investment in its biological control is warranted.

ACKNOWLEDGMENTS

This research was supported by the Central Tablelands Local Land Services through funding from the Australian Government. Ann Lawrie (RMIT University) provided valuable comments on an earlier version of this paper.

REFERENCES

Adair, R.J. (2014). Tutsan rust compatibility evaluation for NSW populations of *Hypericum androsaemum*. Unpublished report to Hawksbury-Nepean Catchment Management Authority.

Bruzzese, E. and I.G. Pascoe (1992). *Melampsora hypericorum*, a rust fungus with potential in the biological control of tutsan, *Hypericum androsaemum*. Proceedings of the International Weed Control Congress, Melbourne, Vic. (Australia), Weed Science Society of Victoria Inc., 1992.

Carey, A. (2007). Protecting swamp communities in the Blue Mountains. *Australasian Plant Conservation: Journal of the Australian Network for Plant Conservation* 16, 14-16.

Casonato, S. (1998). Biological control of the weed tutsan (*Hypericum androsaemum*) with the rust fungus, *Melampsora hypericorum*. Honours Thesis. Department of Applied Biology and Biotechnology, RMIT, Bundoora.

Casonato, S.G., Lawrie, A.C. and McLaren, D.A. (1999). Biological control of *Hypericum androsaemum* with *Melampsora hypericorum*. Proccedings of the 12th Australian Weeds Conference, eds A.C. Bishop, M. Boersma and C.D. Barnes, pp. 339-42. Tasmanian Weed Society, Hobart, Tasmania.

Department for Environment and Conservation NSW (2006). The vegetation of the Western Blue Mountains. Unpublished report funded by the Hawkesbury. Nepean Catchment Management Authority.

Dodds, P.N., Catanzariti, A.M., Lawrence, G.J. and Ellis, J.G. (2007). Avirulence proteins of rust fungi: penetrating the host-haustorium barrier. *Australian Journal of Agricultural Research* 58, 512-17

Gourlay, H., Rendell, E., Olsen, E., Groetmann, R., Probst, C., Padamsee, M., Bellgard, S. and Goecke, D. (2014). The results of surveys and DNA analysis of tutsan (*Hypericum androsaemum*) and tutsan rust (*Melampsora hypericorum*) in New Zealand and Europe. Proceedings of the XIV International Symposium on Biological Control of Weeds, eds F.A.C. Impson, C.A. Kleinjan and J.H. Hoffman, 2–7 March 2014. Kruger National Park, South Africa.

McLaren, D.A., Bruzzese, E. and Pascoe, I.G. (1997). The potential of fungal pathogens to control *Hypericum* species in Australia. *Plant Protection Quarterly* 12, 81-3.

Nel, T. (2012). How extensive is host resistance in the tutsan–*Melamspora hypericorum* relationship. Honours Thesis. School of Applied Science, RMIT University.

New Zealand Landcare Research (2016). www.landcareresearch.co.nz/publications/newsletters/biological-control-of-weeds/issue-74/tutsan-agents. Cited 17 June 2016.

Parsons, W.T. and Cuthbertson, E.G. (2001). 'Noxious weeds of Victoria', 2nd edition. (CSIRO Publishing, Collingwood).

Williams, K.J., Ford, A., Rosauer, D.F., De Silva, N., Mittermeier, R., Bruce, C., Larsen, F.W. and Margules, C. (2011). Forests of East Australia: the 35th biodiversity hotspot. *In* 'Biodiversity hotspots', pp. 295-310. (Springer).

Is dieback in invasive plants due to a loss of protective microorganisms, or encounters with new pathogens?

Tracey V. Steinrucken[1,3], Andrew Bissett[2], Jeff R. Powell[1], Rieks D. van Klinken[3], Anil K.H. Raghavendra[3] and Matteo Garbelotto[4]
[1] Western Sydney University Locked Bag 1797, Penrith, New South Wales 2751, Australia
[2] CSIRO GPO Box 2583, Brisbane, Queensland 4001, Australia
[3] CSIRO GPO Box 1538, Hobart, Tasmania 7001, Australia
[4] University of California Berkeley, 54 Mulford Hall, Berkeley, CA 94720, United States of America
(tracey.steinrucken@csiro.au)

Summary *Parkinsonia aculeata* L. (parkinsonia) is a Weed of National Significance. It impacts heavily on biodiversity, livestock and pasture management across northern Australia. Dieback in some parkinsonia populations has resulted in levels of control not previously achieved by other methods, and therefore raises the possibility of biological control of parkinsonia, using dieback. The composition of parkinsonia's microbial endophyte communities, or 'phytobiome', suggests that multiple endophyte species may be responsible for dieback. Since dieback does not occur in the native range, we hypothesized that native-range parkinsonia hosts a different phytobiome that 'protects' its host against dieback-causing agents, whereas introduced parkinsonia does not. Hence, dieback may be caused by pathogens encountered by invasive parkinsonia upon its introduction to Australia.

To test this hypothesis we compare microbial DNA from healthy parkinsonia sampled in the USA, to DNA from healthy and dieback-affected parkinsonia in Australia. Using Next-Gen Illumina sequencing we identified species of fungal, bacterial and archaeal endophytes that are unique to the country from which they were sampled. We also found evidence that some of these endophytes are vertically transmitted, but others are not, suggesting that parkinsonia may have lost potentially protective endophytes upon introduction to Australia, thereby reducing its capacity to fight generalist pathogens that cause dieback.

This study describes a novel way to analyse multiple microbial taxa using a single high-throughput method. Although many plant diseases are caused by single putative pathogens, we've shown that parkinsonia dieback may be caused by interactions between endophyte communities. These results will lead to further insights into the biological control of dieback-affected weeds in Australia.

Keywords Endophytes, fungi, bacteria, Illumina sequencing, native-range, phytobiome.

Developing additional herbicide control options for rubber vine (*Cryptostegia grandiflora* R.Br.)

Shane D. Campbell and Dannielle A. Brazier
Department of Agriculture and Fisheries, Tropical Weeds Research Centre, PO Box 187, Charters Towers, Queensland 4820, Australia
(Shane.Campbell@daf.qld.gov.au)

Summary Herbicide control of rubber vine (*Cryptostegia grandiflora* R.Br.), a Weed of National Significance in Australia, can be effectively achieved in most situations using a range of chemicals and various techniques, including basal bark, cut stump, foliar and aerial applications. Nevertheless control in areas with poor access and with native vegetation still poses difficulties for some land managers, particularly doing so in a cost effective manner.

The successful incorporation of low-volume high-concentration herbicide applications into control programs for several woody weeds in recent years prompted further testing of this technique for control of rubber vine. The efficacy of single rates of four herbicides plus a combination of two herbicides was compared against an untreated control. The herbicide treatments tested contained the active ingredients aminopyralid/metsulfuron-methyl (375/300 g a.i. kg^{-1}), metsulfuron-methyl (600 g a.i. kg^{-1}), metsulfuron-methyl + glyphosate (600 + 360 g a.i. kg^{-1}), triclopyr/picloram (300/100 g a.i. kg^{-1}) and triclopyr/picloram/aminopyralid (200/100/25 g a.i. kg^{-1}).

Two years after application, triclopyr/picloram was the only treatment to have given 100% mortality, but statistically it was not significantly different ($P > 0.05$) to aminopyralid/metsulfuron-methyl, metsulfuron-methyl or triclopyr/picloram/aminopyralid, which averaged between 82–91% mortality. The metsulfuron-methyl + glyphosate treatment performed poorly (49% mortality): this appears to be associated with some antagonistic effect given that metsulfuron-methyl on its own performed much better. Based on these results, two more trials have been initiated to refine rates for metsulfuron-methyl and triclopyr/picloram and to compare low-volume high-concentration applications against traditional foliar spraying using the same herbicide (triclopyr/picloram) (in terms of efficacy and cost).

Keywords Chemical, splatter gun, weed.

INTRODUCTION

Rubber vine (*Cryptostegia grandiflora* R.Br.) is one of the most problematic weeds in northern Australia which has led to its declaration throughout Australia, along with its classification as a Weed of National Significance (Doak and Deveze 2004). A large body of research has been undertaken into the impacts, ecology and control of this weed over the last 30 years (Tomley 1995, Grice 1996, Palmer and Vogler 2012). This has resulted in the identification of effective control options for different densities and situations, including the individual and integrated use of biocontrol, chemical, and mechanical techniques and the use of fire (Doak and Deveze 2004, Palmer and Vogler 2012).

In recent years low-volume high-concentration herbicide applications (e.g. splatter guns) have become increasingly used in Queensland as an effective option for controlling some other woody weeds in difficult to access areas, such as lantana (*Lantana camara* L.), bellyache bush (*Jatropha gossypiifolia* L.) and Siam weed (*Chromolaena odorata* (L.) King & Robinson) (Somerville *et al.* 2011, Brooks *et al.* 2014). The use of low-volume high-concentration applications has been tested previously (1980s) on rubber vine using sprinkler sprayers, with two chemicals found to be effective (picloram/2,4-D and dicamba) (Harvey 1987). Despite its effectiveness, this technology was not widely adopted by landholders for control of rubber vine, with a preference for other options such as the use of fire, foliar spraying, cut stump and basal bark applications and ground applied residual herbicides where permissible (Doak and Deveze 2004). With an increased range of herbicides and advances in technology since then, a screening trial commenced in 2014 to evaluate the use of splatter guns for applying low-volume high-concentration herbicide applications on rubber vine.

MATERIALS AND METHODS

A randomised complete block experiment was implemented on a cattle property about 20 km east of Mount Surprise, north Queensland. The site comprised a light to medium density infestation of mainly free standing rubber vine (i.e. not climbing up trees) growing in open eucalypt woodland on basalt derived soil.

There were six treatments in total each replicated three times using a randomised complete block design. Single rates of four herbicides plus a combination of two herbicides were compared against an untreated control (Table 1). Experimental units comprised clusters of 15 medium-sized rubber vine plants. All selected plants were of the shrub form with those growing as vines (i.e. climbing up neighbouring trees) excluded. Selected plants were also a minimum of 4 m apart from each other to minimise the risk of spray drift.

On the day before treatments were implemented, the height, basal diameter and width of the canopy of all tagged rubber vine plants was measured. An estimate of the percentage leaf cover and whether plants were flowering and/or podding was also recorded. The rubber vine plants were on average 2.02 ± 0.03 m high, with a basal diameter of 6.4 ± 0.2 cm and a canopy width of 2.19 ± 0.04 m. Leaf cover averaged 80 ± 2% and 99% and 56% of the plants were flowering and podding, respectively. There was no evidence of a leaf rust (*Maravalia cryptostegiae* (Cummins)) being present at the time.

Treatments were implemented between the 10–12 March 2014. They were applied using a manually operated 'Forestmaster' applicator (N.J. Phillips®) set to deliver 20 mL shots via an 'adjustable cone' nozzle. For each experimental unit, two litres of solution was prepared, which included the selected chemical (Table 1) plus 2 mL per L of the non-ionic wetter/spreader/penetrant Pulse® (Nufarm) (1020 g L^{-1} polyether modified polysiloxane) and 1 mL per L of red Spraymate™ Spray Marker Dye (150 g L^{-1} Rhodamine B).

Spraying was undertaken in accord with the herbicide labels for Picloram + Triclopyr 400 and Stinger™, with 10 mL of mixture applied per metre squared of surface area of the plant. On average, each plant received 36 ± 1.5 mm of herbicide mixture.

RESULTS

Two years after application, triclopyr/picloram was the only treatment to have given 100% mortality, but statistically it was not significantly different ($P > 0.05$) to aminopyralid/metsulfuron-methyl, metsulfuron-methyl or triclopyr/picloram/aminopyralid, which averaged between 82–91% mortality (Figure 1). The metsulfuron-methyl + glyphosate treatment performed poorly (49% mortality). Some mortality (9%) of control plants was recorded but this tended to be smaller plants and was still significantly less than any of the herbicide treatments (Figure 1).

DISCUSSION

In this study, low-volume high-concentration applications of triclopyr/picloram were able to provide 100% mortality of medium sized rubber vine plants that were growing in the shrub form. Although the equipment and herbicide used was different, the findings are consistent with an earlier study by Harvey (1987) who achieved high mortality using a sprinkler sprayer to apply low-volume high-concentration applications of 2,4-D/picloram and dicamba. Surprisingly, the metsulfuron-methyl + glyphosate treatment performed poorly (49% mortality) and appears to be associated with some antagonistic effect given that metsulfuron-methyl on its own performed significantly better (82% mortality).

Since this initial study, two other trials have been initiated to refine rates for triclopyr/picloram and metsulfuron-methyl and to compare low-volume high-concentration applications (in terms of efficacy and cost) against traditional foliar spraying using the same herbicide (triclopy/picloram). In the rate refinement trial, triclopyr/picloram has been included to

Table 1. Herbicide treatments implemented on rubber vine using low-volume high-concentration applications.

Herbicide trade name	Active ingredient (g a.i. kg^{-1} or L^{-1})	Herbicide mix rate	Active ingredient rate (g a.i. L^{-1})
Brush-Off®	metsulfuron-methyl (600 g kg^{-1})	2 g L^{-1}	1.2
Brush-Off® + Weedmaster® Duo	metsulfuron-methyl (600 g kg^{-1}) + glyphosate (360 g L^{-1})	2 g L^{-1} + 100 mL L^{-1}	1.2 + 36
Picloram + Triclopyr 400	triclopyr/picloram (300/100 g L^{-1})	50 mL L^{-1}	15/5
Stinger™	aminopryalid/metsulfuron-methyl (375/300 g L^{-1})	4 g L^{-1}	1.5/1.2
Tordon™ RegrowthMaster	triclopyr/picloram/aminopyralid (200/100/25 g L^{-1})	50 mL L^{-1}	10/5/1.25
Control			

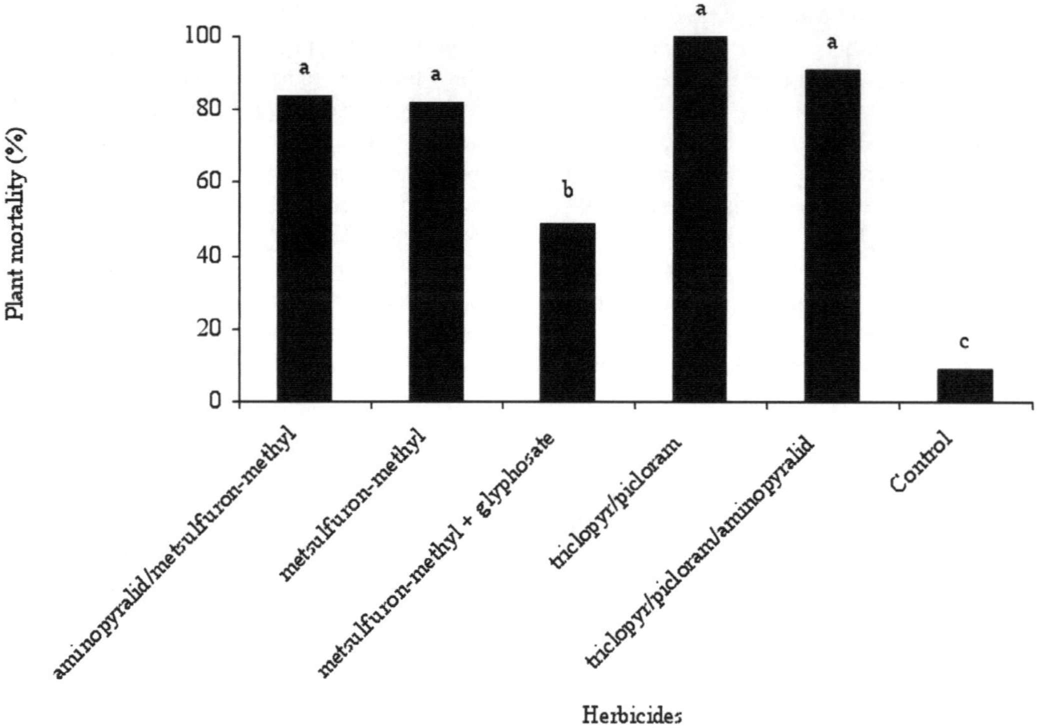

Figure 1. Effect of herbicide treatments on plant mortality (%) of rubber vine 24 months after treatment. Vertical bars with the same letter are not significantly different from each other ($P > 0.05$).

see if lower rates than that used in the current study will still provide high efficacy. Given its low cost, metsulfuron-methyl was also included to determine whether slightly higher rates would increase mortality.

While still in the early stages, preliminary observations of these trials suggest that mortality will be lower than the initial study. Differences in plant density may be a contributing factor and will be explored further if this trend continues.

The level of infection from a leaf rust (*M. cryptostegiae*) will need to be monitored by land managers interested in using low-volume high-concentration applications for control of rubber vine. The rust was released as a biological control agent in Australia in the 1990s and in many areas it has proven very effective at reducing the vigour and reproduction of rubber vine plants (Palmer and Vogler 2012). However, the infection of leaves by the rust has been shown to reduce the efficacy of high-volume foliar spraying (Vitelli and Madigan 1999) and it is likely that low-volume high-concentration applications could be equally affected.

Whilst not formally tested at this stage, anecdotally it does not appear that the vine form of rubber vine (i.e. those plants that are growing up trees) will be able to be effectively controlled using low-volume high-concentration applications. Any plant material that escapes being sprayed, by being out of the reach of the sprayer and the equipment being used, does not appear to be affected allowing the plant to remain alive.

The results in this study are applicable to small to medium sized rubber vine plants that are not climbing up trees and where the infestation is still at a density where the whole surface of individual plants is assessable for spraying. In such situations, the findings from studies on other weeds (lantana and Siam weed) suggest that whilst the amount of herbicide used may be slightly higher (Thompson 2013) than high-volume foliar applications, the time taken to treat plants will be greatly reduced using this technique (Thompson 2013, Brooks *et al.* 2014). The labour efficiencies along with the portability of the application equipment make it an attractive option for difficult to access areas. Furthermore, there is also a wide range of equipment that can be used depending on the size and density of infestations, ranging from inexpensive drench guns to manual, gas-powered backpack and larger ATV-mounted units.

A minor use permit (PER82156) has recently been approved by the Australian Pesticides and Veterinary Medicines Authority (APVMA) for the use of chemicals containing triclopyr (300 g a.i. kg^{-1})/picloram (100 g a.i. kg^{-1}) (e.g. Picloram + Triclopyr 400,

Conqueror®) and triclopyr (300 g a.i. kg^{-1})/picloram (100 g a.i. kg^{-1})/aminopyralid (8 g a.i. kg^{-1}) (Grazon™ Extra) to control rubber vine using low-volume high-concentration applications.

ACKNOWLEDGMENTS

The Authors thank the Saunders family for allowing the trial to be undertaken on their property. We are also grateful to Christopher Crowley and Kirsty Gough for their assistance in establishing and monitoring the trial. Joe Scanlan provided valuable feedback on an earlier draft of this paper.

REFERENCES

Brooks, S.J., Gough, K.L. and Campbell, S.D. (2014). Refining low-volume, high-concentration herbicide applications to control *Chromolaena odorata* (L.) King & Robinson (Siam weed) in remote areas. *Plant Protection Quarterly* 36, 459-67.

Doak, A. and Deveze, M. (2004). 'Rubber vine management: control methods and case studies'. Queensland Department of Natural Resources, Mines and Energy, Brisbane.

Grice, A.C. (1996). Seed production, dispersal and germination in *Cryptostegia grandiflora* and *Ziziphus mauritiana*, two invasive shrubs in tropical woodlands of northern Australia. *Australian Journal of Ecology* 21, 324-31

Harvey, G.L. (1987). Low-volume application of herbicides for control of rubber vine. Proceedings of the eighth Australian Weeds Conference, pp. 49-54.

Palmer, B and Vogler, W. (2012). *Cryptostegia grandiflora* (Roxb.) R.Br. – rubber vine. *In* 'Biological control of weeds in Australia', eds M. Julien, R. McFadyen and J. Cullen, pp. 190-7. (CSIRO Publishing, Melbourne).

Somerville, S., Somerville, W. and Coyle, R. (2011). Regenerating native forest using splatter gun techniques to remove lantana. *Ecological Management and Restoration* 12, 164-74.

Thompson, A. (2013). The importance of accurate surveying for implementing best practice weed control methods whilst working to external constraints. Proceedings of the twelfth Queensland Weed Symposium, pp. 39-42.

Tomley, A.J. (1995). The biology of Australian weeds 26 *Cryptostegia grandiflora* R.Br. *Plant Protection Quarterly* 10, 122-30.

Vitelli, J.S. and Madigan, B.A. (1999). Effect of rust on the efficacy of foliar applied herbicides on rubber vine. Proceedings of the twelfth Australian Weeds Conference, pp. 238-41.

Morpho-physiological responses of two Australian biotypes of parthenium weed (*Parthenium hysterophorus* L.) to soil moisture stress

Ali A. Bajwa[1,2], Bhagirath S. Chauhan[2] and Steve W. Adkins[1,2]
[1] School of Agriculture and Food Sciences, The University of Queensland, Gatton, Queensland 4343, Australia
[2] The Centre for Plant Science, Queensland Alliance for Agriculture and Food Innovation, The University of Queensland, Toowoomba, Queensland 4350, Australia
(a.bajwa@uq.edu.au)

Summary A pot study was conducted to evaluate the morpho-physiological responses of two Australian biotypes (Clermont and Toogoolawah) of parthenium weed (*Parthenium hysterophorus* L.) to 50, 75 and 100% of the soil water holding capacity (WHC). Moisture stress affected growth of both biotypes significantly ($P < 0.05$); however, the growth of Clermont was less affected than Toogoolawah across all soil moisture regimes. Significantly greater numbers of branches, leaves, longer shoots, higher fresh weight, and dry weight were observed for plants growing at 75% of WHC as compared with those growing at 100% of WHC. However, root length was increased in response to the 50% reduction in soil moisture. Physiological attributes including, net photosynthesis and stomatal conductance were also higher for Clermont than Toogoolawah and were at a maximum at 75% of WHC. Parthenium weed maintains good growth under sub-optimal soil moisture and therefore, is able to invade a wider range of environments than expected.

Keywords Biological invasion, parthenium weed, stress tolerance, weed biology.

INTRODUCTION

Parthenium weed is a Weed of National Significance (WoNS) in Australia with high invasive potential and severe impact on agriculture (Shabbir and Adkins 2014, Bajwa *et al.* 2016). It causes huge losses to grazing land productivity, livestock production, native biodiversity and crop production (Adkins and Navie 2006). Parthenium weed is more problematic due to its highly invasive nature. Two Australian biotypes of parthenium weed, Toogoolawah and Clermont (named after the places of their first introductions) differ significantly in their invasiveness. Clermont is highly invasive while Toogoolawah is non-invasive (Navie *et al.* 1996, Hanif *et al.* 2012). So, comparative study of their growth responses may provide better understanding of the invasion mechanism and will have implications for management.

Parthenium weed can tolerate abiotic stresses, which enables it to invade, establish, and even flourish under a wide range of stressful conditions (Kohli *et al.* 2006). Soil moisture is one of the most important factors determining the establishment and spread of an invasive weed species (Bajwa *et al.* 2016). Parthenium weed is known to invade areas with low moisture availability which might be due to its ability to tolerate soil moisture stress (Bajwa *et al.* 2016). Nguyen (2011) reported that under a low soil moisture condition, parthenium weed's life span was reduced by up to 43% while growth was enhanced by 20% due to modifications in the plant's vegetative and reproductive biology. Parthenium weed growth was improved under decreased soil moisture levels along with increased temperature and CO_2 levels (Belgeri 2013). The ability to tolerate the abiotic stresses through physiological regulation could be a strong tool for invasion.

Fifty percent of soil WHC is supposed to be a significant drought which affects growth of several crops and weed species (Nguyen 2011). Only a few studies have been conducted to evaluate the morpho-physiological responses of parthenium weed to soil moisture stress. The results of the present study may enable us to predict features of its invasion potential for different climatic and ecological zones.

MATERIALS AND METHODS

The experiment was carried out in a completely randomized design with four replications per biotype and 100% of WHC was considered as a control (which means total amount of water held by soil under field conditions). Soil moisture levels were maintained by calculating the actual water holding capacity of the soil and by adding the required amount of water at regular intervals. The weight of plants was also considered while maintaining the pot weight in each treatment. Parthenium weed seeds were germinated in Petri dishes in an incubator (25/15°C day/night temperature with 12h/12h photoperiod). The seedlings were transplanted to the soil (heavy clay loam) filled pots (300 mm diameter) kept in a glasshouse after 2 weeks growth. The experiment ran for 85 days till the plants matured and produced seed. Number of leaves and branches were counted at peak vegetative growth (40 days after transplanting). Gas exchange parameters were also measured 40 days after transplanting between 10:00 am and 12:00 pm, on the penultimate leaves (three leaves per plant) using a LI-6400 portable photosynthesis system (LI-COR Inc., Lincoln, NE, USA). Root and shoot lengths, and plant fresh and dry weights were recorded at harvest. The effects of treatments on the variables were statistically analyzed and tested by an analysis of variance (ANOVA). Comparisons among treatment means were made using the Tukey's HSD (honest significant difference) test calculated at $P < 0.05$.

RESULTS

Soil moisture stress affected growth of both Australian biotypes of parthenium weed significantly ($P < 0.05$) (Tables 1, 2). Morphological attributes including, number of leaves, number of branches, root length, shoot length, plant fresh weight, and plant dry weight were affected by different soil moisture levels (Table 2). Overall, the growth of Clermont biotype was more than Toogoolawah biotype at all moisture levels (Table 1). Maximum number of branches, number of leaves and shoot length were observed from plants growing at 75% moisture of soil WHC as compared with those growing at 100% WHC (Table 2). However, 50% WHC caused a significant decline in branching and leaf score of both biotypes as compared with 100%.

Root length was also more for Clermont as compared with Toogoolawah at all moisture levels (Table 1). Maximum root length for both biotypes was observed at severe moisture stress (50% WHC) followed by 75% WHC. Minimum root length was observed at 100% WHC. Whole plant fresh and dry weights were also more for Clermont than Toogoolawah (Table 1). Maximum fresh and dry weights were recorded from plants growing at 75% WHC followed by those at 100% WHC whereas, the minimum fresh and dry weights were observed at 50% WHC for both biotypes (Table 2).

Net photosynthesis and stomatal conductance were higher for Clermont than Toogoolawah across the moisture levels (Table 3). Maximum net photosynthesis and stomatal conductance were observed from plants growing at 75% WHC followed by those growing at 100% WHC while minimum values for these attributes were observed at 50% WHC (Table 3).

DISCUSSION

The ability of parthenium weed to grow at different levels of soil moisture indicates its successful biology to adapt a wide range of environmental conditions. Previously, variable responses of parthenium weed to soil type and quality have also been denoted to its invasiveness (Annapurna and Singh 2003). Vigorous growth of Clermont biotype at various moisture levels as compared with Toogoolawah biotype provides the

Table 1. Effect of biotypes on morphological attributes.

	Biotype	
Attribute	Clermont	Toogoolawah
No. of branches	21a[A]	18b
No. of leaves	63a	53b
Shoot length (cm)	60a	55b
Root length (cm)	23a	21b
Plant fresh weight (g)	122a	114b
Plant dry weight (g)	49a	36b

[A] Individual treatment means in each row followed by different letters are significantly different at $P < 0.05$.

Table 2. Effect of soil moisture on morphological attributes.

	Soil moisture level		
Attribute	WHC50	WHC75	WHC100
No. of branches	15a[A]	29a	19b
No. of leaves	43c	73a	59b
Shoot length (cm)	53c	63a	57b
Root length (cm)	24a	22ab	21b
Plant fresh weight (g)	102c	132a	120b
Plant dry weight (g)	32c	46a	39b

[A] Individual treatment means in each row followed by different letters are significantly different at $P < 0.05$.

Table 3. Effect of biotypes and soil moisture on photosynthesis and stomatal conductance.

Net photosynthesis (mol m^{-2} s^{-1})		
	Clermont	Toogoolawah
WHC50	21.8e[A]	17.3f
WHC75	35.2a	30.8b
WHC100	27.2c	25.0d
Stomatal conductance (m mol m^{-2} s^{-1})		
	Clermont	Toogoolawah
WHC50	0.7cd	0.3d
WHC75	3.4a	1.6b
WHC100	1.2bc	0.9c

[A] Interaction means followed by different letters are significantly different at $P < 0.05$.

reason for profound invasiveness of Clermont (Bajwa *et al.* 2016). Sustaining its growth through morphological adaptations and physiological regulation even at low moisture availability provides parthenium weed an excellent mechanism to reproduce and spread effectively (Nguyen 2011). Maximum growth and photosynthetic activity at 75% of soil WHC is unique and shows that parthenium weed thrives best under slightly less moist conditions. Usually, plants thrive best at optimum moisture which exist at soil WHC but the ability of parthenium weed to grow best at 25% less moisture of soil WHC is an important feature towards moisture stress tolerance. More branching, leaf score and tall stature ensures vigorous growth and physiological activity. Increase in root length at 50% less soil moisture would have helped plants to absorb water efficiently. In addition, higher photosynthetic activity of Clermont biotype might have significant role in its invasiveness.

In conclusion, the Clermont biotype of parthenium weed grew best at the 75% of WHC. Vigorous growth and higher photosynthetic efficiency of the Clermont biotype may explain why it is highly invasive compared to Toogoolawah which is a non-invasive biotype. Although growth and photosynthesis of parthenium weed was reduced at the 50% of WHC, plants still completed their life cycle. The ability to grow vigorously and moderately at 75 and 50% of WHC, respectively, may help parthenium weed to establish in a wide range of edaphic environments. There are certainly many other factors involved in invasion mechanism of these parthenium weed biotypes. Further research to evaluate the individual and interactive effects of other environmental factors including, temperature and CO_2 may unravel the mechanism of parthenium weed invasion. It will not only help to predict future invasions under climate change scenario but also will have pragmatic implications for management of this problematic weed.

ACKNOWLEDGMENTS

This manuscript is part of the PhD research project of Ali Ahsan Bajwa who is thankful to the Australian Government and the University of Queensland, Australia for the provision of an International Postgraduate Research Scholarship (IPRS) and UQ Centennial (UQCent) Scholarship, respectively.

REFERENCES

Adkins, S. and Shabbir, A. (2014). Biology, ecology and management of the invasive parthenium weed (*Parthenium hysterophorus* L.). *Pest Management Science* 70, 1023-9.

Adkins, S.W. and Navie, S.C. (2006). Parthenium weed: a potential major weed for agro-ecosystems in Pakistan. *Pakistan Journal of Weed Science Research* 12, 19-36.

Annapurna, C. and Singh, J.S. (2003). Variation of *Parthenium hysterophorus* in response to soil quality: implications for invasiveness. *Weed Research* 43, 190-98.

Bajwa, A.A., Chauhan, B.S., Farooq, M., Shabbir, A. and Adkins, S.W. (2016). What do we really know about alien plant invasion? A review of the invasion mechanism of one of the world's worst weeds. *Planta* 244, 39-57.

Belgeri, A. (2013). Assessing the impact of parthenium weed (*Parthenium hysterophorus* L.) invasion and its management upon grassland community composition. PhD Thesis, The School of Agriculture and Food Sciences, The University of Queensland, Australia.

Hanif, Z., Adkins, S.W., Prentis, P.J., Navie, S.C. and O'Donnell, C. (2012). Characterization of the reproductive behaviour and invasive potential of parthenium weed in Australia. *Pakistan Journal of Weed Science Research* 18, 767-74.

Kohli, R.K., Batish, D.R., Singh, H.P. and Dogra, K.S. (2006). Status, invasiveness and environmental threats of three tropical American invasive weeds (*Parthenium hysterophorus* L., *Ageratum conyzoides* L., *Lantana camara* L.) in India. *Biological Invasions* 8, 1501-10.

Navie, S.C., McFadyen, R.E., Panetta, F.D. and Adkins, S.W. (1996). The biology of Australian weeds. 27. *Parthenium hysterophorus* L. *Plant Protection Quarterly* 11, 76-88.

Nguyen, T.L.T. (2011). The invasive potential of parthenium weed (*Parthenium hysterophorus* L.) in Australia. PhD Thesis, The School of Agriculture and Food Sciences, The University of Queensland, Australia.

Eradication of red witchweed – it is never going to be easy!

Peter J. Austin, Daniel C. Stampa and Steve Csurhes
Biosecurity Queensland, Department of Agriculture and Fisheries, GPO Box 267, Brisbane, Queensland 4001, Australia
(Peter.Austin@daf.qld.gov.au)

Summary The parasitic weed, red witchweed (RWW) (*Striga asiatica* (L.) Kuntze) was first recorded at Habana, near Mackay in July 2013, the only known infestation to have established in Australia. An eradication response program was endorsed and commenced 1 July 2015 involving a treatment period of three years. A '*targeted adaptive three zone approach*' was adopted, utilising the integration of three treatment methods (planting of false host crops, herbicides and ethylene fumigation) to specific sections, or zones, within each infested property to expedite RWW soil seedbank depletion and minimise emergence.

This presentation will outline the treatment progress made to date and challenges encountered. The planting of the false host, soybean, in the high priority zone was completed in January. 1000 mm of rainfall in the RWW area during January and March 2016, necessitated an active weed control program within the soybean crop. Potential grass hosts amongst the soybean needed to be destroyed before RWW could germinate, establish and set seed. To accelerate RWW soil seed bank run-down, ethylene gas will also be injected into the soil at 1.5 to 2.0 kg ha^{-1} using custom-made equipment during the autumn period. Ethylene gas disperses readily through the soil from the point of injection, saturating the soil profile and initiating germination of all seed in the soil that are in a pre-conditioned state. An integrated field trial (efficacy trial) has also been established on an infested property that will yield adaptive results to help govern the management strategies utilised within the RWW eradication program.

Keywords *Striga asiatica*, red witchweed, eradication, treatment.

Operation witch hunt: conjuring eradication of the parasitic red witchweed plant with trickery and potions

Joseph S. Vitelli[1], Anna M. Williams[1], Natasha Riding[1], Annerose A. Chamberlain[1], Peter Austin[2] and Dan Stampa[2]
[1] Biosecurity Queensland, Department of Agriculture and Fisheries, GPO Box 267, Brisbane, Queensland 4001, Australia
[2] Biosecurity Queensland, DAF, 30 Tennyson Street, Mackay, Queensland 4740, Australia
(anna.williams@daf.qld.gov.au)

Summary An integrated field trial was established near Mackay, Queensland, to test treatment efficacy for eradicating the parasitic weed, red witchweed (RWW) (*Striga asiatica* (L.) Kuntze). Treatments tackling the only known infestation of RWW in Australia include pre and post-emergent herbicides, catch crop, trap crop plants and fumigants. One thousand canisters containing 5800 seed sachets were buried to monitor seed bank decline over time. Each canister contained seed sachets buried at five depths. At six months post-treatment application, 100 canisters were exhumed and a further 100 will be exhumed annually, to evaluate RWW seed viability following annual treatment applications. This trial will determine the treatment options that are most effective at accelerating soil seed bank decline and preventing above-ground emergence of RWW. These results will govern the management strategies utilised within the RWW eradication program.

The six-month retrieval data showed encouraging trends. As seed burial depth increased from 0 cm to 50 cm deep, seed viability increased from 61 to 69%. Treatments utilising herbicides did not significantly differ from the control (maintaining bare ground) treatment (81% and 77% seed viability). The catch crop, sorghum, had the biggest reduction in seed viability (17%), followed by the fumigants, dazomat (38%) and ethylene (39%). Ethylene gas has been used with considerable success in the USA for accelerating soil seed bank decline. Our preliminary data indicate, through the use of a host (false or true) plant and soil fumigant, exhaustion of the soil seed bank is achievable within three to five years.

Keywords *Striga asiatica*, red witchweed, parasitic weeds, eradication, field trial, seed bank.

INTRODUCTION

Striga asiatica (L.) Kuntze, commonly known as red witchweed (RWW), is classified in Queensland as Prohibited Matter in Schedule 1 of the *Biosecurity Act 2014* (the Act). RWW is currently found on six properties at Habana (21°2'60"S, 149°5'60"E) west of Mackay. First recorded growing attached to sugarcane in July 2013, an Eradication Response Plan (ERP) was submitted to the Consultative Committee for Exotic Plant Incursions (CCEPI) on 27 August 2013. The Consultative Committee for Red Witchweed agreed to a 10 year ERP, commencing on 1 July 2015, to be jointly funded by State and Federal Governments and affected industry bodies.

RWW is a small, annual plant that grows attached to the roots of a host plant (Musselman 1980). This obligate parasite (cannot develop independently from a suitable host) draws its nutrients and water from the roots of its host, generally causing the host to wilt, turn yellow and often become stunted. Hosts include a wide range of tropical and sub-tropical grass species, including several agriculturally important crops, including sugar cane (*Saccharum officinarium* L.), sorghum (*Sorghum bicolour* (L.) Moench), corn/maize (*Zea mays* L.), pearl millet (*Pennisetum americanum* (L.) Leeke) and rice (*Oryza sativa* L.) (Parker 2012).

RWW reproduces from seed and a single plant can produce between 90,000 to 450,000 seeds (Bebawi *et al.* 1984b). Bebawi *et al.* (1984a) reported that seeds could remain viable for up to 14 years when buried at a depth of 152 cm. Following seed maturation, seeds require a period of 'after-ripening' of up to two years. After this, seeds require certain soil moisture and temperature conditions for pre-conditioning (≥18°C, for 12 to 21 days) and will only then germinate in the presence of a germination stimulant, usually exuded from a nearby host plant root (Worsham 1987). In the absence of appropriate germination conditions, seeds remain dormant but viable in the soil (Worsham 1987).

Research into effective control strategies for RWW has been undertaken by the USDA (United States Department of Agriculture) and state agricultural organisations in the USA, as well as in Africa and Madagascar with costs exceeding $250 million in the USA alone (Sand 1987). Lessons from the Witchweed Eradication Program in the Carolinas (USA), running since 1957 (Eplee 1992), has provided the blueprint for developing eradication strategies in Queensland.

Eradication efforts in Queensland aim to incorporate techniques that accelerate RWW seed bank decline and prevent the production of new RWW seed. Ethylene-based soil fumigants are reported to trigger seed germination of RWW significantly reducing the time required to exhaust the soil seed bank to within three to five years (Bebawi *et al.* 1984). Catch crops are 'true host' species with roots that exude a stimulant to trigger germination of RWW seeds, supporting the subsequent growth of RWW through to flowering and seed-set. Trap crop plants are 'false host' species with roots that exude a stimulant to trigger germination but do not support attachment and subsequent growth of RWW. Catch crops are destroyed prior to emergence of RWW to prevent soil seed bank recruitment of the crop. This paper reports the findings of the six months exhumation on RWW seed viability following the application of pre- and post-emergent herbicides, catch crops, trap crops and fumigants to sugarcane.

MATERIALS AND METHODS

Experimental site and plant material An integrated control study was established in 2014 on one of the infested RWW properties in Habana (21°3'48"S, 149°4'21"E), Mackay, Queensland. Sugarcane is the predominant commercially viable crop grown locally. The trial will investigate the efficacy of agronomic practices for depleting the RWW seed bank and preventing production of new RWW seed, over a ten year period. The site had been planted to sugar cane and was on its eighth ratoon when RWW was discovered in July 2013. The 1.4 ha site is on an undulating rise with mostly gentle slopes of around 8 to 10% incline, with an abundance of angular stones through the profile. The soil has a coarse textured, non-coherent surface, overlying a narrow band of less permeable clay subsoil and coarser textured decomposing granodiorite.

The wet season in Habana is from December to May, with 81% of the annual rainfall falling during this period (Bureau of Meteorology 2016). The mean (± SEM) annual rainfall during the study period (2014 to 2015) was 1155 ± 243 mm. The number of rainy days per year (where a rainy day is >0.2 mm) averaged 128 ± 8.

Prior to treatment application, the standing sugar cane was bumpered, slashed and ripped in January 2014, with reshooting plants sprayed with glyphosate at 1800 g a.i. ha^{-1}. The destroyed cane was raked into mounds, left to dry and burnt in July 2014. The site was maintained fallow through regular discing, ploughing and periodic spraying with glyphosate and 2,4-D (1125 g a.i. ha^{-1}) until planting. On the 6 August 2014, 25 of the 80 plots were furrowed, fertilised with Granam at 170 kg ha^{-1} and planted with stalks of sugar cane cultivar Q208. Row spacings were 1.5 m apart and ~100 m in length. The planter seed cane was sourced from Mackay Area Productivity Services.

Experimental design and treatments The field study to determine RWW soil seed bank run down was laid out in a 16 × 5 factorial replicated five times using a randomised complete block design. Factor A was the 16 treatments (two pre- and two post-emergent herbicides, four false and two true hosts, one fertiliser and three fumigants) assigned to the mainplots, and factor B was seed burial depth (0, 10, 20, 30 and 50 cm). A perforated canister housed five seed sachets (one for each depth), with each sachet containing 100 RWW seeds. A total of 1000 canisters containing 5800 seed sachets have been buried in the field with 100 canisters to be exhumed at six months and annually thereafter to test for RWW seed viability following treatment applications. The 16 treatments were:

Bare ground:

Ground maintained bare through applications of 2,4-D (1125 g a.i. ha^{-1}), paraquat (400 g a.i. ha^{-1}) or glyphosate (1800 g a.i. ha^{-1}) as needed;

Control:

Only host plant sugarcane;

Post-emergent herbicides applied to sugarcane:

- dicamba (300 g a.i. ha^{-1}) – applied to sugarcane foliage every 4 to 8 weeks;
- fluroxypyr (300 g a.i. ha^{-1}) – applied to sugarcane foliage every 4 to 8 weeks;

Pre-emergent herbicides applied to sugarcane:

- pendimethalin (1500 g a.i. ha^{-1}) – applied annually in February and mechanically incorporated into soil within 5 days;
- trifluralin (1400 g a.i. ha^{-1}) – applied annually in February to a 7.5 to 15 cm deep furrow;

False host plants: (hand broadcast)

- Soybean (*Glycine max*) – 50 t ha^{-1};
- Cowpea (*Vigna unguiculata* L.) – 90 t ha^{-1};
- Cotton (*Gossypium hirsutum* L. – 15 t ha^{-1};
- Lablab (*Lablab purpureus* L.) – 20 t ha^{-1};

True host plants: (hand broadcast)

- Sorghum – 25 t ha^{-1};
- Corn – 30 t ha^{-1};
- Corn + urea – corn 30 t ha^{-1} plus urea at 450 kg N ha^{-1};

Fumigants:

- Dazomat – applied annually as a granular fumigant at 330 kg a.i. ha^{-1} to bare ground in April and immediately irrigated with 25 L water m^{-2} to activate and form a surface crust;
- Ethephon – applied annually in April as a liquid fumigant at 70 kg a.i. ha^{-1} to bare ground;

• Ethylene gas – Ethylene gas injected into the soil via a handheld probe to a depth of 15 to 30 cm at a rate of 2.0 kg ha^{-1} annually in February.

Seed sachets and canisters All seeds used in this experiment were collected from RWW plants grown within a QC2 quarantine facility at the Ecosciences Precinct, Brisbane. After collection, approximately 500,000 seeds were removed, pooled, sieved and stored at low humidity at 35°C for 12 months as part of the seed maturation process (Worsham 1987). Only seeds >150 mm were used in this experiment. Matured seeds to be buried at 10, 20, 30 and 50cm depth were placed inside 40 mm × 65 mm 62 mm precision woven polyamide (nylon) mesh tubes (SAATIFIL PA 62/40 PW WH) with the ends sealed using self-adhesive nylon tape (PSP Spinnaker Repair Tape). Surface seeds were placed inside 40 mm × 65 mm 63 mm stainless steel woven mesh sachets and the ends sealed with a hard setting metallic filler (Selleys Metallic Fix Cement Filler). Approximately 100 seeds were removed from the seed-stock, photographed (for counting) then transferred to each sachet containing 4 g of 300 mm sieved river sand to optimise seed soil contact.

The soil canisters were made from 100 mm PVC pipe cut into 550 mm lengths and perforated using a drill press with a 10 mm drill bit. The open space for each canister ranged from 20 to 25%. The sachets were placed at depth within the canisters and filled with 10 mm sieved soil from the study site. Canister burial concluded in December 2014 with treatments commencing in January 2015.

Laboratory procedure Retrieved sachets were stored at low humidity at 35°C until processed. At processing, each sachet was washed to remove exterior dirt. Sachet contents were flushed into a 90 mm sieve, rinsed with water, and transferred to a 100 mL measuring cylinder to which 100 mL of potassium carbonate (K_2CO_3) solution of specific gravity 1.4 was added (Eplee and Norris 1987). The cylinder was placed on a stirring plate for 30 to 45 minutes at 175 rpm to agitate the sample and encourage seeds to float to the surface. When removed from the stirring plate floating seeds were pipetted into a clean 90 mm sieve. Any seeds and debris accumulated on the edge of the cylinder were collected with a cotton bud and washed with water onto a filter paper. The remaining K_2CO_3 was poured through the sieve without disturbing the sand. Seeds collected from the sieve were also added to the filter paper. Sand from the cylinder was further flushed onto the sieve and checked under a microscope for any seeds and seed coats. All collected seeds and seed coats were counted and intact seeds tested for viability. Viability testing was performed using 1% 2,3,5-triphenyltetrazolium chloride (TTC) (Moore 1976). Seeds were immersed in 80 µl TTC and stored in the dark at 35°C for seven to ten days. Embryos turning red to pink in colour were considered viable.

The total number of viable seeds included those that germinated following treatment plus any ungerminated seeds identified as viable after TTC testing. An analysis of variance was performed on arcsine-transformed seed viability data and later back-transformed. Standard error of the mean (s.e.m) was used to identify differences between treatments.

RESULTS

A significant treatment effect was observed on seed viability at six months post treatment application ($P<0.0001$, Figure 1). The most effective treatments at reducing seed viability irrespective of seed burial depth were true hosts (44 ± 3% (± s.e.m) of remaining seeds viable) and fumigants (49 ± 3% (± s.e.m), followed by false hosts (69 ± 2% (± s.e.m). No significant difference was observed across the remaining treatments with seed viability remaining high (>77%). Within the broader category of true host sorghum was the most effective treatment with only 17 ± 3% (± s.e.m) of the buried RWW seed remaining viable after six months. The fumigants, dazomat and ethylene were equally effective at accelerating seed run down, with 38 ± 6% (± s.e.m) and 37 ± 4% (± s.e.m) of the buried seed remaining viable. Viability of laboratory stored seed over the same period was 92%.

As seed burial depth increased from 0 to 50 cm deep, irrespective of treatment, seed viability increased from 61 to 69%. However this trend was not significant at the six month period.

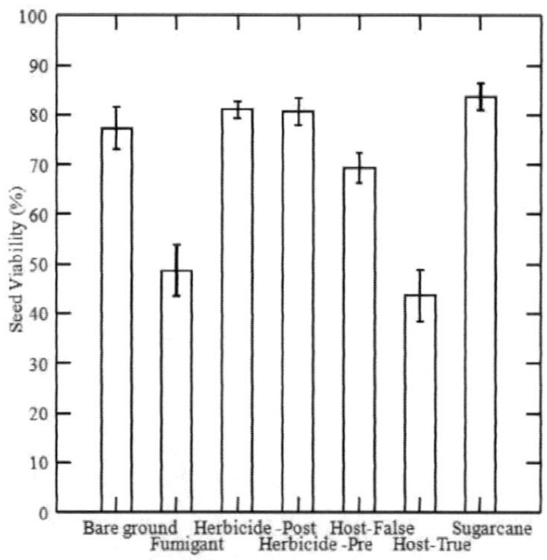

Figure 1. Effect of treatment on seed viability averaged over seed burial depth. Each column is the mean of five replicates (standard error bars shown).

DISCUSSION

Despite the trial being in its infancy (six months), several treatments are having a significant influence at accelerating the RWW seed bank decline. The ability to optimally integrate herbicides, true and false hosts and fumigants will be critical to the successful depletion of RWW seed at infested sites.

The true host crop sorghum reduced RWW seed viability to 17% markedly lower than the bare ground control (77%). Sorghum unfortunately presents a risk to the sugar cane production system, as an additional weed to manage if allowed to seed, with a reported seed longevity of up to 13 years (Burnside *et al.* 1977). Further, if sorghum was left unchecked, RWW would continue to emerge and set seed exacerbating the RWW problem. In West Africa, Murdoch and Kunjo (2003) reported that in systems where *Striga* was allowed to co-exist without chemical control, the soil seed bank of *Striga hermonthica* increased 270% in two wet seasons.

Dazomat is an effective dry formulation fumigant, with treatment application resulting in 3% of surface buried RWW seed remaining viable after six months. Wide scale use of dazomat offers challenges, both in cost ($3777 ha^{-1}) and the need for the product to be soil incorporated with water (25 mm rainfall or irrigation) within a few days after application, when soil temperatures are less than 10°C. The soil injection of ethylene gas saturates the soil profile, initiating germination of all seed that are in a pre-conditioned state, and is very effective at depth with only 35% of RWW seed viable at 50 cm burial after six months. However, the injection of ethylene by hand takes 85 h ha^{-1} to apply. Adoption of ethylene as an eradication tool necessitates the need for fabricating a mechanised system.

In the interim, two Emergency Use Permits PER14357 (http://permits.apvma.gov.au/PER14357.PDF) and PER14361 (http://permits.apvma.gov.au/PER14361.PDF) are currently in place to assist eradication of RWW in Queensland. To date, eradication efforts outside the trial site are incorporating herbicides, the false host soybean and ethylene injection. Findings from our research will allow site specific control strategies to be developed, hastening RWW soil seed bank extirpation.

ACKNOWLEDGMENTS

The authors acknowledge the financial assistance provided by the Queensland Government and CCEPI; land owners Gale and Joe Grech for their patience, understanding and flexibility in dealing with researchers; and Judith Ruhle for crop seeds.

REFERENCES

Bebawi, F.F., Eplee, R.E., Harris, C.E. and Norris, R.S. (1984a). Longevity of witchweed (*Striga asiatica*) seed. *Weed Science* 32, 494-7.

Bebawi, F.F., Eplee, R.E., and Norris, R.S. (1984b). Effects of seed size and weight on witchweed (*Striga asiatica*), seed germination, emergence, host-parasitization. *Weed Science* 32, 202-5.

Burnside, O.C., Wicks, G.A. and Fenster, C.R. (1977). Longevity of shattercane seed in soil across Nebraska. *Weed Research* 17, 139-43.

Eplee, R.E. (1992). Witchweed (*Striga asiatica*) – an overview of management strategies in the USA. *Crop Protection* 11, 3-7.

Eplee, R.E. and Norris, R.S. (1987). Field research techniques. *In* 'Parasitic weeds in agriculture. Volume I. Striga', eds L.J. Musselman, pp. 271-80. (Department of Biological Sciences, Old Dominion University, Norfolk, Virginia).

Moore, R.P. (1976). Tetrazolium seed testing developments in North America. *Journal of Seed Technology* 1, 17-30.

Murdoch, A.J. and Kunjo, E.M. (2003). Depletion of natural soil seedbanks of *Striga hermonthica* in West Africa under different integrated management regimes. *Aspects of Applied Biology* 69, 261-8.

Musselman, L.J. (1980). The biology of *Striga Orobanche* and other root-parasitic weeds. *Annual Review of Phytopathology* 18, 463-89.

Parker, C. (2012). Parasitic weeds: a world challenge. *Weed Science* 60, 269-76.

Sand, P.F. (1987). The American witchweed quarantine and eradication program. *In* 'Parasitic weeds in agriculture. Volume I. Striga', eds L.J. Musselman, pp. 207-23. (Department of Biological Sciences, Old Dominion University, Norfolk, Virginia).

Worsham, A.D. (1987). Germination of witchweeed seeds. *In* 'Parasitic weeds in agriculture. Volume I. Striga', eds L.J. Musselman, pp. 45-61. (Department of Biological Sciences, Old Dominion University, Norfolk, Virginia).

Sterile holoparasite *Cuscuta australis* R.Br. (dodder) spread and its physiological impact on hosts in tropical Brunei Darussalam

Kushan U. Tennakoon[1,2], Quang-Vuong Le[2], Faizah H. Metali[1], Rusydinnah Buntar[1] and Jay F. Bolin[3]
[1] Faculty of Science, Universiti Brunei Darussalam, Jalan Tungku Link, Gadong BE1410, Brunei Darussalam
[2] Institute for Biodiversity and Environmental Research, Universiti Brunei Darussalam, Jalan Tungku Link, Gadong BE1410, Brunei Darussalam
[3] Catawba College, Department of Biology Salisbury, North Carolina and Smithsonian Research Institute, United States of America
(kushan.tennakoon@ubd.edu.bn)

Summary Two proliferative modes of dodders (fertile and sterile) were observed in Brunei Darussalam. The fertile dodder was identified as *C. australis* R.Br. Field studies conducted over six years have revealed except for a few isolated localized populations, almost all *Cuscuta* spp. are exclusively spread by means of perennation and they are widespread along the waterways and low lying areas and show invasive traits. Using a DNA barcoding approach, we found that ITS and *trn*L-F sequences of the sterile dodder were identical to those of *C. australis* and different from *C. campestris* Yunck.

Even though it is widely known that dodders negatively affect host photosynthesis, the information on their impact on hosts in the presence of environmental stress factors (i.e. drought) in the tropics is poorly known. Results of a parallel study showed that the concomitant presence of *C. australis* infection and drought significantly impacted leaf traits, gas exchange (i.e. decreased stomatal conductance, transpiration rates and increased water use efficiency) and quantum yield of Chl *a* fluorescence of one of its common hosts *Mikania micrantha* Kunth.

The presence of a single stress factor (*C. australis* infection or drought), however, only significantly affected leaf traits and gas exchange of host. Results suggested that additive effects of *C. australis* parasitism and drought significantly suppressed the photosynthesis of *M. micrantha* due to both stomatal and non-stomatal limitation of host photosynthesis. This study provides pioneering insights into a novel method of *Cuscuta* spread and varying magnitudes of its impacts on hosts with changes in the abiotic environment.

Keywords Parasitic plants, *Cuscuta*, invasiveness, photosynthesis.

Sucking the lifeblood out of major invasive weeds in Australia

Robert M. Cirocco[1], José M. Facelli[1] and Jennifer R. Watling[2]
[1] DP 312 Benham Building, School of Biological Sciences, The University of Adelaide, Adelaide, South Australia 5005, Australia
[2] Rm NB260 Northumberland Building, Faculty of Health and Life Sciences, Northumbria University, Newcastle upon Tyne NE1 8ST, UK
(robert.cirocco@adelaide.edu.au)

Summary Major invasive weeds cost Australians around four billion dollars annually in addition to incalculable costs to biodiversity. Native parasitic plants may have detrimental effects on performance of invasive weedy hosts by removing resources via 'suckers' called haustoria and thus contribute to their demise. Glasshouse studies have documented severe effects of parasites on invasive species, but the effects of parasites may be highly variable depending on environmental conditions.

We conducted physiological measurements to investigate the effects of the native parasitic vine *Cassytha pubescens* R.Br. on the major invasive weed *Ulex europaeus* L. (gorse) across three field sites in the Mt. Lofty Ranges of South Australia. Photosynthetic performance and nitrogen of gorse were strongly decreased by *C. pubescens* consistently across sites. In addition, at two of the three sites, the parasite had a negative effect on the long-term water-status of gorse and there was also evidence of breakdown in the photosynthetic apparatus of the host in response to infection.

The data indicate that the native parasite negatively affects photosynthesis of gorse by removing large amounts of nitrogen and likely water from the host. Thus, *C. pubescens* shows promise as an effective native biocontrol against major invasive weeds in Australia and if successful, may be used to help restore our native biodiversity.

Keywords Native biocontrol, *Cassytha pubescens*, gorse, nitrogen, parasitic plant-host association, photosynthesis, *Ulex europaeus*, water, Weed of National Significance.

One size does not fit all! Complex issues in managing the weed flora of the Torres Strait Islands

Barbara M. Waterhouse
Northern Australia Quarantine Strategy (NAQS), Department of Agriculture and Water Resources, Box 96 AAC, Cairns International Airport, Queensland 4870, Australia
(barbara.waterhouse@agriculture.gov.au)

Summary The Torres Strait Islands lie on the biogeographic interface between the Australasian and Papuasian floristic regions. The native flora comprises a subset of that found in adjacent northern Australia and New Guinea.

Geomorphologically diverse, the islands were part of a land bridge connecting present day Australia and New Guinea for lengthy intervals throughout much of the Quaternary period. Consequently, the origin of some weedy taxa found there is unclear, and some species now considered introduced on the Australian mainland may in fact be indigenous to the islands.

In recent decades, biosecurity intervention has substantially reduced human-mediated introduction of live plant material southwards from Papua New Guinea into the islands. However, lack of biosecurity controls on northwards-moving cargo almost certainly contributes to the accidental introduction of new taxa. Consequently, naturalised species represent an ever-increasing proportion of the Torres Strait flora, and weeds not previously recorded from the islands are frequently encountered during annual NAQS plant health surveys.

If recognised early enough, recently arrived weeds should be eradicable with relatively little cost or effort. Some localised successes are documented. Decision-making over management and control of other longer established taxa can be more complex and requires respectful consultation with traditional owners. Factors that must be considered include whether the plants have traditional uses as food or medicine or for ceremonial purposes. Localised control of species known to be alternative hosts for newly arrived plant pests or pathogens should also be considered.

Keywords Torres Strait Islands, land bridge, weedy taxa, biosecurity, NAQS, traditional uses, alternative hosts.

Managing weeds for restoration of seabird habitat on Penguin Island, Western Australia

Kate Brown[1], Grazyna Paczkowska[1] and Aurelie Labbe[2]
[1] PO Box Locked Bag 104, Bentley Delivery Centre, Western Australia 6983, Australia
[2] 90 South Street, Murdoch University, Murdoch, Western Australia 6150, Australia
(kate.brown@dpaw.wa.gov.au)

Summary The transition from native shrubs and succulents to a cover of introduced annual grasses following establishment of large colonies of gulls has been reported for floras of small islands across the world. Sometime after 1961 a large colony of silver gulls (*Chroicocephalus novaehollandiae* Stephens) became established on Penguin Island off the coast of Perth. Currently large numbers of silver gulls nest on the island each year over the winter months. They cause extensive trampling, guano deposition and disturbance. The birds are effective carriers of weed seed and in large areas where gulls nest, a cover of nitrogen-loving exotic grasses and annuals has now replaced native shrublands. The vegetation in the north-east section of Penguin Island has historically been important nesting habitat for bridled terns (*Onychoprion anaethetus* Scopoli). In recent years silver gulls have moved into the area and the cover of native shrubs that was important nesting habitat is largely being replaced by a cover of weedy annual grasses in winter/spring and bare ground through summer. In this study we trialled techniques for re-establishing native shrub cover at the site. The objective of all treatments was to replace dense stands of annual weeds with a resilient native vegetation cover that provided bridled tern habitat. Treatments included direct seeding, laying brush of fertile material, planting tube stock and simply removing weeds from plots to assess potential natural regeneration from the soil's seed bank. All treatments were protected from silver gulls by weldmesh cages in the first 12 months of the trial. Results including effectiveness of treatments at establishing a native cover and at displacing exotic annual herbs and grasses will be discussed.

Keywords Silver gulls, annual weeds, restoration, bridled terns, Penguin Island.

Generating insight to underpin improved weed management and consequent protection of Aboriginal sites on the Burrup Peninsula

Vicki Long[1], Ben Fitzpatrick[2], David Pozzari[3], Bruce L. Webber[4,5], Paul B. Yeoh[4] and Craig Bonney[6]
[1] Vicki Long and Associates, PO Box 713, Karratha, Western Australia 6714, Australia
[2] Oceanwise Australia, 4 Cherry Road, Woodlands, Perth, Western Australia 6018, Australia
[3] Pilbara Ports Authority, Mof Road, Burrup Peninsula, Dampier, Western Australia 6714, Australia
[4] CSIRO Land and Water, Private Bag 5, PO Wembley, Western Australia 6913, Australia
[5] School of Plant Biology, The University of Western Australia, 35 Stirling Highway, Crawley, Western Australia 6009, Australia
[6] Murujuga Aboriginal Corporation, PO Box 1544, Karratha, Western Australia 6714, Australia
(vicki.long@hotmail.com vicki.long@hotmail.com)

Summary The Burrup Peninsula, or Murujuga, is located in the Pilbara region, 1500 km north of Perth. This National Heritage listed area has a unique geology and topography of rockpile and block slope terrain surrounded by a marine environment. It supports a mix of arid inland, coastal and remnant Kimberley plant species, over 70 of which are utilized by Aboriginal people. The diverse terrestrial and marine life is of high cultural significance for Aboriginal culture, as evidenced by the globally significant rock-art galleries occurring there. A causeway, built to support industrial development some 53 years ago linked the then Burrup Island to the mainland. Subsequent visitation and development has contributed to the introduction of a diversity of weeds, one of the most threatening being the neotropical vine, stinking passionflower (*Passiflora foetida* L.). This weed is rapidly invading the rock terrain and coastal vegetation, forming dense mats that smother native flora and rock art. The invasion and subsequent control of this species is impacting upon the Aboriginal heritage and natural ecosystem values of the Burrup. Here we document ongoing efforts across multiple collaborations to improve weed management on the Burrup Peninsula. We discuss the results of a six month phenology monitoring study and the implications of those results to date. We will explore possible impacts and also control solutions. We show that this insight may help to make tangible improvements to the on-ground management of weeds to improve outcomes for biodiversity conservation, as well as Aboriginal cultural and historical values across this significant landscape.

Keywords Burrup Peninsula, geology, native vegetation, Aboriginal heritage, plant use, rock art, invasive weeds, *Passiflora foetida*, ecosystem impacts.

Evaluating the implementation of declared pest policies – new ideas for enhancing public value

Michaela Heinson[1,2]
[1] Biosecurity SA, GPO Box 1671, Adelaide, South Australia 5001, Australia
[2] Natural Resources Adelaide and Mount Lofty Ranges, 205 Greenhill Road, Eastwood, South Australia 5063, Australia
(michaela.heinson@sa.gov.au)

Summary Effective systems for managing declared pests improve the private economy and national competitiveness, resulting in enhanced public value. Due to the intangible, complex and shifting nature of public value for strategic control of declared pests, a high level of engagement with stakeholders will always be required to bring insight and intelligence to policy development and implementation. In 2015 a feasibility study for evaluating the implementation of declared pest policies in South Australia was prepared, using survey and interview data from four stakeholder groups: policy writers; regional natural resources staff; community representatives on regional Natural Resources Management Boards; and executives in relevant state agencies. Participating stakeholders suggested sources of data for measuring inputs such as: compliance work; protection orders; inspections; surveillance; formal communication with landholders; human resources; and allocated budgets. Key drivers of change for public value outcomes were analysed, including: expectations for disciplined and consistent enforcement to address non-compliance; variation in political will for each invasive animal and weed; declining budgets; and opportunities to use new technologies to improve efficiency. To improve evaluation systems, recommendations were made for: managing spatial data; communicating the delivery model for implementing policies; further research on social drivers and dynamics; and measuring inputs at the regional level.

Keywords Public value, declared pests, policy, South Australia.

INTRODUCTION

All landholders have a legal responsibility to control declared animals and plants to reduce threats to primary industries, the natural environment and public safety. However these threats are always changing. An evaluation system will help ensure public value is improved and control programs, education, enforcement and compliance are guided by effective policies.

In 2014–15 the eight regional Natural Resources Management (NRM) boards in South Australia collectively invested more than $8 million to coordinate management programs for invasive animals and weeds. NRM boards have the responsibility of setting regional priorities for management programs and engaging with landholders to achieve compliance under the *Natural Resources Management Act 2004* (Government of South Australia 2012). The Department of Environment, Water and Natural Resources (DEWNR) manages control programs, education, enforcement and compliance, delivering the operational activities for NRM boards. Biosecurity SA, in Primary Industries and Regions South Australia (PIRSA) leads the development of declared animal and plant policies and provides technical advice for management programs.

METHODS

A feasibility study for evaluating the implementation of declared pest policies in South Australia was prepared. The three objectives of the study were to: (i) provide recommendations on how to evaluate the implementation of policies; (ii) provide recommendations on engaging with a broader range of stakeholders to design an evaluation system; and (iii) examine key drivers of change for public value outcomes in the management of declared animals and plants. These drivers are analysing political, economic, social, technological, environmental and legal (PESTEL) trends, factors and external dynamics. PESTEL analysis is a business tool used to identify factors that have a combined effect on the key drivers of change for an industry, market or sector (Johnson *et al.* 2005).

Survey and interview data were collected from eighteen participants from four stakeholder groups: policy writers; regional natural resources staff; community representatives on regional Natural Resources Management Boards and executives in relevant state agencies. Participants completed quantitative attitudinal surveys and were interviewed using a semi-structured format which allowed each individual to elaborate on topics of particular interest to them.

The recommendations in the study were developed by the author drawing on survey and interview data, existing plans and reports, and published literature on evaluation, stakeholder engagement and business strategy.

RESULTS AND DISCUSSION

How to evaluate The most common evaluation topics mentioned by participants were (i) input measures, (ii) spatial data, (iii) delivery of regional NRM plans, (iv) outcomes and results, and (v) community engagement. Across the participating stakeholder groups individuals suggested key performance areas for:

- measuring inputs at the regional level (data sources include compliance work, protection orders, inspections, surveillance, formal communication with landholders, human resources and allocated budgets);
- the density and distribution of a species, or group of species, over time;
- delivery of regional NRM plans, including uptake and delivery of pest management plans, and information on assets being protected;
- outcomes and results, such as the actual impacts resulting from management and control activities (before and after monitoring); and
- community engagement, including number of extension activities.

Some of the unique ideas from individuals were also highly valuable in developing recommendations for an evaluation system, including:

- clearly describe and communicate the delivery model for implementing policies;
- use social mapping – a visual method of exploring social dynamics of a community – to help understand how landholders engage and what's important to them;
- measure all activities (across multiple stakeholder groups) for the recently declared plants – to establish 'baseline data' for inputs; and
- stories and case studies that explain benefits and costs in a qualitative way can be more important than detailed, economic cost-benefit analysis.

Engaging a broader range of stakeholders Policy writers, regional natural resources staff; members of regional Natural Resources Management Boards and executives in relevant state agencies are influential in evaluating the implementation of declared pest policies. Nonetheless there are other stakeholders with a high level of interest and influence, particularly individual landholders, industry peak bodies and local government. To design an effective evaluation system these stakeholders should be invited to participate in regional evaluation workshops or via direct consultation for state level peak bodies. Emphasis must be placed on stakeholder analysis and engagement, collaborative approaches, and using holistic rather than linear thinking because the management of declared pests can be inherently complex, unpredictable and open-ended.

Key drivers of change for public value outcomes The survey results for PESTEL trends, factors and dynamics revealed that participants collectively identified 'expectations for disciplined and consistent enforcement to address non-compliance' and 'opportunities to use new technologies to improve efficiency' as the two most significant drivers of change for public value outcomes in the management of declared pests. However, semi-structured interviews with each participant provided better insight and depth to help develop the PESTEL analysis presented in Figure 1 below.

CONCLUSION

The recommendations in the feasibility study address: key performance areas for input measures and spatial data; line-of-sight reporting for relevant regional and state level plans; research on social drivers and dynamics; stakeholder engagement targeting landholders, local government and industry; and an agreed delivery model. Due to the intangible, intricate and shifting nature of public value for strategic management of declared pests, a high level of engagement with stakeholders is required to bring insight and intelligence to policy development and implementation. The ongoing challenge will be to interpret evaluation data to update declared pest policies, while checking that performance measures are still relevant. If key drivers of change for public value outcomes reveal a significant shift in political, economic, social, technological, environmental or legal trends, factors and dynamics the evaluation system should be re-assessed.

ACKNOWLEDGMENTS

I thank the individual Natural Resources Management Board members, Executive Directors in PIRSA and DEWNR, regional Natural Resources staff and my colleagues in Biosecurity SA who anonymously and generously completed surveys and were interviewed during this project.

REFERENCES

Government of South Australia (2012). Our place. Our future. State Natural Resources Management Plan South Australia 2012–2017. Adelaide.

Johnson, G., Scholes, K. and Whittington, R. (2005). 'Exploring corporate strategy'. Prentice Hall, London.

POLITICAL FACTORS

Political will varies greatly for each declared pest.

The transition of regional Natural Resources staff from NRM boards to DEWNR has reduced ownership of coordinated management of declared pests.

ECONOMIC FACTORS

Budgets are declining.

Maintaining market access and productivity does not have a direct impact on creating jobs.

Case studies that demonstrate costs and benefits of invasive animal and weed control can be influential; however it can be difficult to quantify the value of keeping an area pest-free.

SOCIAL FACTORS

Abutting land-uses have different goals and resources for invasive animal and weed control. Some landholders view what happens on public land as the benchmark.

Persistence wanes over long timeframes to achieve coordinated control, however some landholders do persevere committing their own resources.

TECHNOLOGICAL FACTORS

Use of mobile phones, tablets and other hand-held devices to gather data and provide quick access to information is increasing.

Advances in detection methods (e.g. Environmental DNA) may support innovative surveillance in the near future.

Lack of technology to reduce soil seed-banks for weeds is inhibiting effort.

ENVIRONMENTAL FACTORS

Perceptions vary regarding environmental impacts and agricultural impacts, and the provision of support for landholders.

The impact of climate change on risks associated with declared pests is an increasing concern.

LEGAL FACTORS

Lack of disciplined and consistent enforcement actions to address non-compliance for declared pests is weakening coordinated programs (a compliance reform project in DEWNR is in progress).

Figure 1. PESTEL analysis examining key drivers of change for public value outcomes in the management of declared pests.

Marathon weed management: How can we effectively connect weed research with policy planning and on-ground action for long-term NRM results?

Keith B. Ferdinands[1], Samantha A. Setterfield[2] and Jonathan Vea[3]
[1] Research Institute for Environment and Livelihoods, Charles Darwin University, Northern Territory 0909, Australia
[2] School of Earth and Environment, University of Western Australia (M004), Perth, Western Australia 6009 Australia
[3] Northern Territory Department of Sport and Recreation, GPO Box 1448, Darwin, Northern Territory 0801, Australia
(keith.ferdinands@nt.gov.au)

Summary It has long been recognised that weed management issues require long-term, strategic investment and well-coordinated efforts. In sport this would be talked about as something like running a marathon. In contrast, there appears too few examples of long-term investment in weed management and where investment occurs it is often seeking to achieve short-term results. We reviewed an example of natural resource management in Litchfield National Park, Northern Territory. In this Park, biodiversity conservation and weed management efforts are designed using high levels of collaboration and seek to improve the balance of short-term and long-term weed management results. Based on these examples, we suggest that in many cases in the Northern Territory (NT), we have the policies, plans, tools, systems and experts to do (and/or continue) marathon-like weed management efforts. The big challenge is not how to do long-term strategic management of priority weeds, it is convincing decision makers and funding bodies that there are a variety of benefits to be gained from doing so.

Keywords National Parks, environmental weeds, policy and planning, WoNS, EPBC.

INTRODUCTION

With short funding cycles and changing priorities, natural resource management actions, of which weed management remains a key element, are often, in sporting terms, a 'sprint', with the emphasis on quick results via short-term projects. In contrast, most weed management priorities require long-term, marathon-like responses to achieve or maintain successful weed management outcomes. Typically, long-term, evidence-based and adaptive weed management policies and plans are needed and strategic partnerships formed to design and implement them. Where, to use a sporting analogy, this sprint versus marathon mismatch occurs, it becomes difficult for planners, land managers, scientists and Park Rangers to find ways to design implement or maintain effective NRM plans.

Litchfield National Park in the Top End of the Northern Territory provides an excellent example of the challenges to protect nationally and internationally significant park values from a number of threats, of which weeds are a high priority. In Litchfield, managing the threats posed by weeds, particularly invasive grasses, continues to be a management priority.

In this paper, we argue that the planning and management approaches used in Litchfield offers the greatest likelihood of long-term NRM outcomes, including weed management. This is because the approach used has the following key elements:

(a) Policy, planning, management, research and on-ground actions are developed and implemented in a coordinated and collaborative fashion.
(b) Weed management is linked to conservation values (biodiversity, cultural, social) and threats (fire, feral animals).
(c) Imperfections and knowledge gaps are accepted and adaptive management strategies included.
(d) Outcomes and success or failure are reviewed over both short-term and long-term time frames.

We outline how weed management and weed management research has been incorporated into the management of Litchfield National Park. We will also outline how weed research has contributed to an Integrated Conservation Strategy (ICS).

MATERIALS AND METHODS

NT National Parks and weed management Litchfield National Park is managed by the Parks and Wildlife Commission NT (PWCNT) who classify Litchfield as a Class 1 Biodiversity Park. Situated 120 km south of Darwin, Litchfield is an iconic Northern Territory attraction with more than 300,000 visitors in 2014. For all NT, Class 1 Biodiversity Parks, an ICS and an Annual Action Plan is developed and reviewed annually (PWCNT 2015). The planning and management process is directed by the Conservation Action Planning (CAP) process developed by the Nature

Conservancy (TNC 2007), resulting in an ICS. This approach is widely used nationally and internationally. The PWCNT planning process is being applied to several NT parks, and was recently completed for Litchfield National Park. We therefore use the development process for the Litchfield Park ICS to illustrate how protecting conservation values, threat management and weed management can be linked.

RESULTS

The elements of the NRM approach employed by Litchfield National Park are summarised in Table 1.

Project teams comprising planners, park rangers, relevant experts, traditional owners, key stakeholders and researchers were involved in all of these elements. This was done to ensure that evidence based, adaptive management approach, subject to implementation reality checks (e.g. resources) occurs.

Identifying and prioritising threats to key conservation values in Litchfield National Park The initial steps of the ICS process (Table 1) identified the key conservation values (Table 2), then ranked and prioritised the threats to these values (Tables 3).

Table 1. Planning and management elements.

1	Identify, group by function, define practical indicators and prioritise key conservation values. Develop agreed standards to categorise the health of the grouped conservation values based on defined indicators. Develop conservation targets to determine management success based on agreed standards.
2	Identify and prioritise threats to values to define management objectives.
3	Evaluate strategic options by aligning priority conservation targets and prioritised threat objectives.
4	Define management programs based on the strategic priorities to reach conservation targets and threat objectives.
5	Design and implement monitoring and reporting program based on the defined health indicators.
6	Determine and implement simple annual action plans to meet priority conservation targets and prioritised threat objectives.
8	Establish a governance body to review annual results, refine further conservation strategies (5 years) and frame a consistent adaptive management response over successive years.

Conservation values, threats and health: invasive grass The highest priority threats to Litchfield Park are fire and invasive grasses. Fire is a frequent occurrence in the savanna region (Scott *et al.* 2012). There is high awareness in northern Australia about the threat of invasive grasses, particularly gamba grass (*Andropogon gayanus* Kunth.) to the conservation, cultural and visitor values of the region. The evidence is the result of decades of research demonstrating impacts on ecosystem structure and function, particularly by altering fire regimes (Flores *et al.* 2005, Brooks *et al.* 2010, Setterfield *et al.* 2013, Adams and Setterfield 2016). In addition, the NT's weed risk assessment process (Setterfield *et al.* 2010, Setterfield *et al.* 2015) underpinned evidence of the very high risk posed by gamba grass and other invasive grasses. The interaction between gamba grass, other invasive grasses and their ability to alter fire regimes (Rossiter *et al.* 2003,

Table 2. Key values in Litchfield Park identified through the ICS process.

	Key value
1	Visitor safety, visitor, cultural and heritage assets
2	Sandstone plateaus
3	Monsoon rainforest and swamps
4	Melaleuca woodlands
5	Lowlands and alluvial plains

Table 3. The top fourteen priority threats to Litchfield identified through the ICS process.

Priority	Threat to Litchfield	Threat ranking
1	Wildfire	Extreme
2	Gamba grass	Extreme
3	Inappropriate fire regimes	Very high
4	Arson	Very high
5	Mission grass	High
6	Mimosa	Medium
7	Humidicola	Medium
8	Pigs	Low
9	Buffalo	Low
10	Cattle	Low
11	Olive hymenachne	Low
12	Other weeds	Low
13	Cane toads	Low
14	Cats	Very high

Setterfield *et al.* 2013) and increase risk to the Park's visitor values and conservation goals was therefore reflected in the high priorities for these threats in Table 2.

Realistic, evidence based management targets Guided by the conservation values and threats identified through the ICS, the management actions required to protect or maintain these values were identified. Park Rangers had a lead role in identifying the appropriate actions because they have the knowledge and skills in on-ground management as well as knowledge of what is a realistic approach. For NT parks, 5-year realistic management action targets are set. For gamba grass, outcomes of research projects on the current distribution and predicted spread of gamba grass (Adams and Setterfield 2016) directly informed the Litchfield ICS management goals (NERP 2015).

Realistic, evidence based outcomes Because the values being managed and the management methods are sometimes imperfectly understood, monitoring and evaluation is very important. Annual reporting of management actions occurs for Class 1 NT Parks. State of the Park reports are produced to evaluate and report on the state of health of conservation values (PWCNT 2015).

DISCUSSION

Litchfield National Park, like many protected areas in the region, has ongoing weed management problems. With the exception of the rare scenario of a very high-risk weed that is (a) has a high feasibility of eradication, partly due to being identified early and with a short maintenance of a propagule-bank and (b) triggers a rapid and successful response, we have to apply long term (marathon-like) weed management programs. While the value and logic of strategic long-term management responses is not debated, it does not always occur.

This paper outlined the process being used by NT Parks for planning long-term management of all risks to park values – development of an ICS using a CAP approach. The planning and management process in Litchfield, developed using a combination of existing policies and related planning tools, research publications and other survey and monitoring data and reports, along with expert knowledge throughout the process achieved a lot with limited resources.

The CAP process and resulting ICS provides a template for the development, implementation, review and refinement of a planning approach very well suited to both short and long term weed management challenges. Collaborative research aligned with policy and planning requirements continues to play an important role in improving all the elements of the CAP/ICS. In theory, done well and maintained for the long term, park planners, weed managers and researchers can all use the planning process to try and achieve a variety of NRM outcomes.

While the CAP process and linked ICS's may have been under-utilised in weed management in Australia, the underlying principles outlined here are not new. Existing examples of many of these principles include: the Weeds of National Significance (WONS) program (Thorp and Lynch 2000, DoE 2007), the CRC for Australian Weed Management (2001–2008) and the National Land and Water Resource Audit documents (e.g. NLWRA 2002) provide examples of these principles and also highlight the challenge of maintaining well designed, coordinated long-term efforts to manage weeds.

CONCLUSIONS

The CAP process and related integrated conservation strategies, provide another way to strategically pick which weed marathons to run, as well as engage in the occasional sprint. In theory a win-win for the doers and the funders! Incorporating these new tools, while reconsidering the existing tools and methods, may assist weed managers and policy makers to review the need for a long-term, cross-tenure strategic weed management approach.

ACKNOWLEDGMENTS

This paper, like the management plans and conservation strategies referred to, are only possible because the combined long-term efforts and expertise of policy makers, planners, land managers and Park Rangers, traditional owners and researchers. Many thanks to the staff of the Flora and Fauna Division for the flora and fauna survey data and advice that is a key element of ICS's. Special thanks to the Rangers of Litchfield National Parks whose local knowledge and on ground management and monitoring skills and efforts underpin the success of the park management described.

REFERENCES

Adams, V.M. and Setterfield, S.A. (2013). Estimating the financial risks of *Andropogon gayanus* to greenhouse gas abatement projects in northern Australia. *Environmental Research Letters* 8. doi: 10.1016/j.biocon.2012.01.064.

Adams, V.M. and Setterfield, S.A. (2016). Approaches to strategic risk analysis and management of invasive plants: lessons learned from managing gamba grass in northern Australia. *Pacific Conservation Biology* 22, 189-200. doi:10.1071/PC15041

Anon. (2006). 'HB 294-2006 National Post-Border Weed Risk Management Protocol'. (Standards Australia International Ltd., Sydney, Standards New Zealand, Auckland and CRC Australian Weed Management, Adelaide).

Australian Government Department of the Environment (2007). Australian Weeds Strategy – A national strategy for weed management in Australia. Natural Resource Management Ministerial Council Australian Government Department of the Environment and Water Resources, Canberra ACT.

Brooks, K.J., Setterfield, S.A. and Douglas, M.M. (2010). Exotic grass invasions: applying a conceptual framework to the dynamics of degradation and restoration in Australia's tropical savannas. *Restoration Ecology* 18, 188-97.

Flores, T.A., Setterfield, S.A. and Douglas, M.M. (2005). Seedling recruitment of the exotic grass *Andropogon gayanus* (Poaceae) in northern Australia. *Australian Journal of Botany* 53(3), 243-9.

Land and Water Resources Audit (2002). Australia's Natural Resources: 1997–2002 and beyond.

NERP (2015). Northern Australia Resources Program. 'Video. Research helps tackle Gamba Grass in Litchfield National Park'. https://vimeo.com/137781537?lite=1. (Northern Australia National Environmental Research Programme, Darwin).

PWCNT (2014). 'Litchfield National Park Report Card 2014'. https://parksandwildlife.nt.gov.au/__data/assets/pdf_file/0009/249390/Litchfield_22-May-2015.pdf. (PWCNT, Darwin).

PWCNT (2015). 'Litchfield National Park Draft Plan of Management' (PWCNT, Darwin). https://parksandwildlife.nt.gov.au/__data/assets/pdf_file/0007/249082/Litchfield-National-Park-draft.pdf

Rossiter, N.A., Setterfield, S.A., Douglas, M.M. and Hutley, L.B. (2003). Testing the grass-fire cycle: alien grass invasion in the tropical savannas of northern Australia. *Diversity and Distributions* 9, 169-76.

Scott, K., Setterfield, S.A., Douglas, M., Parr, C.L., Schatz, J. and Andersen, A. (2012). Does long-term fire exclusion in an Australian tropical savanna result in biome shift? A test using the re-introduction of fire. *Austral Ecology* 37, 693-711.

Setterfield, S.A, Ferdinands, K., Rossiter-Rachor, N., and Douglas, M. (2010). The Northern Territory's Weed Risk Management System: an application of the national post-border WRM Protocol. *Plant Protection Quarterly* 25(2), 86-8.

Setterfield, S.A., Rossiter-Rachor, N.A., Douglas, M.M., Wainger, L., Petty, A.M., Barrow, P., Shepherd, I.J. and Ferdinands, K.B. (2013). Adding fuel to the fire: the impacts of non-native grass invasion on fire management at a regional scale. PLoS ONE 8, e59144.

Setterfield, S.A., Ferdinands, K., Rossiter-Rachor, N.A. and Elliott, L. (2015). 'Northern Territory Weed Risk Management System User Guide.' (Department of Land Resource Management, Darwin)

Thorp, J.R. and Lynch, R. (2000). Determination of Weeds of National Significance. (National Weeds Strategy Executive Committee, Launceston).

TNC (2007). 'Conservation Action Planning Handbook: Developing strategies, taking action and measuring success at any scale'. (The Nature Conservancy, Arlington, VA).

Relative effort calculation as a guide for regional weed investment

Scott A. Charlton
Department of Primary Industries, 161 Kite Street, Orange, New South Wales 2800, Australia
(scott.charlton@dpi.nsw.gov.au)

Summary Biosecurity NSW has developed a risk-based funding process called the Relative Effort Calculator. This model considers weed risk potential as well as providing for Local Government's capacity to deliver basic compliance obligations - such as property inspections and surveillance.

The model analyses the comparative 'relative effort' of delivering weed management across NSW. Considerations include the relative difficulty of inspecting properties of different scales and the associated costs of travelling to visit those properties. The model also accounts for the influence of terrain on the costs of doing weed management and compares regional variation.

The third component of the model is a pathway analysis of where weeds are most likely to impact NSW. This part of the model assesses historical weed occurrence, climate modelling and the abundance of known vectors such as livestock movements, rivers, roads and fodder production.

The Relative Effort Calculator is potentially a powerful new tool to increase performance across the NSW and presents several benefits for weeds management including:

- Targeting weed resources where they are most needed,
- Quantifying comparative effort required across regions,
- Establishing a monetary value for what has traditionally been an in-kind contribution from Local Government,
- Providing the basis for a 'Compliance Standard' and performance measurement, and
- Creating a competitive and transparent management environment within Local Land Services regions.

Keywords Resource allocation, terrain modelling, weed risk.

Why the polluter pays principle is not a policy panacea for weedy but commercially valuable plant species either

Stephen B. Johnson
Weed Research Unit, New South Wales Department of Primary Industries, Locked Bag 6006, Orange, New South Wales 2800, Australia
(stephen.johnson@dpi.nsw.gov.au)

Summary Accurate assessment and prioritisation of the risks and benefits posed by conflict plant species is needed; these being plants that are both weedy and commercially valuable. When such species are introduced and become invasive, (*sensu* Richardson *et al.* 2000), conflict can arise. This conflict often occurs between those who derive benefits or positive impacts from the plant (e.g. commercial producers) and those who pay for the damage these plants cause when their negative impacts are realised (e.g. environmental custodians; in effect, us the tax-payer). Policy actions and outcomes become mired in this conflict.

Conflict species policy has traditionally been informed by: legislative/regulatory approaches such as noxious weed laws; and economic analyses/tools such benefit cost analyses. These approaches are problematic for various reasons. The polluter pays principle has been proposed to overcome these problems. Broadly, the application of this principle to conflict species would mean that the entity responsible for introducing and/or producing the problem would be responsible for paying for the damage done to the environment/s it invades.

This paper discusses the application of the polluter pays principle to plant invasions. Subsequently, it proposes a fundamentally different process or system of evaluation of the risks and benefits posed by conflict species for informing policy and management responses.

Keywords Conflict or contentious species, Weed Risk Management, negative and positive impacts.

INTRODUCTION

Commercially valuable plant species can become weeds, that is invasive (Richardson *et al.* 2000). This occurs after they are introduced to new areas, cultivated, naturalise away from areas of cultivation, and become weeds (Bennett and Virtue 2004, Ferdinands *et al.* 2011, Johnson 2012).

Policies for the management of these species have focused on legislative/regulatory approaches such as noxious weed laws (e.g. Arcioni 2003, Johnson 2013); and economic analyses/tools such benefit cost analyses (e.g. Ferdinands *et al.* 2010). The disadvantages of both approaches have been discussed elsewhere (Le Gal 2011, Johnson 2012). A different approach is needed.

A potentially larger range of market-based mechanisms have more recently been investigated (e.g. Martin and Le Gal 2010, Le Gal 2011, 2012). Among this alternate regulatory mix is the polluter pays principle. Theoretically, the application of this principle will mean that industry, upon introducing or producing an environmental problem, will be responsible for paying for the damage the problem causes. Applied to plant invasions, it means that industry pays for the negative impacts that are realised after the introduction of such a species.

The polluter pays principle has received recent popular attention as one of four key strategies to reduce future problems with the introduction of new pasture species (Driscoll *et al.* 2014). Although a useful contribution to the literature, that analysis overlooked earlier, superordinate literature examining the polluter pays principle (e.g. Knowler and Barbier 2005, Barbier and Knowler 2006). In particular, those earlier authors examined the introduction of a 'Pigo[u]vian tax' consistent with the principles of an "introducers' pay" tax. This paper discusses the concepts of an "introducers' pay" tax. It then discusses if the introduction of an "introducers' pay" tax should be added the suite of invasive species management practices. In light of this, the paper suggests an alternate process for the evaluation of the risks and benefits from conflict species to inform policy and management.

A PIGO[U]VIAN TAX AND THE ISSUE OF EXTERNALITIES

Pigo[u]vian tax The original arguments for a Pigouvian tax were proposed nearly a century ago (Pigou 1920). (This paper uses the less common spelling 'Pigouvian' throughout this paper). In essence, a Pigouvian tax is a market-based tax that is levied on any activity that generates negative externalities, that is an externality not internalised in the market price. Such a tax is intended to correct an inefficient market outcome, that is a negative externality/ies. It addresses these by introducing, taxing and/or setting a social cost.

The sale of alcohol was used to illustrate the application of this tax (Pigou 1920). The sale of alcohol, unfortunately, often necessitates an increase in policing and prisons, because of the crime associated with that sale. The private benefit accruing to the owners of liquor outlets is large when compared to the negative social impacts that such sales often cause society more generally. Pigou (*ibid.*) suggested that individuals receiving benefits had no incentive to pay for the flow on costs that came to be associated with their business (e.g. the policing and prisons) and that this was the reason most countries taxed alcohol-related businesses.

The legislative management of weeds and the issue of externalities It is useful to consider externalities more generally at this point. In particular, it is important to consider the externalities that are addressed in legislative/regulatory approaches to weed management, such as those surrounding noxious weed laws. Although varying in type, the experience of externalities in weed management is similar to the example above.

To revise, in general, an externality occurs when the actions of individuals impose unintended impacts on others (Blackmore 2008). That reference further states:

> '*Externalities occur as both external costs and external benefits. People who have external costs imposed on them are not compensated for their loss, while people who receive an external benefit are not required to pay for their good fortune... The existence of externalities may justify governments making laws to correct the cause of the externality for the benefit of the community as a whole*';

and with respect to externalities encountered in weed management:

> '*When weeds spread to new areas, the landholders in the newly infested areas suffer economic losses because of the weed spread. These losses may include the cost of control work, loss of production, loss of biodiversity or loss of an amenity value. The landholders are not compensated for these losses by the owner of the land from where the weed spread.*'

For this reason, state and territory governments in Australia attempt to manage the problems caused by negative externalities related to weed management by making laws to require land managers to control particular weeds before they spread further (Blackmore *ibid.*).

The implementation of a market-based instrument such as a Pigouvian tax that is consistent with the "introducers' pay" concept would also seek to address this negative externality. That is, it seeks to address the costs of weed spread to affected landholders. Application of a Pigouvian tax **assumes that the social costs of the negative externality can be calculated, and applied in monetary terms**. While calculation of the costs that apply to this negative externality are fundamental to the concept of an "introducers' pay" or pollution tax, there is often less attention paid to: i. doing so (excepting Knowler and Barbier 2005, Barbier and Knowler 2006); and, if these costs are able to be calculated with any reliability; then ii. how such concepts may be implemented. Testing to see if a pollution tax can be calculated, and then examining how it can be implemented to weed management seem to be crucial first steps before application. Accordingly, these are discussed in this paper.

TESTING THE POLLUTER PAYS CONCEPT

Calculating a polluter pays tax Complete knowledge of a market, while desirable, is not possible (Knowler and Barbier 2005, Barbier and Knowler 2006). Complete knowledge of the costs incurred by all affected landholders for weed management is similarly desirable, but unattainable (Johnson 2014). In particular, the negative impacts of weeds are either partly (de Wit *et al.* 2001), or mostly unmeasured or unknown (Hughey *et al.* 2003, Heikkila 2010). This introduces significant uncertainty into any analysis.

Having said this, using the introduction of a desirable but invasive plant species through nurseries as an example, a key assumption is that:

> '*Economic theory suggests that an unregulated nursery industry would* ***not*** *take into account the risk of invasion posed by the exotic species that it markets...as if there was* ***no risk of invasion***'
> (Knowler and Barbier 2005, emphasis mine).

This assumption is sensible because in many cases the introduction of the plant species does not result in naturalisation out of a 'restricted or controlled environment' (Barbier and Knowler 2006): an important corollary of this fact is that not all plants introduced to a new environment naturalise, nor become invasive (e.g. Johnson 2014).

In contrast, the policy dilemma Knowler and Barbier (2005) identified is to allow the industry selling a plant that will become invasive through nurseries, and that industry to grow to an optimal size but no further (to maximise the profits and positive impacts and to minimise the weedy and negative impacts). Internalising the risk to industry via a Pigouvian tax or charge will restrict the number of outlets selling the plant to an optimal number instead (Knowler and Barbier 2005, Barbier and Knowler 2006). The magnitude of the tax will be dependent upon many factors including: the economic (including environmental) conditions when the invasion occurs; the sensitivity of the risk of invasion to the introduction of further sellers (that is potential propagule sources); not to mention the invasion rate. Using historical data on the introduction of *Tamarix* spp. (Knowler and Barbier 2005) and *Lythrum salicaria* L. (Barbier and Knowler 2006) into North America, they suggest that the: i. 'problem of plant invasives initiated by commercial operations is amenable to standard economic analysis and solutions…by the introduction…of an "introducers' pay tax"' (Barbier and Knowler 2006); ii. mere presence of a risk of invasion from a plant species does not mean it is socially optimal to prevent sales of the species; iii. optimal number of nurseries selling the plant is always higher when there is no risk of invasion (but the marginal risk of invasion was important in influencing the outcome); iv. higher the risk of invasion, the lower number of nurseries and the higher the rate of tax was needed; v. value of the introducers' pays tax was sensitive to the profitability of the plant species; and vi. plant species should not be sold when the risk of invasion exceeds a threshold (they appropriately suggest pre-border screening of such species in these cases): the Australian government currently operates such a system (e.g. Pheloung *et al.* 1999).

The introduction of a polluter pays tax is theoretically possible, but only after making a series of assumptions, sometimes self-reported as 'heroic' (Knowler and Barbier 2005). In addition, the solution values derived 'cannot be used for policy purposes' (Barbier and Knowler 2006). A notable statement oft repeated was that reliable data to support these analyses was either not available (Knowler and Barbier 2005, Barbier and Knowler 2006), or required considerable temporal and economic investment (Johnson 2012).

Application of a polluter pays tax As outlined above, unfortunately, and unless specific detailed data sets are available for species, it is very unlikely that a pollution tax can be applied as a policy-based instrument for managing conflict species.

A FUNDAMENTALLY DIFFERENT WAY

The policy dilemma posed by any conflict species is to balance the risk of invasion (potential negative impacts) presented by its introduction and sale against the economic benefits to industry and society that derive from its introduction (potential positive impacts). As demonstrated above, and earlier (Johnson 2012), doing so in a data-variable environment remains a key challenge.

In light of this, a fundamentally different process or system of evaluation of the risks and benefits posed by conflict species is needed, one that is: scientifically rigorous; repeatable (particularly when used by assessors of variable skill levels); best-practice; and defensible. Ideally it should also be: user-friendly; able to assess a range of different species; able to make use of scientific data when available, any observations where necessary, but not be totally dependent on these; allow for assumptions to be made, and reference material used to be explicitly stated; able to be applied to a range of different scales; and, importantly, allow for revision when new information becomes available.

Semi-quantitative systems that systematically examine the negative impacts of invasive species are widely used (e.g. Johnson 2009, Downey *et al.* 2010, Auld 2012, Auld and Johnson 2014, Steel and Johnson 2015). It is proposed that such negative impact assessments form an integral part of a future system of evaluation (Johnson 2012). A set of positive impact assessments is needed, probably similar to those proposed elsewhere (Johnson 2012). Consultation with key economic, environmental and societal stakeholders, including government and industry, will allow for joint implementation of policy and management responses (Johnson 2010).

ACKNOWLEDGMENTS

Information, comments and ideas from many colleagues have helped distil the ideas in this paper.

REFERENCES

Arcioni, E. (2003). Out damned weeds! Weed management in Australia – keeping them at bay. *Australasian Journal of Natural Resources Law and Policy* 8, 75-122.

Auld, B. (2012). An overview of pre-border weed risk assessment and post-border weed risk management protocols. *Plant Protection Quarterly* 27, 105-11.

Auld, B.A. and Johnson, S.B. (2014). Invasive alien plant management. CAB reviews: Perspectives in Agriculture, Veterinary Science, Nutrition and Natural Resources 9(37), 12 pp.

Barbier, E. and Knowler, D. (2006). Commercialization decisions and the economics on introduction. *Euphytica* 148, 151-64.

Bennett, S.J. and Virtue, J.G. (2004). Salinity mitigation versus weed risks – can we resolve the conflicts of interest in introducing new plants? *Australian Journal of Experimental Agriculture* 44, 1141-56.

Blackmore, P. (2008). Noxious weeds or just obnoxious? Primefact 254. (NSW DPI, [New South Wales Department of Primary Industries] Orange).

3 pp. http://www.dpi.nsw.gov.au/__ data/ assets/ pdf_file/0017/210590/Noxious-weeds-or-just-obnoxious.pdf

de Wit, M.P., Crookes, D.J. and van Wilgen, B.W. (2001). Conflicts of interest in environmental management; estimating the costs and benefits of a tree invasion. *Biological Invasions* 3, 167-78.

Downey, P.O., Johnson, S.B., Virtue, J.G. and Williams, P.A. (2010). Assessing risk across the spectrum of weed management. CAB Reviews: Perspectives in Agriculture, Veterinary Science, Nutrition and Natural Resources 5(38), 15 pp.

Driscoll, D.A., Catford, J.A., Barney, J.N., Hulme, P.E., Inderjit, Martin, T.G., Pauchard, A., Pyšek, P., Richardson, D.M., Riley, S. and Visser, V. (2014). New pasture plants intensify invasive species risk. *PNAS* [Proceedings of the National Academy of Sciences of the United States of America] 111, 16622-7.

Ferdinands, K.B., Setterfield, S.A., Clarkson, J.R., Grice, A.C. and Friedel, M.H. (2010). Embedding economics in weed risk management to assess contentious plants. Proceedings of the 17th Australasian Weeds Conference, ed. S.M. Zydnebos, pp. 482-5. (New Zealand Plant Protection Society, Christchurch).

Ferdinands, K., Virtue, J., Johnson, S.B. and Setterfield, S.A. (2011). 'Bio-insecurities': managing demand for potentially invasive plants in the bioeconomy. *Current Opinion in Environmental Sustainability* 3, 43-9.

Heikkila, J. (2010). Economics of biosecurity across levels of decision-making: a review. *Agronomy for Sustainable Development* 31, 119-38.

Hughey, K.F.D., Cullen, R. and Moran, E. (2003). Integrating economics into priority setting and evaluation in conservation management. *Conservation Biology* 17, 93-103.

Johnson, S.B. (2009). 'New South Wales Weed Risk Management System'. (Industry and Investment New South Wales, Orange).

Johnson, S.B. (2010). A way to weigh dread weeds – a policy framework to estimate the costs and benefits of commercially valuable invasive species. Proceedings of the Global Biosecurity Conference, pp. 122-3.

Johnson, S.B. (2012). Economic tools ≠ policy actions. Why benefit cost analysis are not a policy panacea for weedy but commercially valuable plant species. Proceedings of the 18th Australasian Weeds Conference, ed. V. Eldershaw, pp. 195-8. (Weed Society of Victoria, Melbourne).

Johnson, S.B. (2013). Some weeds have no boundaries. What are the next steps we need to take with species that jump the fence? Proceedings of the 17th Biennial New South Wales Weeds Conference, ed. H. Wu, pp. 61-9, (Weed Society of New South Wales, Corowa).

Johnson, S.B. (2014). The emerging weed challenge of managing native plant species: what are we doing in New South Wales? Proceedings of the 19th Australasian Weeds Conference, ed. M. Baker, pp. 227-30. (Weed Society of Tasmania, Hobart).

Knowler, D. and Barbier, E. (2005). Importing exotic plants and the risk of invasion: are market-based instruments adequate? *Ecological Economics* 52, 341-54.

Le Gal, E. (2011). What might happen if we thought differently about weed risk and risk management? The potential of innovations in laws and policies to address weeds infestations: an illustration with the case study of second-generation biofuels crops. Proceedings of the 16th Biennial New South Wales Weeds Conference, pp. 44-56. (New South Wales Department of Primary Industries, Orange).

Le Gal, E. (2012). The effects of institutional path dependence, political dynamics and transaction costs on the potential for 'smart' regulatory innovation: An illustration with the biofuel weed risk case study. *Australasian Journal of Natural Resources Law and Policy* 15(2), 219-54.

Martin, P. and Le Gal, E. (2010). Concepts for industry co-regulation of bio-fuel weeds. *IUCN* [International Union for Conservation of Nature] *Academy of Environmental Law. E-Journal* (1). 13 pp.

Pigou, A.C. (1920). 'The economics of welfare'. (Macmillan, London).

Pheloung, P.C., Williams, P.A. and Halloy, S.R. (1999). A weed risk assessment model for use as a biosecurity tool evaluating plant introductions. *Journal of Environmental Management* 57, 239-51.

Richardson, D.M., Pysek, P., Rejmanek, M., Barbour, M.G., Panetta, F.D. and West, C.J. (2000). Naturalization and invasion of alien plants: concepts and definitions. *Diversity and Distributions* 6, 93-107.

Steel, J. and Johnson, S.B. (2015). Managing biosecurity risk. Prioritisation of pest animals: Lessons that can be learnt from the Australian Weed Risk experience. *Plant Protection Quarterly* 30, 40-50.

New and revised plant declarations in South Australia

David A. Cooke, Michaela A. Heinson and John G. Virtue
Biosecurity SA, GPO Box 1671, Adelaide, South Australia 5001, Australia
(david.cooke@sa.gov.au)

Summary South Australia has conducted a comprehensive review of plant declarations under the *Natural Resources Management Act 2004*. Various sections of this Act allow for prohibition of sale and road transport of declared plants, legal requirements for landholder control and formal notification of their presence. Declarations are based on risk assessment and policies adopted for each species. The review has occurred over four phases with 155 weed policies revised or newly developed. Policies cover broad goals and objectives, state risk assessment and regional management actions.

There are currently 140 declared plants in SA with another five currently under consideration. These are alisma (*Alisma lanceolatum* With.), coastal tea-tree (*Leptospermum laevigatum* (Gaertn.) F.Muell.), dune onionweed (*Trachyandra divaricata* (Jacq.) Kunth), giant reed (*Arundo donax* L.) and parrot feather (*Myriophyllum aquaticum* (Vell.) Verdc.).

Keywords Declaration, legislation, review.

INTRODUCTION

Declared plants are plants that are regulated under the *Natural Resources Management Act 2004* due to their weed threat to South Australia's primary industries, natural environments and/or public safety. The Minister for Sustainability, Environment and Conservation may declare a weed when some of the powers of the Act are needed to implement one or more of the eight Natural Resources Management (NRM) boards' regional strategic plans to manage the weed. Declaration provides a legislative tool to limit the establishment and spread of a plant, thereby reducing future costs from its weed impacts and control costs. Plant species are declared under various sections of the NRM Act relating to:

- Movement – The declared plant must not be moved on a public road (e.g. as a cutting, seed or potted specimen). Inadvertent movement of the plant on animals, soil, vehicles, machinery or produce may also be prohibited.
- Sale – The declared plant must not be sold at any outlet including nurseries, pet shops and market stalls. Sale of any animal, soil, vehicle, machinery or produce contaminated with the plant may also be prohibited.
- Notification – The presence and locations of the declared plant must be reported to the regional NRM Authority by the owner of the land.
- Control – Land owners are required to take action to destroy or control certain declared plant species present on their property. NRM Authorities are also responsible for controlling these declared plants on road reserves, and may have the power to recover costs of control from the adjoining landowners.

The first comprehensive review of plants declared under weeds legislation in South Australia in 20 years, as described in Heinson *et al.* (2014), commenced in 2010. The review was split into phases and is now in its fourth and final phase. For each phase a revised or new, state-level policy was prepared for a plant. Elements of the policy document include a succinct management plan with dot-points for outcome/s, objective/s and implementation; a table consolidating regional management actions based on a regional risk assessment (Virtue 2010); a declaration statement with sections of the NRM Act which apply; a state-level risk assessment; and synonymy.

A 'plant' as declared under the Act may mean a defined subset of a species, a whole species, two or more species or rarely a whole genus (Cooke 2016). Its circumscription may be modified by the exclusion of particular cultivars assessed as having negligible weed risk, or by restricting the declaration to 'feral' individuals growing outside cultivation. Hybrids between a declared plant and a non-declared plant fall outside the declaration.

This paper summarises changes to the schedule of declared plants arising from the review. The schedule and policies for each declared plant are available on the Biosecurity South Australia website (www.pir.sa.gov.au/biosecurity/weeds_and_pest_animals).

PLANTS REMAINING ON THE SCHEDULE

New policies have been adopted for 104 plants that remain declared on the schedule. However, the specific declaration has changed for many plants. A common change has been uniform declaration of a species across NRM regions, removing anomalies between local government areas. The plants which are declared as notifiable has been rationalised to those that are not known to be or are rarely established in a region or the state. Some plants are now declared for control by land owners in fewer regions, due to being widespread and/or posing a lesser weed risk than other plants.

NEW PLANTS DECLARED

Since the review commenced, 32 new species have been added to the schedule of declared plants (Table 1). This includes the twelve additional Weeds of National Significance that were listed in 2012.

The majority of newly declared plants are environmental weeds with impacts in terrestrial or freshwater natural ecosystems. Many of these were included in the Nursery and Garden Industry of South Australia's 'Grow Me Instead' booklet and hence there was already widespread awareness of their invasive properties. Most newly added environmental weeds are not uniformly declared for control across the state, with this being a decision of regional NRM boards. For example, gazania (*Gazania* spp.) is declared for control in two of the eight NRM regions. Additions to the list include non-indigenous Australian natives such as bluebell creeper (*Billardiera fusiformis* and *B. heterophylla*) and sweet pittosporum (*Pittosporum undulatum*).

Only a few agricultural weeds have been added to the list, reflecting the range of weed control measures available in farming systems to minimise weed impacts. These include the crop weed blue mustard (*Chorispora tenella*) and the pasture weeds apple of Sodom (*Solanum linnaeanum*) and spiny rush (*Juncus acutus*). Buffel grass (*Cenchrus ciliaris* and *C. pennisetiformis*), despite being a widely grown fodder plant in northern Australia, is now declared in South Australia due to the unacceptable biodiversity and fire risk it poses. Most of the newly declared plants were initially proposed by regional NRM boards following regional risk assessments and consultation.

PLANTS NO LONGER DECLARED

Ten plants have been removed from the schedule of declared plants during the review (Table 2).

Nine of the plants now removed from declaration were declared in the 19th or 20th centuries due to concerns about their impact on agricultural or pastoral lands. However, changing farming systems and weed control methods have meant that such species pose less weed risks. The routine use of Group B herbicides enables ready control of soursob (*Oxalis pes-caprae*), small-fruited pheasants eye (*Adonis microcarpa*) and onion weed (*Asphodelus fistulosus*). Three thistle species have been removed from declaration (Table 2) as they are readily controlled by Group I herbicides and regional management programs no longer require enforced control.

Wild carrots had been declared solely to enable carrot seed producers to contract with the former Animal and Plant Control Boards to control wildling carrots, a threat to the genetics of their crops, on public roadsides. This practice is no longer necessary.

PLANTS PROPOSED FOR DECLARATION

Five species are currently under consideration for declaration in Phase 4 of the review:

Alisma (*Alisma lanceolatum*) is an emergent water plant with large broad leaves and herbaceous flowering stems from a short underwater rhizome. It can be mistaken for a native *Alisma* found in the same areas that has been confused with the European *A. plantago-aquatica* L. but may be an undescribed species (J. Conran, pers. comm).

Coastal tea-tree (*Leptospermum laevigatum*) from eastern Australia is an invader of near-coastal native vegetation. Even in its native range it may behave as a seral species as habitats change, encroaching into heath from neighbouring, stabilized, calcarenite dunes, usually after a fire that temporarily raises the pH of acidic heathland soils (Burrell 1981). It was introduced to South Australia as a native ornamental/amenity plant, and now makes incursions into various coastal and near-coastal vegetation from plantings. Many of the public consider coastal tea-tree to be a native of SA, and it can be confused with native Myrtaceae such as *Leptospermum coriaceum* (F.Muell.) Cheel (green tea-tree).

Dune onionweed (*Trachyandra divaricata*) is a sandbinding perennial of coastal front dunes, introduced to Australia from southern Africa. It spreads by seed when dry plants break off and are rolled along beaches by the wind. It can be toxic to livestock by causing photosensitisation (Huxtable *et al.* 1987).

Giant reed (*Arundo donax*) is a large perennial grass growing on stream edges and wetlands, native to a wide region of Eurasia. It resembles the common reed *Phragmites australis* (Cav.) Trin. ex Steud. but is a larger plant. Giant reed is sterile but vegetative dispersal mainly occurs when stem or rhizome fragments are moved during flood events. It is a potential invader in the southern parts of South Australia as it tolerates a wide range of conditions and climates with moderate to high annual rainfall. Nonetheless, the draft policy recognises its potential for biofuel production and proposes measure to manage this risk in production systems (Virtue *et al.* 2010).

Parrot feather (*Myriophyllum aquaticum*) is a submerged aquatic plant introduced for use in ponds and aquaria. It interferes with water flow in streams,

Table 1. Plants added to the declaration 2012–2015.

Scientific name	Common name	Comments
Acer negundo L.	box elder	exemption of a male cultivar is proposed
Amelichloa brachychaeta (Godr.) Arriaga & Barkworth	espartillo	
Andropogon gayanus Kunth	gamba grass	
Anredera cordifolia (Ten.) Steenis	Madeira vine	
Asparagus aethiopicus L.	ground asparagus	the cultivar 'Myersii' is exempt
Asparagus africanus Lam.	ornamental asparagus	
Asparagus plumosus Baker	climbing asparagus fern	
Asparagus scandens Thunb.	asparagus fern	
Billardiera fusiformis Labill. *Billardiera heterophylla* (Lindl.) L.W.Cayzer & Crisp	bluebell creeper	two species that may eventually become synonyms
Casuarina glauca Sieber ex Spreng. *Casuarina obesa* Miq.	swamp oaks	two similar species; exemption of a sterile cultivar is proposed
Cenchrus ciliaris L. *Cenchrus pennisetiformis* Steud.	buffel grass	two species with similar ecology and weed risk
Cenchrus setaceus (Forssk.) Morrone	fountain grass	including all cultivars
Chorispora tenella (Pall.) DC.	blue mustard	
Coprosma repens A.Rich.	mirror bush	
Dipogon lignosus (L.) Verdc.	dolichos pea	
Dolichandra unguis-cati (L.) L.G.Lohmann	cats claw creeper	
Erica arborea L.	tree heath	
Erica baccans L.	berry heath	
Fraxinus angustifolia Vahl	desert ash	the cultivar 'Raywood' is exempt
Gazania Gaertn., all species	gazania	some cultivars are exempt
Genista linifolia L.	flaxleaf broom	
Hieracium aurantiacum L.	orange hawkweed	
Jatropha gossypiifolia L.	bellyache bush	
Juncus acutus L.	spiny rush	
Orbea variegata (L.) Haw.	carrion flower	
Pittosporum undulatum Vent.	sweet pittosporum	
Polygala myrtifolia L.	polygala	including all cultivars
Retama monosperma Webb & Berthel. *Retama raetam* (L.) Boiss.	white weeping brooms	two very similar species
Rhamnus alaternus L.	Italian buckthorn	
Rubus laudatus A.Berger	Bundy blackberry	
Senecio madagascarensis Poir.	fireweed	
Solanum linnaeanum Hepper & P.-M.L.Jaeger	apple of Sodom	
Tamarix parviflora DC. *Tamarix ramosissima* Ledeb.	tamarisks	two species with similar ecology and weed risk
Zantedeschia aethiopica (L.) Spreng.	arum lily	including all cultivars

Table 2. Plants removed from declaration 2012–2015.

Scientific name	Common name
Adonis microcarpa DC.	small-fruited pheasant's eye
Asphodelus fistulosus L.	onion weed
Carduus tenuiflorus Curt.	slender thistle
Cirsium vulgare (Savi.) Tenore	spear thistle
Daucus carota L.	wild carrot
Hirschfeldia incana (L.) Lagr.-Foss.	Buchan weed
Oxalis pes-caprae L.	soursob
Picnomon acarna (L.) Cassini	soldier thistle
Reseda phyteuma L.	rampion mignonette
Sclerolaena birchii (F.Muell.) Domin	galvanised burr

recreational freshwater fishing and use of rivers for recreation (Parsons and Cuthertson 2001); and competes with native aquatic plants for habitat. Parrot feather resembles some native *Myriophyllum* species.

CONCLUSION

Pending the outcome of Phase 4, it is likely that the schedule will consist of 145 declared plants at the conclusion of the review.

ACKNOWLEDGMENTS

We thank the many South Australians who have contributed information and comments to the review of declared plants.

REFERENCES

Burrell, J. (1981). Invasion of coastal heaths of Victoria by *Leptospermum laevigatum* (J.Gaertn.) F.Muell. *Australian Journal of Botany* 29 (6), 747-64.

Cooke, D. (2016) Plant names and Boolean logic. *NRM Science Conference, Adelaide, 13-15 April 2016.* Abstract at https://www.researchgate.net/publication/301894366_Plant_names_and_Boolean_logic (Accessed 22 June 2016).

Heinson, M., Cooke, D., Virtue, J., Michelmore, P., Patrick, G., Drew, J., Wiadrowski, R., Honan, I., Elson, P., Roberts, G. and Gurney, T. (2014). Community engagement for reviewing declared weed policies: how South Australia encourages meaningful input into decision making processes. Proceedings of the 19th Australasian Weeds Conference, ed. M. Baker, pp. 29-32. Tasmanian Weed Society, Hobart.

Huxtable, C., Chapman, H., Main, D., Vas, D., Pearse, B, and Hilbert, B. (1987). Neurological disease and lipofuscinosis in horses and sheep grazing *Trachyandra divaricata* (Branched onion weed) in south Western Australia. *Australian Veterinary Journal* 64 (1), 105-8.

Parsons, W. and Cuthbertson, E. (2001). 'Noxious weeds of Australia', 2nd edition. CSIRO Publishing, Melbourne.

Virtue, J.G. (2010). South Australia's Weed Risk Management System. Proceedings of 2nd International Weed Risk Assessment Workshop. *Plant Protection Quarterly* 25 (2), 75-9.

Virtue, J., Reynolds, T., Malone, J., Preston, C. and Williams, C. (2010). Managing the weed risk of *Arundo donax* L. Proceedings of the 17th Australasian Weeds Conference, ed. S.M. Zydenbos, pp. 176-9. New Zealand Plant Protection Society, Christchurch, New Zealand.

The risk of *Acer pseudoplatanus* (sycamore maple) to plant biodiversity in the Blue Mountains, New South Wales

Robin J. Adair[1], Huw Evans[2], Michaela Jones[3] and Trish Kidd[3]
[1] Australis Biological, PO Box 151, Bittern, Victoria 3918, Australia
[2] PO Box 3109, Bowenfels, New South Wales 2790, Australia
[3] National Parks and Wildlife Service, PO Box 330, Oberon, New South Wales 2787, Australia
(robin.adair@australisbiological.com.au)

Summary *Acer pseudoplatanus* L. is invasive in the Blue Mountains, New South Wales and a threat to the Greater Blue Mountains World Heritage Area. Native plant richness is reduced by 75% in mature stands of *A. pseudoplatanus*, where most main life-forms are reduced in cover-abundance. After removal of *A. pseudoplatanus*, native plant richness and cover increases, but with limited re-establishment of native upper canopy species. Control strategies for *A. pseudoplatanus* need to focus on limiting the rate of spread by eliminating satellite infestations, and containing existing infestations. Treatment of seedling banks and saplings prior to killing mature trees may improve efficacy and effectiveness of control programs.

Keywords Invasions, risk assessment, biodiversity, weed control, species richness.

Chemical and physical methods of controlling wild radish in lupins

Mohammad Amjad[1], Abul Hashem[1] and Glen Riethmuller[2]
Department of Agriculture and Food Western Australia
[1] Northam, Western Australia 6401, Australia
[2] Merredin, Western Australia 6415, Australia
(mohammad.amjad@agric.wa.gov.au)

Summary To minimise the impact of herbicide resistance within Western Australian (WA) Grainbelt, integrated weed management (IWM) strategies employing both chemical and non-chemical weed control options need to be incorporated into the lupin production systems. The effect of on-row and inter-row physical and chemical weed control options on wild radish and performance of a lupin crop sown in wide rows were investigated on grower's field at Ballidu WA. Chemical and non-chemical treatment combinations were applied at pre- and post-emergent to manage wild radish in wide row lupin. On-row herbicide was applied using the knapsack sprayer immediately after the sowing. Sweep points (180 mm) were used for inter-row cultivation at 5-leaf stage or at flowering of lupin crop. The shielded sprayer with Tee jet (TM DG95015EVS) was used for inter-row spray treatments at the two stages of lupins. On-row herbicide application and inter-row cultivation and shielded spray combinations reduced wild radish numbers compared to standard control treatment (simazine) in 2015. However, reduction in lupin grain yield was also observed in some treatments due to impact of inter-row cultivation and shielded spray application. These chemical and non-chemical radish control options in 2015 look promising and need to be tested across seasons and environments to improve the control wild radish and minimise grain yield losses for improving the efficiency of cropping systems involving lupin while minimising the impact of herbicide resistance within WA Grainbelt. Alternative weed control methods are necessary to sustain lupin production systems in Australia.

Keywords Chemical and non-chemical (physical and mechanical) methods, wild radish, shielded sprays, lupin grain yield.

Training and nudging weed practitioners towards better on-ground results

Kate Blood[1] and Bec James[2]
[1] Department of Environment, Land, Water and Planning, PO Box 7, Beaufort, Victoria 3373, Australia
[2] Department of Environment, Land, Water and Planning, Private Bag 15, Ferntree Gully Delivery Centre, Victoria 3156, Australia
(kate.blood@delwp.vic.gov.au)

Summary Asking weed practitioners what they need to do their work better has influenced delivery of capability building activities for the Weeds at the Early Stage of Invasion (WESI) project in Victoria. Surveys of government staff were carried out from 2013 to 2015 using one-on-one and group interviews, and on-line surveys (Survey Monkey). Pilot training activities were undertaken and evaluation carried out. From the information received, a capability plan was completed, including training and awareness. The plan aims to build the capability of staff with responsibilities for invasive species to use the package of WESI tools.

Three priority audiences were identified:

- people authorising the work;
- people designing the work;
- people doing the work on-ground.

Competencies and risks were prioritised to determine the skills and awareness topics to focus on, including:

- invasive plants (especially their identification);
- impacts;
- hazards;
- biology and ecology;
- hygiene;
- reporting weeds;
- conducting investigations, e.g. pathway analysis;
- carrying out monitoring and evaluation; and
- survey techniques, e.g. field search and delimiting surveys.

The training plan includes session topics that meet the greatest need and address the highest risk.

The accompanying awareness plan includes the preferred topics (e.g. weed identification) and the preferred ways in which to access information: on-line; direct email; and through social media e.g. Facebook, Yammer, Twitter. The awareness plan recommends the regular nudging of staff to keep using the WESI tools via these media.

Keywords Awareness, capability building, competencies, early invaders, nudging, risk, skills, training.

Assessing weed risk at the local scale

Kate Blood[1], F. Dane Panetta[2] and Bec James[3]
[1] Department of Environment, Land, Water and Planning, PO Box 7, Beaufort, Victoria 3373, Australia
[2] Faculty of Veterinary and Agricultural Sciences, The University of Melbourne, Parkville, Victoria 3010, Australia
[3] Department of Environment, Land, Water and Planning, Private Bag 15, Ferntree Gully Delivery Centre, Victoria 3156, Australia
(kate.blood@delwp.vic.gov.au)

Summary Determining the risk posed by a potential environmental weed is generally very time consuming. To assist public land managers, the Victorian Government has compiled the Victorian Weed Risk Assessment scores and the ratings from the Environmental Weed Advisory lists into one database, called the *Victorian Environmental Weed Risk Database*. This Excel file has a number of active worksheets. It can be used by land managers to compile a weed list for their patch and then compare scores and ratings for different species. If none is available, the land manager can use the *Environmental Weed Risk Screen* to determine whether the plant should be surveyed to determine its extent or if it should simply be monitored.

The screen was developed to assess the risk posed by a plant that has been either newly detected, or has been known to be present for some time but has never been assessed. It has been designed to identify the plants that pose the highest threat to biodiversity, particularly at the local scale, e.g. national park, conservation reserve, State forest.

By sorting the spreadsheet compiled for the specific public land parcel, the manager can start to both prioritize weeds (based on the comparative risk they pose) and determine what species should be delimited. The database is one part of a decision making framework for early invaders, developed by the *Weeds at the Early Stage of Invasion* project, and available at www.delwp.vic.gov.au/early-invaders.

Keywords Early invaders, environmental weeds, invasive species, risk assessments, risk ratings, risk screen.

Increasing foliar herbicide options for controlling *Clidemia hirta* (L.) D.Don

Simon J. Brooks[1], Stephen D. Setter[2], Kirsty L. Gough[1] and Melissa J. Setter[2]
[1] Department of Agriculture and Fisheries, PO Box 187, Charters Towers, Queensland 4820, Australia
[2] Department of Agriculture and Fisheries, PO Box 20, South Johnstone, Queensland 4859, Australia
(stephen.setter@daf.qld.gov.au)

Summary *Clidemia hirta* (L.) D.Don. (Koster's curse) is a highly invasive, tropical understory shrub that was discovered in north Queensland in 2001. It was an eradication target species at a single location until the discovery of a second infestation in July 2015. The eradication program mostly involved manually uprooting entire plants; however the discovery of the second infestation signalled a need for more control options to accompany changes in managing this weed by a broader range of stakeholders. The second infestation provided sufficient plants to test eight high-volume foliar herbicides at one or two rates. Three of these herbicides were also applied using a low-volume foliar spraying technique, and comparisons were made of the effectiveness, volume and amounts of active ingredients applied using high and low-volume methods. Herbicide damage was assessed one, three and seven months after treatment. High and low-volume rates of treatments containing 600 g L^{-1} of triclopyr, 333 g L^{-1} fluroxypyr, and 300 g L^{-1} triclopyr + 8 g L^{-1} aminopyralid + 100 g L^{-1} picloram were effective. Treatments with 600 g kg^{-1} metsulfuron-methyl, 300 g kg^{-1} metsulfuron-methyl + 375 g kg^{-1} of aminopyralid and 140 g L^{-1} fluroxypyr +10 g L^{-1} aminopyralid was also effective at both high volume rates tested. Treatment with 625 g L^{-1} of 2,4-D amine was less effective, and results for treatments with 360 g L^{-1} of glyphosate were more variable, than the preceding treatments. This trial identified several effective active ingredients to assist in managing a broader *C. hirta* incursion.

Keywords Rainforest, low-volume application.

INTRODUCTION

Clidemia hirta (L.) D.Don (Koster's curse) is a bushy shade tolerant tropical shrub that was the subject of a national cost-shared weed eradication program after it was discovered near Julatten in 2001 (Breaden *et al.* 2012). This was the only known infestation in Australia until a second was discovered in July 2015, in the Wooroonooran National Park, approximately 130 km south of Julatten. As a result of this second infestation the eradication of *C. hirta* was abandoned and alternative management approaches are now being developed.

During the eradication program, most *C. hirta* was controlled by hand removing the entire plants and root system. However, this technique may be less feasible for treating larger or denser infestations. In its invasive range *C. hirta* occurs across a broad range of vegetation, soil, climate types and invades a wide variety of native and agricultural land types (Breaden *et al.* 2012). Beyond the current control activities amongst native forests, there may be future requirements to treat *C. hirta* in areas such as roadsides, riparian zones, steep slopes, grazing areas, walking tracks and plantations. The management of *C. hirta* will also require treatments suitable for a broader range of stakeholders to apply in a variety of situations, including timely responses to new detections.

Multiple overseas studies have tested a range of foliar herbicides for control of *C. hirta*. Herbicides containing the active ingredient triclopyr have proven highly effective over a range of application rates (Faiz (1993), Hafiz *et al.* (2014), Motooka (2003) and Teoh *et al.* (1982)). Fluroxypyr was more than 90% effective on *C. hirta* (Faiz 1993, Kuan *et al.* 1993), as was a metsulfuron-methyl based herbicide (Faiz 1993). Combinations such as metsulfuron-methyl + fluroxypyr (MSIRI 2010), metsulfuron-methyl and paraquat (Hafiz *et al.* (2014) have been found to be effective while other combinations such as aminopyralid + triclopyr, glyphosate + triclopyr (Hafiz *et al.* (2014) and Ismail *et al.* (2014)) and 2,4-D + picloram (Faiz 1993) were less effective. Individually, active ingredients such as glyphosate, dicamba, MCPA and 2,4-D were relatively ineffective in controlling *C. hirta* at the rates tested by Faiz (1993), Ismail *et al.* (2014), Motooka (2003) and Teoh *et al.* (1982).

The discovery of a new infestation in 2015 provided the opportunity to test the most promising foliar chemicals under local conditions, particularly those containing triclopyr, fluroxypyr and metsulfuron-methyl. Glyphosate and 2,4-D were also included to test whether efficacy may be better at different application rates than those tested previously. Based on its success on other weeds, some low-volume, high-concentration (splatter gun) applications were also included as an alternative to physical control and in situations where there is insufficient access for high-volume foliar spraying equipment and to target weeds amongst desirable vegetation (Brooks *et al.* 2014).

MATERIALS AND METHODS

The trial was located on an abandoned forestry track in the Misty Mountains section of the Wooroonooran National Park, near, 17° 42' S, 146° 41' E and 710 m above sea level. There were thousands of *C. hirta* plants spread over several hectares in the surrounding upland complex mesophyll vine forest of very wet rainfall zones (Regional ecosystem 7.8.2a: Queensland Herbarium 2015).

A randomized complete block experiment incorporating 19 herbicide treatments and three replications was established and measured at the site on the 13 October 2015. Plots were 2 × 2 m, 4 × 1 m or 2.5 × 2.5 m in size to create a total area of 12 or 14.5 m² per treatment.

Prior to implementation of treatments, the number of *C. hirta* plants and leaders was recorded in each plot, along with the mean and maximum height of plants. The average height of *C. hirta* was 87 cm. A Campbell

Table 1. Herbicide treatment details. Costs are based on local prices of 5 L of liquid herbicide or 200 g of granular herbicide and applied at trial volumes.

Product	Active ingredients	Active rate	Amount active (g L⁻¹)	Active applied (g ha⁻¹)	Application rate (L ha⁻¹)	Cost of herbicide ($ ha⁻¹)
Amicide® Advance 700	2,4-D (salts)	700 g L⁻¹	2.80	7933	2833	181
Brush-Off®	metsulfuron-methyl	600 g kg⁻¹	0.06	200	2807	58
Brush-Off®	metsulfuron-methyl	600 g kg⁻¹	0.12	337	3333	137
Garlon™ 600	triclopyr	600 g L⁻¹	1.02	2648	2596	150
Garlon™ 600	triclopyr	600 g L⁻¹	1.98	5610	2833	318
Grazon™ Extra	triclopyr picloram aminopyralid	300 g L⁻¹ 100 g L⁻¹ 8 g L⁻¹	1.05 0.35 0.03	2726 909 73	2596	480
Grazon™ Extra	triclopyr picloram aminopyralid	300 g L⁻¹ 100 g L⁻¹ 8 g L⁻¹	1.98 0.66 0.05	6105 2035 163	3083	1075
Hotshot™	fluroxypyr aminopyralid	140 g L⁻¹ 10 g L⁻¹	1.01 0.07	2856 204	2833	690
Hotshot™	fluroxypyr aminopyralid	140 g L⁻¹ 10 g L⁻¹	2.02 0.14	5712 408	2833	1379
Starane™ Advanced	fluroxypyr	333 g L⁻¹	1.00	2914	2917	621
Starane™ Advanced	fluroxypyr	333 g L⁻¹	2.00	5661	2833	1207
Stinger™	aminopyralid metsulfuron-methyl	375 g kg⁻¹ 300 g kg⁻¹	0.08 0.06	213 170	2833	120
Stinger™	aminopyralid metsulfuron-methyl	375 g kg⁻¹ 300 g kg⁻¹	0.15 0.12	463 370	3083	262
Weedmaster® DUO®	glyphosate	360 g L⁻¹	3.60	10200	2833	436
Weedmaster® DUO®	glyphosate	360 g L⁻¹	7.20	20400	2833	873
Garlon™ 600*	triclopyr	600 g L⁻¹	16.80	3584	213	203
Starane™ Advanced*	fluroxypyr	333 g L⁻¹	19.98	4129	207	882
Grazon™ Extra*	triclopyr picloram aminopyralid	300 g L⁻¹ 100 g L⁻¹ 8 g L⁻¹	15.00 5.00 0.40	3300 1100 88	220	581

* Low-volume applications.

Scientific weather station was also placed in a canopy gap on site to record climatic conditions on an hourly basis for the duration of the trial.

Fifteen high-volume and three low-volume applications, plus an untreated control were deployed (Table 1). Five mL L^{-1} of Uptake™ spraying oil was added to the high-volume treatments and 2 mL L^{-1} of Pulse® penetrant was added to the low-volume treatments. Red Spraymate™ Spray Marker Dye (150 g L^{-1} Rhodamine B) was added to all herbicide mixtures at a rate of 1 mL L^{-1} of solution. Fifteen litres of each high-volume treatment (Table 1) was prepared to treat *C. hirta* with a backpack sprayer and hand wand, and the amount of herbicide mixture used was estimated for each treatment.

For the low-volume treatments, one litre of solution was prepared for each treatment and 9 to 12 × 8 mL 'shots' were applied per plot (Table 1) with a manually operated 'Forestmaster' applicator.

Treatments were applied between 9:30 am and 3:30 pm on the 27 October 2015, a sunny day with hourly temperatures between 19.3 and 28.2°C and humidity between 51.3 and 84.5%. The wind was from the south south-east averaging 0.8 km h^{-1} with gusts below 5.8 km h^{-1} during application.

Post treatment assessments were undertaken 30 (26 November 2015), 94 (29 January 2016) and 218 (1 June 2016) DAT (days after treatment) using a herbicide damage rating score of 1 (undamaged) to 9 (rotten) for each stem (Brooks *et al.* 2012) and recording any plant regrowth. Stem damage scores were averaged per plot. Total *C. hirta* seedlings per plot were counted 218 DAT. Plots were photographed at each assessment. Regular visits were undertaken in 2016 to remove flower buds from the control plots and some herbicide treated plots with lower mortality as noted in Table 2. Live plants were removed when

Table 2. Assessments ranked by 218 DAT mean leader damage score, where 1 is unaffected and 8 appears dead. Means followed by the same letters are not significantly different from each other (P >0.05).

Active ingredients and (amounts) in (g L^{-1})	Initial plants m^{-2}	Initial stems m^{-2}	% Plant mortality 218 DAT	Mean stem damage score 218 DAT
aminopyralid (0.15) metsulfuron-methyl (0.12)	6.83	9.50	100.0[a]	8.00[a]
fluroxypyr (1)	5.17	6.50	100.0[a]	8.00[a]
fluroxypyr (2)	5.42	8.50	100.0[a]	8.00[a]
fluroxypyr (19.98)*	4.42	8.08	100.0[a]	8.00[a]
fluroxypyr (2.02) aminopyralid (0.14)	6.00	8.58	100.0[a]	8.00[a]
metsulfuron-methyl (0.06)	6.58	7.50	100.0[a]	8.00[a]
triclopyr (15) picloram (5) aminopyralid (0.4)*	5.58	6.33	100.0[a]	8.00[a]
triclopyr (1.02)	4.30	5.29	100.0[a]	8.00[a]
triclopyr (1.98)	7.67	11.08	100.0[a]	8.00[a]
triclopyr (16.8)*	7.75	10.42	100.0[a]	8.00[a]
triclopyr (1.98) picloram (0.66) aminopyralid (0.05)	7.42	9.50	100.0[a]	8.00[a]
fluroxypyr (1.01) aminopyralid (0.07)	6.05	7.42	100.0a	8.00[a]
metsulfuron-methyl (0.12)	8.17	10.36	99.58[a]	7.98[a]
triclopyr (1.05) picloram (0.35) aminopyralid (0.03)	6.00	7.33	98.72[ab]	7.96[a]
aminopyralid (0.08) metsulfuron-methyl (0.06)	5.08	7.25	96.31[ab]	7.70[ab]
glyphosate (7.2)**	5.25	6.83	90.08[ab]	7.61[ab]
glyphosate (3.6)**	4.42	5.83	87.11[b]	7.17[b]
2,4-D (2.8)**	6.00	7.75	17.20[c]	2.17[c]
Control**	3.58	6.83	0.00[d]	1.00[d]
s.e.m	2.96	3.19	4.18	0.20
l.s.d statistic			12.00	0.57

* Low-volume applications. ** Flowers removed from these treatments between 94 and 218 DAT.

the trial concluded (218 DAT). Genstat 16th Edition (VSN International) was used to undertake analysis of variance and Fisher's protected least significant difference (l.s.d) test identified which treatments differed from each other.

RESULTS AND DISCUSSION

Pre-treatment and final post-treatment assessments are shown in Table 2. As plants were still standing 218 DAT no leaders were scored as category 9, but the category 8 leaders were considered dead with no regrowth recorded under warm (daily min = 19°C and daily max temps = 27.5°C), wet (1620 mm rain) and humid conditions for the duration of the trial.

Twelve of the high and low-volume treatments showed a high degree of *C. hirta* control (Table 2).

There were no flowers, regrowth or live plants in the low-volume treated plots. The cost and amount of active ingredients applied via low-volume sprays fall between the two high-volume rates of the same herbicides (Table 1) and provide an effective alternative for sites where high-volume applications are difficult (Table 2).

Triclopyr was one of the cheapest and consistently effective active ingredients, which parallels the effectiveness reported in the overseas trials mentioned above. The lower rates tested provide the same efficacy for a lower cost than the more concentrated and expensive high-volume applications.

Granular products containing metsulfuron-methyl were the cheapest treatments applied in this trial and plot photos suggest they were among the treatments with more native seedlings. However, the presence of *C. hirta* biotypes resistant to metsulfuron-methyl in Malay plantations (Ramadzan *et al.* 2012) indicates that metsulfuron-methyl should not be repeatedly applied as a sole active ingredient.

Lower mean plot mortality, stem regrowth, flowers and live leaders were recorded in the 2,4-D and some glyphosate treated plots (Table 2). With high seed production from a single fruit (Breaden *et al.* 2012) the failure of these treatments to prevent flowering renders them unsuitable. This trial and the overseas studies suggest that glyphosate-based herbicides may not provide complete *C. hirta* control particularly on roadsides and in riparian areas.

Clidemia hirta seedlings recorded in most plots and were no significant differences between treatments ($P > 0.05$) in the seedling counts 218 DAT. The treatment of *C. hirta* foliage prevented any conclusions being drawn on pre-emergent effects of picloram, aminopyralid or metsulfuron-methyl active ingredients. Particular care would be needed if applying these herbicides near desirable vegetation; in these situations herbicides containing only triclopyr or fluroxypyr may be preferred.

This trial identified alternative active ingredients applied using high and low-volume techniques to control *C. hirta*. These herbicides can be considered by land managers in conjunction with resources, preferences, land use, access and permits.

ACKNOWLEDGMENTS

We thank Chris Roach, Mick Jeffery and Kim Erbacher for assistance in identifying and accessing the site and Joe Scanlan, Shane Campbell, Wayne Vogler for commenting on the manuscript.

REFERENCES

Breaden, R.C., Brooks, S.J. and Murphy, H.T. (2012). The biology of Australia weeds. 59. *Clidemia hirta* (L.) D.Don. *Plant Protection Quarterly* 27, 3-18.

Brooks, S.J., Gough, K.L. and Campbell, S.D. (2012). Testing the efficacy of low volume herbicide applications on *Chromolaena odorata. Pakistan Journal of Weed Science Research* 18, 61-71.

Brooks, S.J., Gough, K.L. and Campbell, S.D. (2014). Refining low-volume, high-concentration herbicide applications to control *Chromolaena odorata* (Siam weed) in remote areas. *Plant Protection Quarterly* 29, 71-7.

Faiz, M.A.A. (1993). Control of *Clidemia hirta* in mature rubber areas. Proceedings II of the 14th Asian Pacific Weed Science Society Conference, eds B.J. Wilson and J.T. Swarbrick, Brisbane, Australia, pp. 72-6.

Hafiz, A., Edison, P. and Sengli Damanik, B.J. (2014). Efficacy of selected herbicides in single and mixed on *Clidemia hirta* (L.) D.Don. in oil palm plantation. *Agroecotechnology* 2, 1578-83. In Indonesian (English Abstract).

Kuan, C.Y., Ahmad Faiz, M.A. and Nang, Y.Z. (1993). Weed management in rubber. II. Control of noxious weeds with herbicides. *Planters' Bulletin* 216, 80-6.

MSIRI (2010). Herbicide treatment for control of *Clidemia hirta* (Watook) in sugar cane. Mauritius Sugar Industry Research Institute. Recommendation sheet. No 178.

Motooka, P. (2003). Summaries of herbicide trials for pasture, range, and non-cropland weed control 2002. College of Tropical Agriculture and Human Resources, University of Hawai'i Manoa, 4 pp.

Queensland Herbarium (2015). Regional Ecosystem Description Database (REDD). Version 9.0 (April 2015) (DSITI: Brisbane).

Ramadzan, A.M.N., Ismail, B.S. and Chuah, T.S. (2012). A preliminary report on the potential resistance of a soapbush ('*Clidemia hirta'* (L.) D.Don) biotype to metsulfuron-methyl in an oil palm plantation in Jerantut, Malaysia. *Plant Protection Quarterly* 27, 64-8.

Teoh, C.H., Toh, P.Y. and Khairudin, H. (1982). Chemical control of *Asystasia intrusa* (Bl.), *Clidemia hirta* (Don.) and *Elettariopsis curtisii* (Bak.) in rubber and oil palm plantations. Proc of the International Conference on Plant Protection in the Tropics, eds K.L. Heong, B.S. Lee, T.M. Lim, C.H. Teoh and Y. Ibrahim, pp. 479-510. (The Malaysian Plant Protection Society, Kuala Lumpur).

Narrow row spacing suppresses weeds more effectively than wide row spacing in mungbean

Bhagirath Singh Chauhan
The Centre for Plant Science, Queensland Alliance for Agriculture and Food Innovation (QAAFI), Gatton/Toowoomba, Queensland 4343/4350, Australia
(b.chauhan@uq.edu.au)

Summary Mungbean, *Vigna radiata* (L.) Wilczek, is a nutritionally important food and feed legume crop. Its production in Australia is concentrated in the northern grains regions; characterised by summer dominant rainfall. In general, the crop is grown in wide rows of up to 1 m and therefore, it is prone to heavy weed infestation. Weeds can be controlled using herbicides; however, continuous use of herbicides with similar modes of action could result in the evolution of resistance in weeds. Therefore, weed management tactics, which can make crops more competitive to weeds, are required. A field trial was conducted in the summer season of 2015 at the research farm of the University of Queensland, Gatton, Queensland, to determine the effects of row spacing (25, 50, and 75 cm) and weed infestation period [no weeds (weed-free); weeds grown up to 3 and 6 weeks after crop planting; and weeds grown up to crop harvest] on weed growth and mungbean yield. There were 60–70% fewer weeds emerging three weeks after crop planting in 25 to 50 cm rows, compared to 75 cm rows. In the weed-free condition, as well as in low to moderate levels of weed infestation, the grain yield of mungbean was lower when grown on 75 cm rows.

Keywords Non-chemical weed control, row spacing, seeding rate, crop competition.

Oats tolerance to grass selective herbicides

Harmohinder S. Dhammu
Department of Agriculture and Food, Western Australia (DAFWA), PO Box 483, Northam, Western Australia 6401, Australia
(harmohinder.dhammu@agric.wa.gov.au)

Summary Three field trials under weed free conditions were conducted during 2006, 2007 and 2015 at Katanning to determine the tolerance of new oat varieties to a range of grass selective herbicides. Trifluralin at 960 and 1920 g ha^{-1}, and chlorsulfuron at 15 g ha^{-1} were included in all the trials, whereas pendimethalin at 594 and 1188 g ha^{-1}, s-metolachlor + prosulfocarb at 300 g + 2000 g and 600 g + 4000 g ha^{-1} and diuron 1000 g ha^{-1} were included in the trials during 2006 and 2007 only. Metolachlor at 1440 g and 2880 g ha^{-1}, and diuron + metolachlor at 1000 g + 1440 g ha^{-1} were tested during 2006, whereas s-metolachlor at 960 and 1920 g, and diuron + s-metolachlor at 500 g + 960 g ha^{-1} were tested during 2007. Terbuthylazine at 1050 g and 2100 g ha^{-1}, terbuthylazine + trifluralin at 1050 g + 960 g and 2100 g + 1920 g ha^{-1} and diuron + s-metolachlor at 500 g + 480 g and 1000 g + 960 g were tested during 2015. The trials were laid out in criss-cross design having 3–5 oat varieties with three replications. Grain yield was recorded across all the trials, but hay yield was recorded during 2015 only.

All the pre-emergent treatments yielded at par with untreated control plots across all the oat varieties except s-metolachlor + prosulfocarb at 600 g + 4000 g ha^{-1} reduced grain yield of Mitika during 2007. Terbuthylazine 2100 g + trifluralin 1920 g ha^{-1} reduced hay yield only of Kojanup significantly. Chlorsulfuron 15 g ha^{-1} applied at Z12–Z13 reduced grain yield of four out of five oat varieties during 2015.

Keywords Oat, herbicides, grass weeds, grain yield, hay yield.

Interaction of shallow seeding and herbicides in wheat

Harmohinder S. Dhammu
Department of Agriculture and Food, Western Australia (DAFWA), PO Box 483, Northam, Western Australia 6401, Australia
(harmohinder.dhammu@agric.wa.gov.au)

Summary Two field trials under weed free conditions were conducted during 2015 at Mullewa and Katanning Research Facilities to determine the tolerance of commonly grown wheat varieties to herbicides. The trials were laid out in criss-cross design having six wheat varieties and 11–13 herbicide treatments with three replications.

Interaction of shallow seeding, good soil moisture and soil residual herbicides terbuthylazine 1050 g and 2100 g ha^{-1}, terbuthylazine 1050 g + trifluralin 1440 g ha^{-1}, and pyroxasulfone 100.3 g + trifluralin 1440 g ha^{-1} caused significant yield reduction across all the wheat varieties at Mullewa on a sandy loam/loam soil. A mixture of sulfosulfuron 18.75 g ha^{-1} with trifluralin 1440 g ha^{-1} also registered significant yield loss in Bonnie Rock, Calingiri, Corack and Eagle Rock at Mullewa. In contrast, only the higher rate of terbuthylazine caused significant yield reduction in Calingiri, Mace and Yitpi at Katanning where varieties were sown at optimum depth of 3–4 cm on a sandy loam soil having good soil moisture.

Pre-emergent trifluralin 1440 g ha^{-1} followed by post-emergent s-metolachlor 300 g + prosulfocarb 2000 g ha^{-1} at Z12–Z13 caused significant yield loss in Bonnie Rock and Corack at Mullewa, but this sequential application was tolerated well by all the varieties at Katanning.

Metribuzin 75 g + carfentrazone 18 g + MCPA amine 250 g ha^{-1} and its double the rate applied at Z13–Z14 caused significant yield reduction across all the wheat varieties at Mullewa. However, at Katanning only Cobra, Mace, Trojan and Yitpi registered significant yield loss with the higher rate of this mixture.

Keywords Wheat, herbicides, seeding depth, grain yield.

Arid land invasive weed *Salvia verbenaca* L. (wild sage): investigation into seedling emergence, soil seedbank, allelopathic effects, and germination

R.L. Fisher, S.K. Florentine and M.E. Westbrooke
Centre for Environmental Management, Faculty of Science and Technology, Federation University Australia, Mount Helen, Victoria 3350, Australia
*Correspondence to: S.K. Florentine, Centre for Environmental Management, Faculty of Science and Technology, Federation University Australia, PO Box 663, Ballarat, Victoria 3350, Australia
(s.florentine@federation.edu.au)

Summary *Salvia verbenaca* L., wild sage (Lamiaceae) is native to western and southern Europe. In Australia *S. verbenaca* is considered an environmental weed with high invasiveness often growing on disturbed sites: tracks, roadsides and around earth tanks. Little is known about the ecology of this invasive species in Australia. Therefore, our objectives in this study were to determine: (i) seedling emergence in relation to sowing depth, (ii) density of soil seed bank, (iii) allelopathic effects, (iv) favourable conditions for seed germination of three month old and nine years old seeds, and (v) seed longevity.

Of four burial depths, only surface sown seeds germinated and survived. From the soil seedbank, *S. verbenaca* density was 343 ± 198 m^{-2}. For the allelopathic effects of *S. verbenaca*, neither the extract nor the leachate inhibited germination of *L. sativa*; however, the growth of the radicle of seedlings was shown to decrease. Germination was highest at 20°C under 12 h light/12 h dark. Burial of seeds in the field prior to germination reduced viability over time. Germination of three month old seed was significantly less than nine year old seed.

In conclusion, it is clear that *S. verbenaca* functions as a casual weed in arid and semi-arid environments of Australia. Understanding the plants' ecological characteristics in this study will help us take appropriate control measures for this species.

Keywords *Salvia verbenaca,* invasive weed, germination, allelopathy.

INTRODUCTION

Salvia verbenaca L. (wild sage) is a perennial herb, native to western and southern Europe, belonging to the family Lamiaceae (Claben-Bockoff and Speck 2000, Huang and van Staden 2002). In Australia *S. verbenaca* is considered an environmental weed with high invasiveness (DSE 2004). However, little is known about the ecology of this species in Australia. Therefore, the objectives of this study were to determine: (i) seedling emergence in relation to sowing depth, (ii) density of soil seed bank, (iii) allelopathic effects, (iv) favourable conditions for seed germination of three month and nine year old seed, and (v) seed longevity.

METHODS

Study site Nanya Research Station (E529680, N6335320) is located in south western New South Wales. The seasonal distribution of rainfall is even, but annual variation between seasons is high. Highest summer temperatures occur during February (mean daily maximum = 32°C; minimum = 16°C) and mild winter temperatures are experienced during July (mean daily maximum = 15°C; minimum = 5°C) (BOM 2013).

Seed collection Seed was obtained by collecting whole plants from dense infestations, in November 1996 and November 2004. Seeds were collected as seed capsules dried out, then stored in a labelled, air tight bottle and kept at room temperature until used.

Effects of seed burial The seeds were tested for germination at four different burial depths; 0, 2, 5 and 10 cm. Sterilised sand was used for the germination medium and four lots of twenty-five seed were placed in pots at each depth. The pots were placed on a bench in a glass house (20–25°C), watered twice a day and monitored for germination for 50 days.

Density and viability of soil seed bank Thirty two soil samples were collected from nine *S. verbenaca* infested sites in February 2005. At each site, four soil samples were collected along belt transects (Pavone and Reader 1982). Samples were taken from randomly selected points within plots and removed with a shovel to a depth of 15 cm (Florentine and Westbrooke 2005).

After being mixed thoroughly to ensure uniform distribution of seeds, each sample was spread evenly in labelled, paper towel-lined punnets, and placed in trays on benches in a glasshouse. All trays were watered twice a day and the number of emerged *S. verbenaca* seedlings was recorded. The germinable seed bank density was calculated as the number of *S. verbenaca* seeds that germinated under favourable water and temperature conditions within 90 days (Florentine and Westbrooke 2005).

Allelopathy: effect of exposure of seed to varying concentration of extract and leachate To determine the effect of metabolites produced by *S. verbenaca* on the germination and growth of other species, seeds of *Lactuca sativa* L. (Lettuce cv. Great Lakes; Yates seeds) were germinated in two different solutions; an aqueous extract and a leachate of *S. verbenaca*.

The aqueous extract was prepared from dried leaves and foliage of *S. verbenaca*, by placing 30 g of ground material in a 500 mL flask with 240 mL of deionised water and shaking the flask for two hours. The leachate was prepared by soaking 30 g of fresh leaves in 240 mL of deionised water for 24 hours. Both solutions were filtered and refrigerated until used.

Lactuca sativa seeds were germinated at a constant temperature of 20°C in four different concentrations of each extract; 0% (Control), 6%, 9% and 12.5%. To test for the effects of each solution on growth, the length of the radicle was measured three days after it had emerged from each seed.

Investigation of the temperature and light requirements of three month old and nine year old seeds Four lots of 25 seeds were surface sterilized with 6% sodium hypochlorite for one minute and placed in labelled Petri dishes that were lined with two moistened filter papers. Petri dishes were placed in incubators set at 20, 25, 30 and 35°C, under a 12 h light/12 h dark regime. For the 24 h dark treatment the dishes were wrapped in aluminium foil. All samples were kept moist until the experiment was terminated, after 50 days. Seeds were checked every 24 hours, with germinants counted and removed.

Germination of three month old and nine year old seeds after burial in the field To assess viability of seed following burial, twenty five small, fine-mesh nylon bags containing nine year-old and three month-old seed were buried in the field and retrieved over the duration of a year. Twenty five seeds were placed into each bag and the bags were buried in groups of five at a depth of 5 cm deep (Grigulis *et al.* 2001). Bags were removed periodically; at one week, three months, five months, and seven months after burial. After retrieval, seeds were sorted and assessed under a microscope for damage and germinated as per the optimum temperature and light requirements.

RESULTS

Seed burial Of the four burial depths, only surface sown seed germinated and survived. 52% of the seeds that germinated did so within eight days and the remainder over the next eight days (16 days in total). No seedlings emerged, from either the nine year old or the three month old seed, from the 2, 5 or 10 cm burial depths.

A significant difference (F 5.45, P = 0.04) was observed between the mean number of seedlings that emerged from nine year old seeds (3.18 ± 0.97) and three month old seeds (1.92 ± 1.42).

Soil seedbank A total of 2233 individuals belonging to ten species germinated from the soil seed bank. Of these, 64% were native species and 36% were exotic. The frequency of *S. verbenaca* was higher than any other species (24.74%), with a mean density across all collection sites of 343 ± 198 seedlings m^{-2}.

Between sites, the density of *S. verbenaca* differed significantly (F 4.01, P = 0.01) with the most significant (P <0.01) difference being attributed to seedling density at a single site, 1830 ± 61.85 seedlings m^{-2}. The seedling density for the other sites ranged from 0 to 655 seedlings m^{-2}.

Allelopathy Full germination was recorded for *L. sativa* seeds exposed to both the extract and leachate. The difference between the mean radicle length for each solution type was significant (F 37.67, P = 0.01). Growth of the radicle was inhibited by both the extract and the leachate, with the extract treatment having a stronger inhibitory effect (1.00 ± 0.50 mm) than the leachate (1.29 ± 0.27 mm).

Concentration of solution was the most significant influence on radicle growth overall, with the difference between the 0% and 12.5% concentrates (mean diff. 0.95 mm) and the lowest difference was between the 9% and 12.5% concentrates (mean diff. 0.18 mm).

Optimum temperature and light requirements There was no overall significant difference in the mean period of final germination (F 2.45, P >0.12) and the mean period for final germination was relatively short. Three month old seed germinated within 4.63 ± 1.07 days and nine year old seed germinated within 3.74 ± 1.27 days. Final germination varied according to age of seed (F = 12.00, P = 0.01) and was significantly higher for nine year old seeds (1.84 ± 0.22 seeds) compared to three month old seeds (1.62 ± 0.59 seeds).

The response to temperature was significant (F 6.12, P = 0.001) with the highest germination observed at 20°C under 12 h light/12 h dark (1.89 ± 0.26 days).

A significant difference (F 53.66, $P = 0.001$) was also observed between light treatments, due to lower germination under the 24 h dark treatment (1.49 ± 0.50 days).

Germination of seeds after burial in the field Viability of seed after burial declined for both ages of seed, with an overall total of 41% of three month old seed germinating and 7.4% of nine year old seed germinating under all temperature treatments. However, germination was observed to increase in seeds retrieved after periods of rainfall and lower temperatures, in July and September.

DISCUSSION

The soil surface appears to provide the best conditions for germination and survival of seedlings, suggesting that this species takes full advantage of available moisture and light. Seeds of this small size may require too much energy for seedlings to reach the surface from any depth of burial (Fenner and Thompson 2005). Alternately, it is likely that there is a light requirement, which may prevent a flush of germination occurring even though temperatures are favourable (Myers and Kitchen 1992). Seeds that are buried will become a part of the seed bank. As long as the seed remains viable, future disturbance events would bring the seed to the surface allowing for re-establishment of the species over time, thus protecting the population.

In conclusion, it is clear that *S. verbenaca* is functioning as a casual weed in arid and semi-arid environments. Understanding the plants' ecological characteristics in a study such as this at an early rather than late stage in the invasion will help us to take appropriate control measures of this species.

REFERENCES

BOM (2013). Australian climate zones – all climate classes. (Bureau of Meteorology, Australia). Available http://www.bom.gov.au/climate/averages/#climatemaps, Accessed 4th May 2016.

Claben-Bockoff, R. and Speck, T. (2000). Diversity and evolution in *Salvia* – presentation of a new research project. Vitex. A newsletter for Lamiaceae and Verbenaceae research. London.

DSE (2004). EVC/Bioregion benchmark for vegetation quality assessment. Victorian Riverina bioregion. (Department of Sustainability and Environment, Melbourne).

Fenner, M. and Thompson, K. (2005). 'The ecology of seeds' (Cambridge University Press, Cambridge).

Florentine, S.K. and Westbrooke, M.E. (2005). Invasion of the noxious weed *Nicotiana glauca* R.Graham after an episodic flooding event in the arid zone of Australia. *Journal of Arid Environments* 60, 534-45.

Grigulis, K., Sheppard, A.W., Ash, J.E. and Groves, R.H. (2001). The comparative demography of the pasture weed *Echium plantagineum* between its native and invaded ranges. *Journal of Applied Ecology* 38, 281-90.

Huang, L.D. and van Staden, J. (2002). *Salvia chamelaegnea* can be micropropagated and its callus induced to produce rosmarinic acid. *South African Journal of Botany* 68, 177-80.

Myers, S.E. and Kitchen, S.G. (1992). Cyclic dormancy in the short-lived perennial *Penstemon palmeri*. *The Journal of Ecology* 80, 115-22.

O'Connor, P.J., Smith, S.E. and Smith, F.A. (2004). Arbuscular mycorrhizas influence diversity and structure in a semi-arid plant community. (Department of Soil and Water, Centre for Plant Root Symbioses, University of Adelaide, Australia).

Pavone, L.V. and Reader, R.J. (1982). The dynamics of seed bank size and seed state of *Medicago lupilina*. *Journal of Ecology* 70, 537-47.

Metazachlor (Butisan™ Herbicide): new herbicide for the control of annual ryegrass (*Lolium rigidum* Gaud.) and other weeds in canola (*Brassica napus* L.) in Australia

Ian Francis and Marco Montagna
BASF Australia Ltd, Level 12, 28 Freshwater Place, Southbank, Victoria 3006, Australia
(marco.montagna@basf.com)

Summary Metazachlor (Butisan™ Herbicide) belongs to the chemical class of chloroacetamides which inhibit the synthesis of very long chain fatty acids (VLCFAI). Metazachlor is a pre- and post-emergence herbicide that enters target primarily through monocot roots/hypocotyls and dicot cotyledons during germination and early emergence. It has, therefore, its best herbicidal effectiveness during early development stages of the weeds. The efficacy and crop safety of metazachlor was evaluated in twenty-five trials conducted during the 2014 and 2015 growing seasons across the Southern canola growing regions of Australia at 750 to 1876 g a.i. ha^{-1} rates applied at the incorporate by sowing (IBS) or post-sowing pre-emergence (PSPE) timings. Efficacy data showed greater annual ryegrass (*Lolium rigidum* Gaud.) control than trifluralin and propyzamide as IBS application. PSPE applications, however, showed lower efficacy when compared to the IBS applications. Metazachlor at 938 g a.i. ha^{-1} also showed good activity on wild oat (*Avena fatua* L.) and wireweed (*Polygonum aviculare* L.). Methazachlor was demonstrated to be safe when applied to canola as an IBS treatment. PSPE applications however exhibited lower levels of safety with some reduction in establishment and early biomass. No significant crop phytotoxicity effects were recorded of across all treatments. Six re-crop trials were undertaken and included major cereal and pulse crops. These trials demonstrated that a twelve month re-crop interval may be established for metazachlor applied at 750 to 1876 g a.i. ha^{-1} rates.

Keywords Metazachlor, canola, annual ryegrass.

Evaluation of fitness cost associated with different EPSPS variants in *Eleusine indica*

Heping Han[1], Martin M. Vila-Aiub[1,2], Adam Jallalulin[1], Qin Yu[1] and Stephen B. Powles[1]
[1] Australian Herbicide Resistance Initiative, 35 Stirling Highway, Nedlands, Western Australia 6009, Australia
[2] Facultad de Agronomía, Departamento de Ecología, Universidad de Buenos Aires (UBA), Viamonte 430, Buenos Aires 1053 Argentina
(heping.han@uwa.edu.au)

Summary Our previous studies have identified the single EPSPS mutation (P106S) and the double mutation (TIPS) in glyphosate resistant *Eleusine indica* (L.) Gaertn. Further research has been conducted estimating the fitness of these two EPSPS mutations. The wild-type (WT) and homozygous P106S and TIPS mutants were isolated from a single resistant population, and fitness traits (e.g. vegetative and reproductive growth, seed production) in the presence and absence of rice competition evaluated in the glasshouse and outdoor conditions. The results showed that while conferring moderate resistance at the glyphosate field rates, P106S mutants displayed no fitness costs compared to WT. In contrast, the TIPS mutants while endowing a high level glyphosate resistance suffered significant fitness penalty with up to 50% reduction in vegetative growth and 69% in seed production, especially in competition with rice. The severe fitness penalty associated with homozygous TIPS mutants is likely due to disturbance of the shikimate pathway, as a 50-fold greater tissue shikimate accumulation was observed in the TIPS than the WT and P106S plants. In addition, we have observed that non-homozygous TIPS mutants (e.g. plants containing one TIPS and one P106S alleles instead two TIPS alleles) displayed favourable traits (e.g. conferring sufficient level of glyphosate resistance at no fitness costs). Lack of fitness costs and even a fitness advantage of the P106S mutation certainly explains its higher frequency observed in glyphosate resistant weed species. In the reality, as observed, the frequency of homozygous TIPS mutants is extremely low, and outnumbered by heterozygous TIPS mutants, in the absence of glyphosate selection.

Keywords *Eleusine indica*, fitness benefit, fitness cost, EPSPS, glyphosate resistance.

Emerging summer weeds are spreading far and wide within Western Australian Grainbelt – a GRDC Survey

Abul Hashem, Catherine Borger, Mohammad Amjad, Cameron Wild, Alex Douglas, Sally Peltzer, John Moore, Harmohinder Dhammu, Barbara Sage, Paul Bartlett, Mario D'Antuono and Greg Doncon
Department of Agriculture and Food, Western Australia
(abul.hashem@agric.wa.gov.au)

Summary Weeds on the roadside today are the potential weeds of crop tomorrow. A GRDC funded survey determined the incidence and density of roadside summer weeds in the Western Australian (WA) Grainbelt from February to April 2015. Sites were selected approximately every 10 km, where weeds were visible on the roadside. At each site, weed species were identified along a 20–50 m long transect. The survey identified 91 species at 1138 sites evenly distributed throughout the Grainbelt and 26 weed species with a total frequency of greater than 1%. The most common species were African lovegrass, fleabane, windmill grass and wild radish with a frequency of 11, 9, 7 and 6% respectively. African lovegrass was evenly distributed throughout the regions of Grainbelt. Fleabane incidence was the greatest in the northern agricultural region. Windmill grass was more prevalent in the central and south-east agricultural regions. Wild radish was most prevalent in the northern and central agricultural regions. Couch and stinking lovegrass were also major weeds in the northern and southern agricultural zones of WA. Feathertop Rhodes grass was common to the central and southern agricultural regions. Other invasive weeds such as matricaria, stinkwort, roly poly, prickly paddy melon and prickly lettuce were also prevalent on the WA roadside with greater frequency in the central agriculture region. This survey highlights those weed species that are spreading and may require further research/development of management programs for optimal control.

Keywords Roadside survey, emerging summer weed, incidence, density, distribution.

Can narrow row spacing in cotton reduce impact of weeds on crop yield?

Nadeem Iqbal[1], Steve Adkins[1] and Bhagirath S. Chauhan[2]
[1] School of Agriculture and Food Sciences (SAFS), The University of Queensland, Gatton 4343, Queensland, Australia
[2] Queensland Alliance for Agriculture and Food Innovation (QAAFI), The University of Queensland, Toowoomba 4350, Queensland, Australia
(b.chauhan@uq.edu.au)

Summary Cotton is the most important fibre crop of the world and covers an area of 2.5% of world arable land. Weeds are one of the major threats to the successful production of cotton, which can severely reduce its yield. In Australia, cotton is generally grown on 1 m rows. A study was conducted at the Gatton farm of the University of Queensland to evaluate the impact of row spacing (50 cm and 100 cm) and weed infestation period [no weeds (weed-free); weeds growing up to 3, 6, 9, and 12 weeks after planting; and weeds growing up to crop harvest) on the yield of glyphosate-tolerant cotton. The experiment was laid out in a split-plot design with crop rows as the main plots and weed infestation periods as the sub-plots. Each treatment was replicated thrice. Results revealed a marked difference in weed density and weed biomass between the two spacings. Lower weed density and weed biomass were observed in the narrow row spacing.

Keywords Cotton, weed control, herbicides, row spacing, weed crop interference.

Performance of alternative herbicides to glyphosate on weeds in glyphosate-tolerant cotton

Nadeem Iqbal[1], Steve Adkins[1] and Bhagirath S. Chauhan[2]
[1] School of Agriculture and Food Sciences (SAFS), The University of Queensland, Gatton 4343, Queensland, Australia
[2] Queensland Alliance for Agriculture and Food Innovation (QAAFI), The University of Queensland, Toowoomba 4350, Queensland, Australia
(b.chauhan@uq.edu.au)

Summary Today, the world uses more cotton than any other fibre. It is a leading cash crop in the world and contributes directly in the world's economy. Weeds are one of the most important biological constraints to the successful production of cotton which can severely reduce its yield. The risk of development of herbicide resistance in weeds has increased in glyphosate-tolerant cotton. A study was conducted at the Gatton farm of the University of Queensland to evaluate the performance of different herbicides on weed control in glyphosate-tolerant cotton. The experiment was laid out in a randomised complete block design with three replications. Different herbicide treatments were weedy (control), glyphosate applied once, glyphosate applied twice, metolachlor, glyphosate + metolachlor, pendimethalin, glyphosate + pendimethalin, and glyphosate + haloxyfop. Significant differences were observed in herbicide efficacy. Soil residual herbicides pendimethalin and metolachlor provided effective weed control as compared glyphosate alone.

Keywords Cotton, herbicide, glyphosate, weed control.

Buffel grass (*Cenchrus ciliaris* L.), an early invader in Victoria

Bec James[1], Kate Blood[2] and Ruth Raleigh[3]
[1] Department of Environment, Land, Water and Planning, Private Bag 15, Ferntree Gully Delivery Centre, Victoria 3156, Australia
[2] Department of Environment, Land, Water and Planning, PO Box 7, Beaufort, Victoria 3373, Australia
[3] Department of Economic Development, Jobs, Transport and Resources, 7 Taylor Street, Epsom, Victoria 3351, Australia
(rebecca.james@delwp.vic.gov.au)

Summary Buffel grass (*Cenchrus ciliaris*) is an invasive grass, with a long-lived deep root system, that can reproduce quickly and out-compete native vegetation. It poses a significant fire risk, having an ability to alter the intensity, frequency and extent of fires, changing vegetation structure and composition, and threatening the habitat of native fauna.

Buffel grass was first recorded in Victoria in 2014. A collaborative inter-agency approach has been implemented in north-west Victoria to tackle the threat it poses to some of the State's highest biodiversity assets. Key activities include undertaking delimiting surveys to determine the extent of infestations, recording and treatment of known infestations, and the promotion of awareness of buffel grass in Victoria.

The Weeds at the Early Stage of Invasion (WESI) project has used buffel grass as a pilot to test a number of the tools that have been developed for early invaders.

Keywords Buffel grass, early invader, delimiting survey.

History of buffel grass in Victoria Buffel grass was actively distributed for pasture from 1910, initially in Western Australia, then more widely (Reynolds and Shepherd 2012). Buffel grass was first recorded and reported in Victoria (excluding the cultivated specimen at Merbein in 1941), in April 2014 by Darren Schmitke from South Australia.

Shortly after, a Buffel Grass Teleconference Group was formed with participants from various groups within the Department of Environment, Land, Water and Planning (DELWP) to discuss the issue and plan how to manage the weed.

The Weeds at the Early Stage of Invasion (WESI) Project team became involved and undertook the first of many delimiting surveys that have been performed in the Mallee, starting with the largest of the buffel grass infestations along the Sturt Highway.

Ecology and distribution Buffel grass is an invasive deep-rooted perennial grass from Africa, Asia and India. It is widespread in central and northern Australia and is invading New South Wales and South Australia (Anon. 2008).

In Victoria the distribution of buffel grass is largely restricted to north west of the State in the Mallee. A distribution map has been developed; please contact the first author for a copy. The largest infestation can be found along the Sturt Highway between the South Australian border and the Mildura airport. The most southern recording is as far south as Bridgewater, which appears to be an outlier. The remaining infestations are largely located along the Murray Valley, Mallee and Calder Highways, in the north of the state.

Buffel grass is dispersed by wind, water and animals, aided by its fluffy bristles (Reynolds and Shepherd 2012). It spreads along drainage lines and roads and spread is assisted by slashers, vehicle draughts and the movement of soil.

Threats to the Victorian Mallee Buffel grass poses a significant threat to native Mallee habitats. South Australia's Buffel Grass Strategic Plan 2012–2017 (Reynolds and Shepherd 2012) indicates the *Environment and Protection and Biodiversity Conservation Act 1999* (EPBC) listed endangered Mallee emu-wren (*Stipiturus mallee*) and vulnerable Malleefowl (*Leiopoa ocellata*) as species that are considered to be threatened by the occurrence of buffel grass. Both species are found in the Victorian Mallee. In addition, food sources and habitat for other native fauna can also be altered by the occurrence of buffel grass (Anon. 2008).

Buffel grass also has rapid regrowth and high biomass that can alter the intensity, frequency and extent of fires and can change vegetation structure and composition (Anon 2008).

A relatively continuous flammable ground layer can be formed with dry foliage that can carry extensive and intense fires in a landscape (Anon. 2008). When moisture is sufficient this grass can recover its biomass very rapidly and can burn partly green, therefore fire can be carried at much shorter intervals in native understorey (Anon. 2008). It can also dominate the ground layer, displacing native grasses and other plants.

Interagency and interstate collaboration Since the discovery of the first known infestations of buffel grass in Victoria there has been a joint collaborative effort amongst all land management agencies within the Mallee to tackle buffel grass.

A Buffel Grass Strategic Action Plan 2015–2020 was been developed in conjunction with the DELWP, VicRoads, Parks Victoria, and other agencies to promote and facilitate actions to manage Buffel grass strategically within Victoria. It is currently in draft form, and is in the process of being formalised.

In addition to Victorian collaboration, there has been interstate collaboration with South Australian colleagues who are dealing with buffel grass, providing invaluable resources and advice on identification, treatment and control of this species.

Delimiting surveys Since the first buffel grass plant was formally recorded, a number of delimiting surveys have been conducted to determine the full extent of the spread of this invasive grass. The WESI project has developed a package of tools, including a set of six 'how to' guides to help public land managers work out the highest priority early invaders to eradicate locally. One of the guides described the process of 'how to conduct a delimiting survey' (James and Blood 2016).

The WESI project has used buffel grass as a pilot project to test and refine the delimiting survey process in the field.

DELWP, Parks Victoria and VicRoads staff have participated in a number of delimiting surveys across the Mallee. During these surveys specimens have also been collected and submitted to the National Herbarium of Victoria.

Treatment and control of buffel grass The infestations located to date are mainly restricted to roadsides. VicRoads have been vigilant in locating and recording infestations and have implemented a treatment plan to reduce the cover and spread of buffel grass along roadsides. In addition, areas surrounding the infestations, such as National Parks and State Forest, are being monitored by DELWP and Parks Victoria for any spreading buffel grass plants.

Plan for the future Moving into the future, continual searching for buffel grass in and around infested areas, particularly in high biodiversity areas such as Murray Sunset National Park and Hattah Kulkyne National Park is the highest priority.

There will be continued collaboration between agencies to manage the threat and impact of buffel grass in the Mallee. Further works will be dependent on resourcing and possible project funding submissions.

ACKNOWLEDGMENTS

A big thank you to everyone who has been involved in the management of buffel grass. A special thank you to the following: Tim Mayr, Ruth Raleigh, Amy Groch (DELWP); Peter Woods and Natasha Kennedy (VicRoads); Matt Baker, David Christian (Parks Victoria), Troy Bowman (Biosecurity South Australia).

REFERENCES

Anon. (2008). Weed management guide: buffel grass (*Cenchrus ciliaris*). CRC for Australian Weed Management.

James, R. and Blood, K. (2016). Looking for weeds: delimiting survey guide. A guide for planning and undertaking delimiting surveys for weeds at the early stage of invasion on public land in Victoria. Department of Environment, Land, Water and Planning, Victoria.

Randall, R.P. (2002). 'A global compendium of weeds'. ISBN 0958743983. (R.G. and F.J. Richardson, Melbourne).

Reynolds, T. and Shepherd, B. (2012). South Australia buffel grass strategic plan 2012–2017. A plan to reduce the weed threat of buffel grass in South Australia. South Australian Government.

Towards maximum utilization of the noxious root parasitic weeds, *Striga* spp.

Masaki Kuse[1], Reem Gaddal[2], Abdel Gabar Babiker[3] and Yukihiro Sugimoto[1]
[1] Graduate School of Agricultural Science, Kobe University, 1-1 Rokkodai, Nada, Kobe, Japan
[2] College of Agricultural Studies, Sudan University of Science and Technology, Shambat, Khartoum North, Sudan
[3] National Center for Research, Mohamed Nagib Street, Deposit 57 St. Amarat, Khartoum, Sudan
(kuse@eagle.kobe-u.ac.jp)

Summary The witchweeds (*Striga* spp.), root parasitic angiosperms predominant in sub-Saharan Africa, are major constraints to cereals production including the main food crops sorghum, millet and maize. The parasites induced to germinate by host-derived chemical signals, attach, penetrate the host roots, establish connection with the vascular system and remain subterranean for 6–8 weeks during which period they inflict most of their damage. Our group has recently demonstrated that chemical induction of germination of *Striga* seed, in absence of or away from host roots, is a practical method of control. The method, commonly referred to as 'suicidal germination', is a powerful tool to evade crop damage by the parasite. Post-emergence control methods including hand-pulling, albeit curtail seed bank replenishment, are not acceptable to farmers as early damage by the parasite preclude financial returns. Our goal is to promote *Striga* spp. control and curtail their spread through commercial utilization of seedlings. Metabolome analysis of *Striga* showed pre-dominance of optically active flavonoids and terpenoids that have high commercial values as sources of fine chemicals and lead compounds of medical importance. Further analysis indicated the presence of other biologically useful compounds including the famous vermifuge santonine. Challenges have been continued to establish simple and efficient methods to isolate these compounds and maximize utilization of *Striga* spp. as medicinal plants.

Keywords Metabolome analysis, *Striga*, parasitic weeds, secondary metabolites

Central Highlands Eden – Working along the invasive species curve continuum

Sally Lambourne and Mark Edmonds
Department of Environment, Land, Water and Planning (DELWP), Private Bag 15, Ferntree Gully Delivery Centre, Victoria 3156, Australia
(sally.lambourne@delwp.vic.gov.au)

Summary The Central Highlands Eden (CHE) project focuses on the detection and removal of high risk weeds that threaten the biodiversity values of the Central Highlands of Victoria. Active since 2012, the project area comprises almost 400,000 ha of mostly public land, including Baw Baw and Yarra Ranges national parks, Moondarra and Bunyip state parks and all interconnecting state forest.

In collaboration with local land managers, project staff undertake regular mapping of all active treatment areas and in response to disturbance such as fire. An increasingly complex picture of weed distribution is forming that aids in decision-making processes. Weeds at the early stage of invasion, for example, are delimited following detection to determine feasibility of eradication. There are a handful of species, restricted to isolated infestations where the risk of reintroduction is low, for which the aim is project-wide eradication. In contrast, treatment of widespread weeds is restricted to areas of recognised biodiversity value.

Working along the weed invasion curve continuum CHE aims to:

- Maintain ecosystem resilience in the absence of weeds,
- Achieve local eradication of weeds at the early stage of invasion before they can become naturalised,
- Contain those species whose range is actively expanding, and
- Protect high value biodiversity asset areas from the impacts of all high risk weeds.

Funded through the Victorian State Government Weeds and Pests on Public Land (WPPL) program, the Central Highlands Eden project is delivered by DELWP in partnership with Parks Victoria, VicRoads, Melbourne Water, alpine resorts, West Gippsland CMA and other land managers.

Combination of Arylex (halauxifen-methyl) + fluroxypyr for broadleaf weed control in winter cereals and fallow

Christopher O. Love, Colin L. Plater, Michael D. Lines, Gregory S. Wells and Peter J. Nott
Dow AgroSciences Australia Limited, 20 Rodborough Road, Frenchs Forest, New South Wales 2086, Australia
(love@dow.com)

Summary Pixxaro™ Herbicide# with Arylex™ active is a new combination herbicide for the control of broadleaf weeds in wheat, barley and triticale as well as knockdown control of broadleaf weeds, when mixed with glyphosate, in fallow situations.

Pixxaro™ (16.25 g a.e. L^{-1} halauxifen + 250 g a.e. L^{-1} fluroxypyr + 16.25 g a.i. L^{-1} cloquintocet-mexyl) at 200 mL ha^{-1} gives 100% control of seedling Mexican poppy (*Argemone mexicana* L.) and deadnettle (*Lamium amplexicaule* L.) up to the four leaf stage. Pixxaro™ at 300 mL ha^{-1} controls seedling flax-leaf fleabane (*Conyza bonariensis* (L.) Cronquist), dense-flowered fumitory (*Fumaria densiflora* DC.), common poppy (*Papaver somniferum* L.) and small-flowered mallow (*Malva parviflora* L.) up to the six leaf stage. Pixxaro™ at 400 mL ha^{-1} controls seedling common sowthistle (*Sonchus oleraceus* L.) and black bindweed (*Fallopia convolvulus* (L.) A.Love) up to the four leaf stage.

In fallow situations, Pixxaro™ at 300 to 400 mL ha^{-1}, when tank mixed with glyphosate at 450 to 720 g a.i. ha^{-1}, controls common sowthistle and red pigweed (*Portulaca oleracea* L.) to 15 cm diameter.

Keywords Arylex, fluroxypyr, wheat, barley, fallow.

INTRODUCTION

Wide window of application of broadleaf herbicides in winter cereals and short plant-back flexibility of broadleaf herbicides tank mixed with glyphosate in fallow are key customer attributes in the Northern Grain Region (NGR). Many broadleaf weeds have multiple germinations during the season and growers want to apply herbicides when weeds appear in their winter cereal crops or fallow. All picloram and aminopyralid based herbicides have a restricted time of application in the crop up to Z31 plus a longer restriction for safe replant of sensitive crops compared to fluroxypyr. Therefore, if the weeds germinate after the Z31 spray window or weather prevents application occurring prior to this crop growth stage window, then the benefits of using picloram and aminopyralid are lost for that season.

The key broadleaf weeds in cereals and fallow in the NGR are flax-leaf fleabane, common sowthistle and black bindweed as they can persist in the crop and into the fallow after harvest. There are currently limited herbicides available that provides flexibility of application, control of these three main broadleaf weeds in winter cereals plus allow for the short plant-back flexibility to following crops when tank mixed with glyphosate in fallow.

Pixxaro™ is currently registered in Canada (Dow AgroSciences 2015) and combines two Group I herbicides, Arylex™ (subclass arylpicolinate) and fluroxypyr (subclass pyridine), that allows a wide application window in cereals from three leaf (Z13) to full flag leaf emergence (Z39) under Australian conditions.

MATERIALS AND METHODS

For the winter cereal trials, seven small plot trials with a randomized complete block design (three or four replicates) were undertaken on the Darling Downs, Queensland and Breeza, Northern New South Wales in 2015. For the fallow trials, eight small plot trials with a randomized complete block design (three or four replicates) were undertaken in Queensland, New South Wales, South Australia and Western Australia in 2015/16.

Treatments were applied with a small plot (LPG powered) precision sprayer operating at 200–230 kPa applying 100 L ha^{-1} water volume in a single pass at 6 km h^{-1} through flat fan air induction nozzles (coarse spray quality) fitted to a 2 or 3 m hand-held spray boom. Uptake™ Spraying Oil at 500 mL 100 L^{-1} was added to each treatment in the winter cereal trials except Tordon™ 242 Cereal Herbicide (26 g a.i picloram potassium salt + 420 g a.i. MCPA potassium salt L^{-1}). For the fallow trials, a non-ionic surfactant was added to all treatments whilst ammonium sulphate was also added to treatments containing glyphosate.

Final assessment as percent visual control of broadleaf weeds was taken between 55 and 65 days after application (DAA) for the cereal trials and 27 and 57 DAA for the fallow trials. Commercially acceptable control was set at greater than 90 percent, which indicates death of the majority of plants with the remaining plants stunted with no ability to produce viable seed.

Results from individual trials have been analysed using Tukey's means test with a 95% confidence interval.

Tolerance Crop selectivity was also assessed as percent visual injury for each crop at each site. There was no significant crop injury observed with any treatment during the trials (data not presented).

RESULTS AND DISCUSSION

Pixxaro™ at 200 mL ha^{-1} gave 100 percent control of deadnettle and Mexican poppy which was equivalent to Paradigm™ Herbicide with Arylex™ active (200 g a.i. florasulam + 200 g a.i. halauxifen kg^{-1}) at 25 g ha^{-1} + LVE 600 MCPA at 500 mL ha^{-1}, but was significantly better than the other standard herbicides in the trials (Table 1).

Pixxaro™ at 300 mL ha^{-1} gave excellent control of flax-leaf fleabane (Table 2). Control was equivalent to Paradigm™ at 25 g ha^{-1} + LVE 600 MCPA at 500

Table 1. Average control of deadnettle and Mexican poppy with herbicides in 2015. Note that the data highlighted in orange indicates the proposed label rate for each weed.

		Trial number	152012HG	152017HG	154006AS
		Location	Breeza	Breeza	Greymare
		State	NSW	NSW	Qld
		Spray date	6-May-15	31-Jul-15	20-May-15
		Situation	Wheat	Fallow	Oats
		Weed	Deadnettle	Deadnettle	Mexican poppy
		Weed stage	2–9 leaf	4 leaf	1–6 leaf
		Density (m^{-2})	10	10	15
		Soil moisture	Moist	Moist	Moist
		Assessment (DAA)	62	59	65
No.	**Treatment**	**Rate (ha^{-1})**	**% Visual control**		
1	Untreated	–	0	0	0
2	Pixxaro	200 mL	100 a	100 a	100 a
3	Pixxaro LVE 600 MCPA	200 mL 500 mL	100 a	–	100 a
4	Pixxaro	300 mL	100 a	100 a	100 a
5	Pixxaro LVE 600 MCPA	300 mL 500 mL	100 a	100 a	100 a
6	Pixxaro	400 mL	100 a	100 a	100 a
7	Pixxaro LVE 600 MCPA	400 mL 500 mL	100 a	100 a	98 a
8	Paradigm LVE 600 MCPA	25 g 500 mL	100 a	100 a	100 a
9	Starane Advanced LVE 600 MCPA	300 mL 500 mL	53 b	26 b	85 b
10	Tordon 242	1 L	32 c	9 c	70 c
		LSD	6.1	5.6	5.8
		CV	3.96	5	4.19

mL ha^{-1}, but was significantly better than Starane™ Advanced (300 g a.e. fluroxypyr L^{-1}) at 300 mL ha^{-1} plus LVE 600 MCPA at 500 mL ha^{-1} and Tordon™ 242 at 1 L ha^{-1}.

Pixxaro™ at 400 mL ha^{-1} gave excellent control of common sowthistle in three out of four trials which was equivalent to all standards (Table 3).

For trial 154009AS (Table 3 and 4), there was a frost on the morning of application and there was no significant rainfall for the two weeks before and four weeks after application, so the weeds were considered under stress. As a result, the control was less than ideal. This situation can happen regularly and can only be reinforced in the restraints section of the label to avoid treating weeds that are under stress.

Pixxaro™ at 400 mL ha^{-1} gave excellent control of black bindweed in two out of three trials which was equivalent to two of the three standards (Table 4).

Summer fallow Pixxaro™ at 300 mL ha^{-1} plus glyphosate gave acceptable control of common sow-thistle in fallow compared to glyphosate alone and fluroxypyr + glyphosate (Table 5).

Table 2. Average control of flax-leaf fleabane with herbicides in 2015. Note that the data highlighted in orange indicates the proposed label rate.

		Trial number	154007AS	152017HG	152029HG
		Location	Warra	Breeza	Breeza
		State	Qld	NSW	NSW
		Spray date	27-May-15	31-Jul-15	10-Sep-15
		Situation	Barley	Fallow	Wheat
		Weed	Fleabane	Fleabane	Fleabane
		Weed stage	cots – 3 leaf	8 leaf	8 leaf
		Density (m^{-2})	20	4	0.5
		Soil moisture	Moist	Moist	Moist
		Assessment (DAA)	55	59	60
No.	**Treatment**	**Rate (ha^{-1})**	**% Visual control**		
1	Untreated	–	0	0	0
2	Pixxaro	200 mL	90 ab	100 a	100 a
3	Pixxaro LVE 600 MCPA	200 mL 500 mL	91 ab	–	100 a
4	Pixxaro	300 mL	94 a	100 a	100 a
5	Pixxaro LVE 600 MCPA	300 mL 500 mL	97 a	100 a	100 a
6	Pixxaro	400 mL	93 a	100 a	100 a
7	Pixxaro LVE 600 MCPA	400 mL 500 mL	98 a	100 a	100 a
8	Paradigm LVE 600 MCPA	25 g 500 mL	97 a	98 a	100 a
9	Starane Advanced LVE 600 MCPA	300 mL 500 mL	77 bc	45 b	100 a
10	Tordon 242	1 L	72 c	49 b	95 b
		LSD	9.0t	10.8	2.0
		CV	8.44	8.84	1.36

Table 3. Average control of common sowthistle with herbicides in 2015. Note that the data highlighted in orange indicates the proposed label rate.

No.	Treatment	Rate (ha^{-1})	152012HG	154008AS	154009AS	152017HG
		Trial number	152012HG	154008AS	154009AS	152017HG
		Location	Breeza	Tara	Millmerran	Breeza
		State	NSW	Qld	Qld	NSW
		Spray date	6-May-15	2-Jun-15	15-Jul-15	31-Jul-15
		Situation	Wheat	Wheat	Barley	Fallow
		Weed	Milk thistle	Milk thistle	Milk thistle	Milk thistle
		Weed stage	4 leaf	2–4 leaf	4–6 leaf	8 leaf
		Density (m^{-2})	3	1	0.5	0.5
		Soil moisture	Moist	Moist	Frost	Moist
		Assessment (DAA)	62	58	68	59
No.	Treatment	Rate (ha^{-1})	% Visual control			
1	Untreated	–	0	0	0	0
2	Pixxaro	200 mL	87 a	91 a	65 bc	100 a
3	Pixxaro	200 mL	85 a	99 a	63 bc	–
	LVE 600 MCPA	500 mL				
4	Pixxaro	300 mL	100 a	97 a	50 bc	100 a
5	Pixxaro	300 mL	100 a	97 a	50 bc	100 a
	LVE 600 MCPA	500 mL				
6	Pixxaro	400 mL	100 a	98 a	68 bc	100 a
7	Pixxaro	400 mL	100 a	97 a	65 bc	100 a
	LVE 600 MCPA	500 mL				
8	Paradigm	25 g	96 a	98 a	57 bc	94 a
	LVE 600 MCPA	500 mL				
9	Starane Advanced	300 mL	67 a	97 a	50 bc	100 a
	LVE 600 MCPA	500 mL				
10	Tordon 242	1 L	100 a	93 a	82 ab	100 a
		LSD	20.9	5.4	19.4	13.8
		CV	12.63	3.89	19.95	9.98

Pixxaro™ at 400 mL ha^{-1} was unreliable for control of common sowthistle in fallow due to the size of the weeds treated and lack of crop competition. Further work is underway on this use pattern.

Conclusion Pixxaro™ combines the weed spectrum of Arylex™ and fluroxypyr to allow wide window of application in cereals with a short plant-back interval.

ACKNOWLEDGMENTS

The authors would like to thank Dow AgroSciences for release of this information for publication.

REFERENCES

Dow AgroSciences. (2015). Pixxaro Herbicide Arylex Active – Technical Bulletin. Dow AgroSciences website: http://www.dowagro.com/en-ca/canada/product-finder/herbicides/pixxaro.

™ Trademark of The Dow Chemical Company ('Dow') or an affiliated company of Dow.

\# Registration pending in Australia.

Table 4. Average control of black bindweed with herbicides in 2015. Note that the data highlighted in orange indicates the proposed label rate.

		Trial number	154009AS	152017HG	152029HG
		Location	Millmerran	Breeza	Breeza
		State	Qld	NSW	NSW
		Spray date	15-Jul-15	31-Jul-15	10-Sep-15
		Situation	Barley	Fallow	Wheat
		Weed	Bindweed	Bindweed	Bindweed
		Weed stage	2–4 leaf	4 leaf	4 leaf
		Density (m^{-2})	10	2	10
		Soil moisture	Frost	Moist	Moist
		Assessment (DAA)	68	59	60
No.	**Treatment**	**Rate (ha^{-1})**	**% Visual control**		
1	Untreated	–	0	0	0
2	Pixxaro	200 mL	65 bcd	100 a	82 ab
3	Pixxaro LVE 600 MCPA	200 mL 500 mL	71 abc	–	71 b
4	Pixxaro	300 mL	73 abc	100 a	88 ab
5	Pixxaro LVE 600 MCPA	300 mL 500 mL	74 abc	100 a	95 a
6	Pixxaro	400 mL	79 ab	100 a	97 a
7	Pixxaro LVE 600 MCPA	400 mL 500 mL	79 ab	100 a	94 a
8	Paradigm LVE 600 MCPA	25 g 500 mL	58 cd	100 a	45 c
9	Starane Advanced LVE 600 MCPA	300 mL 500 mL	70 abc	100 a	90 ab
10	Tordon 242	1 L	88 a	100 a	83 ab
		LSD	11.5	16.8	15.6
		CV	10.93	11.71	13.31

Table 5. Average control of common sowthistle with herbicides in summer fallow in 2015/16. Note that the data highlighted in orange indicates the proposed label rate.

		Trial number	164001AS	152017CP	166001MW
		Location	Moola	Lockhart	Esperance
		State	Qld	NSW	WA
		Spray date	22-Dec-15	31-Dec-15	22-Dec-15
		Situation	Fallow	Fallow	Fallow
		Weed	Milk thistle	Milk thistle	Milk thistle
		Weed stage	Elongated	15 cm	20 cm
		Density (m^{-2})	25	5	20
		Soil moisture	Moist	Dry	Moist
		Assessment (DAA)	29	28	34
No.	**Treatment**	**Rate (ha^{-1})**	**% Visual control**		
1	Untreated	–	0	0	0
2	Pixxaro	300 mL	53 cde	83 ab	83 a
3	Pixxaro	400 mL	65 cd	90 a	84 a
4	Roundup® CT	1.4 L	88 ab	90 a	94 a
5	Pixxaro	300 mL	86 ab	92 a	93 a
	Roundup CT	1.4 L			
6	Starane Advanced	300 mL	93 a	92 a	95 a
	Roundup CT	1.4 L			
		LSD	22.6	12	30.6
		CV	14.11	5.76	15.19

Investigations into the effects of elevated carbon dioxide and drought on the growth and physiology of carpet weed (*Galenia pubescens* Eckl. & Zeyh.)

Ako H. Mahmood[1], Singarayer K. Florentine[1], Nimesha Fernando[1], Wendy Wright[1], Grant Palmer[1], David McLaren[2] and Jim Sillitoe[3]

[1] Centre for Environmental Management, Faculty of Science and Technology, Federation University, Mount Helen, PO Box 663, Victoria 3350, Australia

[2] Department of Economic Development, Jobs, Transport and Resources, Victorian AgriBiosciences Centre, Bundoora, Victoria 3086, Australia

[3] Research Services, Federation University, Mount Helen, Victoria 3350, Australia

(akomahmood4@gmail.com)

Summary The present study aimed to examine the interactive effects of elevated atmospheric CO_2 concentration and drought stress on the growth and some of the physiological processes of *Galenia pubescens.* Photosynthetic rate of plants increased under elevated CO_2 concentration, however drought caused significant reduction in net photosynthetic rate by (45% in 400 ppm CO_2) and (27% in 700 ppm CO_2) after five days simulating the drought treatment when compared with well-watered plants. Plants grown under elevated CO_2 level and well-watered produced a greater biomass (17.5 ± 0.5 g per plant) compared to the plants which were grown under the ambient CO_2 concentration.

Keywords CO_2 concentrations, drought, weeds.

INTRODUCTION

Plant invasion can be developed or hindered as a result of numerous changes in the climate and ecological processes. Changes in climate include; change in precipitation pattern, a rise in temperature and CO_2 levels and deposition of nitrogenous material, and fire regime changes that can lead to changes in resource availability (Miri *et al.* 2012, Sheppard and Stanley 2014).

There has been a report that air CO_2 concentrations may be increased from the present-day 380 to 550 ppm by 2050 (Franks, 2013) and afterward to surpass 700 to 1000 ppm before the end of the 21st Century (Dietz and Stern 2009, IPCC 2015) This increase in atmospheric CO_2 concentration will affect plant growth in a variety of ways (Dukes 2011), as it can have a direct impact upon photosynthesis and metabolism (Plett *et al.* 2015).

The future potential of characteristics of *Galenia pubescens* under an elevated atmospheric CO_2 concentration and drought is unknown. Hence, the objective of this study was to assess the effects of elevated carbon dioxide and water availability on the physiology and growth on the exotic weed *G. pubescens* in order to improve management methods.

MATERIALS AND METHODS

Seedlings establishment Mature *G. pubescens* seeds were collected from several populations of mature plants in the Werribee region (37° 49' 5.63" S, 144° 34' 58.77" E), and cleaned seeds were placed in an airtight container until use.

In the glasshouse twenty plastic pots (13 cm diameter and 13 cm high) were prepared as follows: for soil moisture movement and aeration, the sterilised industrial, seed-free sand soil was placed to a depth of 1 cm lined with paper towel. One kg of well-mixed dry sterilised soil was placed into each pot. On the top of each pot ten *G. pubescens* seeds were sown. Pots were labelled and placed into large white trays (44 × 36 × 7 cm) to facilitate watering from below, which ensured minimal disturbance of seeds and later, germinants. One month old *G. pubescens* seedlings (4–6 leaf growth stage), were thinned to one per pot and transferred to CO_2 chambers.

Setting-up CO_2 chambers and growth conditions Two identical *Steriudium e2400* model chambers were used for this experiment. Each chambers was set-up with a different CO_2 concentration. One of the chambers was set at an ambient atmospheric CO_2 concentration (400 ppm), simulating the present concentration of CO_2 in the atmosphere. The second chamber was placed at an elevated atmospheric CO_2 concentration (700 ppm), simulating the likely future concentration of atmospheric CO_2 (by the end of this century) as predicted in climate change scenarios (Franks 2013).

Thereafter, 10 pots (randomly chosen) were placed in one growth chamber (with a CO_2 concentration of 400 ppm). Another 10 pots (randomly chosen) were placed into the other growth chamber (with a CO_2 concentration of 700 ppm). All pots were watered to field capacity based on their need for water. One week prior to the taking of measurements in each CO_2 chamber, half of the plants from each chamber (randomly chosen) were kept well-watered and the other half were allowed to dry out slowly until visible wilting occurred.

Physiological parameters The net photosynthetic rate (μmol CO_2 m^{-2} s^{-1}), transpiration rate (mmol H_2O m^{-2} s^{-1}) and stomatal conductance (gs; mol H_2O m^{-2} s^{-1}) were measured using a portable gas exchange system (Li-6400XT, Li-Cor, Lincoln, NE, USA), following methods described by Le *et al.* (2016). The water use efficiency (WUE), was determined by dividing the net photosynthetic rate value by the transpiration rate value (Shabbir *et al.* 2014).

Plant biomass After the completion of the trial, approximately at the end of the vegetative stage and the beginning of flowering, the following morphological parameters were measured: The number of branches that were present on each plant was recorded. Then the plants were harvested from the soil surface and the stems and leaves were separated and kept in labelled paper bags and oven dried at 70°C for 48 hours. The roots were removed from the soil and washed thoroughly, were put in labelled paper bags separately and oven dried at 70°C for 48 hours. The dry leaves were added to the dry branches, to determine individual dry biomass. The root/shoot ratio was measured by dividing the root dry weight to the branch dry weight.

Data analysis In order to statistically account for any potential influence of CO_2 concentration and drought treatments on plant responses, a randomized complete design was used with two treatments and five replications. Data were analysed using a general linear model (GLM). The GLM was set up with two atmospheric CO_2 concentrations, two irrigation levels and interaction as the main factor. An analysis was then undertaken using an adjusted sum of squares approach using 95.0% confidence intervals. A two-way analysis of variance tests was run for each trait separately to detect differences caused by each factor using the Statistix 8.1 Regd package. Means were compared by Tukey's test at a 5% level of probability.

RESULTS

Figure 1 shows the effects of atmospheric CO_2 concentration and water availability of the net photosynthesis rate of *G. pubescens* seedlings at 60 days after the placement of plants in CO_2 chambers, and measurements were taken four times (48 hours were between each recorded date). It should be noted that the plants under elevated CO_2 concentration recorded significantly ($F = 9.60$, $P < 0.05$) increases in the mean net photosynthesis rates on all recorded dates (21.3 μmol mol^{-1}) when compared to the plants in ambient atmospheric CO_2 concentration (13.3 μmol mol^{-1}) under well-watered treatment. Whereas moisture stress significantly decreased the rate of net photosynthesis, the decrease was greater in plants that were grown under elevated CO_2 conditions.

The *G. pubescens* plants were not significantly different ($F = 1.68$, $P = 0.2132$) in their production of dry biomass by changing atmospheric CO_2 concentrations. However, the plants were significantly different between well-watered and drought treatments in producing dry biomass (data not shown). Similarly, elevated CO_2 concentration did not have a significant effect on root shoot ratio; however, data showed that the ratio was slightly decreased by increasing CO_2 concentration in both well-watered and drought treatments. Regarding dry root weight, data showed no significant difference ($F = 2.26$, $P = 0.152$) among plants under both CO_2 levels and water availability treatment. Dry leaf and dry shoot weight were also found not significantly affected by changing CO_2 levels; in contrast, water availability significantly affected dry leaf and dry shoot weight. The number of branches per plant was found not to be significantly affected by changing atmospheric CO_2 concentration. However, the number of branches per plant was significantly different under well-watered and drought treatments.

DISCUSSION

Elevated CO_2 is beneficial for *G. pubescens* growth under well-watered conditions. The photosynthesis rate and the percentage of dry biomass were increased subsequently, whereas under drought conditions, elevated CO_2 markedly showed reductions in the photosynthesis rate and dry biomass of the plants. Similar results were obtained by Erice *et al.* (2007), who reported that plant biomass increased under elevated CO_2 under well-watered conditions but not under drought. CO_2 concentration had no significant effects on the leaf, shoot, and root dry weight in either CO_2 concentrations. These responses are in contrast with those of many herbaceous plants (Farrar and Williams 1991, Ziska 2002).

It can be concluded, that this study supported the view that the growth of *G. pubescens* will be significantly enhanced in a future climate with an enriched atmospheric CO_2 concentration, producing taller plants with a greater biomass, and presumably with a higher seed output under well-watered conditions.

These changes will have important implications for management of this noxious weed in the future.

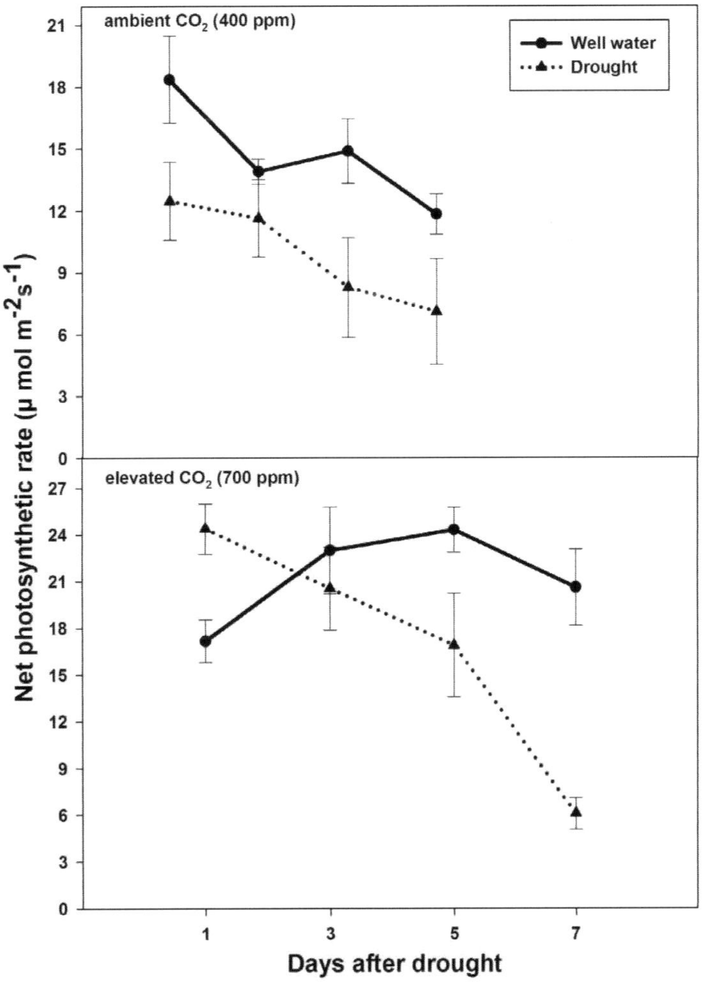

Figure 1. The effect of atmospheric CO_2 concentrations (a) ambient CO_2 (b) elevated CO_2 of the net photosynthetic rate (µmol $mol^{-2}s^{-1}$) of *G. pubescens* plants growing under well-watered and drought conditions. Vertical bars show standard errors of the mean from five replications.

ACKNOWLEDGMENTS

Special thanks are extended to the Centre for Environmental Management, Faculty of Science and Technology at Federation University, Australia for the provision of financial support. We also thank Melbourne University for allowing us to use Li-6400XT, Li-Cor. We acknowledge Wendy Cloke for providing laboratory equipment. We also thank Agnieszka Wujeska-Klause for technical help and Mr Allan Adair for reviewing this manuscript.

REFERENCES

Dietz, S. and Stern, N. (2009). Note on the timing of greenhouse gas emissions reductions: a final rejoinder to the symposium on "the economics of climate change: the stern review and its critics". *Review of Environmental Economics and Policy* 3, 138-40.

Dukes, J.S., Chiariello, N.R., Loarie, S.R. and Field, C.B. (2011). Strong response of an invasive plant species (*Centaurea solstitialis* L.) to global environmental changes. *Ecological Applications* 21, 1887-94.

Erice, G., Aranjuelo, I., Irigoyen, J.J. and Sánchez-Díaz, M. (2007). Effect of elevated CO_2, temperature and limited water supply on antioxidant status during regrowth of nodulated alfalfa. *Physiologia Plantarum* 130, 33-45.

Farrar, J.F. and Williams, M.L. (1991). The effects of increased atmospheric carbon dioxide and temperature on carbon partitioning, source-sink relations and respiration. *Plant, Cell and Environment* 14, 819-30.

Franks, P.J., Adams, M.A., Amthor, J.S., Barbour, M.M., Berry, J.A., Ellsworth, D.S. and Norby, R.J. (2013). Sensitivity of plants to changing atmospheric CO_2 concentration: from the geological past to the next century. *New Phytologist* 197, 1077-94.

Intergovernmental Panel on Climate Change (IPCC). (2015). 'Climate change 2014: mitigation of climate change, Volume 3'. (Cambridge University Press).

Le, Q.V., Tennakoon, K.U., Metali, F., Lim, L.B.L. and Bolin, J.F. (2016). Ecophysiological responses of mistletoe *Dendrophthoe curvata* (Loranthaceae) to varying environmental parameters. *Journal of Tropical Forest Science* 28, 59.

Miri, H.R., Rastegar, A. and Bagheri, A.R. (2012). The impact of elevated CO_2 on growth and competitiveness of C3 and C4 crops and weeds. *European Journal of Experimental Biology* 2, 1144-50.

Morgan, J.A., Mosier, A.R., Milchunas, D.G., LeCain, D.R., Nelson, J.A., and Parton, W.J. (2004). CO_2 enhances productivity, alters species composition, and reduces digestibility of shortgrass steppe vegetation. *Ecological Applications* 14, 208-19.

Plett, J.M., Kohler, A., Khachane, A., Keniry, K., Plett, K.L., Martin, F. and Anderson, I.C. (2015). The effect of elevated carbon dioxide on the interaction between *Eucalyptus grandis* and diverse isolates of *Pisolithus* sp. is associated with a complex shift in the root transcriptome. *New Phytologist* 206, 1423-36.

Sheppard, C.S., Burns, B.R. and Stanley, M.C. (2014). Predicting plant invasions under climate change: are species distribution models validated by field trials? *Global Change Biology* 20, 2800-14.

Ziska, L.H. (2002). Influence of rising atmospheric CO_2 since 1900 on early growth and photosynthetic response of a noxious invasive weed, Canada thistle (*Cirsium arvense*). *Functional Plant Biology* 29, 1387-92.

Factors affecting the germination ecology of turnip weed (*Rapistrum rugosum*) in the Northern Grain Region of Australia

Sudheesh Manalil and Bhagirath Singh Chauhan
Queensland Alliance for Agriculture and Food Innovation (QAAFI), The University of Queensland, Gatton/Toowoomba, Queensland 4343/4350, Australia
(s.manalil@uq.edu.au)

Summary Turnip weed (*Rapistrum rugosum* (L.) All.) is one of the rapidly emerging weeds in the Northern Grain Region (NGR) of Australia. Limited studies are available on the germination and emergence of turnip weed from the NGR. There are also areas in the NGR with varying rainfall and climatic conditions. The maternal environment can be a major determinant that decides the germination and dormancy characteristics of weeds. Therefore, two weed populations were chosen from Gatton and St George, high and low rainfall areas, respectively, to study the pattern of germination and emergence. Under the laboratory environment, lack of germination was observed in the freshly harvested seeds with siliquae intact. However, when the silique was removed, germination improved to 91 and 87% from the populations from Gatton and St George, respectively. In order to understand the level of seed bank persistence, seeds were placed in permeable nylon bags and buried at three depths: 0 (surface), 2 and 10 cm. The seeds were exhumed after three months, at the time of sampling, 2% of seeds from both the populations were germinated at 0 cm, and no germination was observed from seeds buried at 2 and 10 cm depths. Seed decay varied from 42 to 81% across the depths. Under the laboratory environment, germination was low at 0 cm (19–21%) compared to 2 cm (58–67%) and 10 cm (54–57%) depths. The responses of this weed population under varying temperatures and light conditions are under examination. In addition, the effects of pH, osmotic potential, salinity, and depth of burial on the germination of this weed are being examined. The information gained in this study would lead to a better understanding of the germination and emergence of this weed in the NGR, thereby supplement the existing knowledge, and help to frame the appropriate weed management tactics.

Saflufenacil (Sharpen® WG Herbicide): new crop topping herbicide for the control of seed set of wild radish (*Raphanus raphanistrum* L.) in wheat (*Triticum aestivum* L.)

Marco Montagna and Ian Francis
BASF Australia Ltd, Level 12, 28 Freshwater Place, Southbank, Victoria 3006, Australia
(marco.montagna@basf.com)

Summary Saflufenacil (Sharpen® WG Herbicide) is a broadleaf herbicide of the Protoporphyrinogen Oxidase Inhibitors (PPOI) mode of action group. Saflufenacil was tested as a potential crop topping herbicide to be used in wheat (*Triticum aestivum* L.) crops to reduce the seed set in wild radish (*Raphanus raphanistrum* L.). Eighteen small plot field trials were conducted between 2013 and 2015 in Victoria, South Australia and Western Australia. The efficacy and crop safety of saflufenacil was compared against 2,4-D, glyphosate, diquat and triasulfuron. The applications were made between flowering and maturity stage of the crop and when the wild radish was at early to late flowering stage. The crop topping applications of saflufenacil significantly reduced the number of set pods when compared to the untreated control. Even though the number of seeds per pod was not affected by the applications. Moreover, the germination tests of wild radish seeds collected after an earlier application of saflufenacil showed significant reductions in viable seeds. The combination of reduced number of pods and then reduced germination of those pods that were present at harvest provides a significant reduction in wild radish seed that is able to contribute to the weed seed population in a field for following seasons. Low level of phytotoxicity observed on the wheat, and generally no significant negative effect on grain yield, showed that saflufenacil has an excellent technical fit in early crop topping applications.

Keywords Saflufenacil, crop topping, seed set, wild radish, wheat.

Metabolic profiling for benzoxazinoids in weed-suppressive and early vigour wheat genotypes

James M. Mwendwa[1], Paul A. Weston[1], Inge Fomsgaard[2], Bente B. Laursen[2], William B. Brown[1], Hanwen Wu[1], Greg Rebetzke[3], Jane C. Quinn[1] and Leslie A. Weston[1]
[1] Graham Centre for Agricultural Innovation (NSW Department of Primary Industries and Charles Sturt University), Wagga Wagga, New South Wales 2678, Australia
[2] Department of Agroecology, Aarhus University, Forsøgsvej 1, DK-4200 Slagelse, Denmark
[3] CSIRO Plant Industry, P.O. Box 1600, Australian Capital Territory 2601, Australia
(jmwendwa@csu.edu.au)

Summary Replicated wheat (*Triticum aestivum* L.) cultivar trials were performed in commercial fields in moderate to low rainfall zones at Wagga Wagga (572 mm) and Condobolin (449 mm) NSW in 2014 and 2015. At each experimental site, crop and/or weed growth were monitored at selected growth stages including tillering, vegetative, grain filling, harvest and post-harvest to the crop. In addition, shoots, roots, rhizoplane and bulk rhizosphere soil samples were collected. All shoot and root samples were extracted in methanol using an automated Buchi high pressure extractor while soil samples were extracted using a rotary shaking method. Extracts were profiled for unique secondary plant products including benzoxazinoids (BXs) using liquid chromatography coupled to mass spectrometry.

Metabolic profiling of wheat cultivar shoots, roots, and soils resulted in detection of up to 14 BXs including BX glycosides and other metabolites of interest. Both qualitative and quantitative differences in BXs were observed and were cultivar and location dependent. Plant part and rhizosphere location (distance from root) also impacted BX concentration. Additional metabolic profiling is now underway and will provide crucial information regarding crop metabolism and biosynthesis of metabolites associated with weed suppression in commercial wheat cultivars.

Keywords Weed suppression, metabolomics, residue, competition, resource allocation.

INTRODUCTION

Benzoxazinoids are important allelochemicals present in wheat, barley and rye, and their suppressive effects on weeds, pest and diseases are of great interest in sustainable agriculture (Bertholdsson *et al.* 2012). BXs are known to play important roles in plant defence against herbivory, in plant interactions including allelopathy (Wu *et al.* 2000, Macías *et al.* 2007), and are considered to be natural pesticides (Tanwir *et al.* 2013). When produced in excessive amounts, BXs are stored in the form of glucosides which are enzymatically converted to aglycone forms under stress conditions (Tanwir *et al.* 2013). BXs comprise a number of chemical groups, the most important being lactams and hydroxamic acids (Figure 1).

Precise metabolic profiling of allelochemicals in the plant and at the same time in the soil rhizosphere could provide strong insight into the release of bioactive metabolites following incorporation of plant material or living root exudates into the soil (Krogh *et al.* 2006, Weston *et al.* 2015). Therefore, we evaluated selected diverse Australian wheat cultivars for their ability to suppress annual weeds under field conditions. At the same time, we also assessed their respective plant parts and rhizosphere soils for the presence of bioactive allelochemicals associated with weed suppression over two growing seasons.

MATERIALS AND METHODS

Establishment of field trials In 2014 and 2015, field trials were sown at two locations in NSW: Condobolin, considered to be a low rainfall (mean 449 mm) region and Wagga Wagga, a moderate rainfall (mean 572 mm) region. Plots were seeded with six replications in a randomised complete block design. Eleven wheat cultivars representing four major genetic families of winter wheat commercially grown in Australia were selected for evaluation, plus one cultivar of winter cereal rye, *Secale cereale* L., as a known weed-suppressive control.

In 2015, two additional cultivars (Trojan and Federation) were included. Trojan is a recently released cultivar and Federation is an older heritage cultivar bred and released in 1901 and widely used until 1970. Soils were typical Aeolian fine red clays with the clay content varying with soil depth.

Cultivar sampling and sample extraction Shoots, roots, rhizoplane and bulk rhizosphere soil from around the roots were sampled at four strategic times based on crop growth stage which included early (tillering), vegetative (stem elongation), flowering and

Benzoxazolinones		**Lactams**			**Hydroxamic acids**		
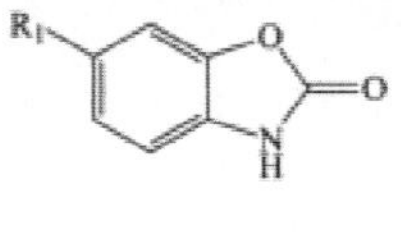		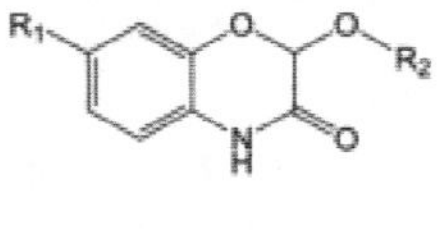			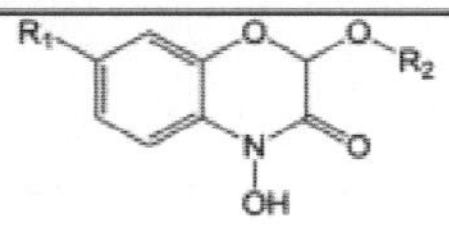		
R_1		**R_1**	**R_2**		**R_1**	**R_2**	
H	BOA	H	H	HBOA	H	H	DIBOA
OCH_3	MBOA	H	Glc	HBOA-Glc	H	Glc	DIBOA-Glc
		OCH_3	H	HMBOA	OCH_3	H	DIMBOA
		OCH_3	Glc	HMBOA-Glc	OCH_3	Glc	DIMBOA-Glc
		H	Glc-Hex[a]	HBOA-Glc-Hex	H	Glc-Hex[a]	DIBOA-Glc-Hex

Figure 1. Chemical structures of the BXs most commonly found in cereal grains and bakery products. BOA, benzoxazolin-2-one; MBOA, 6-methoxy-benzoxazolin-2-one; HBOA, 2-hydroxy-1,4-benzoxazin-3-one; HMBOA, 2-hydroxy-7-methoxy-1,4-benzoxazin-3-one; HBOA-Glc, 2-β-D-glucopyranosyloxy-1,4-benzoxazin-3-one; HMBOA-Glc, 2-β-D-glucopyranosyloxy-7-methoxy-1,4-benzoxazin-3-one; HBOA-Glc-Hex, double-hexose derivative of HBOA; DIBOA, 2,4-dihydroxy-1,4-benzoxazin-3-one; DIMBOA, 2,4-dihydroxy-7-methoxy-1,4-benzoxazin-3- one; DIBOA-Glc, 2-β-D-glucopyranosyloxy-4-hydroxy-1,4-benzoxazin-3-one; DIMBOA-Glc, 2-β-D-glucopyranosyloxy-4-hydroxy-7-methoxy- 1,4-benzoxazin-3-one; DIBOA-Glc-Hex, double-hexose derivative of DIBOA. [a] Structure not fully elucidated (Adhikari *et al.* 2015, Tanwir *et al.* 2013).

maturity (grain fill) growth stages. Approximately 5 g of sampled shoots and roots tissue were collected in a factorial experiment. Samples (taken from 4 replicates × 7 cultivars × 2 locations × 4 sample types) were each extracted in methanol using an automated high pressure extractor (Weston *et al.* 2015). Rhizoplane and bulk soil samples were extracted by shaking as per Krogh *et al.* (2006). Following extraction, samples were filtered (0.22 µm filter) before storage in amber HPLC vials. Metabolic profiling of the extracts for targeted secondary metabolites was accomplished using an LC-MS/MS Q Trap 4500 Mass spectrometer (AB SCIEX) and validated using an Agilent 6410 LC-MS QQQ (Agilent).

Secondary metabolite analyses Analytical standards for 10 BXs and 2 provisionally characterised BXs, as listed in Table 1, were gifted by the Fomsgaard lab in Denmark. Analytical grade methanol and acetonitrile were obtained from Rathburn (Walkerburn, Scotland). Glacial acetic acid was obtained from Baker (Griesheim, Germany). HPLC water used for the solvent mobile phase was purified with a Milli-Q Millipore purification system.

LC-MS/MS Detection and analysis – Q-Trap 4500 LC-MS/MS Methanolic extracts were further diluted in methanol by 4–400 times depending on sample type. The diluted samples were analysed using electrospray ionization and multiple reaction monitoring (MRM) mode using an Applied Biosystems 4500 Q Trap LC-MS (Nærum, Denmark). The chromatographic column used for the BX analysis was a Synergi Polar RP-80A with a 250 × 2 mm internal diameter and a 4 µm particle size (Phenomenex). The instrument and the compound-dependent parameters were optimised using analytical standards (Adhikari *et al.* 2015). Metabolites were analysed using negative ion mode in conjunction with optimal instrument parameters.

Nitrogen was used as a collision and source gas. The instrument and compound-dependent MS/MS parameters were optimised using flow injection analysis of individual authentic compounds. Two mobile phases (A, 7% acetonitrile in water; and B, 78% acetonitrile in water, each containing 20mM of acetic acid) were used in a linear gradient system as follows: 0–1 min, 92:08; 1–3 min, 90:10; 3–13 min, 30:70; 13–14 min, 10:90; 14–16 min, 10:90; 16–17 min, 100:0; 17–23 min, 100:0 of A:B at a flow rate of 300 µL min^{-1}

Table 1. Identification (based on molecular weight, MRM transition ion pair, and retention time on chromatographic column) of 10 known benzoxazinoids and 2 provisionally characterised benzoxazinoids analysed by Q Trap 4500 LC-MS/MS. The standards were mixed in groups of fours.

Standard	Acronym	Mass	m/z Values		Retention time/ min
			Q1 Mass (Da)	Q3 Mass (Da)	
A	BOA	135.12	133.901	41.933	12.98
A	MBOA	165.15	163.930	148.867	13.82
A	HBOA	165.15	163.916	107.929	10.69
A	HMBOA	195.17	193.902	122.901	11.78
C	HBOA-Glc	327.2867	325.9	163.9	8.79
C	HMBOA-Glc	357.3127	355.971	193.9	10.11
C	DIBOA	181.1455	179.888	133.976	10.80
C	DIMBOA	211.1715	210.1	164.3	14.00
L	DIBOA-Glc	343.2861	341.900	133.800	8.40
L	DIMBOA-Glc	373.3121	372.000	148.800	10.28
L	DIBOA-Glc-hex	505.43	504.000	133.900	4.90
L	HBOA-Glc-hex	489.43	488.100	163.900	4.90

with an injection volume of 10 µL. Twelve BXs were identified based on comparisons of mass spectra and the selectivity of the LC separation producing matching peaks at the predetermined retention times for the authentic standards listed in Table 1.

Calibration curves - Analyst AB SCIEX Calibration curves for the 12 BX standards listed in Table 1 were prepared by serial dilution of the pure standard compounds. Initially, a 10 ppm standard solution was diluted in 25% acetonitrile to the desired final concentrations. The concentrations used for the standard curves ranged from 0.05 to 200 ng L^{-1} (200, 100, 50, 25, 12.5, 6.25, 1.56, 0.39, 0.19, 0.95 and 0.05) and calibration curves were fitted using quadratic regression with Analyst software (v. 1.5) from AB SCIEX. The data points were weighted by 1/x (r >0.99 in all cases). The standard curve was generated by plotting the area of the integrated peak (y-axis) as a function of concentration (x-axis).

LC-MS/MS Detection and analysis – Agilent 6410 Triple Quad LC-MS/MS The sample extracts were diluted with aqueous solvent by 1 to 5 times and were analysed via electrospray ionization and multiple reaction monitoring (MRM) mode using an Agilent 6410 Triple Quad LC-MS/MS (Agilent Technologies). The C_{18} column (Phenomenex Synergi Polar RP-80A) possessed a 250 × 2.0 mm internal diameter and a 4 µm particle size (Phenomenex Australia Pty Ltd). The instrument and the compound-dependent parameters were optimised using analytical standards for BOA and MBOA. Negative mode ionization was used in conjunction with the following instrument parameters: curtain gas, 11 psi; temperature, 350°C; ion source gas 1, 60 psi; ion source gas 2, 60 psi; interface heater, on; collision gas, medium; and ion spray voltage, −4500 V.

The compounds were grouped according to dwell time, and the MRM mass transitions were assessed for each molecule at the corresponding time of elution. The instrument and compound-dependent MS/MS parameters were optimised using flow injection analysis of individual authentic standards. Mobile phases were as follows: A, 7% acetonitrile in water; and B, 78% acetonitrile in water, each containing 20mM of acetic acid, in a linear gradient system (0–1 min, 92:08; 1–4 min, 90:10; 4–14 min, 30:70; 14–15 min, 10:90; 15–24 min, 10:90; 24–25 min, 100:0; 25–28 min, 100:0 of A:B) at a flow rate of 300 µL min^{-1} with an injection volume of 10 µL. The 12 BXs were identified based on comparisons of mass spectra with known analytical standards and corresponding retention times as listed in Table 2. The scanning time increased from 23 to 28 minutes per sample.

RESULTS AND DISCUSSION

Metabolic profiling of wheat secondary metabolites from roots, rhizoplane and bulk rhizosphere soil using LC-MS Q Trap targeted analysis of BXs resulted in detection of all the BX compounds evaluated. The BXs were identified at varying concentrations, depending on cultivar and plant part or soil type analysed (Table 3).

Table 2. Retention time (Rt) comparison between the AB SCIEX Q Trap 4500 and Agilent Triple Quad 6410 LC-MS/MS identification of the key benzoxazinoid metabolites in the tissues of wheat cultivars and rye scanned previously in the Q Trap (based on molecular weight, MRM transition ion pair, and retention time on chromatographic column). ND* means not detected in the samples used.

Metabolite	Mass	m/z Values		Q Trap 4500 Rt (min)	Triple Quad Rt (min)
		Q1 Mass (Da)	Q3 Mass (Da)		
BOA	135.12	133.901	41.933	12.98	11.75
MBOA	165.15	163.930	148.867	13.82	12.48
HBOA	165.15	163.916	107.929	10.69	9.56
HMBOA	195.17	193.902	122.901	11.78	10.45
HBOA-Glc	327.2867	325.9	163.9	8.79	7.17
HMBOA-Glc	357.3127	355.971	193.9	10.11	8.62
DIBOA	181.14548	179.888	133.976	10.80	ND*
DIMBOA	211.17146	210.1	164.3	14.00	ND*
DIBOA-Glc	343.2861	341.900	133.800	8.40	7.39
DIMBOA-Glc	373.3121	372.000	148.800	10.28	9.01
DIBOA-Glc-hex	505.43	504.000	133.900	4.90	5.33
HBOA-Glc-hex	489.43	488.100	163.900	4.90	3.99

Table 3. Detection of targeted secondary metabolites (BXs) in cereal rye (*Secale cereale*) and wheat (*Triticum aestivum* L.) cultivar (Gregory and Wedgetail) shoot and root tissues, rhizoplane (root surface) and bulk soil surrounding field-grown living roots sampled in July 30, 2014 as assessed with AB SCIEX Q Trap 4500 LC-MS/MS. The root sample results for Wedgetail are not presented as they have not been analysed yet. The <LDL means lower than detection limit.

Metabolite	Cereal rye				Gregory wheat				Wedgetail Wheat		
	Shoots	Roots	Rhizo	Soil	Shoots	Roots	Rhizo	Soil	Shoots	Rhizo	Soil
HBOA-Glc-Hex	0.43	0.08	<LDL	<LDL	0.94	0.02	<LDL	<LDL	0.05	<LDL	<LDL
DIBOA-Glc-Hex	1.85	0.00	<LDL	<LDL	0.01	<LDL	<LDL	<LDL	0.02	<LDL	<LDL
DIBOA-Glc	1477.97	0.02	0.033	<LDL	0.91	0.00	<LDL	<LDL	0.34	<LDL	<LDL
DIMBOA-Glc	7.75	6.65	0.003	<LDL	389.55	2.36	0.002	<LDL	5.08	0.007	0.010
HBOA	1.02	3.63	0.016	<LDL	0.06	1.17	0.002	<LDL	<LDL	0.001	<LDL
BOA	2.85	0.66	0.027	0.008	0.09	4.99	0.003	0.002	0.01	0.001	<LDL
MBOA	0.98	625.12	2.797	0.047	104.13	151.03	0.258	0.010	0.05	0.070	0.265
HMBOA	0.04	3.27	0.018	0.003	1.67	3.63	0.015	<LDL	0.01	0.003	0.003
DIMBOA	2.65	0.38	<LDL	<LDL	0.54	5.20	<LDL	<LDL	<LDL	<LDL	<LDL
DIBOA	314.97	0.06	0.004	<LDL	0.52	45.45	0.003	0.015	0.03	0.002	0.011
HBOA-Glc	5.32	2.69	0.037	<LDL	5.03	5.44	0.001	0.004	6.41	<LDL	<LDL
HMBOA-Glc	0.68	13.02	0.013	<LDL	23.30	5.23	0.006	0.338	3.70	0.016	0.011

Using chromatographic conditions described in the methodology, and the use of cereal rye as a positive control (rye extracts are known to contain high concentrations of BXs), well resolved peaks for the 12 BX compounds evaluated were obtained with comparable retention times using both mass spectrometers and similar chromatography methods (Table 2).

Limit of detection (LOD) was determined by considering signal to noise (S/N) ratio of 3:1 for the strongest mass transition with respect to the background noise for each metabolite and sample type. The detection limit is the lowest quantity of a substance that can be distinguished from the absence of that substance; a blank value (Browne and Whitcomb 2010). Both instruments showed a similar level of sensitivity with relatively minor differences observed over metabolites surveyed (data not shown).

Results obtained showed reasonable sensitivity and the presence of substantial quantities of nearly all metabolites profiled in both shoot and root tissues, as well as bulk and rhizosphere soils. By profiling the level of these metabolites in soils surrounding living root systems, the role of BXs in weed suppression will be further examined. Future results using numerous wheat genotypes could provide strong insight into the resulting availability of these metabolites following their release by exudation and incorporation of plant material into the soil (Krogh *et al.* 2006), as well as the complex interplay between plants and their associated rhizosphere microorganisms, an area which is relatively understudied (Weston *et al.* 2015).

Targeted metabolic profiling in soil and plant tissues will provide important physiological information regarding crop competitive traits and biosynthesis and activity of related allelochemicals that may be important in long-term weed suppression in crop. Currently, profiling of large sample data sets from 2014 and 2015 growing seasons are underway, with differences between season, location, cultivar and plant part the subject of further analysis.

ACKNOWLEDGMENTS

We acknowledge the financial support of the Grains Research and Development Corporation (Projects UCS 00020, 00022, and 00023) and technical support of Graeme Heath, Shamsul Haque and Dom Skoneczny with respect to data collection, sampling and analyses.

REFERENCES

Adhikari, K.B., Tanwir, F., Gregersen, P.L., Steffensen, S.K., Jensen, B.M., Poulsen, L.K. and Fomsgaard, I.S. (2015). Benzoxazinoids: cereal phytochemicals with putative therapeutic and health-protecting properties. *Molecular nutrition and food research 59* (7), 1324-38.

Bertholdsson, N.O., Andersson, S.C. and Merker, A. (2012). Allelopathic potential of *Triticum* spp., *Secale* spp. and *Triticosecale* spp. and use of chromosome substitutions and translocations to improve weed suppression ability in winter wheat. *Plant Breeding* 131 (1), 75-80.

Browne, R.W. and Whitcomb, B.W. (2010). Procedures for determination of detection limits: application to high-performance liquid chromatography analysis of fat-soluble vitamins in human serum. *Epidemiology* 21 (4), S4-S9.

Krogh, S.S., Mensz, S.J., Nielsen, S.T., Mortensen, A.G., Christophersen, C. and Fomsgaard, I.S. (2006). Fate of benzoxazinone allelochemicals in soil after incorporation of wheat and rye sprouts. *Journal of Agricultural and Food Chemistry* 54, 1064-74.

Macías, F.A., Molinillo, J.M., Varela, R.M. and Galindo, J.C. (2007). Allelopathy—a natural alternative for weed control. *Pest Management Science* 63 (4), 327-48.

Tanwir, F., Fredholm, M., Gregersen, P.L. and Fomsgaard, I.S. (2013). Comparison of the levels of bioactive benzoxazinoids in different wheat and rye fractions and the transformation of these compounds in homemade foods. *Food Chemistry* 141 (1), 444-50.

Weston, L.A., Skoneczny, D., Weston, P.A. and Weidenhamer, J.D. (2015). Metabolic profiling: an overview—new approaches for the detection and functional analysis of biologically active secondary plant products. *Journal of Allelochemical Interactions* 1, 15-27.

Wu, H., Haig, T., Pratley, J., Lemerle, D. and An, M. (2000). Distribution and exudation of allelochemicals in wheat (*Triticum aestivum*). *Journal of Chemical Ecology* 26 (9), 2141-54.

Effects of the invasive grass *Andropogon gayanus* Kunth (gamba grass) on soil nitrogen dynamics and the soil microbial community

Natalie A. Rossiter-Rachor[1], Karen Gibb[1], Levente Bodrossy[2], Bronwyn Holmes[2], Damien McMaster[1] and Samantha A. Setterfield[1,3]
[1] Charles Darwin University, Darwin, Northern Territory 0909, Australia
[2] CSIRO Oceans and Atmosphere, Hobart, Tasmania 7000, Australia
[3] University of Western Australia, Crawley, Western Australia 6009, Australia
(natalie.rossiter@cdu.edu.au)

Summary The African grass *Andropogon gayanus* Kunth (gamba grass) is a serious threat to savanna ecosystems across northern Australia. The success of this high biomass grass, growing on low nitrogen (N) status savanna soils, may, in part, be due to the documented ability of gamba grass to alter the soil microbial community via allelopathy. Suppressing the activity of ammonium oxidising community reduces nitrification and conserves N in the ecosystem. This mechanism could play a key role in the invasion success and competitive superiority of gamba grass in Australian savannas.

We examined the soil ammonia oxidiser (archaea and bacteria), and N-fixing communities following invasion, as well as soil N relations. Soil inorganic N availability (NO_3^- and NH_4^+) under native grass and gamba grass stands were not significantly different. This was reflected in the microbial communities, with no difference in *amoA* gene copy number between the native and gamba grass soils. While gene abundances were similar, the identities of some *amoA* phylotypes between native grass and gamba grass soils were different. The N-fixing bacterial community, as measured by *nifH* amplicons sequencing, provided new information about N-fixing bacteria in these soils, however there were no measurable differences between native and gamba grass soils. The dominant N fixers were *Streptomyces*, *Actinomyces*, *Candidatus, Methylomirabilis* and *Pseudomonas.*

Keywords *Andropogon gayanus*, ammonia oxidiser, exotic grasses, invasive alien grasses, nitrogen cycling genes, nitrification, savanna.

INTRODUCTION

Gamba grass (*Andropogon gayanus* Kunth) invasion is a key threat to savanna ecosystems across northern Australia. Gamba grass dramatically transforms the structure and function of invaded ecosystems; producing 10 times more biomass than native grasses; altering fires regimes, water, nutrient and carbon cycling, and increasing savanna tree mortality (Rossiter *et al.* 2003, Rossiter-Rachor *et al.* 2008, 2009, Setterfield *et al.* 2010, 2013). These impacts contributed to gamba grass being declared a weed across all northern states of Australia (in 2008), declared a Key Threatening Process under the EPBC Act (in 2010), and being declared a weed of national significance (WoNS; in 2012). Despite these declarations, gamba grass is now widely distributed throughout native savannas, with the invaded area expanding rapidly, covering approximately 10,000–15,000 km^2 in the Northern Territory (Adams and Setterfield 2016).

The paradox of a high biomass grass growing on low N status soils has led researchers to question if gamba grass productivity could be sustained in the long-term, or would heavily invaded sites experience a 'run down' or decline in soil N, resulting in reduced plant vigour, and productivity. While there is no evidence of long-term declines in total soil N pools on gamba grass sites, invasion has been associated with significant increases in soil ammonium availability, accompanied by decreases in soil nitrate availability, compared to native grasses (Rossiter-Rachor *et al.* 2009). These results led to speculation that: (1) gamba grass may inhibit the process of nitrification, as it does in its native range in Africa, by altering the soil microbial community, and/or (2) gamba grass may stimulate N fixation, again by altering the soil microbial community (Rossiter-Rachor *et al.* 2009).

Gamba grass is one of several African grasses known to release allelopathic nitrification inhibitors from their roots, which suppress the activity of ammonium oxidising bacteria (Lata *et al.* 2004, Subbaro *et al.* 2012). Nitrification (the oxidation of ammonium to nitrate) is a two-step process. The first step being mediated by ammonia oxidisers (AO), comprising ammonia oxidising bacteria (AOB) and archaea (AOA). Suppressing this process, and retaining N in the ecosystem as ammonium reduces the potential for N losses via leaching, and denitrification (Subbaro *et al.* 2012). The capacity of plants to inhibit nitrification is likely to be an adaptation mechanism for the conservation and efficient use of N in ecosystems with low N availability, such as savannas (Abbadie and Lata 2006). Field experiments have shown that grasses that inhibit nitrification exhibit greater productivity in above-aground biomass than those that lack this ability (Lata *et al.* 2004). We speculate this mechanism could play a key role in the invasion and persistence of gamba grass in the low N savannas of Australia, as it could allow gamba grass to regulate soil N relations in the invaded ecosystem and outcompete native species for N (Rossiter-Rachor *et al.* 2009). It is therefore *critical* that we understand the impact of gamba grass invasion on soil microbial communities, particularly those involved in nitrification (ammonium oxidising bacteria or archaea), and N fixation (N-fixing bacteria). We aimed to quantify changes in the soil microbial community that occur in response to gamba grass invasion, and determine if altered microbial communities were accompanied by changes in soil N relations.

MATERIALS AND METHODS

Study site Soil microbial communities and soil N were sampled at three paired plots near Batchelor, approximately 100 km south of Darwin, Northern Territory, Australia. The overstorey was dominated by *Eucalyptus miniata* (Cunn. ex Schauer) and *E. tetrodonta* (F.Muell), with a grass understory dominated by native perennial grasses *Alloteropsis semialata* (R.Br.) Hitchc. and perennial *Sorghum* spp. Each paired plot consisted of one native grass plot, and an adjacent gamba grass invaded plot.

Sample collection Soil samples were collected in April 2015 at the end of the wet/growing season. Six replicate soil cores (7.5 × 15 cm) were taken within each native and gamba grass plot. Soil cores were placed in a snap-lock polyethylene bag and stored on ice until return to the lab, where approximately 0.25 g of soil was taken for nucleic acid extraction. Soil genomic DNA was extracted using a 50 prep PowerLyzer PowerSoil DNA Isolation Kit (MB-12855-50) according to the manufacturer's instructions. A subsample of soil was taken for determination of gravimetric moisture content (weighed, dried at 105°C for 48 h, and reweighed). The remainder was dried at 40°C for 48 h, ground, and analysed for Total N (% N), NO_3^- and NH_4^+ (mg kg^{-1}) on a Carlo Erba analyser.

Soil microbial community The soil ammonia oxidising community in native and gamba plots was measured using two approaches. The first approach was a qPCR analysis whereby DNA was extracted from soil and a portion of the ammonia oxidizing gene from archaea (A-*amoA*) and bacteria (B-*amoA*) was amplified in separate experiments and gene copy numbers measured (following Abell *et al.* 2011). These data provide a relative measure of ammonia oxidation gene abundance in native and gamba grass soil.

The second approach was a microarray analysis whereby ammonia oxidizing gene probes were used in hybridisation assays (following Abell *et al.* 2012). The bank of probes used in these assays provides a broader measure of the diversity of ammonia oxidisers in this system, than is possible by qPCR. This is a semi-quantitative approach which can be used to measure patterns and compare the ammonia oxidizer community in native versus gamba grass soils.

For the N-fixing community we used next generation sequencing targeting the N-fixing (*nif*) gene. These data give a semi quantitative measure of abundance in the form of sequence counts and they provide a measure of diversity so we can determine what N-fixers are dominant in these soils.

Soil N availability We measured soil inorganic N availability *in situ* in native and gamba grass plots using mixed ion-exchange resin bags (Dowex-MR3; Sigma, St. Louis, Missouri, USA) following Rossiter-Rachor *et al.* (2009). These data provide an index of plant NH_4^+ and NO_3^- availability, and were related back to the microbial communities present.

RESULTS AND DISCUSSION

Ammonia oxidising gene copy number – qPCR Ammonia oxidation in this system was overwhelmingly dominated by archaea ammonia oxidisers (AOA) with 16 times more archaeal oxidisers than bacterial nitrifiers (AOB) (Table 1).

The MDS of communities based on total gene copy number as measured by qPCR (irrespective of the identity of the ammonia oxidisers) showed no clear pattern between native and gamba grass soils (Figure 1).

Table 1. Gene copy numbers of archaeal nitrifiers (AOA) and bacterial nitrifiers (AOB) in native grass and gamba grass plots (mean ± standard deviation; n = 6).

Plot type	AOA gene copy numbers per ng DNA	AOB gene copy numbers per ng DNA
Native grass	14,923 ± 642	894 ± 52
Gamba grass	14,979 ± 699	887 ± 50

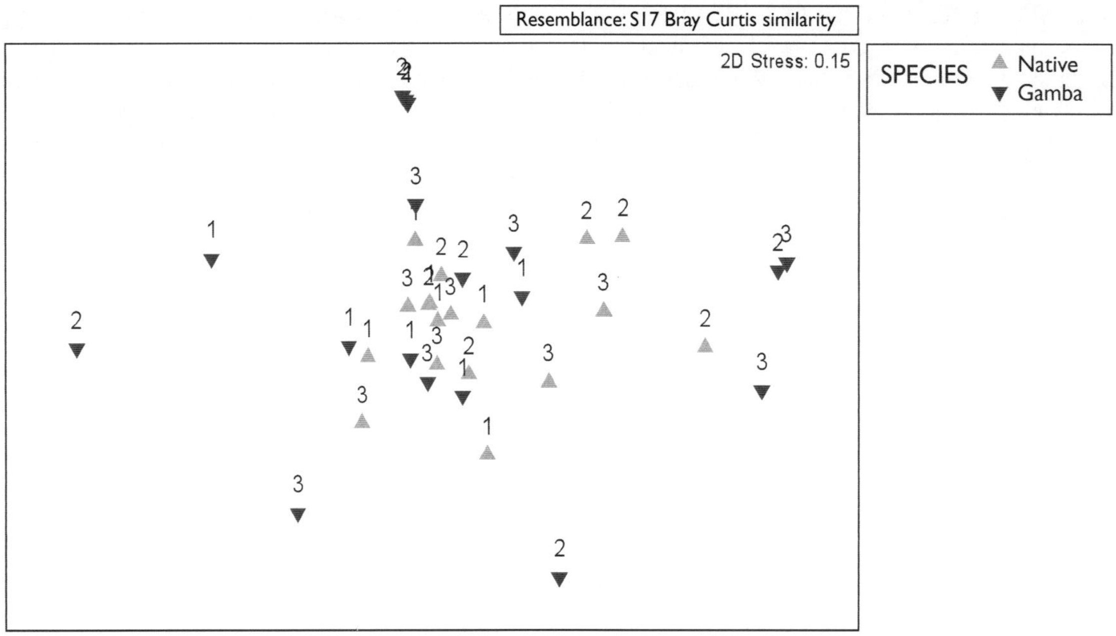

Figure 1. MDS plot of the ammonia oxidising microbial community size in native and gamba grass plots (A-*amoA* and B-*amoA* gene copy numbers per plot). Replicates not pooled.

Ammonia oxidising gene diversity – microarray The MDS of communities based on the presence or absence of AOA phylotypes derived from the microarray probes (a measure of genetic diversity of the AOA genes, Figure 2) showed no clear pattern between native and gamba grass plots. In fact, no significant community level patterns were measured between habitats due to the large sample variability, particularly

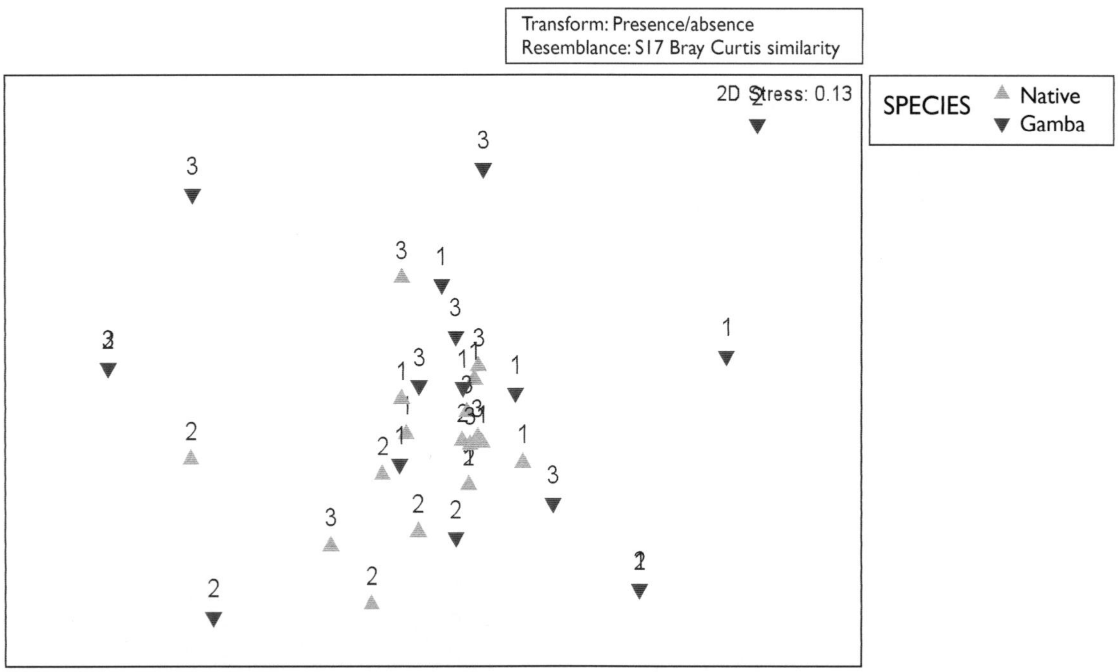

Figure 2. MDS plot of the archaeal ammonia oxidiser (AOA) community structure in native and gamba grass plots (ANOSIM global R >0.05).

within gamba samples. Despite this, the MDS showed that some gamba samples were separate and therefore different from the main cluster (Figure 2). This suggests that while there may not be any major differences compared to native, there are some unique aspects of the AOA community for some gamba samples.

The microarray data showed that the DNA from both native and gamba grass plots hybridized to 45 AOA genes across all samples; 21 distinct to gamba plots, eight distinct to native plots and 16 common to both native and gamba plots (Figure 3).

N-fixing community composition derived from the next generation sequencing nifH gene data After removing rare sequences, eleven N-fixing genera were detected across most samples (data not shown). Only one genus (*Shinella*) was unique to gamba plots. *Streptomyces*, *Actinomyces*, *Candidatus Methylomirabilis* and *Pseudomonas* were dominant across all sites.

Soil N There was no significant differences in soil N parameters between native and gamba grass plots (P >0.05; Table 2). Our results directly contrast with a previous study by Rossiter-Rachor *et al.* (2009), at a different savanna location, which found significantly increased soil ammonium availability, and decreased soil nitrate availability, as a result of gamba grass invasion. We speculate that the difference between the two studies may be due to differences in soil conditions, or the age of the gamba grass tussocks and their root density. Lata *et al.* (2000, 2004) found that the impact of the African grass *Hyparrhenia diplandra* on nitrification varied significantly with site, and also root density, and did not always result in a depressive effect on nitrification.

In conclusion, soil N in native and gamba grass plots was not significantly different, and this was reflected in the microbial communities. There was no evidence of higher ammonia concentrations in gamba grass soils. Subsequently, there was no evidence for suppression of ammonia oxidation as measured by *amoA* gene copy number. Although gene abundances were similar, there were differences in the identities of *amoA* phylotypes between native and gamba grass soil. This is for subsequent investigations. There was also no evidence for stimulation of N fixation as measured by *nifH* sequence counts. N fixers dominant in these systems: *Streptomyces*, *Actinomyces*, *Candidatus*, *Methylomirabilis* and *Pseudomonas*.

ACKNOWLEDGMENTS

This project was funded by a 2015 Charles Darwin University EHSE Faculty Grant. We thank the landowner at Batchelor for access to the study plots.

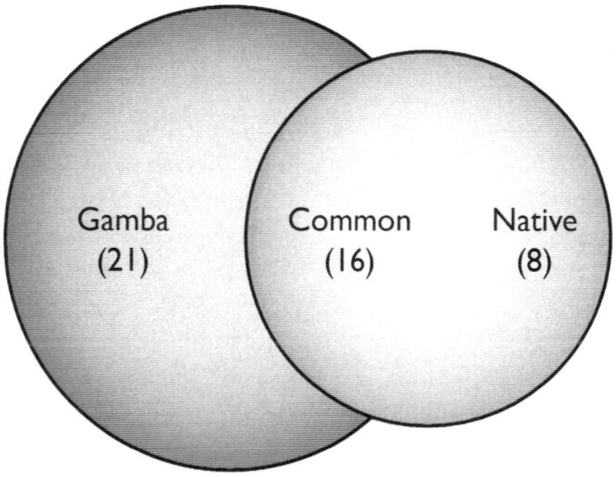

Figure 3. Venn diagram illustrating the number of the ammonia oxidising archaeal genes (a proxy for ammonia oxidising archaea) detected within and between each of the grass types.

Table 2. Soil chemical parameters in native and gamba grass plots (mean ± SE).

Soil chemistry	Native grass soil	Gamba grass soil
Gravimetric water content (%)	6.25 (0.16)	6.58 (0.15)
NH_4–N (mg kg^{-1})	1.77 (0.10)	1.70 (0.11)
NO_3–N (mg kg^{-1})	0.28 (0.04)	0.27 (0.05)
Total N (% N)	0.10 (0.005)	0.09 (0.004)
Ammonium availability (ng NH_4^+ g^{-1} resin day^{-1})	972 (16)	1068 (40)
Nitrate availability (ng NO_3^- g^{-1} resin day^{-1})	1177 (98)	1055 (118)

REFERENCES

Abell, G.C., Banks, J., Ross, D.J., Keane, J.P., Robert, S.S., Revill, A.T. and Volkman, J.K. (2011). Effects of estuarine sediment hypoxia on nitrogen fluxes and ammonia oxidizer gene transcription. *FEMS Microbiology Ecology* 75 (1), 111-22.

Abell, G.C., Robert, S.S., Frampton, D.M., Volkman, J.K., Rizwi, F., Csontos, J. and Bodrossy, L. (2012). High-throughput analysis of ammonia oxidiser community composition via a novel, *amoA*-based functional gene array. *PloS one* 7(12), e51542.

Abbadie, L., and Lata, J.C. (2006). Nitrogen dynamics in the soil-plant system. *In* Lamto, eds L. Abbadie *et al.*, pp. 277-97. (Springer, New York).

Adams, V. and Setterfield, S. (2016). Approaches to strategic risk analysis and management of invasive plants: lessons learned from managing gamba grass in northern Australia. *Pacific Conservation Biology* 22, 189-200.

Lata, J.C., Guillaume, K., Degrange, V., Abbadie, L. and Lensi, R. (2000). Relationships between root density of the African grass *Hyparrhenia diplandra* and nitrification at the decimetric scale: an inhibition–stimulation balance hypothesis. *Proceedings of the Royal Society of London B: Biological Sciences* 267, 595-600

Lata, J.C., Degrange, V., Raynaud, X., Maron, P.A., Lensi, R., and Abbadie, L. (2004). Grass populations control nitrification in savanna soils. *Functional Ecology* 18 (4), 605-11.

Rossiter-Rachor, N.A., Setterfield, S.A., Douglas, M.M., Hutley, L.B. and Cook, G.D. (2008). *Andropogon gayanus* (gamba grass) invasion alters fire-mediated nitrogen losses in the tropical savanna of Northern Australia. *Ecosystems* 11, 77-88.

Rossiter, N.A., Setterfield, S.A., Douglas, M.M. and Hutley, L.B. (2003). Testing the grass-fire cycle: alien grass invasion in the tropical savannas of northern Australia. *Diversity and Distributions* 9, 169-76.

Rossiter-Rachor, N.A., Setterfield, S.A., Douglas, M.M., Hutley, L.B., Cook, G.D. and Schmidt, S. (2009). Invasive *Andropogon gayanus* (gamba grass) is an ecosystem transformer of nitrogen relations in Australian savanna. *Ecological Applications* 19 (6), 1546-60.

Setterfield, S.A., Rossiter-Rachor, N.A., Hutley, L.B., Douglas, M.M. and Williams, R.J. (2010). Turning up the heat: the impacts of *Andropogon gayanus* (gamba grass) invasion on fire behaviour in northern Australian savannas. *Diversity and Distributions* 16 (5), 854-61.

Setterfield, S.A., Rossiter-Rachor, N.A., Douglas, M.M., Wainger, L., Petty, A.M., Barrow, P., Shepherd, I.J. and Ferdinands, K.B. (2013). Adding fuel to the fire: the impacts of non-native grass invasion on fire management at a regional scale. *PLOS One* 8 (5), e59144.

Subbarao, G.V., Sahrawat, K.L., Nakahara, K., Ishikawa, T., Kishii, M., Rao, I., Hash, C.T., George, T.S., Srinivasa Rao, P., Nardi, P., Bonnett, D.G., Berry, W., Suenaga, K. and Lata, J.C. (2012). Biological nitrification inhibition—a novel strategy to regulate nitrification in agricultural systems. *Advances in Agronomy* 114 (1), 249-301.

Managing *Mimosa pigra* in Kakadu National Park: 30 years of flood, sweat and cheers

Samantha Setterfield[1], Damien McMaster[2], Calvin Murakami[3], Frederick Hunter[3], Buck Salau[3] and Anthony Mann[3]
[1] University of Western Australia, Crawley, Western Australia 6009, Australia
[2] Charles Darwin University, Darwin, Northern Territory 0909, Australia
[3] Kakadu National Park, PO Box 71 Jabiru, Northern Territory 0886, Australia
(samantha.setterfield@uwa.edu.au)

Summary Kakadu National Park is Australia's premier protected area and one of the few World Heritage areas listed for both its natural and cultural values. Kakadu contains vast areas of seasonally inundated wetlands that support an outstanding abundance of biodiversity, particularly birds and fish. The wetlands provide critical resources for the Indigenous landowners and are also a major tourist attraction. Their listing under the Ramsar Wetlands Convention reflects the international importance of Kakadu's wetlands. Unfortunately, these wetlands are under significant threat from a range of weeds including *Mimosa pigra* L. (mimosa), a Weed of National Significance. Outbreaks of mimosa were first discovered in the early 1980s. The response in the Park was rapid, appropriately resourced, consistent over time and well-monitored. The Park's mimosa management program has now been running for over 30 years, and has resulted in the removal of over 300,000 mimosa plants from the Park. As a result, Kakadu's World Heritage listed floodplains contain no large infestations and are a major contrast to neighbouring floodplains outside the Park which have been extensively invaded. This poster outlines the aims and approaches of the management programme that has been in place for thirty years ago, and will outline the key factors underpinning its success.

Keywords Kakadu, *Mimosa pigra*, Weed of National Significance, best practice weed management.

Practicality of the suicidal germination approach for controlling *Striga hermonthica* (Delile) Benth.

Yukihiro Sugimoto[1], Hiroaki Samejima[1], Abdel Gabar Babiker[2], Hirosato Takikawa[1] and Mitsuru Sasaki[1]
[1] Graduate School of Agricultural Science, Kobe University, 1-1 Rokkodai, Nada, Kobe, Japan
[2] College of Agricultural Studies, Sudan University of Science and Technology, Shambat, Khartoum North, Sudan
(yukihiro@kobe-u.ac.jp)

Summary Purple witchweed (*Striga hermonthica* (Delile) Benth.), Orobanchaceae, is an obligate root parasitic weed of important cereal crops. The parasite is a copious seed producer and a huge seed bank develops soon after the onset of the initial infestation. To germinate a *Striga* seed requires a pre-treatment in a moist warm environment and a subsequent exposure to an exogenous stimulant. One approach to reduce the seed bank is to artificially induce germination of the seeds in the absence of or away from the host roots. A newly developed germination stimulant for *S. hermonthica* designated as T-010 was selected as a model stimulant, in principle, owing to its ease of preparation, and evaluated for efficacy in greenhouse and field experiments under artificial *Striga* infestation. Formulated T-010 applied at 0.1, 1 and 10 kg a.i. ha^{-1} to potted soil containing *S. hermonthica* seeds, previously conditioned by judicious irrigation, reduced *Striga* emergence by 94–100%. Results of a field trial showed that formulated T-010, at the same rates as for the pot experiment, delayed and reduced *Striga* emergence by 33% and increased sorghum shoot and head dry weight by 18.7–40.2% and 187–241%, respectively. These findings demonstrated, for the first time, the technical feasibility of suicidal germination for controlling *S. hermonthica*. Optimising structure, formulation, and application protocol of germination stimulants should be the main goals for further improvement of the technology.

Keywords Parasitic weeds, *Striga hermonthica*, suicidal germination, strigolactone analogue.

Risks and impacts of environmental weeds to Vietnam's forests

Tuyet T.A. Truong[1,2], Margaret E. Andrew[1], Giles Hardy[1], Bernard Dell[1] and Thu Q. Pham[3]
[1] Murdoch University, Western Australia
[2] Thai Nguyen University of Agriculture and Forestry, Vietnam
[3] Vietnam Academy of Forest Science
(t.truong@murdoch.edu.au)

Summary In the past decades, the forests in Vietnam have been under increasing pressure from disturbance, which has increased the chance of invasion of both non-native and native weeds. However, impacts of weeds on the forests have received little attention in Vietnam. *Microstegium* spp. are aggressive invasive grasses; dense infestations may prevent forest regeneration. *Microstegium* impacts are most well-known in its non-native range in North America, but our preliminary observations suggest that it may also be a native invasive species threatening forests in Vietnam, where it may be especially problematic as forest regeneration features prominently in environmental management as well as socioeconomic development goals. With a view to determining the impacts of *Microstegium ciliatum* (Trin.) A.Camus to forest regeneration and developing management recommendations, the study will conduct a three-year field experiment in Cuc Phuong, Vietnam's oldest national park, investigating the response of the woody species community to experimental removal of *Microstegium*. This presentation presents the experimental design and the results of the initial survey (May–June 2016). Invasion level of *Microstegium* and species abundance and composition of woody species (all size classes) in invaded plots under different canopy cover will be indicated. The relationship between invasion level and species abundance and composition, therefore, will also be evaluated.

Keywords Cuc Phuong National Park, environmental weeds, invasive species impacts, forest regeneration, *Microstegium*, species abundance, species richness, woody species.

Detoxication of acetochlor phytotoxicity for rice induced by 5-methoxypsoralen

Lifeng Wang[1,2], Lifeng Hu[2], Xiaomao Zhou[1,2], Qiong Peng[1,2], Zuren Li[1,2], Lianyang Bai[1,2] and Chenzhong Jin[2]
[1] Biotechnology Research Center, Hunan Academy of Agricultural Sciences, Changsha 410125, China
[2] Collaborative Innovation Center for Field Weeds Control, Loudi, Hunan 417000, China
(wanglifeng112358@163.com)

Summary *Rhizoma et Radix Notopterygii*, derived from the dried rhizome/roots of *Notopterygium incisum* Ting. ex H.T.Chang or *N. forbesii* Boiss, is often used as a traditional medicine in China. However, there were few papers reporting its extensional use in agricultural production. In this study, we found that its extract 5-methoxypsoralen (5-MOP) can effectively reduce phytotoxicity of acetochlor (A) on rice-seedling, in a growth chamber test. We observed that the shoot height, root length, and fresh biomass of A-treated rice seedlings were about 37, 25, 35 percent lower than that of water-treated rice seedlings (control), respectively. However, these physiological indexes could be recovered to 91.7%, 92.5%, and 92.2% in the 5-MOP + A treatments, comparing with control groups, respectively. The emergence rate can be enhanced to about 81% in the 5-MOP + A treatment, comparing with 63% in the A treatment. And this 5-MOP + A mixture treatment can also elevate the root activity to higher than 79.0% of the value for the non-treated control. Furthermore, we observed that the glutathione S-transferase (GST) activity has a significant improvement in rice seedlings during the detoxification process of acetochlor induced by 5-methoxypsoralen, which could plays a leading role for detoxification of acetochlor. Our findings suggest that 5-MOP have the potential to be applied as safener for rice.

Keywords 5-methoxypsoralen, rice, safener, acetochlor, herbicide phytotoxicity.

***** THIS PAPER HAS BEEN WITHDRAWN AT THE REQUEST OF THE AUTHOR *****
24 August 2016

Aggregating weed data to improve on-ground outcomes for management: a case study in the Pilbara

Bruce L. Webber[1,2], Kathryn L. Batchelor[1], Stephen J. van Leeuwen[3] and John K. Scott[1,4]
[1] CSIRO Land and Water, Private Bag 5, PO Wembley, Western Australia 6913, Australia
[2] School of Plant Biology, The University of Western Australia, 35 Stirling Highway, Crawley, Western Australia 6009, Australia
[3] Western Australia Department of Parks and Wildlife, Dick Perry Avenue, Kensington, Western Australia 6151, Australia
[4] School of Animal Biology, The University of Western Australia, 35 Stirling Highway, Crawley, Western Australia 6009, Australia
(bruce.webber@csiro.au)

Summary The Pilbara region of Western Australia is experiencing a significant increase in human activity in remote areas with high biodiversity value. The resultant increase in biosecurity risk and threat from weed invasions associated with this change requires greater efforts for control. However, existing knowledge on weed distributions in the Pilbara is extremely fragmented and most of the data are not accessible. The timely development of a weed database could, therefore, help with the identification and prioritisation of management options that span tenure boundaries and that could have the greatest benefit for on-ground outcomes. Although the majority of the weeds of the Pilbara are relatively well known, in the last 20 years a huge volume of flora and vegetation reports have been produced by botanical consultants employed by the mining industry for the purpose of obtaining environmental approvals. We have examined these reports, as well as a range of other weed databases from the Chichester and Fortescue IBRA sub-regions for information on weed distributions. From this information we will develop a weed risk assessment for a prioritised set of weed species that will contribute to an overall invasive plant management strategy for the Pilbara. A major product of the project will be a publically available, comprehensive weed database that can be efficiently updated and expanded, providing extractable data for use as a weed management tool.

Keywords Pilbara, flora, weeds, database, aggregation, risk management.

The exceptional success achieved with biological control of *Cylindropuntia fulgida* (Engelm.) F.M.Knuth var. *mamillata* (Schott ex Engelm.) Backeberg using the cochineal, *Dactylopius tomentosa* Costa (Cholla biotype), in South Africa

Travor Xivuri[1], Helmuth Zimmermann[2], Hildegard Klein[3] and Pierre van den Berg[4]
[1] South African National Biodiversity Institute, National Invasive Species Programme: Early Detection and Rapid Response, in the Eastern, Northern and Western Cape
[2] South African National Biodiversity Institute, Pretoria, South Africa or Helmuth Zimmermann & Associates (Central), Pretoria, South Africa
[3] Agricultural Research Council, Plant Protection Research Institute, Pretoria, South Africa
[4] 34 Pentz Street, Heidelberg (Western Cape Province), South Africa
(t.xivuri@sanbi.org.za)

Summary *Cylindropuntia fulgida* (Engelm.) F.M.Knuth var. *mamillata* (Schott ex Engelm.) Backeberg, also known as the boxing glove cactus, is a garden escapee which has invaded large areas in the Northern Cape, Western Cape and Limpopo provinces of South Africa and also in parts of Australia. It is a monstrose form of the chain-fruit cholla (*C. fulgida* (Engelm.) F.M. Knuth var. *fulgida*) and is native to south-western USA and north-western Mexico.

A biotype of the cochineal, *Dactylopis tomentosus* Costa was collected from a closely related cholla, known as *C. cholla* A.Weber in Baja California Sur which was subsequently introduced into quarantine in South Africa for host-specificity studies. This cochineal biotype, also known as the Cholla biotype, is highly host-specific and shows a high preference for both *C. fulgida* varieties. The first releases were made on *C. fulgida* var. *fulgida* in 2008 followed by further releases on the boxing glove cactus in 2012. The results are spectacular and invasions are usually brought under biological control within three years. Best results are achieved in hot and dry areas. Several boxing glove infestations are now being monitored to demonstrate population declines. Two institutions and private landowners are involved in the manual dispersal of the insect to all infestations. The dramatic outcome of this project could be the result of a new association effect which is based on the avoidance of evolved interspecific homeostasis typical of old associations.

Metabolic herbicide resistance gene discovery in wild oat

Qin Yu[1], Heping Han[1], Lianyang Bai[2] and Stephen Powles[1]
[1]Australian Herbicide Resistance Initiative (AHRI), School of Plant Biology, The University of Western Australia, 35 Stirling Highway Crawley, Western Australia 6009
[2] Biotechnology Research Center, Hunan Academy of Agricultural Sciences, Changsha 410125, China
(qin.yu@uwa.edu.au)

Summary Weedy plant species that have evolved resistance to herbicides due to enhanced metabolic capacity to detoxify herbicides (metabolic herbicide resistance) are a major issue threatening herbicide sustainability and global crop production. Metabolic enzymes (e.g. cytochrome P450 monooxygenase (P450), glycosyl transferase (GT), glutathione S-transferase, GST)) are major super families involved in herbicide metabolism. However, progress in biochemical and molecular understanding of these enzymes in endowing metabolic resistance has been very slow until recently, especially with rapid advance in next generation RNA-sequencing. Our recent work have identified a number of wild oat populations resistant to ACCase-inhibiting herbicides diclofop-methy, fenoxaprop and pinoxaden. We have isolated, characterised, and purified wild oat lines only possessing metabolic resistance, and these are valuable materials for metabolic herbicide resistance gene discovery. The RNA samples have been prepared from individually phenotyped most resistant or susceptible plants, and RNA-sequencing will be performed using Illumina Hiseq4000 with PE 150 and total 35 M reads. Differentially expressed contigs relevant to herbicide metabolism (e.g. P450, GT, GST) and gene regulation (e.g. transcription factors) pathways will be identified and validated using RNA-seq samples and additional sample from the resistant (R) and susceptible (S) wild oat lines. Candidate genes will be amplified from the R and S plants, sequences compared for structural changes and expressed in *planta* for functional characterisation.

Keywords Diclofop, fenoxaprop, metabolic resistance, RNA-sequencing, wild oats.

AUTHOR INDEX

AUTHOR INDEX

AUTHOR INDEX

AUTHOR INDEX

Weed Society of Western Australia (WSWA)
Department of Agriculture
Baron-Hay Court
South Perth 6151
Australia

ISBN 978-1-5108-3408-8

9 781510 834088

CURRAN ASSOCIATES INC.
proceedings.com

Underwater Intervention (UI 2023)

New Orleans, Louisiana, USA
29 November - 1 December 2023

UNDERWATER INTERVENTION